GEOMETRY AND PHYSICS

Geometry and Physics

A Festschrift in Honour of Nigel Hitchin

Edited by

JØRGEN ELLEGAARD ANDERSEN
ANDREW DANCER
OSCAR GARCÍA-PRADA

VOLUME I

OXFORD
UNIVERSITY PRESS

OXFORD
UNIVERSITY PRESS

Great Clarendon Street, Oxford, OX2 6DP,
United Kingdom

Oxford University Press is a department of the University of Oxford.
It furthers the University's objective of excellence in research, scholarship,
and education by publishing worldwide. Oxford is a registered trade mark of
Oxford University Press in the UK and in certain other countries

First Edition published in 2018
Impression: 1

Published in the United States of America by Oxford University Press
198 Madison Avenue, New York, NY 10016, United States of America

British Library Cataloguing in Publication Data
Data available

Library of Congress Control Number: 2018940609

Two volume set ISBN 978-0-19-880200-6
Volume one ISBN 978-0-19-880201-3
Volume two ISBN 978-0-19-880202-0

DOI: 10.1093/oso/9780198802006.001.0001

Printed and bound by
CPI Group (UK) Ltd, Croydon, CR0 4YY

PREFACE

In September 2016, a conference was held in three venues (Aarhus, Oxford, Madrid) to mark Nigel Hitchin's seventieth birthday and to honour his far-reaching contributions to geometry and mathematical physics.

The three locations chosen are all places with which Nigel has been closely associated. He has played a key role in the development of the Centre for Quantum Geometry of Moduli Spaces (QGM) at Aarhus and the Instituto de Ciencias Matemáticas at Madrid. He has spent much of his career at Oxford, first as a student working with Michael Atiyah and Brian Steer, then as University Lecturer and Fellow of St Catherine's College and, from 1997, as Savilian Professor of Geometry.

In these volumes, we have included a wide selection of articles based on talks given in the conference, as well as articles by mathematicians (Simon Donaldson, Carlos Simpson, Shing-Tung Yau) whose work has been closely associated with Nigel's interests but who were unable to attend in person.

While the talks span a wide range of geometric topics, reflecting the breadth of Nigel's contributions, the meeting in Aarhus had a particular focus on Riemannian geometry and quantization, while that in Madrid was particularly concerned with the themes of Higgs bundle moduli spaces and generalized geometry.

In more detail, the volumes include the following topics. The articles by Donaldson, by Gauduchon and by Conti, Madsen and Salamon deal with issues of special holonomy (the variational approach to G_2-holonomy, Kähler and hyperkähler geometry and cohomogeneity one constructions for quaternionic structures, respectively). Christian Bär's article with Sebastian Hannes deals with the Dirac operator, the subject of much of Nigel's early work, but this time in Lorentzian signature.

One of Nigel's more recent contributions has been the concept of generalized complex structures—this rapidly growing area is the subject of the articles by Behrens, Cavalcanti, Klaasse and by Gualtieri. Generalized complex geometry is closely intertwined with Poisson geometry, the subject of the paper by Brent Pym and Travis Schedler. Poisson geometry also links to the theory of Lie bialgebras, a subject explored in another article by Merkulov and Willwacher. Both generalized complex and Poisson structures can, of course, be viewed as generalizations of symplectic structures, another area where Nigel has worked extensively. Symplectic geometry and moment maps occur in the articles by Jeffrey and Mracek and by Hurtubise, Jeffrey, Rayan, Selick and Weitsman.

One of Nigel's most celebrated papers is his 1987 Proceedings of the LMS article 'The self-duality equations on a Riemann surface', which introduced the concept of Higgs bundles and their moduli spaces. These carry a hyperkähler structure, one of whose complex structures (the Dolbeault model) comes from the Higgs bundle viewpoint, while the others (de Rham or Betti model) are related to a description in terms of flat bundles/local systems.

These complex structures have very different properties—the paper by Simpson explores the idea of transferring the Hitchin fibration, which lives on the Dolbeault side, to the Betti/de Rham picture.

In recent years, these moduli spaces have become central to the study of the geometric Langlands programme. This (roughly) posits a correspondence between local systems and D-modules on the moduli space of principal bundles for the Langlands dual group. Work by Gukov, Kapustin and Witten interprets this as a manifestation of a physics duality on the Hitchin space, using its double nature as a space of local systems and a space of Higgs bundles. The paper by Teschner looks at a concrete quantization approach to the geometric Langlands correspondence in the case where the local system has an oper structure. Quantization issues also arise (in particular, the Hitchin connection) in the paper by Ellegaard Andersen and Rasmussen. Another key link between geometry and mathematical physics has been the idea of mirror symmetry; in the context of Higgs moduli spaces, this is explored in the article of by Hausel, Mellit and Pei. Dedushenko, Gukov and Putrov's article in the current volume explores a physical approach to 4-manifold invariants.

An insight of Nigel's original paper was that restricting to flat bundles corresponding to representations valued in a real form of the complex group gave a generalization of the classical Teichmüller space. This theme is pursued in the articles by Collier, by Bradlow, García-Prada, Gothen and Heinloth, by García-Prada and Ramanan and by Mundet i Riera.

The paper by Baraglia, Biswas and Schaposnik investigates the Brauer group of Higgs bundle moduli spaces. In Nigel's original paper, the Higgs fields were holomorphic, but it has become apparent that many interesting spaces can be obtained by allowing poles—this is dealt with in the papers by Boalch and by Chekhov, Mazzocco and Rubtsov. The paper by Forni and Goldman links the theory of moduli spaces (in the de Rham model) to dynamical systems via the action of the mapping class group. Biquard's article deals with Higgs bundles for an infinite-dimensional group and also addresses quantization questions. The wide range of subjects dealt with here is a testament to the incredibly rich and multifarious structure of the Hitchin moduli space.

Although Nigel is usually viewed as a differential geometer, of course much of his work, from the ADHM construction in the 1970s to recent work on Higgs bundles, has a strong algebraic flavour. Three papers in these proceedings have a pure algebraic geometric theme—those by Bogomolov, Fu and Tschinkel (on elliptic curves in positive characteristic), by Mori and Prokhorov (on extremal curve germs in threefolds) and by Grushevsky, Hulek and Tommasi, with an appendix by Dutour Sikirić (on the topology of partial compactifications of the moduli space of Abelian varieties).

Two papers focused more on mathematical physics. These are the papers by de la Ossa, Larfors and Svanes (on heterotic strings and manifolds with G_2 structure), and by Collins, Xie and Yau (on deformed Hermitian–Yang–Mills equations and mirror symmetry).

The organizing committee of the conference comprised: Luis Álvarez-Cónsul (Madrid), Jørgen Ellegaard Andersen (Aahus), Steve Bradlow (Urbana), Andrew Dancer (Oxford), Oscar García-Prada (Madrid), Frances Kirwan (Oxford), Henrik Pedersen (Odense), Yat Sun Poon (UC Riverside) and Andrew Swann (Aarhus).

The conference was supported by a wide range of funding agencies, including the Clay Mathematics Institute, the London Mathematical Society, EPSRC (through the Symmetries and Correspondences grant), QGM Aarhus, the Carlsberg Foundation, the Madrid Instituto de Ciencias Matemáticas (via the Excellence Grant Severo Ochoa), the EU-IRSES project 'Moduli' and the NSF-funded network Geometric Structures and Representation Varieties (GEAR). We thank all these bodies for their generous support.

We extend our thanks to the other members of the organizing committee, to all the administrative staff who made sure the conferences ran so smoothly, to Gil Cavalcanti and Laura Schaposnik for providing the photographs, to the speakers and contributors to this volume, and finally, of course, to Nigel himself.

Jørgen Ellegaard Andersen
Andrew Dancer
Oscar García-Prada

Laura Schaposnik

CONTENTS

PART II: VOLUME II

PART I

Volume I

1

Boundary Value Problems for the Lorentzian Dirac Operator

CHRISTIAN BÄR
SEBASTIAN HANNES

Institut für Mathematik, Universität Potsdam, Karl-Liebknecht-Str. 24-25, 14476 Potsdam, Germany, Email: baer@math.uni-potsdam.de, shannes@math.uni-potsdam.de

Dedicated to Nigel Hitchin on the occasion of his seventieth birthday

ABSTRACT. On a compact globally hyperbolic Lorentzian spin manifold with smooth space-like Cauchy boundary, the (hyperbolic) Dirac operator is known to be Fredholm when Atiyah–Patodi–Singer boundary conditions are imposed. In this paper, we investigate to what extent these boundary conditions can be replaced by more general ones and how the index then changes. There are some differences to the classical case of the elliptic Dirac operator on a Riemannian manifold with boundary.

1. Introduction

The Atiyah–Singer index theorem [1] for elliptic operators on closed manifolds is one of the central mathematical discoveries of the twentieth century. It contains famous classical results such as the Gauss–Bonnet theorem, the Riemann–Roch theorem and Hirzebruch's signature theorem as special cases and has numerous applications in analysis, geometry, topology and mathematical physics. For instance, it has been used in [14] to obtain a topological obstruction to the existence of metrics with positive scalar curvature, and a refinement of the index theorem was employed in [12] to show that, on many manifolds,

Date: 14 July, 2017.
2010 *Mathematics Subject Classification*. 58J20, 58J45.
Key words and phrases. Dirac operator, globally hyperbolic Lorentzian manifold, Fredholm pair, Dirac–Fredholm pair, index theorem.

Bär, C., Hannes, S., *Boundary Value Problems for the Lorentzian Dirac Operator*. In: *Geometry and Physics: A Festschrift in Honour of Nigel Hitchin*, Jørgen Ellegaard Andersen, Andrew Dancer, Oscar García-Prada (Eds): Oxford University Press (2018). DOI: 10.1093/oso/9780198802006.003.0001

a change of metric in a neighbourhood of a point will alter the dimension of the space of harmonic spinors. This contrasts with the space of harmonic forms whose dimensions are given topologically by the Betti numbers.

The index theorem for compact manifolds with boundary, by Atiyah, Patodi and Singer [2], requires the introduction of suitable nonlocal boundary conditions which are based on the spectral decomposition of the operator induced on the boundary. An exposition of the most general boundary conditions which one can impose in order to obtain a Fredholm operator can be found e.g. in [4].

While an analogue of the Atiyah–Singer index theorem for Lorentzian manifolds is unknown and not to be expected, one for Lorentzian manifolds with space-like boundary has been found recently [6]. More precisely, we consider the (twisted) Dirac operator on a spatially compact globally hyperbolic manifold. It is supposed to have boundary consisting of two disjoint smooth space-like Cauchy hypersurfaces. The Dirac operator is now hyperbolic rather than elliptic but the operator induced on the boundary is still self-adjoint and elliptic so that Atiyah–Patodi–Singer boundary conditions still make sense. It was shown in [6] that, under these boundary conditions, the Dirac operator becomes Fredholm and its index is given formally by the same geometric expression as in the Riemannian case. As an application, the chiral anomaly in algebraic quantum field theory on curved space-times was computed in [7].

In the present paper, we investigate more general boundary conditions which turn the hyperbolic Dirac operator into a Fredholm operator. There are similarities with and differences from the Riemannian case. It was already observed in [7] that the boundary conditions complementary to the Atiyah–Patodi–Singer boundary conditions, the anti-Atiyah–Patodi–Singer boundary conditions, also give rise to a Fredholm operator. This is false in the Riemannian case. On the other hand, we show that the conditions described in [3, 4] for the Riemannian case work in the Lorentzian setting only under an additional assumption. One can think of these boundary conditions as graph deformations of the Atiyah–Patodi–Singer boundary conditions plus finite dimensional modifications. The finite-dimensional modifications work just the same in the Lorentzian setting but the deformations need to be small, either in the sense that the linear maps whose graphs we are considering are compact or in the sense that they have sufficiently small norm. We show by example that these conditions cannot be dropped.

The paper is organized as follows. In the next section, we summarize what we need to know about Dirac operators on Lorentzian manifolds. The most important fact is well-posedness of the Cauchy problem. In the third section, we discuss some functional–analytic topics concerning Fredholm pairs. In the last section, we combine everything and consider various examples of boundary conditions giving rise to Fredholm operators and compute their index.

2. The Dirac operator on Lorentzian manifolds

We collect a few standard facts on Dirac operators on Lorentzian manifolds. For a more detailed introduction to Lorentzian geometry, see e.g. [9, 16]; for Dirac operators on semi-Riemannian manifolds, see [5, 8].

2.1. *Globally Hyperbolic Manifolds*

Suppose that M is an $(n+1)$-dimensional oriented time-oriented Lorentzian spin manifold, with n odd. We use the convention that the metric of M has signature $(-+\cdots+)$.

A subset $\Sigma \subset M$ is called a *Cauchy hypersurface* if every inextensible time-like curve in M meets Σ exactly once. If M possesses a Cauchy hypersurface, then M is called *globally hyperbolic*. All Cauchy hypersurfaces of M are homeomorphic. We assume that M is *spatially compact*, i.e. the Cauchy hypersurfaces of M are compact.

If $\Sigma_0, \Sigma_1 \subset M$ are two disjoint smooth and space-like Cauchy hypersurfaces with Σ_0 lying in the past of Σ_1, then M can be written as

$$M = \mathbb{R} \times \Sigma \tag{1}$$

such that $\Sigma_0 = \{0\} \times \Sigma$, $\Sigma_1 = \{1\} \times \Sigma$ and each $\Sigma_t = \{t\} \times \Sigma$ is a smooth space-like Cauchy hypersurface. The metric of M then takes the form $\langle\cdot,\cdot\rangle = -N^2\,dt^2 + g_t$ where $N : M \to \mathbb{R}$ is a smooth positive function (the lapse function) and g_t is a smooth 1-parameter family of Riemannian metrics on Σ; see [10, Thm 1.2] (and also [15, Thm 1]).

2.2. *Spinors*

Let $SM \to M$ be the complex spinor bundle on M endowed with its invariantly defined indefinite inner product $(\cdot,\cdot)$. Denote Clifford multiplication with $X \in T_pM$ by $\gamma(X) : S_pM \to S_pM$. It satisfies

$$\gamma(X)\gamma(Y) + \gamma(Y)\gamma(X) = -2(X,Y)$$

and

$$(\gamma(X)u, v) = (u, \gamma(X)v)$$

for all $X, Y \in T_pM$, $u, v \in S_pM$ and $p \in M$.

Let $e_0, e_1, \ldots, e_n$ be a positively oriented Lorentz-orthonormal tangent frame. Then Clifford multiplication with the volume form $\Gamma = \mathrm{i}^{n(n+3)/2}\gamma(e_0)\cdots\gamma(e_n)$ satisfies $\Gamma^2 = \mathrm{id}_{SM}$. This induces the eigenspace decomposition $SM = S^RM \oplus S^LM$ for the eigenvalues ± 1 into right-handed and left-handed spinors. Since the dimension of M is even, $\Gamma\gamma(X) = -\gamma(X)\Gamma$ for all $X \in TM$. In particular, S^RM and S^LM have equal rank and Clifford multiplication by tangent vectors reverses handedness.

Now let $\Sigma \subset M$ be a smooth space-like hypersurface. Denote by ν be the past-directed time-like vector field on M along Σ with $\langle \nu, \nu \rangle \equiv -1$, which is perpendicular to Σ. The restriction of S^RM or S^LM to Σ can be naturally identified with the spinor bundle of Σ, i.e. $S^RM|_\Sigma = S^LM|_\Sigma = S\Sigma$. The spinor bundle of Σ carries a natural positive definite scalar product $\langle\cdot,\cdot\rangle$ induced by the Riemannian metric of Σ. The two inner products are related by $\langle\cdot,\cdot\rangle = (\gamma(\nu)\cdot,\cdot)$.

Clifford multiplication $\gamma_\Sigma(X)$ on $S\Sigma$ corresponds to $\mathrm{i}\gamma(\nu)\gamma(X)$ under this identification. Note that $\gamma(\nu)^2 = \mathrm{id}$. Clifford multiplication on Σ is skew-adjoint because

$$\begin{aligned}\langle \gamma_\Sigma(X)u,v\rangle &= (\gamma(\nu)\mathrm{i}\gamma(\nu)\gamma(X)u,v) = (\mathrm{i}\gamma(X)u,v) = -(u,\mathrm{i}\gamma(X)v)\\ &= -(u,\gamma(\nu)\gamma_\Sigma(X)v) = -(\gamma(\nu)u,\gamma_\Sigma(X)v) = -\langle u,\gamma_\Sigma(X)v\rangle .\end{aligned}$$

2.3. *The Dirac Operator*

Let $E \to M$ be a Hermitian vector bundle with a compatible connection ∇^E. Sections of the vector bundles $V^R := S^R M \otimes E$ and $V^L := S^L M \otimes E$ are called right-handed (resp. left-handed) twisted spinors (or spinors with coefficients in E). The inner product $(\cdot,\cdot)$ on $S^R M$ and the scalar product on E induce an (indefinite) inner product on V^R, again denoted by $(\cdot,\cdot)$. When restricted to a space-like hypersurface the scalar product $\langle\cdot,\cdot\rangle$ on $S^R M$ and the one on E induce a (positive definite) scalar product on V^R, again denoted by $\langle\cdot,\cdot\rangle$.

Let $D : C^\infty(M;V^R) \to C^\infty(M;V^L)$ be the Dirac operator acting on right-handed twisted spinors. In terms of a local Lorentz-orthonormal tangent frame $e_0, e_1, \ldots, e_n$ this operator is given by

$$D = \sum_{j=0}^{n} \varepsilon_j \gamma(e_j)\nabla_{e_j},$$

where $\varepsilon_j = \langle e_j, e_j\rangle = \pm 1$ and ∇ is the connection on V^R induced by the Levi–Civita connection on $S^R M$ and ∇^E. The Clifford multiplication $\gamma(X)$ of a tangent vector $X \in T_pM$ on a twisted spinor $\phi \otimes e \in V^R_p = S^R_p M \otimes E_p$ is to be understood as acting on the first factor, $\gamma(X)(\phi\otimes e) = (\gamma(X)\phi)\otimes e$. The Dirac operator is a hyperbolic linear differential operator of first order.

Along a smooth space-like hypersurface $\Sigma \subset M$ with past-directed unit normal field ν, the Dirac operator can be written as

$$D = -\gamma(\nu)\left(\nabla_\nu + \mathrm{i}A_\Sigma - \frac{n}{2}H\right), \tag{2}$$

where H is the mean curvature of Σ with respect to ν, and A_Σ is the elliptic twisted Dirac operator of the Riemannian manifold Σ.

2.4. *The Cauchy Problem*

Now we fix two smooth space-like Cauchy hypersurfaces Σ_1 and Σ_0 in M. We assume that Σ_0 lies in the chronological past of Σ_1. Then we consider the region M_0 ‘in between’ Σ_0 and Σ_1, more precisely, $M_0 = J^+(\Sigma_0)\cap J^-(\Sigma_1)$, where J^- and J^+ denote the causal past and future, respectively. Since M is spatially compact, the region M_0 is a compact manifold with boundary, the boundary being the disjoint union of Σ_0 and Σ_1.

For any compact space-like hypersurface $\Sigma \subset M$, we define the L^2-scalar product for $u, v \in C^\infty(\Sigma, V^R)$ by

$$(u,v)_{L^2} = \int_\Sigma \langle u,v\rangle \,\mathrm{dA},$$

where dA denotes the volume element of Σ induced by its Riemannian metric. Recall that the inner product $\langle\cdot,\cdot\rangle$ on V^R is positive definite. The completion of $C^\infty(\Sigma,V^R)$ w.r.t. the L^2-norm will be denoted by $L^2(\Sigma,V^R)$.

Similarly, using an auxiliary positive definite scalar product on V^L, we can define $L^2(M_0, V^L)$, where we integrate against the volume element induced by the Lorentzian metric on M. By compactness of M_0, different choices of auxiliary scalar products on V^L will give rise to equivalent L^2-norms. Hence $L^2(M_0, V^L)$ is unanimously defined as a topological vector space.

Finally, we complete $C^\infty(M_0, V^R)$ w.r.t. the L^2-graph-norm of D,

$$\|u\|_{FE}^2 = \|u\|_{L^2}^2 + \|Du\|_{L^2}^2,$$

and obtain the 'finite-energy' space $FE(M_0, V^R)$.

Now the Dirac operator obviously extends to a bounded operator $D : FE(M_0, V^R) \to L^2(M_0, V^L)$. It can be checked that the restriction map $C^\infty(M_0, V^R) \to C^\infty(\Sigma, V^R)$ extends uniquely to a bounded operator $\mathrm{res}_\Sigma : FE(M_0, V^R) \to L^2(\Sigma, V^R)$ if Σ is a space-like Cauchy hypersurface.

In these function spaces, the Cauchy problem is well posed:

Theorem 2.1 *Let $\Sigma \subset M_0$ be a smooth space-like Cauchy hypersurface. Then the mapping*

$$\mathrm{res}_\Sigma \oplus D : FE(M_0, V^R) \to L^2(\Sigma; V^R) \oplus L^2(M_0, V^L)$$

is an isomorphism of Hilbert spaces. □

In particular, we get well-posedness of the Cauchy problem for the homogeneous Dirac equation:

Corollary 2.2 *For any smooth space-like Cauchy hypersurface $\Sigma \subset M_0$, the restriction mapping*

$$\mathrm{res}_\Sigma : \{u \in FE(M_0, V^R) \mid Du = 0\} \to L^2(\Sigma, V^R)$$

is an isomorphism of Hilbert spaces. □

For details, see [6].

2.5. *The Wave Evolution*

Applying Corollary 2.2 to $\Sigma = \Sigma_0$ and to $\Sigma = \Sigma_1$, we can define the *wave evolution operator*

$$Q : L^2(\Sigma_0, V^R) \to L^2(\Sigma_1, V^R) \tag{3}$$

by the commutative diagram

$$\begin{array}{ccc} & \{u \in FE(M_0, V^R) \mid Du = 0\} & \\ {}^{\mathrm{res}_{\Sigma_0}}\swarrow\cong & & \cong\searrow^{\mathrm{res}_{\Sigma_1}} \\ L^2(\Sigma_0, V^R) & \xrightarrow{\quad Q \quad} & L^2(\Sigma_1, V^R) \end{array}$$

By construction, Q is an isomorphism. One can check that Q is actually unitary, that it restricts to isomorphisms $H^s(\Sigma_0, V^R) \to H^s(\Sigma_1, V^R)$ for all $s > 0$ and that it extends to isomorphisms $H^s(\Sigma_0, V^R) \to H^s(\Sigma_1, V^R)$ for all $s < 0$. Here H^s denote the corresponding Sobolev spaces. As a consequence, Q maps $C^\infty(\Sigma_0, V^R)$ to $C^\infty(\Sigma_1, V^R)$.

In fact, well-posedness of the Dirac equation also holds for smooth sections, i.e.

$$\mathrm{res}_\Sigma \oplus D : C^\infty(M_0, V^R) \to C^\infty(\Sigma; V^R) \oplus C^\infty(M_0, V^L) \tag{4}$$

is an isomorphism of Fréchet spaces.

3. Fredholm Pairs

In this section we collect a few functional-analytic facts which will be useful later.

Definition 3.1 Let H be a Hilbert space and let $B_0, B_1 \subset H$ be closed linear subspaces. Then (B_0, B_1) is called a *Fredholm pair* if $B_0 \cap B_1$ is finite dimensional and $B_0 + B_1$ is closed and has finite codimension. The number

$$\mathrm{ind}(B_0, B_1) = \dim(B_0 \cap B_1) - \dim(H/(B_0 + B_1))$$

is called the *index* of the pair (B_0, B_1).

We list a few elementary properties of Fredholm pairs. For details, see [13, Ch. 4, § 4].

Remark 3.2 (1) The pair (B_0, B_1) is Fredholm if and only if (B_1, B_0) is a Fredholm pair and, in this case,

$$\mathrm{ind}(B_0, B_1) = \mathrm{ind}(B_1, B_0).$$

(2) The pair (B_0, B_1) is Fredholm if and only if $(B_0^\perp, B_1^\perp)$ is a Fredholm pair and, in this case,

$$\mathrm{ind}(B_0, B_1) = -\mathrm{ind}(B_0^\perp, B_1^\perp).$$

(3) Let $B_0' \subset H$ be a closed linear subspace with $B_0 \subset B_0'$ and $\dim(B_0'/B_0) < \infty$. Then (B_0, B_1) is a Fredholm pair if and only if (B_0', B_1) is a Fredholm pair and, in this case,

$$\mathrm{ind}(B_0', B_1) = \mathrm{ind}(B_0, B_1) + \dim(B_0'/B_0).$$

The following lemma reformulates the concept of Fredholm pairs in terms of orthogonal projections. For a proof, see e.g. [11, Lemma 24.3]. Here and henceforth, the orthogonal projection onto a closed subspace V of a Hilbert space H will be denoted by

$$\pi_V : H \to V.$$

Lemma 3.3 *Let $B_0, B_1 \subset H$ be closed linear subspaces. Then (B_0, B_1) is a Fredholm pair of index k if and only if*

$$\pi_{B_1^\perp}|_{B_0} : B_0 \to B_1^\perp$$

is a Fredholm operator of index k. In this case, we have $\ker\left(\pi_{B_1^\perp}|_{B_0}\right) = B_0 \cap B_1$ *and* $\mathrm{coker}\left(\pi_{B_1^\perp}|_{B_0}\right) \cong B_0^\perp \cap B_1^\perp$. □

Let E, F, H_0 and H_1 be Hilbert spaces and let $L : E \to F$, $r_0 : E \to H_0$ and $r_1 : E \to H_1$ be bounded linear maps. We assume that

$$r_j \oplus L : E \to H_j \oplus F$$

is an isomorphism for $j = 0$ and for $j = 1$. Then r_j restricts to an isomorphism $\ker(L) \to H_j$. We define the isomorphism $Q : H_0 \to H_1$ by the commutative diagram

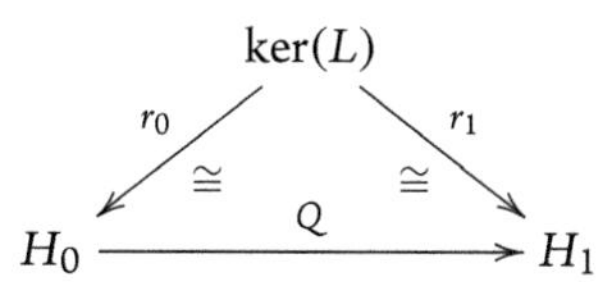

Proposition 3.4 *Assume that $r_0 \oplus r_1 : E \to H_0 \oplus H_1$ is onto. Let $B_j \subset H_j$ be closed linear subspaces. Then the following are equivalent:*

(i) The pair $(B_0, Q^{-1}B_1)$ is Fredholm of index k;

(ii) The pair (QB_0, B_1) is Fredholm of index k;

(iii) The operator $(\pi_{B_0^\perp} \circ r_0) \oplus (\pi_{B_1^\perp} \circ r_1) \oplus L : E \to B_0^\perp \oplus B_1^\perp \oplus F$ is Fredholm of index k;

(iv) The restriction $L : \ker(\pi_{B_0^\perp} \circ r_0) \cap \ker(\pi_{B_1^\perp} \circ r_1) \to F$ is a Fredholm operator of index k.

Proof (a) The equivalence of (i) and (ii) is clear because $Q : H_0 \to H_1$ is an isomorphism.

(b) Proposition A.1.(iv) in [3] with $P = L$ states that (iii) is equivalent to the operator

$$((\pi_{B_0^\perp} \circ r_0) \oplus (\pi_{B_1^\perp} \circ r_1))|_{\ker L} : \ker L \to B_0^\perp \oplus B_1^\perp$$

being Fredholm of index k. Since $r_1|_{\ker L} = Q \circ r_0|_{\ker L}$, this is again equivalent to

$$\widetilde{L} : H_0 \to B_0^\perp \oplus B_1^\perp, \quad h_0 \mapsto (\pi_{B_0^\perp}(h_0), \pi_{B_1^\perp}(Q(h_0))),$$

being Fredholm of index k. With respect to the splittings, $\widetilde{L} : H_0 = B_0^\perp \oplus B_0 \to B_0^\perp \oplus B_1^\perp$ takes the operator matrix form

$$\widetilde{L} = \begin{pmatrix} 1_{B_0^\perp} & 0 \\ \pi_{B_1^\perp} Q|_{B_0^\perp} & \pi_{B_1^\perp} Q|_{B_0} \end{pmatrix}.$$

Now $\widetilde{L}$ is Fredholm if and only if it is an invertible modulo compact operator. This is equivalent to $\pi_{B_1^\perp} Q|_{B_0}$ being invertible modulo compact operators, i.e. being Fredholm. By deformation invariance of the index, we have

$$\operatorname{ind}(\widetilde{L}) = \operatorname{ind}\begin{pmatrix} 1_{B_0^\perp} & 0 \\ \pi_{B_1^\perp} Q|_{B_0^\perp} & \pi_{B_1^\perp} Q|_{B_0} \end{pmatrix} = \operatorname{ind}\begin{pmatrix} 1_{B_0^\perp} & 0 \\ 0 & \pi_{B_1^\perp} Q|_{B_0} \end{pmatrix} = \operatorname{ind}\big(\pi_{B_1^\perp} Q|_{B_0}\big).$$

This shows that (iii) is equivalent to $\pi_{B_1^\perp}|_{QB_0}$ being a Fredholm operator of index k. The equivalence of (ii) and (iii) now follows with Lemma 3.3.

(c) The equivalence of (iii) and (iv) is Proposition A.1.(iv) in [3] with

$$P = (\pi_{B_0^\perp} \circ r_0) \oplus (\pi_{B_1^\perp} \circ r_1) : E \to B_0^\perp \oplus B_1^\perp.$$

Note that P is onto because $r_0 \oplus r_1 : E \to H_0 \oplus H_1$ is onto by assumption. □

4. Boundary Value Problems for the Dirac Operator

We return to our twisted Dirac operator D on a Lorentzian manifold as introduced in Section 2. Let $B_0 \subset L^2(\Sigma_0, V^R)$ and $B_1 \subset L^2(\Sigma_1, V^R)$ be closed subspaces. Denote by $Q : L^2(\Sigma_0, V^R) \to L^2(\Sigma_1, V^R)$ the wave evolution operator as defined in (3). Combining Theorem 2.1 and Proposition 3.4, we get

Theorem 4.1 *The following are equivalent:*

(i) The pair $(B_0, Q^{-1}B_1)$ is Fredholm of index k;

(ii) The pair (QB_0, B_1) is Fredholm of index k;

(iii) The operator

$$(\pi_{B_0^\perp} \circ \mathrm{res}_{\Sigma_0}) \oplus (\pi_{B_1^\perp} \circ \mathrm{res}_{\Sigma_1}) \oplus D : FE(M_0, V^R) \to B_0^\perp \oplus B_1^\perp \oplus L^2(M_0, V^L)$$

is Fredholm of index k;

(iv) The restriction

$$D : \ker(\pi_{B_0^\perp} \circ \mathrm{res}_{\Sigma_0}) \cap \ker(\pi_{B_1^\perp} \circ \mathrm{res}_{\Sigma_1}) \to L^2(M_0, V^L)$$

is a Fredholm operator of index k. □

Definition 4.2 If these conditions hold, then we call (B_0, B_1) a *Dirac–Fredholm pair* of index k.

Condition (iv) in Theorem 4.1 means that we consider the Dirac equation $Du = f$ subject to the boundary conditions $u|_{\Sigma_0} \in B_0$ and $u|_{\Sigma_1} \in B_1$. We now look at concrete examples.

4.1. (Anti-) Atiyah–Patodi–Singer Boundary Conditions

For any subset $I \subset \mathbb{R}$, denote by $\chi_I : \mathbb{R} \to \mathbb{R}$ its characteristic function. Denote the Riemannian Dirac operators for the boundary parts by $A_0 := A_{\Sigma_0}$ and $A_1 := A_{\Sigma_1}$ and then compare (2). Since A_0 and A_1 are self-adjoint elliptic operators on closed Riemannian manifolds, they have real and discrete spectra. We consider the spectral projector $\chi_I(A_0) : L^2(\Sigma_0, V^R) \to L^2(\Sigma_0, V^R)$ and similarly for A_1. This is the orthogonal projector onto the sum of the A_0-eigenspaces to all eigenvalues contained in I. Its range will be denoted by $L^2_I(A_0) := \chi_I(A_0)(L^2(\Sigma_0, V^R)) \subset L^2(\Sigma_0, V^R)$ and similarly for A_1.

Now the *Atiyah–Patodi–Singer boundary conditions* correspond to the choice $B_0 = L^2_{(-\infty,0)}(A_0)$ and $B_1 = L^2_{(0,\infty)}(A_1)$. It was shown in [6, Thm 7.1] that the Dirac operator subject to these boundary conditions is Fredholm. In other words, the pair $(B_0, B_1) = (L^2_{(-\infty,0)}(A_0), L^2_{(0,\infty)}(A_1))$ is Fredholm and its index is given by

$$\mathrm{ind}(B_0, B_1) = \int_{M_0} \widehat{\mathrm{A}}(\nabla) \wedge \mathrm{ch}(\nabla^E) + \int_{\partial M_0} \mathcal{T} - \frac{h(A_0) + h(A_1) + \eta(A_0) - \eta(A_1)}{2}.$$

Here $\widehat{\mathrm{A}}(\nabla)$ is the $\widehat{\mathrm{A}}$-form built from the curvature of the Levi–Civita connection ∇ on M, and $\mathrm{ch}(\nabla^E)$ is the Chern character form of the curvature of ∇^E. The form $\mathcal{T}$ is the corresponding transgression form. In particular, the boundary integral vanishes if the given metric and connections have product structure near the boundary. Moreover, $h(A)$ denotes the dimension of the kernel of an operator A and $\eta(A)$, its η-invariant. See [6] for details.

By Remark 3.2, the complementary boundary conditions $(\mathrm{APS}_0(0)^\perp, \mathrm{APS}_1(0)^\perp) = (L^2_{[0,\infty)}(A_0), L^2_{(-\infty,0]}(A_1))$, the *anti-Atiyah–Patodi–Singer boundary conditions*, are also Fredholm and the index has the opposite sign,

$$\mathrm{ind}(B_0^\perp, B_1^\perp) = -\int_{M_0} \widehat{\mathrm{A}}(\nabla) \wedge \mathrm{ch}(\nabla^E) - \int_{\partial M_0} \mathcal{T} + \frac{h(A_0) + h(A_1) + \eta(A_0) - \eta(A_1)}{2}.$$

4.2. *Generalized Atiyah–Patodi–Singer Boundary Conditions*

For $a \in \mathbb{R}$, we define $\mathrm{APS}_0(a) := L^2_{(-\infty,a)}(A_0) \subset L^2(\Sigma_0, V^R)$ and $\mathrm{APS}_1(a) := L^2_{(a,\infty)}(A_1) \subset L^2(\Sigma_1, V^R)$. Since cutting the spectrum of the boundary operator at 0 is somewhat arbitrary, we may want to fix $a_0, a_1 \in \mathbb{R}$ and consider the boundary conditions $(B_0, B_1) = (\mathrm{APS}_0(a_0), \mathrm{APS}_1(a_1))$. These boundary conditions are known as *generalized Atiyah–Patodi–Singer boundary conditions.* Since all eigenvalues of the Riemannian Dirac operators A_{Σ_i} are of finite multiplicity, these boundary conditions differ from the Atiyah–Patodi–Singer boundary conditions only by finite dimensional spaces, more precisely,

$$\begin{aligned} \mathrm{APS}_0(a_0)/\mathrm{APS}_0(0) &\cong L^2_{[0,a_0)}(A_0) \quad \text{for } a_0 \geq 0, \\ \mathrm{APS}_0(0)/\mathrm{APS}_0(a_0) &\cong L^2_{[a_0,0)}(A_0) \quad \text{for } a_0 < 0, \end{aligned}$$

and similarly for APS_1. Remark 3.2 implies that generalized Atiyah–Patodi–Singer boundary conditions also form a Dirac–Fredholm pair. Setting

$$W_0 := \begin{cases} L^2_{[0,a_0)}(A_0), & a_0 \geq 0, \\ L^2_{[a_0,0)}(A_0), & a_0 < 0, \end{cases} \quad \text{and} \quad W_1 := \begin{cases} L^2_{(0,a_1]}(A_1), & a_1 \geq 0, \\ L^2_{(a_1,0]}(A_1), & a_1 < 0, \end{cases}$$

we get for their index

$$\begin{aligned}
&\operatorname{ind}(\mathrm{APS}_0(a_0),\mathrm{APS}_1(a_1))\\
&\quad = \operatorname{ind}(\mathrm{APS}_0(0),\mathrm{APS}_1(0)) + \operatorname{sgn}(a_0)\dim(W_0) - \operatorname{sgn}(a_1)\dim(W_1)\\
&\quad = \int_{M_0} \widehat{\mathrm{A}}(\nabla)\wedge \operatorname{ch}(\nabla^E) + \int_{\partial M_0} \mathcal{T} - \frac{h(A_0)+h(A_1)+\eta(A_0)-\eta(A_1)}{2}\\
&\qquad + \operatorname{sgn}(a_0)\dim(W_0) - \operatorname{sgn}(a_1)\dim(W_1).
\end{aligned}$$

In other words, the correction terms in the index formula are given by the total multiplicity of the eigenvalues of A_j between 0 and a_j.

4.3. *Boundary Conditions in Graph Form*

In the previous section, we discussed certain finite dimensional modifications of the Atiyah–Patodi–Singer boundary conditions, which lead to corrections in the index formula. Now we introduce continuous deformations of Dirac–Fredholm pairs, leaving the index unchanged. Formally, the definition coincides with that of D-elliptic boundary conditions for the elliptic Dirac operator on *Riemannian* manifolds, as introduced in [3, Def. 7.5 and Thm 7.11].

Definition 4.3 We call a pair (B_0, B_1) of closed subspaces $B_i \subset L^2(\Sigma_i, V^R)$ *boundary conditions in graph form* if there are L^2-orthogonal decompositions

$$L^2(\Sigma_i, V^R) = V_i^- \oplus W_i^- \oplus V_i^+ \oplus W_i^+, \qquad i = 0,1, \tag{5}$$

such that

(i) W_i^+, W_i^- are finite dimensional;

(ii) $W_i^- \oplus V_i^- = L^2_{(-\infty,a_i)}(A_i)$ and $W_i^+ \oplus V_i^+ = L^2_{[a_i,\infty)}(A_i)$ for some $a_i \in \mathbb{R}$;

(iii) There are bounded linear maps $g_0 : V_0^- \to V_0^+$ and $g_1 : V_1^+ \to V_1^-$ such that

$$\begin{aligned}
B_0 &= W_0^+ \oplus \Gamma(g_0),\\
B_1 &= W_1^- \oplus \Gamma(g_1),
\end{aligned}$$

where $\Gamma(g_{0/1}) := \{v + g_{0/1}v \mid v \in V_{0/1}^{\mp}\}$ denotes the graph of $g_{0/1}$.

Remark 4.4 In the setting of Definition 4.3, we have

$$V_i^- \oplus V_i^+ = \Gamma(g_i) \oplus \Gamma(-g_i^\dagger),$$

where both decompositions are orthogonal. With respect to the splitting $V_i^{\mp} \oplus V_i^{\pm}$, the projections onto $\Gamma(g_i)$ are given by

$$\begin{pmatrix} \mathrm{id} & 0 \\ g_i & 0 \end{pmatrix}\begin{pmatrix} \mathrm{id} & -g_i^\dagger \\ g_i & \mathrm{id} \end{pmatrix}^{-1} = \begin{pmatrix} (\mathrm{id}+g_i^\dagger g_i)^{-1} & (\mathrm{id}+g_i^\dagger g_i)^{-1}g_i^\dagger \\ g_i(\mathrm{id}+g_i^\dagger g_i)^{-1} & g_i(\mathrm{id}+g_i^\dagger g_i)^{-1}g_i^\dagger \end{pmatrix} \tag{6}$$

(see [3, Lemma 7.7 and Remark 7.8]). Here $g_i^\dagger : V_i^\pm \to V_i^\mp$ denotes the adjoint linear map.

The next proposition shows that deforming Atiyah–Patodi–Singer boundary conditions for the Dirac operator to a graph preserves Fredholm property as well as the index.

Proposition 4.5 *Let $a_0, a_1 \in \mathbb{R}$ be given. Then there exists an $\varepsilon > 0$ such that, for any bounded linear maps $g_i : \mathrm{APS}_i(a_i) \to \mathrm{APS}_i(a_i)^\perp$, the pair $(\Gamma(g_0), \Gamma(g_1))$ is a Dirac–Fredholm pair of the same index as $(\mathrm{APS}_0(a_0), \mathrm{APS}_1(a_1))$, provided*

(A) g_0 *or* g_1 *is compact or*

(B) $\|g_0\| \cdot \|g_1\| < \varepsilon$.

Proof We put $\tilde{g}_1 := Q^{-1} \circ g_1 \circ Q$. Now $(\Gamma(g_0), \Gamma(g_1))$ is a Dirac–Fredholm pair of index k if and only if $(\Gamma(g_0), Q^{-1}\Gamma(g_1)) = (\Gamma(g_0), \Gamma(\tilde{g}_1))$ is a Fredholm pair of index k. By Lemma 3.3, this is equivalent to

$$\pi_{\Gamma(g_0)}|_{\Gamma(-\tilde{g}_1^\dagger)} : \Gamma(-\tilde{g}_1^\dagger) \to \Gamma(g_0)$$

being a Fredholm operator of index $-k$. Since the maps $(\mathrm{id} + g_i) : \mathrm{APS}_i(a_i) \to \Gamma(g_i)$ and $(\mathrm{id} - g_i^\dagger) : \mathrm{APS}_i(a_i)^\perp \to \Gamma(-g_i^\dagger)$ are isomorphisms, this is equivalent to

$$L : Q^{-1}\mathrm{APS}_1(a_1)^\perp \xrightarrow{\cong} \Gamma(-\tilde{g}_1^\perp) \xrightarrow{\pi_{\Gamma(g_0)}} \Gamma(g_0) \xrightarrow{\cong} \mathrm{APS}_0(a_0)$$

being Fredholm of index $-k$. We write $B_0 := \mathrm{APS}_0(a_0)$ and $B_1 := Q^{-1}\mathrm{APS}_1(a_1)$. Using (6), we see that L takes the form

$$\begin{aligned} L &= \begin{pmatrix} \mathrm{id} & 0 \end{pmatrix} \begin{pmatrix} (\mathrm{id} + g_0^\dagger g_0)^{-1} & (\mathrm{id} + g_0^\dagger g_0)^{-1} g_0^\dagger \\ g_0(\mathrm{id} + g_0^\dagger g_0)^{-1} & g_0(\mathrm{id} + g_0^\dagger g_0)^{-1} g_0^\dagger \end{pmatrix} \begin{pmatrix} \pi_{B_0}|_{B_1} & \pi_{B_0}|_{B_1^\perp} \\ \pi_{B_0^\perp}|_{B_1} & \pi_{B_0^\perp}|_{B_1^\perp} \end{pmatrix} \begin{pmatrix} -\tilde{g}_1^\dagger \\ \mathrm{id} \end{pmatrix} \\ &= \begin{pmatrix} (\mathrm{id} + g_0^\dagger g_0)^{-1} & (\mathrm{id} + g_0^\dagger g_0)^{-1} g_0^\dagger \end{pmatrix} \begin{pmatrix} -(\pi_{B_0}|_{B_1})\tilde{g}_1^\dagger + \pi_{B_0}|_{B_1^\perp} \\ -(\pi_{B_0^\perp}|_{B_1})\tilde{g}_1^\dagger + \pi_{B_0^\perp}|_{B_1^\perp} \end{pmatrix} \\ &= (\mathrm{id} + g_0^\dagger g_0)^{-1}(-\underbrace{(\pi_{B_0}|_{B_1})\tilde{g}_1^\dagger}_{=:\alpha} + \underbrace{\pi_{B_0}|_{B_1^\perp}}_{=:\beta} - \underbrace{g_0^\dagger(\pi_{B_0^\perp}|_{B_1})\tilde{g}_1^\dagger}_{=:\gamma} + g_0^\dagger(\underbrace{\pi_{B_0^\perp}|_{B_1^\perp}}_{=:\delta})). \end{aligned}$$

The operators α and δ are compact by [7, Lemma 2.6]. If g_0 or g_1 (and hence $g_0^\dagger$ or $\tilde{g}_1^\dagger$) is compact, then γ is compact as well. Since $\beta = \pi_{B_0}|_{B_1^\perp}$ is a Fredholm operator of index $\mathrm{ind}(B_0^\perp, B_1^\perp) = -\mathrm{ind}(B_0, B_1)$, the same is true for L.

If $\|g_0\| \cdot \|g_1\| < \varepsilon$, then

$$\|\gamma\| \le \|g_0^\dagger\| \cdot \|\pi_{B_0^\perp}|_{B_1}\| \cdot \|\tilde{g}_1^\dagger\| = \|g_0\| \cdot \|\pi_{B_0^\perp}|_{B_1}\| \cdot \|g_1\| < \varepsilon$$

is small so that $\beta - \gamma$ (and hence L) is again a Fredholm operator of the same index as β. □

Remark 4.6 In Example 4.11 we will see that Conditions (A) and (B) cannot be dropped. Without these assumptions, boundary conditions in graph form do not give rise to a Fredholm operator in general.

Remark 4.7 Proposition 4.5 can be applied in particular if $g_0 = 0$ or $g_1 = 0$. Thus, in the setting of the proposition, $(\Gamma(g_0), \mathrm{APS}_1(a_1))$ and $(\mathrm{APS}_0(a_0), \Gamma(g_1))$ are Dirac–Fredholm pairs of index $\mathrm{ind}(\mathrm{APS}_0(a_0), \mathrm{APS}_1(a_1))$.

If $g : B \to B^\perp$ is a bounded linear map and $\widetilde{B} \subset B$, then $\Gamma(g)/\Gamma(g|_{\widetilde{B}}) \cong B/\widetilde{B}$. Combining Section 4.2 and Proposition 4.5 we get the following result.

Corollary 4.8 *Let (B_0, B_1) be boundary conditions in graph form with g_0 or g_1 compact or $\|g_0\| \cdot \|g_1\|$ sufficiently small. Then (B_0, B_1) is a Dirac–Fredholm pair and its index is given by*

$$\begin{aligned}\mathrm{ind}(B_0, B_1) = \mathrm{ind}(L^2_{(-\infty,a_0)}(A_0), L^2_{[a_1,\infty)}(A_1)) + \dim(W_0^+) - \dim(W_0^-)\\ + \dim(W_1^-) - \dim(W_1^+).\end{aligned}$$

4.4. Local Boundary Conditions

Suppose we have subbundles $E_j \subset V^R|_{\Sigma_j}$. Then the boundary condition $B_j = L^2(\Sigma_j, E_j) \subset L^2(\Sigma_j, V^R)$ is called a *local boundary condition*. The following proposition relates local boundary conditions to graph deformations of generalized Atiyah–Patodi–Singer boundary conditions, as discussed in Section 4.3.

Proposition 4.9 *Let $E_j \subset V^R|_{\Sigma_j}$ be a subbundle and $B_j = L^2(\Sigma_j, E_j)$. Then the following are equivalent:*

(i) There exists an L^2-orthogonal splitting

$$L^2(\Sigma_j, V^R) = V_j^- \oplus W_j^- \oplus V_j^+ \oplus W_j^+$$

and bounded linear maps $g_j : V_j^\mp \to V_j^\pm$ as in Definition 4.3 such that $B_j = W_j^\pm \oplus \Gamma(g_j)$;

(ii) For every $x \in \Sigma_j$ and $\xi \in T_x\Sigma_j$, $\xi \neq 0$, the projection $\pi_{(E_j)_x} : (V^R|_{\Sigma_j})_x \to (E_j)_x$ restricts to an isomorphism from the sum of eigenspaces to the negative (for $j = 0$) or positive (for $j = 1$) eigenvalues of $i\sigma_{A_j}(\xi)$ onto $(E_j)_x$.

Proof First note that the fibrewise projections $\pi_{(E_j)_x}$ yield a projection map

$$P_j : L^2(\Sigma_j, V^R) \to L^2(\Sigma_j, E_j).$$

Then (ii) is equivalent to the operator

$$P_j - \pi_{\mathrm{APS}_j(a)^\perp} : L^2(\Sigma_j, V^R) \to L^2(\Sigma_j, V^R)$$

being Fredholm for some (and then all) $a \in \mathbb{R}$, as stated in [3, Thm 7.20].

First we show that (i) implies (ii). Since the sum of a Fredholm operator and a finite rank operator is again Fredholm, we can assume that $W_j^- = W_j^+ = 0$, $V_j^- = \mathrm{APS}_j(0)$ and $V_j^+ = \mathrm{APS}_j(0)^\perp$. With respect to the splitting $L^2(\Sigma_j, V^R) = V_j^- \oplus V_j^+$, we then have by (6)

$$\pi_{\mathrm{APS}_j(0)^\perp} = \begin{pmatrix} 0 & 0 \\ 0 & \mathrm{id} \end{pmatrix} \quad \text{and} \quad P_j = \begin{pmatrix} \mathrm{id} & 0 \\ g_j & 0 \end{pmatrix} \begin{pmatrix} \mathrm{id} & -g_j^\dagger \\ g_j & \mathrm{id} \end{pmatrix}^{-1}$$

and hence

$$P_j - \pi_{\mathrm{APS}_j(0)^\perp} = \begin{pmatrix} \mathrm{id} & 0 \\ 0 & -\mathrm{id} \end{pmatrix} \begin{pmatrix} \mathrm{id} & -g_j^\dagger \\ g_j & \mathrm{id} \end{pmatrix}^{-1}$$

is an isomorphism.

Now we show that (ii) implies (i). We construct the decomposition in (i) and the map g_j for $j = 0$. The case $j = 1$ is analogous with the labels $+$ and $-$ interchanged.

Assuming that $P - \pi_{\mathrm{APS}_0(0)^\perp}$ is Fredholm, we set

$$\begin{aligned} W_0^+ &:= B_0 \cap \mathrm{APS}_0(0)^\perp, & V_0^+ &:= (W_0^+)^\perp \cap \mathrm{APS}_0(0)^\perp, \\ V_0^- &:= \pi_{\mathrm{APS}_0(0)}(B_0), & W_0^- &:= (V_0^-)^\perp \cap \mathrm{APS}_0(0). \end{aligned}$$

We then have $V_0^+ \oplus W_0^+ = L^2_{[0,\infty)}(\Sigma_0, V^R)$ and $V_0^- \oplus W_0^- = L^2_{(-\infty,0)}(\Sigma_0, V^R)$. Furthermore, for $w \in W_0^+$, we have

$$(P_0 - \pi_{\mathrm{APS}_0(0)^\perp})(w) = w - w = 0,$$

so W_0^+ is contained in the kernel of $P_0 - \pi_{\mathrm{APS}_0(0)^\perp}$ and is hence finite dimensional. For $w \in W_0^-$, we observe

$$(w, P_0 x - \pi_{\mathrm{APS}_0(0)^\perp} x)_{L^2} = (w, \pi_{\mathrm{APS}_0(0)} P_0 x)_{L^2} - (w, \pi_{\mathrm{APS}_0(0)^\perp} x)_{L^2} = 0 - 0 = 0$$

for all $x \in L^2(\Sigma_0, V^R)$. Thus, W_0^- is contained in the orthogonal complement of the range of $P_0 - \pi_{\mathrm{APS}_0(0)^\perp}$ and is hence finite dimensional.

Setting $U_0 := (W_0^+)^\perp \cap B_0$, we have $B_0 = W_0^+ \oplus U_0$, and $\pi_{\mathrm{APS}_0(0)}|_{U_0} : U_0 \to V_0^-$ is an isomorphism. Now the composition

$$g_0 : V_0^- \xrightarrow{(\pi_{\mathrm{APS}_0(0)}|_{U_0})^{-1}} U_0 \xrightarrow{\pi_{\mathrm{APS}_0(0)^\perp}} V_0^+$$

is a bounded linear map with $U_0 = \Gamma(g_0)$. Then we have $B_0 = W_0^+ \oplus \Gamma(g_0)$, which concludes the proof. □

Example 4.10 Let G be a smooth field of unitary involutions of V^R along $\Sigma_0 \cup \Sigma_1$ which anti-commutes with A_j, $j = 0, 1$. Then G anti-commutes with all $\sigma_{A_j}(\xi)$, $\xi \in T^*\Sigma_0 \cup T^*\Sigma_1$. We split $V^R|_{\Sigma_j} = E_j^+ \oplus E_j^-$ into the sum of (± 1)-eigenspaces for G.

Let $\xi \in T_x\Sigma_j$, $\xi \neq 0$. W.l.o.g., we assume $|\xi| = 1$. Then $(\mathrm{i}\sigma_{A_j}(\xi))^2 = 1$ and $\mathrm{i}\sigma_{A_j}(\xi)$ has only the eigenvalues ± 1. Since G and $\sigma_{A_j}(\xi)$ anti-commute, $(E_j^+)_x$, $(E_j^-)_x$ and the (± 1)-eigenspaces for $\mathrm{i}\sigma_{A_j}(\xi)$ all have the same dimension, namely half the dimension of $(V^R)_x$. The projections onto $E_j^\pm$ are given by $\pi_{(E_j^\pm)_x} = \frac{1}{2}(1 \pm G)$.

Now assume that $\mathrm{i}\sigma_{A_j}(\xi)u = -u$ and that $\pi_{(E_j^\pm)_x} u = 0$. Then

$$0 = \mathrm{i}\sigma_{A_j}(\xi)\pi_{(E_j^\pm)_x} u = \pi_{(E_j^\mp)_x} \mathrm{i}\sigma_{A_j}(\xi)u = -\pi_{(E_j^\mp)_x} u$$

and hence

$$u = \pi_{(E_j^\pm)_x} u + \pi_{(E_j^\mp)_x} u = 0.$$

Thus, both $\pi_{(E_j^+)_x}$ and $\pi_{(E_j^-)_x}$ are injective when restricted to the (-1)-eigenspace of $\mathrm{i}\sigma_{A_j}(\xi)$. For dimensional reasons, these restrictions form an isomorphism onto $(E_j^\pm)_x$. Thus, Proposition 4.9 applies to both $E_j = E_j^+$ and $E_j = E_j^-$. Hence, $(L^2(\Sigma_0, E_0^\pm), L^2(\Sigma_1, E_1^\pm))$ are boundary conditions in graph form. These boundary conditions are known as *chirality conditions.*

The description of the chirality boundary conditions in Definition 4.3 is as follows: $V_0^- = L^2_{(-\infty,0)}(A_0)$, $W_0^+ = \ker(A_0) \cap L^2(\Sigma_0, E_0^\pm)$, $V_0^+ = L^2_{(0,\infty)}(A_0)$, $W_0^- = \ker(A_0) \cap L^2(\Sigma_0, E_0^\mp)$, $g_0 = \pm G$ and similarly for Σ_1.

In order to conclude from Proposition 4.5 that the chirality conditions yield a Fredholm operator, one would need to verify condition (A) or (B). Condition (A) is not satisfied but, for (B), this is not clear. So let us look at a special case.

Example 4.11 Let h be a fixed Riemannian metric on the closed spin manifold Σ and equip $M = \mathbb{R} \times \Sigma$ with the 'ultrastatic' metric $-dt^2 + h$. Then (2) simplifies to

$$D = -\gamma(\nu)(\nabla_\nu + \mathrm{i}A) = -\gamma(\nu)\Big(-\frac{\partial}{\partial t} + \mathrm{i}A\Big),$$

where A is the Dirac operator on (Σ, h). If we now solve the equation $Du = 0$ with initial condition $u(0,\cdot) = u_0$, where u_0 is an eigenspinor for A, $Au_0 = \lambda u_0$, then the solution is given by $u(t,x) = e^{\mathrm{i}\lambda t}u_0(x)$. We choose $\Sigma_j = \{j\} \times \Sigma$. Then $Qu_0 = e^{\mathrm{i}\lambda}u_0$.

Specializing even further, we put $\Sigma = S^1$ and choose h such that Σ has length 1. The one-dimensional sphere has two spin structures. For the so-called trivial spin structure, the Dirac operator A has the eigenvalues $\lambda = 2\pi k$, where $k \in \mathbb{Z}$. Thus, $e^{\mathrm{i}\lambda} = 1$ and hence $Q = \mathrm{id}$, where we have identified $L^2(\Sigma_0, V^R) = L^2(S^1, SS^1) = L^2(\Sigma_1, V^R)$ and $A_0 = A = A_1$. Now choose

$$V_0^- = L^2_{(-\infty,0)}(A), \quad V_0^+ = L^2_{(0,\infty)}(A), \qquad W_0^- = \ker(A), \qquad W_0^+ = 0,$$
$$V_1^- = L^2_{(-\infty,0)}(A), \quad V_1^+ = L^2_{(0,\infty)}(A), \qquad W_1^- = 0, \qquad W_1^+ = \ker(A).$$

Let $g_0 : L^2_{(-\infty,0)}(A) \to L^2_{(0,\infty)}(A)$ be an isomorphism of Hilbert spaces and put $g_1 = g_0^{-1}$. Then

$$B_0 = \Gamma(g_0) = \{v + g_0 v \mid v \in L^2_{(-\infty,0)}(A)\} = \{g_1 w + w \mid w \in L^2_{(0,\infty)}(A)\} = \Gamma(g_1) = B_1.$$

Now $(B_0, Q^{-1}B_1) = (B_0, B_1) = (B_0, B_0)$ is clearly not a Fredholm pair. This shows that, in Proposition 4.5, Conditions (A) and (B) cannot be dropped. Without these assumptions, boundary conditions in graph form do not give rise to a Fredholm operator in general.

Example 4.12 Finally, we want to point out that one also obtains Dirac–Fredholm pairs (B_0, B_1) if B_0 is finite dimensional and B_1 has finite codimension, or vice versa. According to Theorem 4.1, the Dirac operator with these boundary conditions is Fredholm with index $\dim(B_0) - \mathrm{codim}(B_1)$.

REFERENCES

[1] Michael F Atiyah and Isadore M. Singer, *The index of elliptic operators on compact manifolds* Bull. Amer Math. Soc. **69**(1963), no. 3, 422–33.

[2] Michael F. Atiyah, Vijay K. Patodi and Isadore M. Singer, *Spectral asymmetry and Riemannian geometry. I* Math. Proc. Cambridge Philos. Soc. 77 (1975), no. 1, 43–69.

[3] Christian Bär and Werner Ballmann, *Boundary value problems for elliptic differential operators of first order* in: H.-D. Cao and S.-T Yau (eds), Surveys in differential geometry, Vol. 27, Int. Press, Boston, MA, 2012, pp. 1–78.

[4] Christian Bär and Werner Ballmann, *Guide to boundary value problems for Dirac-type operators* in: W Ballmann et al. (eds), Arbeit- stagung Bonn 2013, Prog. Math. 319, Birkhäuser Cham, 2016, pp. 43–80.

[5] Christian Bär, Paul Gauduchon and Andrei Moroianu, *Generalized cylinders in semi-Riemannian and Spin geometry,* Math. Z. **249** (2005), no. 3, 545–80.

[6] Christian Bär and Alexander Strohmaier, *An index theorem for Lorentzian manifolds with compact space-like Cauchy boundary* to appear in Amer. J. Math arXiv:1506.00959 (2015).

[7] Christian Bär and Alexander Strohmaier, *A rigorous geometric derivation of the chiral anomaly in curved backgrounds* Comm. Math. Phys. **347** (2016), no. 3, 703–21.

[8] Helga Baum, *Spin-Strukturen und Dirac-Operatoren über pseudoriemannschen Mannigfaltigkeiten* TeubnerTexte zur Mathematik, Vol. 41, BSB B. G. Teubner Verlagsgesellschaft, Leipzig, 1981.

[9] John K. Beem, Paul E. Ehrlich and Kevin L. Easley *Global Lorentzian geometry* second edn, Monographs and Textbooks in Pure and Applied Mathematics, Vol. 202, Marcel Dekker Inc., New York, 1996.

[10] Antonio N Bernal and Miguel Sánchez, *Further results on the smoothability of Cauchy hypersurfaces and Cauchy time functions* Lett. Math. Phys. **77** (2006), no. 2, 183–97.
[11] Bernhelm Booß-Bavnbek and Krzysztof P Wojciechowski, *Elliptic boundary problems for Dirac operators* Mathematics: Theory and Applications, Birkhäuser, Boston, 1993.
[12] Nigel Hitchin, *Harmonic spinors* Adv. Math. **14** (1974), no, 1, 1–55.
[13] Tosio Kato, *Perturbation theory for linear operators* Classics in Mathematics, Springer-Verlag, Berlin, 1995. Reprint of the 1980 edition.
[14] André Lichnerowicz, *Spineurs harmoniques* C. R. Acad. Sci. Paris **257** (1963), no. 7, 7–9.
[15] Olaf Müller, *A note on invariant temporal functions* Lett. Math. Phys. **106** (2016), no. 7, 959–71.
[16] Barrett O'Neill, *Semi-Riemannian geometry: With applications to relativity* Pure and Applied Mathematics, Vol. 103, Academic Press, Inc., New York, 1983.

2

Torsion of Elliptic Curves and Unlikely Intersections

FEDOR BOGOMOLOV[1,2]
HANG FU[1]
YURI TSCHINKEL[1,3]

[1] *Courant Institute of Mathematical Sciences, N.Y.U. 251 Mercer Str., New York, NY 10012, U.S.A. Email: fu@cims.nyu.edu*
[2] *National Research University Higher School of Economics, Russian Federation, AG Laboratory, HSE, 7 Vavilova Str., Moscow, Russia, 117312, Email: bogomolo@cims.nyu.edu*
[3] *Simons Foundation, 160 Fifth Av., New York, NY 10010, U.S.A., Email: tschinkel@cims.nyu.edu*

ABSTRACT. We study effective versions of unlikely intersections of images of torsion points of elliptic curves on the projective line.

To Nigel Hitchin, with admiration.

Introduction

Let k be a field of characteristic $\neq 2$, and $\bar{k}/k$ an algebraic closure of k. Let E be an elliptic curve over k, presented as the double cover

$$\pi : E \to \mathbb{P}^1,$$

ramified in four points, and $E[\infty] \subset E(\bar{k})$ the set of its torsion points. In [1] we proved:

Theorem 1 *If E_1, E_2 are non-isomorphic elliptic curves over $\bar{\mathbb{Q}}$, then*

$$\pi_1(E_1[\infty]) \cap \pi_2(E_2[\infty])$$

is finite.

Key words and phrases. Elliptic curves, torsion points, fields.

Bogomolov, F., Fu, H., Tschinkel, Y., *Torsion of Elliptic Curves and Unlikely Intersections.* In: *Geometry and Physics: A Festschrift in Honour of Nigel Hitchin*, Jørgen Ellegaard Andersen, Andrew Dancer, Oscar García-Prada (Eds): Oxford University Press (2018). DOI: 10.1093/oso/9780198802006.003.0002

Here, we explore effective versions of this theorem, specifically, the size and structure of such intersections (see [5] for an extensive study of related problems). We expect the following universal bound:

Conjecture 2 (Effective Finiteness–EFC-I) *There exists a constant $c > 0$ such that, for every pair of non-isomorphic elliptic curves E_1, E_2 over $\mathbb{C}$, we have*

$$\pi_1(E_1[\infty]) \cap \pi_2(E_2[\infty]) < c.$$

We say that two subsets of the projective line

$$S := \{s_1, \ldots, s_n\}, \quad S' := \{s'_1, \ldots, s'_n\} \subset \mathbb{P}^1(\bar{k})$$

are projectively equivalent, and write $S \sim S'$, if there is a $\gamma \in \mathrm{PGL}_2(\bar{k})$ such that (modulo permutation of the indices) $s_i = \gamma(s'_i)$, for all i.

Let E be an elliptic curve over k, $e \in E$ the identity, and

$$\begin{array}{ccc} E & \xrightarrow{\iota} & E \\ x & \mapsto & -x \end{array}$$

the standard involution. The corresponding quotient map

$$\pi : E \to E/\iota = \mathbb{P}^1$$

is ramified in the image of the 2-torsion points of $E(\bar{k})$. Conversely, for

$$r := \{r_1, r_2, r_3, r_4\} \subset \mathbb{P}^1(\bar{k}),$$

the double cover

$$\pi_r : E_r \to \mathbb{P}^1$$

with ramification in r defines an elliptic curve; given another such r', the curves E_r and $E_{r'}$ are isomorphic (over $\bar{k}$) if and only if $r \sim r'$, in particular, the image of 2-torsion determines the elliptic curve, up to isomorphism.

Let $E_r[n] \subset E_r(\bar{k})$ be the set of elements of order *exactly* n, for $n \in \mathbb{N}$. The behaviour of torsion points of other small orders is also simple:

$$\pi_r(E_r[3]) \sim \{1, \zeta_3, \zeta_3^2, \infty\},$$

where ζ_3 is a non-trivial third root of 1, and

$$\pi_r(E_r[4]) \sim \{0, 1, -1, i, -i, \infty\}.$$

In particular, up to projective equivalence, these are *independent* of E_r. However, for all $n \geq 5$, the sets $\pi_r(E_r[n])$, modulo $\mathrm{PGL}_2(\bar{k})$, do depend on E_r, and it is tempting to inquire into the nature of this dependence.

In this note, we study $\pi_r(E_r[n])$, for varying curves E_r and varying n. Our goal is to establish effective and uniform finiteness results for intersections

$$\pi_r(E_r[n]) \cap \pi_{r'}(E_{r'}[n']), \quad n, n' \in \mathbb{N},$$

for elliptic curves $E_r, E_{r'}$, defined over k. We formulate several conjectures in this direction and provide evidence for them.

The next step is to ask, given elliptic curves $E_r, E_{r'}$ over $\bar{k}$, when is

$$r \subset \pi_{r'}(E_{r'}[\infty])?$$

We modify this question as follows: which minimal subsets $\tilde{L} \subset \mathbb{P}^1(\bar{k})$ have the property

$$r \subset \tilde{L} \quad \Rightarrow \pi_r(E_r[\infty]) \subset \tilde{L}?$$

The sets $\tilde{L}$ carry involutions, obtained from the translation action of the 2-torsion points of E on E, which descends, via π, to an action on $\mathbb{P}^1$ and defines an embedding of $\mathbb{Z}/2 \oplus \mathbb{Z}/2 \hookrightarrow \mathrm{PGL}_2(\bar{k})$. It is conjugated to the standard embedding of $\mathbb{Z}/2 \oplus \mathbb{Z}/2$, generated by involutions

$$z \mapsto -z \quad \text{and} \quad z \mapsto 1/z$$

acting on $\tilde{L}$. This observation is crucial for the discussion in Section 4, where we prove that, modulo projectivities, $L := \tilde{L} \setminus \{\infty\}$ are fields.

Acknowledgements: The first author was partially supported by the Russian Academic Excellence Project '5-100', a Simons Fellowship and the EPSRC programme grant EP/M024830. The second author was supported by the MacCracken Program offered by New York University. The third author was partially supported by NSF grant 1601912.

1. Generalities

Let $j : \mathcal{E} \to \mathbb{P}^1$ be the standard universal elliptic curve, with j the j-invariant morphism. Consider the diagram

$$\begin{array}{ccc}
E_\lambda & \xrightarrow{\iota} & P_\lambda \\
\cap & & \cap \\
\mathcal{E} & \xrightarrow{\iota} & \mathcal{P} \\
{\scriptstyle j}\downarrow & & \downarrow{\scriptstyle j} \\
\mathbb{P}^1 & = & \mathbb{P}^1
\end{array}$$

assigning to each fibre $E_\lambda := j^{-1}(\lambda)$ the quotient $P_\lambda := \pi(E_\lambda) \simeq \mathbb{P}^1$, by the involution ι : $x \mapsto -x$ on E_λ. (This is well defined even for singular fibres of j.)

Note that $\mathcal{P} \to \mathbb{P}^1$ is a PGL_2-torsor. Taking the fibrewise n-symmetric product

$$P_\lambda \mapsto \mathrm{Sym}^n(P_\lambda),$$

we have the associated PGL_2-torsors

$$j_n : \mathcal{P}_n := \mathrm{Sym}^n(\mathcal{P}) \to \mathbb{P}^1.$$

Taking PGL_2-invariants, we have a canonical projection,

$$\mathrm{Sym}^n(P_\lambda) \to \mathcal{M}_{0,n}(P_\lambda) \simeq \mathcal{M}_{0,n},$$

to the moduli space of n-points on $\mathbb{P}^1$. The associated PGL_2-torsor is trivial; fixing a trivialization, we obtain the morphism

$$\mu_n : \mathcal{P}_n \to \overline{\mathcal{M}}_{0,n}.$$

For every $N \in \mathbb{N}$, we have the modular curve $X(N) \to \mathbb{P}^1$, parametrizing pairs of elliptic curves together with N-torsion subgroups. The involution ι induces an involution on every $X(N)$, so we have the induced quotient

$$X(N) \to Y(N) := X(N)/\iota.$$

Since the family $j : \mathcal{E} \to \mathbb{P}^1$ has maximal monodromy $\mathrm{SL}_2(\mathbb{Z})$, the curves $X(N)$ and $Y(N)$ are irreducible. We have a natural embedding $Y(N) \hookrightarrow \mathcal{P}$. Put

$$Y := \cup_{N \in \mathbb{N}} Y(N)$$

and consider

$$\mathrm{Sym}^n(Y) \hookrightarrow \mathcal{P}_n \to \overline{\mathcal{M}}_{0,n}.$$

Note that $\mathrm{Sym}^n(Y)$ is a union of infinitely many irreducible curves, each corresponding to an orbit of the action of the monodromy group $\mathrm{PGL}_2(\mathbb{Z})$ on the generic fibre of the restriction of j_n to $\mathrm{Sym}^n(Y)$. Let $Y_{n,\omega} \subset \mathrm{Sym}^n(Y)$ be an irreducible component corresponding to a $\mathrm{PGL}_2(\mathbb{Z})$-orbit ω (for the monodromy action, as above). We now formulate conjectures about μ_n, for small n, which guide our approach to the study of images of torsion points.

Conjecture 3 *The map*

$$\mu_4 : Y_{4,\omega} \to \overline{\mathcal{M}}_{0,4} = \mathbb{P}^1$$

is finite surjective, for all but finitely many ω.

Conjecture 4 *The map*

$$(\mu_4, j) : Y_{4,\omega} \to \overline{\mathcal{M}}_{0,4} \times \mathbb{P}^1$$

is a rational embedding, for all but finitely many ω.

Conjecture 5 *The map*

$$\mu_5 : Y_{5,\omega} \to \overline{\mathcal{M}}_{0,5}$$

is a rational embedding, for all but finitely many ω. *Moreover, if for some distinct orbits* ω *and* ω' *the corresponding images* $\mu_5(Y_{5,\omega})$ *and* $\mu_5(Y_{5,\omega'})$ *are curves, then they are different.*

Conjecture 6 *The map*

$$\mu_6 : Y_{6,\omega} \to \overline{\mathcal{M}}_{0,6}$$

is a rational embedding, for all but finitely many ω. *Moreover, if* $\mu_6(Y_{6,\omega})$ *is a curve, then there exist at most finitely many* ω' *such that:*

- $\mu_6(Y_{6,\omega'})$ *is a curve,*
- $\mu_6(Y_{6,\omega}) \cap \mu_6(Y_{6,\omega'}) \neq \emptyset$.

2. Examples and Evidence

We now discuss examples and evidence for the conjectures presented in Section 1.

Example 7 We have:

- $\mu_4(\mathrm{Sym}^4(Y(2))) \simeq \overline{\mathcal{M}}_{0,4} = \mathbb{P}^1$,
- $\mu_4(\mathrm{Sym}^4(Y(3)))$ is a point in $\overline{\mathcal{M}}_{0,4}$.

Consider $\mathrm{Sym}^4(Y(4))$. Note that $\pi(E[4]) = \{0, 1, -1, i, -i, \infty\}$ is an orbit of the symmetric group $\mathfrak{S}_4$, acting on $\mathbb{P}^1$. The pairs

$$(0, \infty), (1, -1), (i, -i)$$

are pairs of stable points for three even involutions in $\mathfrak{S}_4$, and the action of $\mathfrak{S}_4$ is transitive on pairs and inside each pair. There are two different $\mathfrak{S}_4$-orbits of 4-tuples: either the orbit contains two pairs of vertices such as $(0, \infty), (1, -1)$, or a pair and two points from different pairs $(0, \infty), (1, i)$. Thus, $\mathrm{Sym}^4(Y(4))$ has two components which project to different points modulo PGL_2; therefore, there exist exceptional orbits ω such that $\mu_4(Y_{4,\omega})$ is a point.

Lemma 8 *If* $\mu_4(Y_{4,\omega})$ *is a point, then all cross-ratios of 4-tuples of points parametrized by* $Y_{4,\omega}$ *are constant.*

Proof The map μ_4 can be viewed as the composition

$$(\mathbb{P}^1)^4 \xrightarrow{cr} (\mathbb{Z}/2 \oplus \mathbb{Z}/2)\backslash(\mathbb{P}^1)^4/\mathrm{PGL}_2 = \mathbb{P}^1_1 \to \mathfrak{S}_3\backslash\mathbb{P}^1_1.$$

Thus, we have the diagram

$$\begin{array}{ccc} (\mathbb{P}^1)^4 & \xrightarrow{cr} & (\mathbb{Z}/2 \oplus \mathbb{Z}/2)\backslash(\mathbb{P}^1)^4/\mathrm{PGL}_2 \\ \downarrow & & \downarrow \mathfrak{S}_3 \\ \mathfrak{S}_4\backslash(\mathbb{P}^1)^4 & \longrightarrow & \mathfrak{S}_4\backslash(\mathbb{P}^1)^4/\mathrm{PGL}_2. \end{array}$$

Note that any irreducible $Y_{4,\omega}$ lifts to a union of connected components $Y_{4,\omega,i} \subset (\mathbb{Z}/2 \oplus \mathbb{Z}/2)\backslash Y^4$, where the cross-ratio is well defined. Thus, if μ_4 is a rational function of cross-ratio on any 4-tuple of points and if μ_4 is constant, then the cross-ratio is also constant. □

Proposition 9 *There exist orbits ω such that*

$$\mu_4 : Y_{4,\omega} \to \mathbb{P}^1$$

is surjective.

Proof The singular fibre $\mathcal{E}_\infty := j^{-1}(\infty)$ is an irreducible rational curve with one node p_∞. The group scheme $\cup_{d|n}\mathcal{E}[d]$, whose generic fibre is isomorphic to $\mathbb{Z}/n \oplus \mathbb{Z}/n$, specializes to $\{\zeta_n^i\} \subset \mathbb{G}_m = \mathcal{E}_\infty \setminus p_\infty$. Let $\mathcal{E}_\infty[n]$ be the specialization of $\mathcal{E}[n]$; then:

- $\mathcal{E}_\infty[n] \subset \{\zeta_n^i\}$,
- there exists a subgroup scheme $\mathcal{W}_n \simeq \mathbb{Z}/n \subset \mathbb{Z}/n \oplus \mathbb{Z}/n$ in the group scheme of points killed by n, specializing to $\mathcal{E}_\infty$, while the complementary branches specialize to p_∞.

Taking the quotient by ι, we find that $((\mathbb{Z}/n \oplus \mathbb{Z}/n) \setminus \mathbb{Z}/n)/\iota$ specializes to 0 in the fibre $\mathbb{P}^1_\infty$, and all other points specialize to subset in $(\mathbb{Z}/n)/\iota$; the limit depends on the selected direction of specialization.

Assume that we have distinct points $\{z_1, z_2, z_3, z_4\} \subset \pi(E[n])$, for a smooth fibre E of $\mathcal{E}$, such that

$$z_1, z_2 \in W_n/\iota \quad \text{and} \quad z_3, z_4 \notin W_n/\iota.$$

The z_1, z_2 can be specialized to different non-zero points in $\mathcal{E}_\infty/\iota$, and z_3, z_4 will specialize to 0.

Assume that μ_4 is constant, i.e. the cross-ratio is constant. Since z_3, z_4 will specialize to 0, the cross-ratio equals 1. Then

$$(z_1 - z_3)(z_2 - z_4) = (z_2 - z_3)(z_1 - z_4),$$

and

$$z_1(z_3 - z_4) = z_2(z_3 - z_4).$$

Near the special fibre, $z_3 \neq z_4$; thus, $z_1 = z_2$, a contradiction. Thus, on orbits of this type, μ_4 is not constant and hence is surjective. □

3. A Geometric Approach to Effective Finiteness

Let $E := E_r, E' := E_{r'}$ be elliptic curves. Consider the diagram

$$\begin{array}{ccc} C & \longrightarrow & E \times E' \\ \downarrow & & \downarrow \\ \Delta & \longrightarrow & \mathbb{P}^1 \times \mathbb{P}^1 \end{array}$$

where $C \subset E \times E'$ is the fibrewise product over the diagonal $\Delta \subset \mathbb{P}^1 \times \mathbb{P}^1$. If $r \neq r'$, then C has genus ≥ 2. By Raynaud's theorem [4],

$$C(\bar{k}) \cap E[\infty] \times E'[\infty]$$

is finite, since it is the preimage of $\pi(E[\infty]) \cap \pi(E'[\infty]) \subset \Delta$; the latter set is also finite. This finiteness argument appeared in [1].

Consider the curve C occurring in this construction. We have the diagram

$$\begin{array}{ccc} C & \xrightarrow{\sigma} & E \\ {\scriptstyle \sigma'}\downarrow & & \\ E' & & \end{array}$$

where σ, σ' are involutions with fixed points c_1, c_2 and c_1', c_2', respectively. Assume that

$$r \cap r' = \{0, 1, \infty\}.$$

Then the product involution $\sigma\sigma'$ on $C \subset E \times E'$ has fixed points in the six preimages of the points $\{0, 1, \infty\} \subset \Delta_{\mathbb{P}^1} \subset \mathbb{P}^1 \times \mathbb{P}^1$ (diagonally), i.e. it is the hyperelliptic involution. Thus, there is an action of $\mathbb{Z}/2 \oplus \mathbb{Z}/2$ on C, induced by the covering maps π and π'. The curve $C \subset A = E \times E'$ has self-intersection $C^2 = 8$ since it is a double cover of both E and E' and its class is equal to $2(E + E')$:

- If the genus $\mathsf{g}(C) = 2$ (three such points), then the image of C in its Jacobian $J(C)$ has self-intersection 2. Consider the map

$$\nu : J(C) \to A := E \times E'$$

and let n be its degree. The preimage $\nu^{-1}(C) \subset J(C)$ has self-intersection $8n$. On the other hand, its homology class is equal to n translations of C; hence, it has self-intersection $2n^2$ and thus $n = 4$. Moreover, $\ker(\nu) = \mathbb{Z}/2 \oplus \mathbb{Z}/2$, generated by the pairwise differences of preimages of points $\{0, 1, \infty\}$. Thus, $J(C)$ is 4-isogenous to $A := E \times E'$, and $\nu(C)$ is singular, with nodes exactly at the preimages of $\{0, 1, \infty\} \subset \Delta_{\mathbb{P}^1}$. Consider a point $c \in C \subset J(C)$ and assume that $\nu(c)$ has order m with respect to $0 \in A$. Then, c has order m or $2m$ in $J(C)$, with respect to $0 \in J(C)$. Hence, the corresponding curve $Y(m) \subset \mathbb{P}^1 \times \mathbb{P}^1$ (viewed as a moduli space of pairs E, E') is given as an intersection of genus 2 curves containing a point of order m or $2m$, respectively. This is a locus in the moduli space $\mathcal{M}_2$ of genus 2 curves.

- If $\mathsf{g}(C) = 3$ (two such points), then there are three quotients of C which are elliptic curves E_1, E_2, E_3, with involutions $\sigma_i \in \mathbb{Z}/2 \oplus \mathbb{Z}/2$ fixing four points on E_i which are invariant under the hyperelliptic involution given by complement to σ_j. The kernel of

$$\nu_i : J(C) \to E_j \times E_k$$

contains E_i, for $i, j, k \in \{1, 2, 3\}$.

- If $\mathsf{g}(C) = 4$, then C is $C/\sigma_i = E_i, i = 1, 2$ and $C/\sigma_1\sigma_2 = C'$, where $\mathsf{g}(C') = 2$ and there are exactly two ramification points on C'.
- If $\mathsf{g}(C) = 5$, then $C/\sigma_1\sigma_2 = C'$ is a hyperelliptic curve of genus 3, and the covering is an unramified double cover.

Remark 10 Assume that there is $b \in \mathbb{P}^1$ and a subset $S \subset C(\bar{k})$ such that $S + b \subset C \subset E \times E'$. Then

$$\#S \leq 8 = C^2 = C \cap (C + b);$$

hence we have at most eight points $c_i \in \mathbb{P}^1$ such that, for x-coordinates, $c_i +_1 b = c_i +_2 b$, where the summation $+_1$ corresponds to the summation on the first curve, and $+_2$ to that on the second.

Remark 11 The construction can be extended to products of more than two elliptic curves. We may consider

$$\pi := \prod_{i=1}^{r} \pi_i : \mathcal{A} := \prod E_i \to \mathcal{P} := \prod \mathbb{P}^1_i.$$

The ramification divisor of $\pi : \mathcal{A} \to \mathcal{P}$ is a union of products of projective lines. Let $\Delta = \mathbb{P}^1 \subset \mathcal{P}$ be the diagonal; then there exists the canonical identification $\delta_i : \mathbb{P}^1_i \simeq \Delta$. If $p \in \Delta$ is contained in $\delta_i(\pi_i(E_i[\infty]))$, for all i, then the preimage of p in $\mathcal{A}$ is contained in the preimage of the diagonal. This is a curve of genus at least 2, provided there exist E_i, E_j with $r_i \neq r_j$. Then the set of such p is finite. In particular, if E is defined over a number field k, and p is defined over a proper subfield, then p is also in the image torsion points of $\gamma(E)$, where γ is a Galois conjugation. Hence, the existence of torsion points with x-coordinates in a smaller field has a geometric implication.

We expect the following version of Conjecture 2:

Conjecture 12 (Effective Finiteness–EFC-II) *There exists a constant $c > 0$ such that, for every elliptic curve E_r over a number field and every $\gamma \in \mathrm{PGL}_2(\bar{\mathbb{Q}})$ with $\gamma(r) \neq r$, we have*

$$\pi_r(E_r[\infty]) \cap \pi_\gamma(E_\gamma[\infty]) < c.$$

4. Fields Generated by Elliptic Division

In this section, we explore properties of subsets of $\mathbb{P}^1(\bar{k})$ generated by images of torsion points, following closely [1]. For

$$r := \{r_1, r_2, r_3, r_4\} \subset \mathbb{P}^1(\bar{k}),$$

a set of four distinct points, let E_r be the corresponding elliptic curve defined in the introduction. Let

$$\tilde{L}_r \subset \mathbb{P}^1(\bar{k})$$

be the smallest subset such that, for every $E_{r'}$ with $r' \subset \tilde{L}_r$, we have $\pi_{r'}(E_{r'}[\infty]) \subset \tilde{L}_r$.

Theorem 13 [1] *Let k be a number field. For every $a \in k \setminus \{0, \pm 1, \pm i\}$ and*

$$r = r_a := \{a, -a, a^{-1}, -a^{-1}\} \subset \mathbb{P}^1(k), \tag{1}$$

the set

$$L_a := \tilde{L}_{r_a} \backslash \{\infty\}$$

is a field.

At first glance, it is rather surprising that such a simple and natural construction, inspired by comparisons of x-coordinates of torsion points of elliptic curves, produces a field. The conceptual reason for this is the rather peculiar structure of 4-torsion points of elliptic curves: translations by 2-torsion points yields, upon projection to $\mathbb{P}^1$, *two* standard commuting involutions on $\mathbb{P}^1(\bar{k})$, which make it possible to define addition and multiplication on L_a.

We may inquire about the arithmetic and geometric properties of the fields L_a. For $a \in \bar{k}$, we let $k(a) \subset \bar{k}$ denote the smallest subfield containing a. We have:

- For every $a \in \bar{k}$, the field L_a is a Galois extension of $\mathbb{Q}(a)$.
- For every k of characteristic zero, L_a contains $\mathbb{Q}^{ab}$, the maximal abelian extension of $\mathbb{Q}$.

- The field L_ζ, where ζ is a primitive root of order 8, is contained in any field L_a. Indeed, the corresponding elliptic curve E has the ramification subset

$$\{\zeta, \zeta^3, \zeta^5, \zeta^7\},$$

which is projectively equivalent to $\{1, -1, i, -i\} \subset \pi(E[4])$. Since $\pi(E[4])$ projectively does not depend on the curve E, we obtain that $L_\zeta \subset L_a$, for all a. The same holds for L_a where E_a is isomorphic to E_3 (an elliptic curve with an automorphism of order 3).

- The field L_a is *contained* in a field obtained as an iteration of Galois extensions with Galois groups either abelian or $\mathrm{PGL}_2(\mathbb{F}_q)$, for various prime powers q. Is L_a equal to such an extension? As soon as the absolute Galois group is not equal to a group of this type, for example, for a number field k, we have

$$L_a \subsetneq \bar{k}.$$

- Let $a, a' \in \bar{\mathbb{Q}}$ be algebraic numbers such that $\mathbb{Q}(a) = \mathbb{Q}(a')$. Then $L_a = L_{a'}$. Varying $a \in \bar{\mathbb{Q}}$, we obtain a supply of interesting infinite extensions $L_a/\mathbb{Q}$.

The rest of this section is devoted to the proof of Theorem 13.

Proof Let $r_0 := \{0, \infty, 1, -1\}$ and put $L := \tilde{L}_{r_0} \setminus \{\infty\}$. Let

$$\pi := \pi_{r_a} : E_{r_a} \to \mathbb{P}^1$$

be the elliptic curve with ramification in r_a. Since

$$\{0, \infty, \pm 1\} \subset \pi(E_{r_a}[4]),$$

we have $L \subset L_a$, for all a. We first show that L is a field.

Step 1. $L\backslash\{0\}$ is a multiplicative group. Indeed, for any $b \in L\backslash\{0\}$, we have

$$r_0 := \{0, 1, -1, \infty\} = b^{-1} \cdot \{0, b, -b, \infty\} =: r_b$$

and hence

$$L_{r_b} = b \cdot L_{r_0} = b \cdot L.$$

Since $b^{-1}, -b^{-1} \in L$, we also have $\{0, 1, -1, \infty\} \subset b \cdot L$. Thus, $L \subset bL$. Similarly, $L \subset b^{-1} \cdot L$ or $b \cdot L \subset L$, which implies $L = bL$. Thus, for any $a, b \in L$, we have $ab \in L$ and, since the same holds for $ab^{-1}, b \neq 0$, we obtain $L \setminus \{0\} \subset \bar{k}^\times$.

Step 2. Let

$$\mathrm{Aut}_L := \{\gamma \in \mathrm{PGL}_2(\bar{k}) \mid \gamma(\tilde{L}) \subset \tilde{L}\}$$

be the subgroup preserving $\tilde{L}$. It is non-trivial, since it contains $L \setminus \{0\}$ as a multiplicative subgroup, together with the involution $x \mapsto x^{-1}$. Consider

$$\gamma_1 : x \mapsto (x-1)/(x+1).$$

It is an involution with $\gamma_1(\infty) = 1, \gamma_1(0) = -1$, and hence γ_1 comes from $r := \{0, 1, -1, \infty\}$. Thus, it maps L into L, and $\gamma_1 \in \mathrm{Aut}_L$.

Consider any pair of distinct elements $\{b, c\} \subset L$: it can be transformed into $\{0, 1\}$ by an element from Aut_L. If $b \neq 0, \infty$, then, dividing by b, we obtain $\{1, c/b\}$ and $\gamma_1(\{1, c/b\}) = \{0, 1\}$. If $b = 0$ and $c \neq \infty$, then, dividing by c, we obtain $\{0, 1\}$. If $b = 0, c = \infty$, then $\gamma_1(\{0, \infty\}) = \{-1, 1\}$ and we reduce to the first case.

Step 3. L is closed under addition. We show that $\gamma : x \mapsto x + 1$ is contained in Aut_L: by Step 2, there exists a $g \in \mathrm{Aut}_L$ which maps $\{-1, \infty\}$ to $\{0, \infty\}$ and hence $\{-1, 0, \infty\}$ to $\{0, b, \infty\}$, for some $b \in L \setminus \{0\}$. Then $b^{-1}g \in \mathrm{Aut}_L$ maps $\{-1, 0, \infty\}$ to $\{0, 1, \infty\}$ and hence $b^{-1}g(x) = \gamma(x) = x + 1$. Thus, for any $a \in L$, we have $a + b = b(a/b + 1) \in L$, which shows that L is an abelian group.

Now let us turn to the general L_a.

Step 4. Note that $L \subset L_a$ and that L_a is closed under taking square roots. Indeed, for any $a \in L$ and E_r with $r := \{0, 1, a, \infty\}$, we have $\sqrt{a} \in \pi_r(E_r[4])$ and hence $\sqrt{a} \in L_a$. Furthermore, for any $a, b \in L_a$ we have $\sqrt{ab} \in L_a$. Indeed, consider the curve E_r with $r = \{0, a, b, \infty\}$. Then $\sqrt{ab} \in \pi(E_r[4])$, since the involution $z \to ab/z$ is contained in the subgroup $\mathbb{Z}/2 \oplus \mathbb{Z}/2$ corresponding to the 2-torsion on E_r, and its invariant points are in $\pi_r(E_r[4])$. Iterating, we obtain that

$$\sqrt[2^{m-1}]{b_1 \cdots b_m} \in \tilde{L}_a \setminus \{\infty\} \text{ for all } b_i \in \tilde{L}_a \setminus \{\infty\}.$$

Step 5. For all $b \in L_a$, and $c \in L$, we have $\sqrt{b + c} \in L_a$. Indeed, for $c \in L$, we know that there is a solution $d \in L$ of the quadratic equation $d^2 + d + c = 0$. Consider the curve E_r for $r := \{\infty, b, d, d + 1\}$. Then

$$d \pm \sqrt{b - d} \in \pi(E_r[4])$$

and hence $d \pm \sqrt{b - d} \in L_a$. Thus,

$$\sqrt{(\sqrt{b-d} + d)(\sqrt{b-d} - d)} = \sqrt{b - d^2 - d} = \sqrt{b + c} \in L_a.$$

Step. 6. Let $P_m \in L[x]$ be a monic polynomial of degree m and let $b \in L_a$. Then there is an $N(m) \in \mathbb{N}$ such that

$$\sqrt[4^{N(m)}]{P_m(b)} \in L_a.$$

Indeed, we have

$$P_m(b) = c_m + b(c_{m-1} + b(c_{m-2} + \cdots)\cdots).$$

The statement holds for $m = 1$ by Step 4. Assume that it holds for $m - 1$. Then $c_{m-1} + b(c_{m-2} + \cdots) = d^{4^{N(m-1)}}$ for some $d \in L_a$. We can then write

$$P^m(b) = c_m + bd^{4^{N(m-1)}};$$

by taking $t = \sqrt[4^{N(m-1)}]{b}$ and $u_m = \sqrt[4^{N(m-1)}]{c_m}$, we obtain

$$P^m(b) = \prod(t + \zeta^i u_m),$$

where $t \in L_a, u_m \in L$, and ζ^i runs through the roots of unity of order $4^{N(m-1)}$.

By Steps 4 and 5, we obtain that the $4^m 4^{N(m-1)}$-th root of $P_m(b)$ is contained in L_a; thus, the result holds for $N(m) = 4^{N(m-1)}$.

Step 7. Let $b \in L_a$ be any algebraic element over L. Then the field $L(b)$ is a finite extension of L and there is an $n \in \mathbb{N}$ such that any $x \in L(b)$ can be represented as a monic polynomial of b with coefficients in L of degree $\leq n$. For such n, we define a power 4^N such that $\sqrt[4N]{x} \in L_a$, but then any element in $L(b)$ is in L_a. □

Remark 14 In the proof, we have only used points in $\pi(E[4])$. Therefore, for any subset $D \subset \mathbb{N}$ containing 4, we can define $L_{a,D}$ as the smallest subset containing all $\pi(E[n])$ for all $n \in D$ and all elliptic curves obtained as double covers with ramification in $L_{a,D}$. It will also be a field.

For example, if $D = \{3, 4\}$, then $L_{a,D}$ is exactly the closure of L_a under abelian degree 2 and 3 extensions, since $\mathrm{PGL}_2(\mathbb{F}_2) = \mathfrak{S}_3$ and $\mathrm{PGL}_2(\mathbb{F}_3) = \mathfrak{S}_4$ and both groups are solvable with abelian quotients of exponent 3 or 2.

On $(\mathrm{Sym}^4(\mathbb{P}^1(\bar{\mathbb{Q}}))\setminus\Delta)/\mathrm{PGL}_2(\bar{\mathbb{Q}})$, we can define a directed graph structure DGS, postulating that

$$r_z = \{z_1, z_2, z_3, z_4\} \to r_w = \{w_1, w_2, w_3, w_4\}$$

if there is an elliptic curve E' isogeneous to E_{r_z} such that r_w is projectively equivalent to a subset in $\pi(E'[\infty])$. Any path in the graph is equivalent to a path contained in $(\mathrm{Sym}^4(\mathbb{P}^1(L(E)))\setminus\Delta)/\mathrm{PGL}_2(\bar{\mathbb{Q}})$, for some E. The graph contains cycles, periodic orbits and preperiodic orbits, i.e. paths which at some moment end in periodic orbits.

Question 15 *Consider the field $L_0 = L_{r_0}$ for $r_0 = \{0, 1, -1, \infty\}$. Does*

$$(\mathrm{Sym}^4(\mathbb{P}^1(L(E)))\setminus\Delta)/\mathrm{PGL}_2(L(E))$$

consist of one cycle in DGS? Note that any path beginning from r_0 extends to a cycle (in many different ways) since r_0 is PGL_2-equivalent to a 4-tuple of points of order 4 on any elliptic curve.

Remark 16 In Step 7, we used the algebraicity of L_a/L, and we do not know how to extend the proof to geometric fields. What are the properties of L_a in geometric situations, when a is transcendental over k?

We have seen in the proof that the field L_a is closed under extensions of degree 2. We also have:

Lemma 17 *For any $b \in L_a$, we have $\sqrt[3]{b} \in L_a$.*

Proof Consider a curve E_r with $r := \{b, \sqrt{b}, -\sqrt{b}, \infty\}$. Its 3-division polynomial takes the form

$$f_3(x) = 3x^4 - 4bx^3 - 6bx^2 + 12b^2x - 4b^3 - b^2.$$

We can represent it as the product $3\prod_{i=1}^4 (x - x_i)$, where the set $\{x_i\} \subset L_a$ is equal to $\pi(E_r[3])$. The corresponding cubic resolvent is

$$rc(x) := \prod (x - (x_ix_j + x_kx_l)),$$

where $(i,j),(k,l)$ is any splitting into pairs of indices among $1,2,3,4$. In terms of b, we have

$$rc(x) = x^3 + 2bx^2 + 4b^2x/3 + 8b^3/3 - 128b^4/27 + 64b^5/27.$$

Since the set $\{x_i\}$ is projectively equivalent to $\{0, 1, \zeta_3, \zeta_3^2\}$, we can see that the cubic polynomial above has the form $C(x^3 + B)$, for some constants C, B. It can be checked that

$$rc(2b(2x - 1)/3) = (4b/3)^3(x^3 + (b - 1)^2).$$

After a projective map in $\mathrm{PGL}_2(L_a)$, we can transform the elements $x_ix_j + x_kx_l$ into $-\sqrt[3]{(b-1)^2}$. Hence, $-\sqrt[3]{(b-1)^2} \in L_a$, for any $b \in L_a$; since L_a is a field closed under 2-extensions, we obtain the claim. □

This raises a natural

Question 18 *Is L_a is closed under taking roots of arbitrary degree?*

If we add $\mathbb{G}_m$ to the set of allowed elliptic curves, then the answer is affirmative. However, there may exist a purely *elliptic* substitute for obtaining roots of prescribed order.

Corollary 19 *If the $j(E) \in L_a$ then any set $\{b, -b, b^{-1}, -b^{-1}\}$ with $\mu_4(\{b, -b, b^{-1}, -b^{-1}\}) = j(E)$ is contained in L_a. Note that such b are solutions of a cubic equation. Thus, L_a depends only on the curve E and we will write $L(E)$.*

It is also easy to see that $L(E) = L(E')$ if E and E' are isogenous.

5. Intersections

In this section, we present further results concerning the intersections

$$\pi_1(E_1[\infty]) \cap \pi_2(E_2[\infty])$$

for different elliptic curves E_1, E_2 and provide evidence for Effective finiteness conjecture 2.

Proposition 20 *Assume that*

$$\pi_1(E_1[4]) = \pi_2(E_2[4]) = \{0, 1, -1, i, -i, \infty\} \tag{2}$$

and that

$$\#\{\pi_1(E_1[3]) \cap \pi_2(E_2[3])\} \geq 2.$$

Then $r_1 = r_2$, and $E_1 = E_2$.

Proof By our assumption (2), E_i are given by the equation

$$y^2 = x^4 - t_i x^2 + 1.$$

With a_i defined by

$$r_i = \{a_i, -a_i, a_i^{-1}, -a_i^{-1}\},$$

we have

$$t_i = a_i^2 + a_i^{-2}.$$

We assume that $\pi_i(e_i) = a_i$. In this case, points $\pi_i(E_i[3]) \subset \bar{\mathbb{Q}} \subset \mathbb{P}^1$ are the roots of

$$x^4 + 2ax^3 - (2/a)x - 1 = 0 \tag{3}$$

or, equivalently,

$$2x^3a^2 + (x^4 - 1)a - 2x.$$

If $x, y \in \pi_{a_1}(E_{a_1}[3]) \cap \pi_{a_2}(E_{a_2}[3])$, where $x \neq y$, and $a_1 \neq a_2$, then a_1 and a_2 are the roots of $2x^3a^2 + (x^4 - 1)a - 2x$ and of $2y^3a^2 + (y^4 - 1)a - 2y$, which means that their coefficients are proportional:

$$\frac{2x^3}{2y^3} = \frac{x^4 - 1}{y^4 - 1} = \frac{-2x}{-2y}.$$

Then, on the one hand, $x^3/y^3 = x/y$ implies $x^2 = y^2$, and hence $x = -y$, by our assumption that $x \neq y$. On the other hand,

$$x/y = -1 = (x^4-1)/(y^4-1) = 1,$$

a contradiction. □

Given any $x \in \bar{\mathbb{Q}}$, we obtain $a_i = a_i(x)$, $i = 1,2$, which satisfy (3). Then the resulting elliptic curves E_i satisfy (2) and we have

$$\#\{\pi_1(E_1[3]) \cap \pi_2(E_2[3])\} = 1$$

unless

$$(x^4-1)^2 + 16x^4 = x^8 + 14x^4 + 1 = 0 \quad \text{or} \quad x^4 = -7 \pm 4\sqrt{3}.$$

Moreover,

$$\#\{\pi_{a_1}(E_{a_1}[\infty]) \cap \pi_{a_2}(E_{a_2}[\infty])\} = 6 + 4n \geq 10, \tag{4}$$

where 6 is the number of images of common points of order 4 (from 2), and 4 stands for the size of the $(\mathbb{Z}/2 \oplus \mathbb{Z}/2)$-orbit of a point in $\mathbb{P}^1$. However, it may happen that the inequality in (4) is strict.

Example 21 Consider the polynomial $f_5(x,a)$ defined in [2, Thm 18]. Its roots are exactly $\pi_a(E_a[5])$. It has degree 12 with respect to x, and 6 with respect to a. The polynomial $f_3(x,a)$ has degree 2 with respect to a and generically has exactly two solutions, $a_1(z), a_2(z)$, for any given z. We want also $f_5(v, a_i(z)) = 0$ for some v and z. This is equivalent to $f_5(v,a)$ being divisible by $f_3(z,a)$, as polynomials in a. Writing the division with the remainder, we obtain

$$f_5(v,a) = g(a)f_3(z,a) + C(v,z)a + C'(v,z)$$

for some explicit polynomials C and C', which have to vanish. This condition gives an explicit polynomial in u and which is divisible by a high power of u and $(u-1)$. Excluding the trivial solutions $u = 0, 1$, and substituting $t = u^4$, we obtain the equation

$$\begin{aligned} & 32u^{24} + 1369u^{20} + 18812u^{16} + 90646u^{12} + 18812u^8 + 1369u^4 + 32 \\ = \; & 32t^6 + 1369t^5 + 18812t^4 + 90646t^3 + 18812t^2 + 1369t + 32 \\ = \; & t^3\left[32\left(t^3 + \frac{1}{t^3}\right) + 1369\left(t^2 + \frac{1}{t^2}\right) + 18812\left(t + \frac{1}{t}\right) + 90646\right] \end{aligned}$$

Since $t \neq 0$, we have

$$\begin{aligned} = \; & 32\left(t + \frac{1}{t}\right)^3 + 1369\left(t + \frac{1}{t}\right)^2 + 18716\left(t + \frac{1}{t}\right) + 87908 \\ = \; & 32r^3 + 1369r^2 + 18716r + 87908 \\ =: \; & f(r) \end{aligned}$$

Computing the discriminant of this cubic polynomial, we find that it has no multiple roots. Its solutions give rise to pairs u, v such that, for $a_1 := a_1(u), a_2 := a_2(u)$, we have

$$f_5(v, a_i) = f_3(u, a_i) = 0$$

and hence

$$\#\{\pi_{a_1}(E_{a_1}[\infty]) \cap \pi_{a_2}(E_{a_2}[\infty])\} \geq 14.$$

The symmetry of the above equation reduces the problem to a cubic equation with coefficients in $\mathbb{Q}$, followed by a quadratic equation. The roots can be expressed in closed form and hence we get an explicit description for the 24 roots u.

The same scheme can be applied to points of higher order. Indeed, we have a polynomial $f_n(u,x) = 0$ which has increasing degree with respect to u, and the existence of a pair u, v such that $f_n(v,x) = 0$ is divisible by $f_3(u,x)$, depends on the divisibility of $f_n(v,x)$ by $f_3(u,x)$. Using long division, we obtain the two polynomials $C_{0,n}(u,v)$ and $C_{1,n}(u,v)$ so that their common zeroes (u,v) correspond to pairs (u,v) with $f_3(u,x) = 0$ and $f_n(v,x) = 0$ simultaneously.

Example 22 Applying this scheme to points of order 3 and 7 (or 3 and 11, 3 and 13, 3 and 17) we obtain that the corresponding resultant has roots of multiplicity 3, which implies the existence of three points v for a given u with $f_3(u,x) = 0$, and $f_7(v,x) = 0$; hence,

$$\#\{\pi_1(E_1[\infty]) \cap \pi_2(E_2[\infty])\} \geq 6 + 16 = 22.$$

Since we have every reason to expect polynomials $C_{0,n}(u,v)$ and $C_{1,n}(u,v)$ to have an increasing number of intersection points with the growth of n, we are led to the following conjecture:

Conjecture 23 *There is an infinite dense subset of points $a \in \mathbb{P}^1$ such that*

$$\pi_a(E_a[\infty]) \cap \pi_{a_2}(E_{a_2}[\infty]) \geq 22$$

with

$$\pi_a(E_a[3]) \cap \pi_{a_2}(E_{a_2}[3]) \neq 0.$$

Note that, in all such cases, the fields $L_a = L_{a_2}$.

6. General Weierstrass Families

The family of elliptic curves considered in Section 5 is the most promising for obtaining large intersections of torsion points. In this section, we consider other families where the intersections tend to be smaller, following [2].

We consider elliptic curves E_a with the same

$$\pi_a(e_a) = \infty \in \mathbb{P}^1.$$

These are given by their Weierstrass form

$$y^2 = x^3 + a_2x^2 + a_4x + a_6. \tag{5}$$

Using the formulas in, for example, [3, Ch. 3, §2], we write down (modified) division polynomials $f_{n,a}$, whose zeroes are *exactly* $\pi_a(E_a[n])$:

$$f_{n,a}(x) = \sum_{0\le r,s,t,r+2s+3t\le d(n)} c_{r,s,t}(n)a_2^r a_4^s a_6^t x^{d(n)-(r+2s+3t)},$$

where $d(n)$ and the coefficients $c_{r,s,t}(n)$ can be expressed via totient functions $J_k(n)$, with $d(n) = J_2(n)/2$, if $n > 2$, and $d(2) = 3$ (see [2]).

Lemma 24 *Let E_1, E_2 be elliptic curves in generalized Weierstrass form (5) such that, for some $n > 1$, we have*

$$\pi_1(E_1[n]) = \pi_2(E_2[n]).$$

Then $E_1 \simeq E_2$.

Proof The statement is trivial for $n = 2$. For $n > 2$, we have $d(n) \ge 4$; the comparison of division polynomials implies that the terms

$$a_2^r a_4^s a_6^t, \quad r + 2s + 3t \le 3$$

must be equal. For

$$(r,s,t) = (0,0,1),(0,1,0),(1,0,0)$$

we find equality of coefficients a_i for both curves. □

Often, already the existence of non-trivial intersections

$$\#\{\pi_1(E_1[n]) \cap \pi_2(E_2[n])\} \ge 1 \tag{6}$$

leads to the isomorphism of curves E_1, E_2. For example, if both curves are defined over a number field k, and the action of the absolute Galois group G_k on $\pi_1(E_1[n])$ and $\pi_2(E_2[n])$ is transitive, then (6) implies that $E_1 \simeq E_2$. For many, but not all, $n \in \mathbb{N}$, the equality of totient functions $J_2(n) = J_2(m)$, for some $m \in \mathbb{N}$, implies $n = m$.

Example 25 There exist many tuples (m,n) for which

$$J_2(m) = J_2(n) \quad \text{and} \quad J_1(m) \neq J_1(n).$$

For example,

$$J_2(5) = J_2(6) \quad \text{but} \quad J_1(5) = 4, \quad J_1(6) = 2.$$

We also have

$$J_2(35) = J_2(40) = J_2(42), \text{ while } J_1(35) = 24, J_1(40) = 16, J_1(42) = 12.$$

On the other hand, we have

$$J_2(15) = J_2(16) = 192 \quad \text{and} \quad J_1(15) = J_1(16) = 8.$$

These results indicate a relation of our question to Serre's conjecture. He considered the action of the Galois group on torsion points of an elliptic curve E defined over a number field k. If E does not have complex multiplication, then the image of the absolute Galois group G_k is an open subgroup of $\mathrm{GL}_2(\hat{\mathbb{Z}})$, i.e. of finite index.

Conjecture 26 (Serre) *For any number field k, there exists a constant $c = c(k)$ such that, for every non-CM elliptic curve E over k, the index of the image of the Galois group G_k in $\mathrm{GL}_2(\hat{\mathbb{Z}})$ is smaller than c.*

In particular, for $k = \mathbb{Q}$, he conjectured that, for primes $\ell \geq 37$, the image of G_k surjects onto $\mathrm{PGL}_2(\mathbb{Z}_\ell)$. Thus, modulo Serre's conjecture, our conjecture holds for curves defined over $\mathbb{Q}$.

Proposition 27 *Assume that*

$$\pi_1(E_1[n]) = \pi_2(E_2[m]), \quad n \neq m.$$

Then $k(E[n])$ contains $\mathbb{Q}(\zeta_d)$, where $d = \mathrm{lcm}(m,n)$, the least common multiple of m,n.

Proof By Serre, we have

$$\mathbb{Q}(\zeta_n) \subset k(E[n]) \quad \text{and} \quad \mathbb{Q}(\zeta_m) \subset k(E[m])$$

as subfields of index at most 2. □

Corollary 28 *Assume that k does not contain roots of 1 of order divisible by n,m. Then $k(E[n]), k(E[m])$ contain a cyclotomic subfield of index at most 2.*

This provides a strong restriction on intersections of images of torsion points for elliptic curves over $\mathbb{Q}$, or over more general number fields k with this property. This yields a restriction on fields $k(E[n])$, since $(n,m) > 4$, for all (n,m) with $J_2(n) = J_2(m)$.

REFERENCES

[1] Fedor A. Bogomolov and Yuri Tschinkel. Algebraic varieties over small fields. In Umberto Zannier (ed.) *Diophantine geometry*, volume 4 of *CRM Series*, pages 73–91. Edizioni della Normale, Pisa, 2007.

[2] Fedor A. Bogomolov and Hang Fu. Division polynomials and intersection of projective torsion points. *Eur. J. Math.*, 2(3):644–60, 2016.

[3] Anthony W. Knapp. *Elliptic curves*, volume 40 of *Mathematical Notes*. Princeton University Press, Princeton, NJ, 1992.

[4] Michel Raynaud. *Courbes sur une variete abelienne et points de torsion. Invent. Math.*, 71(1):207–33, 1983.

[5] Umberto Zannier. *Some problems of unlikely intersections in arithmetic and geometry*, volume 181 of *Annals of Mathematics Studies*. Princeton University Press, Princeton, NJ, 2012. With appendixes by David Masser.

3

Algebras of Quantum Monodromy Data and Character Varieties

LEONID CHEKHOV[1]
MARTA MAZZOCCO[2]
VLADIMIR RUBTSOV[3]

[1] *Steklov Mathematical Institute of Russian Academy of Sciences and Laboratoire Poncelet, Moscow, Russia, and Michigan State University, East Lansing, USA. Email: chekhov@mi.ras.ru.*
[2] *School of Mathematics, University of Birmingham, Edgbaston, Birmingham, B15 2TT, United Kingdom. Email: m.mazzocco@bham.ac.uk.*
[3] *Université d'Angers, France and Theory Division, ITEP; Moscow, Russia; Email: volodya@univ-angers.fr.*

For the seventieth birthday of Nigel Hitchin

1. Introduction

The classical Riemann–Hilbert problem deals with Fuchsian systems on the Riemann sphere. Let us consider a meromorphic system of first-order ODEs:

$$\frac{d\Psi}{dz} = \sum_{i=1}^{s} \frac{A_i}{z - a_i}\Psi, \tag{1.1}$$

where z is a coordinate on the sphere $\Sigma_{0,s} := \mathbb{P}^1 \setminus \{a_1, \ldots, a_s\}$, where $\{A_1, \ldots, A_s\} \subset \mathfrak{sl}_k(\mathbb{C})$ are constant in z.

The Riemann–Hilbert correspondence is defined by associating to each Fuchsian system its monodromy representation class obtained by considering the analytic continuation of a fundamental matrix $\Psi(z)$ of (1.1) around loops γ_i, $i = 1, \ldots, s$, encircling each singular point a_i:

$$\rho : \pi_1(\mathbb{P}^1 \setminus \{a_1, \ldots, a_s\}, a_0) \to SL_k(\mathbb{C}).$$

Chekhov, L., Mazzocco, M., Rubtsov, V., *Algebras of Quantum Monodromy Data and Character Varieties.* In: *Geometry and Physics: A Festschrift in Honour of Nigel Hitchin*, Jørgen Ellegaard Andersen, Andrew Dancer, Oscar García-Prada (Eds): Oxford University Press (2018). © Oxford University Press. DOI: 10.1093/oso/9780198802006.003.0003

Taking the conjugacy classes, one obtains the Betti moduli space of monodromy representations, or the SL_k-character variety:

$$\mathcal{M}_B = \mathrm{Hom}\left(\pi_1(\Sigma_{0,s}) \to SL_k(\mathbb{C})\right)/_{SL_k(\mathbb{C})}. \tag{1.2}$$

Geometrically, the system (1.1) can be replaced by the meromorphic connection

$$\nabla := \left(\frac{\partial}{\partial z} - A(z)\right) dz = d - \sum_{i=1}^{s} \frac{A_i}{z - a_i} dz$$

on the trivial holomorphic vector bundle $\mathbb{C}^k \times \mathbb{P}^1 \to \mathbb{P}^1 \setminus \{a_1, \ldots, a_s\}$. In this setting, the Riemann–Hilbert correspondence is the isomorphism

$$RH : \mathcal{M}_{DR} \simeq \mathcal{M}_B, \tag{1.3}$$

where $\mathcal{M}_{DR}$ is the de Rham moduli space

$$\mathcal{M}_{DR} = \left\{(\nabla, E), E \to \Sigma_{0,s}\right\}/\mathcal{S}$$

of logarithmic connections ∇ on holomorphic rank k vector bundles E over the Riemann sphere $\Sigma_{0,s}$ with s boundary components, and $\mathcal{S}$ is the gauge group.

In [36], Hitchin proved that this map is a symplectomorphism. To be precise, denoting by $\mathcal{O}_i$ the conjugacy classes of the residues A_i, $i = 1, \ldots, s$, one can endow $\mathcal{M}_{DR}$ with the standard Lie–Poisson structure on $\mathcal{O}_1 \times \cdots \times \mathcal{O}_s \subset \mathfrak{sl}_k \times \mathfrak{sl}_k \times \cdots \times \mathfrak{sl}_k$ obtained by identifying $\mathfrak{sl}_k$ with $\mathfrak{sl}_k^*$. Upon fixing the conjugacy classes, this Poisson structure restricts to the Kostant-Kirillov symplectic structure. On the Betti moduli space $\mathcal{M}_B$, Hitchin considered the Poisson structure constructed by Audin [5] as follows. Consider the Atiyah–Bott symplectic
structure

$$\Omega = \frac{k}{4\pi} \mathrm{Tr} \int_\Sigma \delta A \wedge \delta A$$

on the space $\mathcal{C}onn(\Sigma)$ of all smooth $\mathfrak{g}$-valued connections A (for $\mathfrak{g}$ a simple Lie algebra) on a compact Riemann surface Σ. When no boundaries are present, one replaces the space $\mathcal{C}onn(\Sigma)$ by the space $\mathcal{M}(\Sigma) = \mathcal{M}_0(\Sigma)/\mathcal{S}$, the quotient of the space $\mathcal{M}_0(\Sigma)$ of all flat connections on Σ by the gauge group. Since Σ is closed, the momentum map is just the curvature, so that the space $\mathcal{M}(\Sigma)$ is just a reduced level set of the momentum mapping and thus a symplectic manifold. In the presence of boundaries, the curvature is the momentum map of a smaller group, so one needs to consider a central extension of the group of gauge transformations in order to construct the Poisson structure. In particular, one needs to add a correction term to the Atiyah–Bott symplectic structure:

$$\Omega_c = \frac{k}{4\pi} \mathrm{Tr} \int_{\partial\Sigma} \phi \wedge \delta A, \qquad \mathrm{d}_A \phi = \delta A.$$

The Poisson structure on the Betti moduli space is the result of the Hamiltonian reduction on the zero level of the momentum map associated to $\Omega + \Omega_c$. This Poisson structure coincides with the Goldman bracket on the character variety $\mathcal{M}_B$ [28, 29].

In this paper, we address the question of what happens to this theory if we allow connections with higher-order poles on holomorphic rank k vector bundles E on Riemann surfaces $\Sigma_{g,s}$ of genus g and s boundary components.

This question has been addressed by a number of Hitchin's disciples. In [9] Boalch treated the case of a system with a regular pole and a pole of order 2 at infinity by using the Laplace transform [21, 34, 37] to map it back to the Fuchsian case. More recently, he introduced the notion of wild character variety [10] in which higher-order poles are blown up to produce extra regular poles (at the intersection between a boundary circle and the Stokes directions) and the fundamental group is replaced by the groupoid of closed loops around these extra regular poles. He defined the Poisson structure on the wild character variety by using the quasi-Hamiltonian approach by Alexeev and collaborators [2, 3].

An elegant approach was proposed by Gualtieri, Li and Pym [30]. In this case, they start with the space of meromorphic connections on a smooth curve with a pole divisor D (poles with multiplicity). Let S be the space such that the parallel transport defined by this connection with poles bounded by the divisor D exists and is holomorphic (outside of the poles). They then consider the Lie algebroid $\mathcal{A}$ tangent to S and define the *Stokes groupoid* as the Lie groupoid that integrates this Lie algebroid. They compute this groupoid in the case of the Airy equation and demonstrate that it is the usual pairing groupoid with a twist. A similar computation can be carried out for every case in which total summability works. This result gives a beautiful geometric explanation of Ecalle resummability theory.

In [18], based on the idea of interpreting higher-order poles in the connection as boundary components with *bordered cusps*[1] [17] on the Riemann surface, we introduced the notion of *decorated character variety*. Let us recall this definition here.

Topologically speaking, a Riemann surface $\Sigma_{g,s,n}$ of genus g with s holes and n bordered cusps is equivalent to a Riemann surface $\tilde{\Sigma}_{g,s,n}$ of genus g, with s holes and n marked points $m_1, \dots, m_n$ on the boundaries. Then, one defines *the fundamental groupoid of arcs* $\pi_{\mathfrak{a}}(\Sigma_{g,s,n})$ as the set of all directed paths $\gamma_{ij} : [0,1] \to \tilde{\Sigma}_{g,s,n}$ such that $\gamma_{ij}(0) = m_i$, and $\gamma_{ij}(1) = m_j$, modulo homotopy. The groupoid structure is dictated by the usual path–composition rules. The SL_k decorated character variety is defined as

$$\mathcal{M}^k_{g,s,n} := \mathrm{Hom}\left(\pi_{\mathfrak{a}}(\Sigma_{g,s,n}), SL_k(\mathbb{C})\right) / \textstyle\prod_{j=1}^{n} U_j, \tag{1.4}$$

where U_j is the unipotent Borel subgroup in $SL_k(\mathbb{C})$ (one unipotent Borel subgroup for each bordered cusp).

Our interest in the representation spaces and their interpretation as decorated character varieties goes back to study of the moduli space of monodromy representations for the fundamental group of the 4-holed sphere. This is the SL_2–character variety

[1] We use the term 'bordered cusp', meaning a vertex of an ideal triangle in the Poincaré metric in order to distinguish it from standard cusps (without borders) associated to punctures on a Riemann surface.

$$\mathcal{M}_B = \mathrm{Hom}\left(\pi_1(\Sigma_{0,4}) \to SL_2(\mathbb{C})\right)/_{SL_2(\mathbb{C})}, \tag{1.5}$$

and the above discussed Poisson structure can also be obtained as a reduction of so-called *Korotkin–Samtleben* bracket ([32]) on $\mathrm{Hom}\left(\pi_1(\Sigma_{0,4}) \to SL_2(\mathbb{C})\right)$ (which is, in fact, the quasi-Poisson structure in sense of [2, 3]). The relation with Teichmüller space parametrization and a quantization of this Poisson character variety was proposed by the first two authors in [15].

This Poisson manifold is also known as the monodromy manifold of the linear system corresponding to the Painlevé *VI* equation. Our notion of decorated character variety was motivated by a challenging problem of giving a definition compatible with the confluence operations that give rise to all other Painlevé differential equations–in this case, the *Stokes phenomenon* appears–the solutions in the vicinity of the multiple poles have different asymptotic behaviours in different sectors.

In this paper, we study the Poisson structure on the representation space

$$\mathcal{R}^k_{g,s,n} := \mathrm{Hom}\left(\pi_{\mathfrak{a}}(\Sigma_{g,s,n}), SL_k(\mathbb{C})\right),$$

induced by the Fock–Rosly bracket [26] as explained in [4] (see also [12, 13, 40]) and prove that the quotient by unipotent Borel subgroups giving rise to the decorated character variety (1.4) is a Poisson reduction. More precisely, we consider the Poisson structure on the matrices M that correspond to directed arcs in $\pi_{\mathfrak{a}}(\Sigma_{g,s,n})$. We call these matrices *monodromy data,* because they indeed contain Stokes matrices, connection matrices and standard monodromy matrices of linear systems on first-order ODEs (see Section 5). We treat classical and quantum case simultaneously, thus providing a quantization of the decorated character variety (1.4). It would be interesting to understand the categorical version of our quantization along the lines of the recent papers by Ben-Zvi, Brochier and Jordan [6, 7]–we postpone this to future publications.

In their seminal papers [23], [24], Fock and Goncharov introduced a set of Darboux coordinates for $SL_k(\mathbb{R})$ systems on Riemann surfaces with holes. Nevertheless, to the best of our knowledge, a comprehensive analysis relating the Fock–Goncharov construction to the Fock–Rosly algebras was still missing. Elements of this construction (in the classical case, without references to Poisson or quantum structures) had appeared in papers of Musiker, Schiffler and Williams [38], [39], mostly devoted to establishing connection to cluster algebras; it was there where λ-lengths were identified with upper-left elements of $SL_2(\mathbb{R})$-monodromy data. Shear coordinates associated with decorated bordered cusps were introduced simultaneously and independently by the first two authors in [17] and by Allegretti [1]. A useful technical tool allowing avoiding most difficulties of the standard combinatorial description of structures on Riemann surfaces with holes is that, having at least one bordered cusp, we can consider ideal-triangle partitions of these surfaces with triangles based only at bordered cusps enclosing all holes without cusps in monogons. We then restrict the set of mapping class group (MCG) transformations (and the corresponding cluster mutations) to the generalized cluster algebras introduced in [20]. Upon imposing these restrictions, we can establish an isomorphism between extended shear coordinates and λ-lengths, both enjoying homogeneous Poisson or quantum relations, and explicitly construct the monodromy data of the system (1.4) for $k = 2$. The homogeneous Poisson or quantum relations for shear coordinates then induce classical or quantum Fock–Rosly relations for elements of monodromy data, with the Poisson reduction imposed by the quotient by unipotent Borel subgroups. All matrix elements of all monodromy data are then sign-definite Laurent polynomials either of exponentiated shear coordinates or of λ-lengths.

The paper is organized as follows. In Section 2, we briefly recall the hyperbolic geometry description of Teichmüller spaces of Riemann surfaces with holes and bordered cusps and formulate the main statement of the paper (Theorem 2.5). We then derive the quantum Fock–Rosly-like algebras of monodromy data for SL_2 out of coordinate algebras of the quantum Teichmüller spaces. In Section 3, we study Poisson and quantum R-matrix structures of decorated character varieties for general SL_k-monodromy data, paying special attention to the Poisson reduction due to factorization w.r.t. Borel subgroups. In Section 4, we consider in details three important examples of our construction for general $SL_k(\mathbb{R})$-monodromy data: $\Sigma_{1,s+1,1}$, $\Sigma_{0,2,2}$ and $\Sigma_{0,1,3}$. Finally, in Section 5, keeping in mind the idea of extending the Riemann–Hilbert correspondence to $\mathcal{Z}^{irr}_{g,s} \to \mathcal{M}^k_{g,s,n}$, where $\mathcal{Z}^{irr}_{g,s}$ is a suitably decorated moduli space of irregular connections ∇ on a holomorphic rank k vector bundle $E \to \Sigma_{g,s}$, we propose a formula linking the number n of bordered cusps on $\Sigma_{g,s,n}$ to the irregular type of the connection. Because, as explained in [18], the decorated character variety (1.4) contains the wild character variety as the subalgebra of functions that Poisson commute with the function associated to certain arcs connecting bordered cusps, the space $\mathcal{Z}^{irr}_{g,s}$ is in fact an extension of the one considered by Boalch in his survey on Riemann–Hilbert correspondence in this same issue [11].

Throughout this paper, we use the following notation: we denote by M a monodromy datum, or a matrix in SL_k, and by m_{ij} its elements.

2. Algebras for $SL_2(\mathbb{C})$ Monodromy Data for Surfaces with Bordered Cusps

2.1. *Darboux Coordinates in Dimension 2*

In the case of dimension $k = 2$, the complex dimension of the decorated character variety is $6g - 6 + 3s + 2n$. In this section, we restrict to the real-analytic subvariety

$$\mathcal{R}^{\mathbb{R}}_{g,s,n} := \mathrm{Hom}\left(\pi_{\mathfrak{a}}(\Sigma_{g,s,n}), SL_2(\mathbb{R})\right),$$

and show how to construct real Darboux coordinates. We then complexify them to define Darboux coordinates on the complex representation space $\mathcal{R}^2_{g,s,n}$.

Every bordered cusp is endowed with a *decoration*—a horocycle based at the end of the cusp, which is a point on an absolute. Such a horocycle cuts out an infinite part of a cusp and we consider only parts of arcs that are confined between two horocycles decorating two cusps at which this arc terminates (this can be the same cusp, then the same horocycle).

We split the Riemann surface $\Sigma_{g,s,n}$ into ideal triangles based at bordered cusps; the edges of these triangles are arcs; if there are holes without bordered cusps, these holes by prescription are always enclosed in monogons (obviously bordered by arcs starting and terminating at the same bordered cusp).

For every ideal-triangle decomposition, as above, we consider the dual fat graph $\Gamma_{g,s,n}$, all vertices of which are 3-valent except exactly n 1-valent vertices being in 1-1 relation with bordered cusps. Each (non-directed) edge of $\Gamma_{g,s,n}$ carries an (extended) shear coordinate (denoted by capital Z letters for internal edges and by π_j for edges ending at 1-valent vertices) from the set $\{Z_\alpha, \pi_j\}$; for every hole without cusps, we have an edge dual to the bordering arc and the loop attached to the (inner) end of this edge carrying the coefficient ω ($\omega = e^{P/2} + e^{-P/2}$ for a hole with the perimeter P or $\omega = 2\cos(\pi/n)$ for a $\mathbb{Z}_n$ orbifold point). The shear coordinates $\{Z_\alpha, \pi_j\}$ are either real numbers (in the classical case) or Hermitian operators (in the quantum case) with constant commutation relations determined by the graph $\Gamma_{g,s,n}$. All coefficients ω are Casimirs.

From [20], for any choice of real numbers $\{Z_\alpha, \pi_j\}$ and parameters ω_β, we have a metrizable Riemann surface $\Sigma_{g,s,n}$, and vice versa; for any Poincaré uniformizable $\Sigma_{g,s,n}$, we have a (non-unique) set of $\{Z_\alpha, \pi_j\}$ and ω_β and a graph $\Gamma_{g,s,n}$ determining the glueing of this surface out of ideal triangles using the extended shear coordinates $\{Z_\alpha, \pi_j\}$.

Then, we have an explicit parametrization of the monodromy data in terms of extended shear coordinates. To every edge, we set into correspondence the *edge matrix* $X_A := \begin{pmatrix} 0 & -e^{A/2} \\ e^{-A/2} & 0 \end{pmatrix}$, where $A \in \{Z_\alpha, \pi_j\}$ is the extended shear coordinate of this edge. When an oriented path goes through the corresponding edge (in any direction), we multiply from the left by the edge matrix. When a path turns left or right at 3-valent vertices, we multiply from the left by the corresponding matrices $L = \begin{pmatrix} 0 & 1 \\ -1 & -1 \end{pmatrix}$ and $R = \begin{pmatrix} 1 & 1 \\ -1 & 0 \end{pmatrix}$ and, when a path goes clockwise around the loop containing the hole endowed with the coefficient ω, we multiply from the left by $F_\omega = \begin{pmatrix} 0 & 1 \\ -1 & -\omega \end{pmatrix}$. We always begin with the edge matrix X_{π_i} of the cusp the arc begins with; the last matrix in the product is always the edge matrix X_{π_j} of the cusp at which the arc terminates. In the quantum case, the quantum ordering is the natural ordering in the product, provided we scale the left and right turn matrices: $L \to q^{1/4}L$, $R \to q^{-1/4}R$ (if $[X, Y] = 2\pi i\hbar$, then $e^X e^Y = qe^{X+Y}$, so $q = e^{\pi i\hbar}$).

The thus constructed matrix products are invariant under the quantum MCG transformations (mutations of inner edges including those dual to monogons containing holes without cusps), so we can always reduce a given matrix product to a simpler one: for example, a monodromy datum corresponding to an arc with ends at two different cusps and such that it does not correspond to a bordered arc can be brought to the form $M = X_{\pi_j} L X_Z R X_{\pi_i}$; every monodromy datum of a bordered arc that goes clockwise can be brought to the form $M = X_{\pi_j} L X_{\pi_i}$, and the one going anticlockwise can be brought to the form $M = X_{\pi_j} R X_{\pi_i}$.

We now define the monodromy data for arc-like paths.

Definition 2.1 For a given set of extended shear coordinates $\{Z_\alpha, \pi_j\}$ and coefficients ω_β associated with a spine (fat graph) $\Gamma_{g,s,n}$, the $SL_2(\mathbb{R})$-monodromy data associated to (directed) arcs (directed paths starting and terminating at bordered cusps) are

$$M_{\mathfrak{a}} = X_{\pi_2} L X_{Z_{\alpha_n}} R \cdots L X_{Z_{\alpha_j}} F_{\omega_\beta} X_{Z_{\alpha_j}} R \cdots L X_{Z_{\alpha_1}} R X_{\pi_1}, \tag{2.6}$$

where π_1 and π_2 are the extended shear coordinates of the respective starting and terminating bordered cusps. In the quantum case, the quantum ordering is the natural ordering prescribed by matrix multiplication and we make a scaling $L \to q^{1/4}L$, $R \to q^{-1/4}R$.

Definition 2.2 We define the λ-length of an arc $\mathfrak{a}$ to be the upper-left element of $M_{\mathfrak{a}}$ defined by (2.6) (u.r.(M) in [38], [39] or $\mathrm{tr}_K(M)$ in [17]).

Identifying $\{Z_\alpha, \pi_j\}$ with the extended shear coordinates, the thus defined λ-lengths (in the classical case) are $e^{\ell_{\mathfrak{a}}/2}$, where $\ell_{\mathfrak{a}}$ are actual (signed) lengths of stretched between decorating horocycles parts of geodesic curves that join the corresponding bordered cusps and belong to the same homotopy class as the arc $\mathfrak{a}$. We often just identify these arcs with the corresponding λ-lengths, writing merely $\lambda_{\mathfrak{a}}$.

Lemma 2.1 [17, 19, 20] *The classical and quantum monodromy data $M_{\mathfrak{a}}$ are invariant under MCG transformations (extended cluster mutations) induced by mutations (flips) of inner edges of $\Gamma_{g,s,n}$.*

Definition 2.3 [17] A *complete geodesic lamination (CGL)* is the set of all the edges of all ideal triangles constituting an ideal-triangle decomposition of $\Sigma_{g,s,n}$ with vertices at bordered cusps. We call *algebraic* CGL the collection of all λ-lengths of the elements in the CGL.

Lemma 2.2 [17] *Every algebraic CGL can be identified with a seed of a quantum cluster algebra of geometric type [8, 17]; the corresponding λ-lengths enjoy homogeneous commutation relations among themselves.*

Lemma 2.3 [17, 38, 39] *For the fat graph $\Gamma_{g,s,n}$ dual to the corresponding partition of $\Sigma_{g,s,n}$ into ideal triangles, the relations between λ lengths from the corresponding CGL and exponentiated extended shear coordinates $\{e^{\pm Z_\alpha}, e^{+\pi_j/2}\}$ are 1-1 and monomial in both directions.*

Lemma 2.4 [17] *Every λ-length in every CGL (a seed) is a polynomial from $\mathbb{Z}_+[e^{\pm Z_\alpha}, e^{+\pi_j/2}, \omega_\beta]$ of exponentiated shear coordinates of any given seed. Every quantum λ-length is a Hermitian operator represented by an ordered polynomial from $\mathbb{Z}_+[e^{\pm Z_\alpha}, e^{+\pi_j/2}, \omega_\beta, q^{\pm 1/4}]$.*

Note that Lemmas 2.3 and 2.4 then immediately imply the Laurent and positivity phenomenon for λ-lengths in all seeds.

The main result of this paper is as follows.

Theorem 2.5 *Provided the extended shear coordinates $\{Z_\alpha, \pi_j\}$ enjoy the standard constant commutation or Poisson relations [14, 17], the set of (classical or quantum) monodromy data $M_{\mathfrak{a}}$ determined by formula (2.6) for any (directed) arc $\mathfrak{a}$ from a fixed CGL satisfies the following properties:*

(a) *The (classical or quantum) matrices $M_{\mathfrak{a}}$ corresponding to arcs from the same CGL satisfy R-matrix permutation relations of Fock–Rosly type* (2.8), (2.9), (2.12), (2.13), (2.14), (2.15), (2.16), (2.17), (2.18).

(b) *All elements of every monodromy datum* $M_\mathfrak{a}$ *are polynomials of* $e^{\pm Z_\alpha/2}$, $e^{\pm\pi_j/2}$, $q^{\pm 1/4}$ *and* ω_β *with integers coefficients and, by Lemma 2.3, are sign-definite Laurent polynomials in* λ*-lengths of any given CGL (for any seed of the corresponding cluster algebra of geometric type) and polynomials in* ω_β.

(c) *The Borel subgroup restriction (see (1.4)) is realized by reducing all monodromy data* $M_\mathfrak{a}$ *that correspond to paths between two neighbouring bordered cusps (or the same cusp, if a hole contains only one cusp) that go along the hole boundary, with the hole being of the left of the form, with entries* $m_{\mathfrak{a}_{i,j}} = 0, \forall i+j \geq k+2$. *This restriction is Poisson for any system of monodromy data (and survives the generalization to the case of* $SL_k(\mathbb{R})$*-monodromy data).*

Observe that, thanks to Lemma 2.3 and point (b) of Theorem 2.5, the extended shear coordinates $\{Z_\alpha, \pi_j\}$ can be straightforwardly complexified and therefore the complex representation space $\mathcal{R}^2_{g,s,n}$ is endowed with the same Poisson structure.

In the following subsections, we derive point (a) of the theorem (points (b) and (c) will be clear from construction) in the quantum case (the classical one follows by taking the semi-classical limit). Let us stress that, while the final formulas for the quantum commutation relations have been derived from the Fock–Rosly bracket [26] already in [4] (see also [12, 13, 40]), our approach allows to express all matrices in Darboux coordinates. For this reason, we repeat the derivation of the relations from [4] here.

In the following, we call an arc joining different cusps an *open arc*; we call an arc joining the same cusp a *closed arc*.

2.2. *Basic Relations*

In this subsection, we use our geometric construction to find the two Fock–Rosly commutation relations, from which all other commutation relations can be found using the groupoid property.

The first relation pertains to the case where two different arcs $\mathfrak{a}_i$ and $\mathfrak{a}_j$ start at the same bordered cusp π and then go to the left and to the right, respectively, never colliding again (see Figure 1).

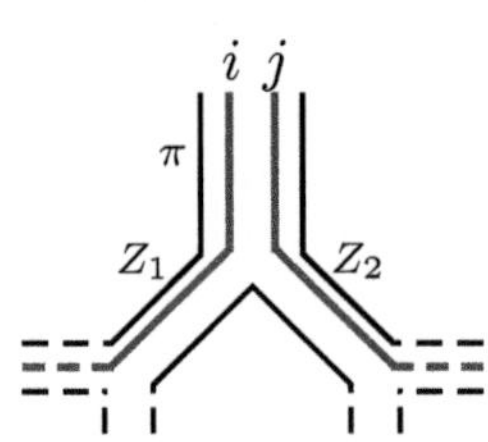

Figure 1 Fat graph on a Riemann surface with at least one cusp of coordinate π. We denote the standard shear coordinates by Z_1 and Z_2. The dashed part is the rest of the Riemann surface.

The matrices corresponding to arcs are

$$M_i = QX_{Z_1}RX_\pi; \qquad M_j = SX_{Z_2}LX_\pi, \qquad i<j, \tag{2.7}$$

where we use the notation that $i<j$ when the arc $\mathfrak{a}_i$ is on the left of the arc $\mathfrak{a}_j$, and the matrices Q and S correspond to portions of the arcs that never intersect the arcs $\mathfrak{a}_j$ and $\mathfrak{a}_i$, respectively, so that

$$[Q,M_j]=0, \quad [S,M_i]=0 \quad \text{and} \quad [S,Q]=0.$$

To deduce the commutation relations of the matrices (2.7), we use the following commutation relations:

$$[Z_2,Z_1]=[Z_2,\pi]=[\pi,Z_1]=2\pi i\hbar,$$
$$e^{Z_1/2}e^{\pi/2}=q^{-1/2}e^{\pi/2}e^{Z_1/2}, \quad e^{Z_2/2}e^{\pi/2}=q^{1/2}e^{\pi/2}e^{Z_2/2}, \quad e^{Z_2/2}e^{Z_1/2}=q^{-1/2}e^{Z_1/2}e^{Z_2/2}.$$

Then, by a direct calculation, we obtain the following relations:

$$\overset{1}{M_i}\overset{2}{M_j}=\overset{2}{M_j}\overset{1}{M_i}R_{12}(q) \qquad i<j, \tag{2.8}$$

$$\overset{1}{M_i}\overset{2}{M_j}R_{12}(q)^T=\overset{2}{M_j}\overset{1}{M_i} \qquad i>j, \tag{2.9}$$

where for any matrix A, $\overset{1}{A}=A\otimes\mathbb{I}$ and $\overset{2}{A}=\mathbb{I}\otimes A$ and $R_{12}(q)$ is the rescaled Kulish–Sklyanin R-matrix; explicitly,

$$R_{12}(q)=\left(\begin{array}{cc|cc} q^{1/2} & 0 & 0 & 0 \\ 0 & q^{-1/2} & q^{1/2}-q^{-3/2} & 0 \\ \hline 0 & 0 & q^{-1/2} & 0 \\ 0 & 0 & 0 & q^{1/2} \end{array}\right). \tag{2.10}$$

Observe that, as long as the arc $\mathfrak{a}_i$ is on the left of the arc $\mathfrak{a}_j$, by MCG transformations, we can always flip edges in our fat graph to match this situation.

We obtain the second basic relation for entries of the *same* monodromy datum corresponding to an open arc: every such matrix (except the case where it borders a hole) can be brought by quantum MCG transformations to the form

$$M:=X_{\pi_2}LX_{Z_1}RX_{\pi_1}, \quad [\pi_1,Z_1]=[\pi_2,Z_1]=2\pi i\hbar,\ [\pi_1,\pi_2]=0,$$

(see Figure 2), thus giving the following commutation relation:

$$R_{12}^{\mathrm{T}}\overset{1}{M}\overset{2}{M}=\overset{2}{M}\overset{1}{M}R_{12}. \tag{2.11}$$

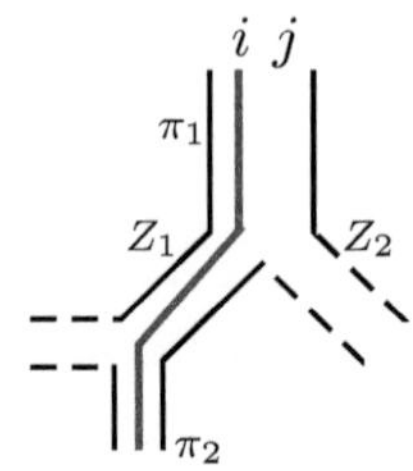

Figure 2 Fat graph on a Riemann surface with two cusps of coordinates π_1 and π_2. We denote the standard shear coordinates by $Z_1, Z_2, \ldots$. The dashed part is the rest of the Riemann surface.

2.3. *Composite Relations*

Here we explain how to obtain all other relations from the basic relations (2.8) and (2.11), using the groupoid property.

Using the basic relation (2.8), we can deduce what happens when two open arcs meet at two different cusps. Let these arcs be oriented in the same way, that is, they originate at the same cusp and end at the same cusp. Inverting orientation corresponds to inverting a matrix. We denote the above two matrices by M_i^j and M_k^l, where i and k are in the source cusp, and j, l in the target cusp; at each cusp, we have a linear ordering of indices originated from orientation of the surface. We can think of $M_i^j = {M^j}^{-1} M_i$, and $M_k^l = {M^l}^{-1} M_k$, where

$$[M^j, M_k] = [M_i, M^l] = 0,$$

and the pairs M_i, M_k and M^j, M^l enjoy the commutation relations (2.8). In this way, we obtain

$$\overset{1}{M}{}_i^j \overset{2}{M}{}_k^l = R_{12} \overset{2}{M}{}_k^l \overset{1}{M}{}_i^j R_{12} \quad \text{for } i < k, j < l, \tag{2.12}$$

$$\overset{1}{M}{}_i^j \overset{2}{M}{}_k^l = R_{12} \overset{2}{M}{}_k^l \overset{1}{M}{}_i^j R_{12}^{-T} \quad \text{for } i > k, j < l, \tag{2.13}$$

$$\overset{1}{M}{}_i^j \overset{2}{M}{}_k^l = R_{12}^{-T} \overset{2}{M}{}_k^l \overset{1}{M}{}_i^j R_{12} \quad \text{for } i < k, j > l, \tag{2.14}$$

$$\overset{1}{M}{}_i^j \overset{2}{M}{}_k^l = R_{12}^{-T} \overset{2}{M}{}_k^l \overset{1}{M}{}_i^j R_{12}^{-T} \quad \text{for } i > k, j > l. \tag{2.15}$$

Consider the case of two monodromy data corresponding to two closed arcs (starting and terminating at the same cusp) having no intersections inside the surface. Then, their four ends can be uniquely enumerated from 1 to 4, depending on the order in which the corresponding arcs enter the cusp; see Figure 3, where the index 1 corresponds to the rightmost thread, and 4 to the leftmost thread. We have three different cases, all of which

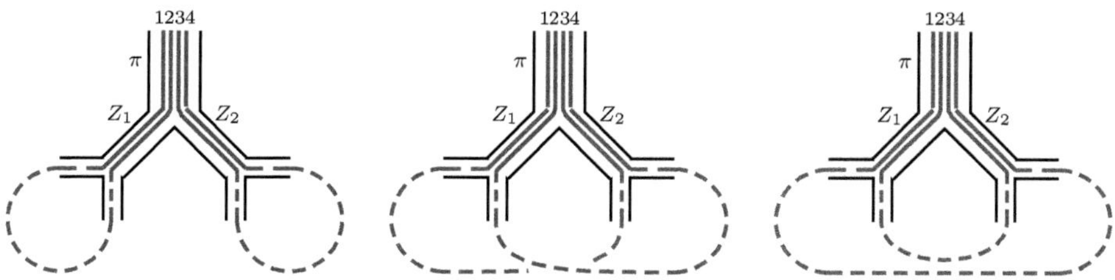

Figure 3 Three cases of non-intersecting geodesic arcs terminating at the same bordered cusp.

can be obtained from the basic relation (2.8); M_{ij} indicates the arc that starts at thread i and terminates at j and which has the structure $M_j^{-1}M_i$, with M_i and M_j from (2.7). In all examples below, we take $i > j$, that is, the corresponding arc goes clockwise along the surface:

$$\overset{1}{M}_{21}R_{12}\overset{2}{M}_{43}R_{12}^{-1} = R_{12}\overset{2}{M}_{43}R_{12}^{-1}\overset{1}{M}_{21}, \tag{2.16}$$

$$\overset{1}{M}_{41}R_{12}^{-\mathrm{T}}\overset{2}{M}_{32}R_{12}^{\mathrm{T}} = R_{12}\overset{2}{M}_{32}R_{12}^{-1}\overset{1}{M}_{41}, \tag{2.17}$$

$$\overset{1}{M}_{31}R_{12}^{-\mathrm{T}}\overset{2}{M}_{42}R_{12}^{-1} = R_{12}\overset{2}{M}_{42}R_{12}^{-1}\overset{1}{M}_{31}. \tag{2.18}$$

The last commutation relation is for the entries of the same matrix M_{ij} corresponding to a closed arc. We can obtain this arc as the product of two arcs: one starts at the same cusp and terminates at another cusp, and the other starts at the second cusp and terminates at the first cusp, going along a different path in the surface, in such a way that the composition gives the arc that starts and terminates at the same cusp. We obtain

$$R_{12}^{\mathrm{T}}\overset{1}{M}_{ij}R_{12}^{-\mathrm{T}}\overset{2}{M}_{ij} = \overset{2}{M}_{ij}R_{12}^{-1}\overset{1}{M}_{ij}R_{12}. \tag{2.19}$$

Note that, for the R-matrix of the form (2.10), we have another, equivalent way of writing the *same* quantum commutation relations for elements of the matrix M_{ij} encoded in (2.19):

$$\overset{1}{M}_{ij}R_{12}^{-\mathrm{T}}\overset{2}{M}_{ij}R_{12}^{\mathrm{T}} = R_{12}\overset{2}{M}_{ij}R_{12}^{-1}\overset{1}{M}_{ij}. \tag{2.20}$$

We are free to use any of relations (2.19), (2.20) at our discretion.

Remark 2.4 That (2.20) is equivalent to (2.19) implies the following commutation relation:

$$R_{12}^{\mathrm{T}}R_{12}\overset{2}{M}_{ij}R_{12}^{-1}\overset{1}{M}_{ij} = \overset{2}{M}_{ij}R_{12}^{-1}\overset{1}{M}_{ij}R_{12}R_{12}^{\mathrm{T}}, \text{ and } R_{12}R_{12}^{\mathrm{T}}\overset{1}{M}_{ij}R_{12}^{-\mathrm{T}}\overset{2}{M}_{ij} = \overset{1}{M}_{ij}R_{12}^{-\mathrm{T}}\overset{2}{M}_{ij}R_{12}^{\mathrm{T}}R_{12}.$$

3. Classical and Quantum R-Matrix Structures of $SL_k(\mathbb{C})$ Monodromy Data

3.1. R-Matrix Relations in the $SL_k(\mathbb{C})$ Case

For generic dimension k, consider the following special case of a trigonometric R-matrix generalizing the Kulish–Sklyanin matrix R_{12} to the case of $SL_k(\mathbb{C})$:

$$R_{12}(q) = q^{-1/k}\left(\sum_{i,j} \overset{1}{e}_{i,i}\overset{2}{e}_{j,j} + \sum_{i}(q-1)\overset{1}{e}_{i,i}\overset{2}{e}_{i,i} + \sum_{j>i}(q-q^{-1})\overset{1}{e}_{i,j}\overset{2}{e}_{j,i}\right) \tag{3.21}$$

Following [4], we use the R-matrix relations (2.8) and (2.11) as well as all 'composite' relations (2.16)–(2.20) to define a bracket operation on the representation space $\mathcal{R}^k_{g,s,n}$. The following theorem shows that actually this is a Poisson structure–we call it *Poisson algebra of monodromy data*:

Theorem 3.1 *For any R-matrix $R_{\alpha\beta}(q)$ that satisfies the QYBE $R_{12}R_{13}R_{23} = R_{23}R_{13}R_{12}$, and such that $R^{\mathrm{T}}_{\alpha\beta} = R_{\beta\alpha}$, relations (2.19) and (2.20) are equivalent and the relations (2.16)–(2.20) satisfy quantum Jacobi property.*

Proof This is a consequence of the fact that this Poisson algebra comes from the Fock–Rosly one. However, we would like to prove it directly, at least in one case, for the sake of completeness: the case in which two entries come from the same monodromy datum and the third one comes from another monodromy datum. We begin with

$$\overset{1}{M}_{21}R_{12}^{-\mathrm{T}}\overset{2}{M}_{21}R_{13}R_{23}\overset{3}{M}_{43}.$$

We then have the following chain of equalities in which we use the basic QYBE $R_{12}R_{13}R_{23} = R_{23}R_{13}R_{12}$ and its derivatives:

$$R^{\mathrm{T}}_{12}R_{23}R_{13} := R_{21}R_{23}R_{13} = R_{13}R_{23}R_{21} := R_{13}R_{23}R^{\mathrm{T}}_{12}$$

and

$$R_{13}R_{12}^{-\mathrm{T}}R_{23}^{-1} = R_{23}^{-1}R_{12}^{-\mathrm{T}}R_{13}$$

(we underline the terms in which we use commutation relations or QYBE); note that every $\overset{i}{M}$ commutes with R_{jk} if $i \neq \{j,k\}$:

$$\begin{aligned}
&\underline{\overset{1}{M}_{21}R_{12}^{-\mathrm{T}}\overset{2}{M}_{21}}R_{13}R_{23}\overset{3}{M}_{43} = R_{12}^{-\mathrm{T}}\overset{2}{M}_{21}R_{12}^{-1}\overset{1}{M}_{21}\underline{R_{12}R_{13}R_{23}}\overset{3}{M}_{43}\\
&= R_{12}^{-\mathrm{T}}\overset{2}{M}_{21}R_{12}^{-1}\overset{1}{M}_{21}R_{23}R_{13}R_{12}\overset{3}{M}_{43} = R_{12}^{-\mathrm{T}}\overset{2}{M}_{21}R_{12}^{-1}R_{23}\underline{\overset{1}{M}_{21}R_{13}\overset{3}{M}_{43}}R_{12}\\
&= R_{12}^{-\mathrm{T}}\overset{2}{M}_{21}\underline{R_{12}^{-1}R_{23}R_{13}}\overset{3}{M}_{43}R_{13}^{-1}\overset{1}{M}_{21}R_{13}R_{12} = R_{12}^{-\mathrm{T}}\overset{2}{M}_{21}R_{13}R_{23}R_{12}^{-1}\overset{3}{M}_{43}R_{13}^{-1}\overset{1}{M}_{21}R_{13}R_{12}\\
&= R_{12}^{-\mathrm{T}}R_{13}\underline{\overset{2}{M}_{21}R_{23}\overset{3}{M}_{43}}R_{12}^{-1}R_{13}^{-1}\overset{1}{M}_{21}R_{13}R_{12}
\end{aligned}$$

$$
\begin{aligned}
&= \underline{R_{12}^{-\mathrm{T}} R_{13} R_{23}} \overset{3}{M}_{43} R_{23}^{-1} \overset{2}{M}_{21} \underline{R_{23} R_{12}^{-1} R_{13}^{-1}} \overset{1}{M}_{21} R_{13} R_{12} \\
&= R_{23} R_{13} R_{12}^{-\mathrm{T}} \overset{3}{M}_{43} R_{23}^{-1} R_{13}^{-1} \underline{\overset{2}{M}_{21} R_{12}^{-1} \overset{1}{M}_{21}} \underline{R_{23} R_{13} R_{12}} \\
&= R_{23} R_{13} R_{12}^{-\mathrm{T}} \overset{3}{M}_{43} \underline{R_{23}^{-1} R_{13}^{-1} R_{12}^{\mathrm{T}}} \overset{1}{M}_{21} R_{12}^{-\mathrm{T}} \overset{2}{M}_{21} R_{13} R_{23} \\
&= R_{23} \underline{R_{13} \overset{3}{M}_{43} R_{13}^{-1} \overset{1}{M}_{21}} R_{23}^{-1} R_{12}^{-\mathrm{T}} \overset{2}{M}_{21} R_{13} R_{23} = R_{23} \overset{1}{M}_{21} R_{13} \overset{3}{M}_{43} \underline{R_{13}^{-1} R_{23}^{-1} R_{12}^{-\mathrm{T}}} \overset{2}{M}_{21} R_{13} R_{23} \\
&= R_{23} \overset{1}{M}_{21} R_{13} R_{12}^{-\mathrm{T}} \underline{\overset{3}{M}_{43} R_{23}^{-1} \overset{2}{M}_{21} R_{23}} = R_{23} \overset{1}{M}_{21} \underline{R_{13} R_{12}^{-\mathrm{T}} R_{23}^{-1}} \overset{2}{M}_{21} R_{23} \overset{3}{M}_{43} \\
&= \overset{1}{M}_{21} R_{12}^{-\mathrm{T}} \overset{2}{M}_{21} R_{13} R_{23} \overset{3}{M}_{43}.
\end{aligned}
$$

We present one more calculation demonstrating Jacobi property for the same matrix M. We use just one form, (2.19), of the commutation relation. We begin with the same expression

$$
\overset{1}{M} R_{12}^{-\mathrm{T}} \overset{2}{M} R_{13}^{-\mathrm{T}} R_{23}^{-\mathrm{T}} \overset{3}{M} = \overset{1}{M} R_{12}^{-\mathrm{T}} R_{13}^{-\mathrm{T}} \overset{2}{M} R_{23}^{-\mathrm{T}} \overset{3}{M}.
$$

We first transform the left-hand side:

$$
\begin{aligned}
&\underline{\overset{1}{M} R_{12}^{-\mathrm{T}} \overset{2}{M}} R_{13}^{-\mathrm{T}} R_{23}^{-\mathrm{T}} \overset{3}{M} = R_{12}^{-\mathrm{T}} \overset{2}{M} R_{12}^{-1} \overset{1}{M} \underline{R_{12} R_{13}^{-\mathrm{T}} R_{23}^{-\mathrm{T}}} \overset{3}{M} \\
&= R_{12}^{-\mathrm{T}} \overset{2}{M} R_{12}^{-1} R_{23}^{-\mathrm{T}} \underline{\overset{1}{M} R_{13}^{-\mathrm{T}} \overset{3}{M}} R_{12} = R_{12}^{-\mathrm{T}} \overset{2}{M} \underline{R_{12}^{-1} R_{23}^{-\mathrm{T}} R_{13}^{-\mathrm{T}}} \overset{3}{M} R_{13}^{-1} \overset{1}{M} R_{13} R_{12} \\
&= R_{12}^{-\mathrm{T}} R_{13}^{-\mathrm{T}} \underline{\overset{2}{M} R_{23}^{-\mathrm{T}} \overset{3}{M}} R_{12}^{-1} R_{13}^{-\mathrm{T}} \overset{1}{M} R_{13} R_{12} = R_{12}^{-\mathrm{T}} R_{13}^{-\mathrm{T}} R_{23}^{-\mathrm{T}} \overset{3}{M} R_{23}^{-1} \overset{2}{M} \underline{R_{23} R_{12}^{-1} R_{13}^{-1}} \overset{1}{M} R_{13} R_{12} \\
&= R_{12}^{-\mathrm{T}} R_{13}^{-\mathrm{T}} R_{23}^{-\mathrm{T}} \overset{3}{M} R_{23}^{-1} \overset{2}{M} R_{13}^{-1} R_{12}^{-1} \overset{1}{M} R_{23} R_{13} R_{12}.
\end{aligned}
$$

We now turn to the right-hand side:

$$
\begin{aligned}
&\overset{1}{M} R_{12}^{-\mathrm{T}} R_{13}^{-\mathrm{T}} \underline{\overset{2}{M} R_{23}^{-\mathrm{T}} \overset{3}{M}} = \overset{1}{M} \underline{R_{12}^{-\mathrm{T}} R_{13}^{-\mathrm{T}} R_{23}^{-\mathrm{T}}} \overset{3}{M} R_{23}^{-1} \overset{2}{M} R_{23} \\
&= R_{23}^{-\mathrm{T}} \underline{\overset{1}{M} R_{13}^{-\mathrm{T}} \overset{3}{M}} R_{12}^{-\mathrm{T}} R_{23}^{-1} \overset{2}{M} R_{23} = R_{23}^{-\mathrm{T}} R_{13}^{-\mathrm{T}} \overset{3}{M} R_{13}^{-1} \overset{1}{M} \underline{R_{13} R_{12}^{-\mathrm{T}} R_{23}^{-1}} \overset{2}{M} R_{23} \\
&= R_{23}^{-\mathrm{T}} R_{13}^{-\mathrm{T}} \overset{3}{M} R_{13}^{-1} R_{23}^{-1} \underline{\overset{1}{M} R_{12}^{-\mathrm{T}} \overset{2}{M}} R_{13} R_{23} = R_{23}^{-\mathrm{T}} R_{13}^{-\mathrm{T}} \overset{3}{M} \underline{R_{13}^{-1} R_{23}^{-1} R_{12}^{-\mathrm{T}}} \overset{2}{M} R_{12}^{-1} \overset{1}{M} R_{12} R_{13} R_{23} \\
&= \underline{R_{23}^{-\mathrm{T}} R_{13}^{-\mathrm{T}} R_{12}^{-\mathrm{T}}} \overset{3}{M} R_{23}^{-1} \overset{2}{M} R_{13}^{-1} R_{12}^{-1} \overset{1}{M} \underline{R_{12} R_{13} R_{23}} \\
&= R_{12}^{-\mathrm{T}} R_{13}^{-\mathrm{T}} R_{23}^{-\mathrm{T}} \overset{3}{M} R_{23}^{-1} \overset{2}{M} R_{13}^{-1} R_{12}^{-1} \overset{1}{M} R_{23} R_{13} R_{12},
\end{aligned}
$$

which coincides with the final expression in transformations of the left-hand side. □

3.2. *Decorated Character Variety*

In order to define a Poisson structure on the decorated character variety

$$
\mathcal{M}_{g,s,n}^{k} := \mathcal{R}_{g,s,n}^{k} / \textstyle\prod_{j=1}^{n} U_j,
$$

we prove that the quotient by unipotent Borel subgroups is a Poisson reduction.

In the $SL_2(\mathbb{C})$ case, we have the following Poisson reduction for monodromy data corresponding to passing along the hole boundary: if M corresponds to a path along the boundary that goes clockwise (the hole is to the left w.r.t. the path direction), then $m_{22} = 0$, that is, the lower-right element vanishes.

For generic $k \geq 2$, we have the following

Lemma 3.2 *Consider the monodromy data $M \in SL_k(\mathbb{C})$ corresponding to paths that go clockwise along boundaries of holes (they may start and terminate at the same cusp if a hole has only one cusp). The reduction*

$$m_{i,j} = 0,\ i+j \geq k+2, \tag{3.22}$$

that is, all its entries below the main anti-diagonal vanish, is a Poisson reduction.

Proof The proof is based on the following observation: if M is a matrix corresponding to a path that is leftmost at the starting cusp and rightmost at the terminating cusp (examples are M_2 in the basic relation (2.8) and M_{41} in (2.17)), then (Poisson or quantum) commutation relations of elements $m_{i,j}$ of this matrix with elements of every other matrix or among themselves are such that every term of the corresponding commutation relation necessarily contains an element $m_{k,j}$ or $m_{i,l}$ with $k \geq i$ and $l \geq j$, that is, $k+j \geq i+j$ and $i+l \geq i+j$. Therefore, imposing a constraint $m_{i,j} = 0$ for $i+j > k+1$ is consistent: commutation relations of such elements with all other elements of algebra automatically vanish. □

Remark 3.1 From a purely algebraic standpoint, one may consider other Poisson reductions; the one for which $m_{i,j} = 0$ for elements below the main anti-diagonal is consistent with factoring out a gauge freedom associated with Borel subgroups at cusps.

The Poisson reduction in Lemma 3.2 is the quotient w.r.t. unipotent Borel subgroups $U_i \subset SL_k(\mathbb{C})$ associated with the cusps; therefore, it endows the decorated character variety $\mathcal{M}^k_{g,s,n}$ with a Poisson structure.

3.3. *Powers of Matrices*

Using the commutation relations (2.16)–(2.20) in the R-matrix form, we obtain the following generalizations of these relations to powers of matrices:

$$\overset{1}{M}{}^{p}_{21} R_{12} \overset{2}{M}{}^{m}_{43} R_{12}^{-1} = R_{12} \overset{2}{M}{}^{m}_{43} R_{12}^{-1} \overset{1}{M}{}^{p}_{21}, \quad p, m \in \mathbb{Z}, \tag{3.23}$$

$$\overset{1}{M}_{41} R_{12}^{-\mathrm{T}} \overset{2}{M}{}^{p}_{32} R_{12}^{\mathrm{T}} = R_{12} \overset{2}{M}{}^{p}_{32} R_{12}^{-1} \overset{1}{M}_{41}, \quad p \in \mathbb{Z}, \tag{3.24}$$

$$\overset{1}{M}_{31} R_{12}^{-\mathrm{T}} \overset{2}{M}_{42} R_{12}^{-1} = R_{12} \overset{2}{M}_{42} R_{12}^{-1} \overset{1}{M}_{31}, \text{ no generalization}, \tag{3.25}$$

$$R_{12}^{\mathrm{T}} \overset{1}{M}{}^{p}_{ij} R_{12}^{-\mathrm{T}} \overset{2}{M}_{ij} = \overset{2}{M}_{ij} R_{12}^{-1} \overset{1}{M}{}^{p}_{ij} R_{12}, \quad p \in \mathbb{Z}, \tag{3.26}$$

$$\overset{1}{M}_{ij}R_{12}^{-\mathrm{T}}\overset{2}{M}{}_{ij}^{p}R_{12}^{\mathrm{T}} = R_{12}\overset{2}{M}{}_{ij}^{p}R_{12}^{-1}\overset{1}{M}_{ij}, \quad p \in \mathbb{Z}. \tag{3.27}$$

Note that, for the *same* matrix M_{ij}, we can take powers of only one of the matrices M_{ij} in relations (2.19) and (2.20) but not powers of both matrices.

3.4. *Semi-Classical limit*

By taking $q = \exp(-i\pi\hbar)$, we can expand the Kulish–Sklyanin matrix R_{12} as

$$R_{12}(q) = (1 + \frac{i\pi\hbar}{k} + \mathcal{O}(\hbar^2))\sum_{i,j}\overset{1}{e}_{ii}\otimes\overset{2}{e}_{jj} + (-i\pi\hbar + \mathcal{O}(\hbar^3))\sum_{i}\overset{1}{e}_{ii}\otimes\overset{2}{e}_{ii}$$
$$+ (-2i\pi\hbar + \mathcal{O}(\hbar^2))\sum_{j>i}\overset{1}{e}_{ij}\otimes\overset{2}{e}_{ji}$$

so that we obtain

$$R_{12}(q) = \overset{1}{1}\otimes\overset{2}{1} + i\pi\hbar r + \mathcal{O}(\hbar^2), \qquad R_{12}(1/q) = \overset{1}{1}\otimes\overset{2}{1} - i\pi\hbar r + \mathcal{O}(\hbar^2),$$

where

$$r = \frac{1}{k}\sum_{i,j}\overset{1}{e}_{ii}\otimes\overset{2}{e}_{jj} - \sum_{i}\overset{1}{e}_{ii}\otimes\overset{2}{e}_{ii} - 2\sum_{j>i}\overset{1}{e}_{ij}\otimes\overset{2}{e}_{ji}. \tag{3.28}$$

Now, using the correspondence principle that $[A^\hbar, B^\hbar] \mapsto i\pi\hbar\{A, B\}$, we can take the semi-classical limits of (2.16):

$$\overset{1}{M}_{21}\overset{2}{M}_{43} + i\pi\hbar\overset{1}{M}_{21}r\overset{2}{M}_{43} - i\pi\hbar\overset{1}{M}_{21}\overset{2}{M}_{43}r = \overset{2}{M}_{43}\overset{1}{M}_{21} + i\pi\hbar r\overset{2}{M}_{43}\overset{1}{M}_{21} - i\pi\hbar\overset{2}{M}_{43}r\overset{1}{M}_{21},$$

so that (2.16) becomes

$$\{\overset{1}{M}_{21}, \overset{2}{M}_{43}\} = -\overset{1}{M}_{21}r\overset{2}{M}_{43} + \overset{1}{M}_{21}\overset{2}{M}_{43}r + r\overset{2}{M}_{43}\overset{1}{M}_{21} - \overset{2}{M}_{43}r\overset{1}{M}_{21}. \tag{3.29}$$

In the same way, (2.17) becomes

$$\{\overset{1}{M}_{41}, \overset{2}{M}_{32}\} = \overset{1}{M}_{41}r^T\overset{2}{M}_{32} - \overset{1}{M}_{41}\overset{2}{M}_{32}r^T + r\overset{2}{M}_{32}\overset{1}{M}_{41} - \overset{2}{M}_{32}r\overset{1}{M}_{41}, \tag{3.30}$$

while (2.18) becomes

$$\{\overset{1}{M}_{31}, \overset{2}{M}_{42}\} = \overset{1}{M}_{31}r^T\overset{2}{M}_{42} + \overset{1}{M}_{31}\overset{2}{M}_{42}r + r\overset{2}{M}_{42}\overset{1}{M}_{31} - \overset{2}{M}_{42}r\overset{1}{M}_{31}, \tag{3.31}$$

and (2.19) becomes

$$\{\overset{1}{M}_{ij}, \overset{2}{M}_{ij}\} = \overset{1}{M}_{ij}r^T\overset{2}{M}_{ij} - r^T\overset{1}{M}_{ij}\overset{2}{M}_{ij} + \overset{2}{M}_{ij}\overset{1}{M}_{ij}r - \overset{2}{M}_{ij}r\overset{1}{M}_{ij}. \tag{3.32}$$

We let $m_{k,l}^{\alpha,\beta}$ denote the (k,l)-element of the matrix $M_{\alpha,\beta}$. For matrix elements, we have the following Poisson relations (in the formulas below, repeated indices imply summations):

$$\{m_{i,j}^{2,1}, m_{k,l}^{4,3}\} = m_{i,s}^{2,1} m_{s,l}^{4,3} \delta_{j,k} \theta(j-s) - m_{i,l}^{2,1} m_{k,j}^{4,3} \theta(j-l) + m_{s,j}^{2,1} m_{k,s}^{4,3} \delta_{i,l} \theta(s-i) \\ - m_{k,j}^{2,1} m_{i,l}^{4,3} \theta(k-i), \tag{3.33}$$

$$\{m_{i,j}^{4,1}, m_{k,l}^{3,2}\} = -m_{i,s}^{4,1} m_{s,l}^{3,2} \delta_{j,k} \theta(s-j) + m_{i,l}^{4,1} m_{k,j}^{3,2} \theta(l-j) + m_{s,j}^{4,1} m_{k,s}^{3,2} \delta_{i,l} \theta(s-i) \\ - m_{k,j}^{4,1} m_{i,l}^{3,2} \theta(k-i), \tag{3.34}$$

$$\{m_{i,j}^{3,1}, m_{k,l}^{4,2}\} = -m_{i,s}^{3,1} m_{s,l}^{4,2} \delta_{j,k} \theta(s-j) - m_{i,l}^{3,1} m_{k,j}^{4,2} \theta(j-l) + m_{s,j}^{3,1} m_{k,s}^{4,2} \delta_{i,l} \theta(s-i) \\ - m_{k,j}^{3,1} m_{i,l}^{4,2} \theta(k-i) + 2 m_{i,s}^{3,1} m_{s,l}^{4,2}, \tag{3.35}$$

$$\{m_{i,j}^{\alpha,\beta}, m_{k,l}^{\alpha,\beta}\} = -m_{i,s}^{\alpha,\beta} m_{s,l}^{\alpha,\beta} \delta_{j,k} \theta(s-j) + m_{s,j}^{\alpha,\beta} m_{k,s}^{\alpha,\beta} \delta_{i,l} \theta(s-i) + m_{k,j}^{\alpha,\beta} m_{i,l}^{\alpha,\beta} (\theta(l-j) \\ - \theta(k-i)), \tag{3.36}$$

where $\theta(k) := \text{sign}(k) + 1$.

Remark 3.2 Note that quantum commutation relations (2.19) and (2.20) have the same semi-classical limit (3.36).

For a monodromy datum M_2^1 corresponding to an arc starting and terminating at different cusps, we have

$$\{m_{2_{i,j}}^1, m_{2_{k,l}}^1\} = m_{2_{i,j}}^1 m_{2_{k,l}}^1 (\theta(i-k) - \theta(j-l)). \tag{3.37}$$

3.5. *Casimirs of the Poisson Algebra of Monodromy Data*

We now address the problem of constructing Casimirs for the Poisson algebra of monodromy data. For technical convenience, we do not impose the restriction $\det M = 1$, although relations (3.33)–(3.36) imply that determinants of all monodromy data corresponding to arcs starting and terminating at the same cusp are central.

Theorem 3.3 *In any system of $SL_k(\mathbb{C})$ monodromy data, for a monodromy datum M corresponding to an arc homeomorphic to circumnavigating a single hole without cusps, all elements* $\operatorname{tr}[M^p]$, $p = 1, \ldots, k$, *are Casimirs of the Poisson algebra of monodromy data.*

Proof The monodromy datum M corresponding to an arc homeomorphic to circumnavigating a single hole without cusps can be identified with the matrix M_{21}, while all other monodromy data can be identified with a matrix denoted by $M_{3,[x]}$ that corresponds to a path starting at the same cusp as M_{21} (to the right of both ends of M_{21}) and terminating at a different cusp (denoted $[x]$). In the quantum case, we have the relation

$$\overset{1}{M}_{21} \overset{2}{M}_{3,[x]} = \overset{2}{M}_{3,[x]} R_{12}^{-1} \overset{1}{M}_{21} R_{12}, \tag{3.38}$$

which admits an immediate generalization to powers of M_{12}:

$$\overset{1}{M}{}^{p}_{21}\overset{2}{M}_{3,[x]} = \overset{2}{M}_{3,[x]}R_{12}^{-1}\overset{1}{M}{}^{p}_{21}R_{12}, \quad p \in \mathbb{Z}. \tag{3.39}$$

Evaluating traces in space 1 in the semi-classical limits of relations (3.23), (3.24), (3.26) and (3.39), we find that the traces of the matrix M_{21}^{n} (or M_{32}^{n}) Poisson commute with elements of all other matrices (and elements of the matrix M_{21} itself), which completes the proof. We do this computation in detail only for relations (3.39); all other cases are completely analogous.

The semi-classical limit of relation (3.39) reads

$$\{m^{p}_{21_{i,j}} m_{3,[x]_{k,l}}\} = -\sum_{s=1}^{n}[m^{p}_{21_{s,j}} m_{3,[x]_{k,s}} \delta_{i,l}\theta(s-i) + m^{p}_{21_{i,l}} m_{3,[x]_{k,j}} \theta(j-l)]. \tag{3.40}$$

Taking the sum over i with $i = j$, we obtain

$$\begin{aligned}\sum_{i=1}^{n}\{m^{p}_{21_{i,i}} m_{3,[x]_{k,l}}\} &= -\sum_{s=1}^{n}\sum_{i=1}^{n}[m^{p}_{21_{s,i}} m_{3,[x]_{k,s}} \delta_{i,l}\theta(s-i) + m^{p}_{21_{i,l}} m_{3,[x]_{k,i}} \theta(i-l)] \\ &= -\sum_{s=1}^{n} m^{p}_{21_{s,l}} m_{3,[x]_{k,s}}\theta(s-l) + \sum_{i=1}^{n} m^{p}_{21_{i,l}} m_{3,[x]_{k,i}}\theta(i-l) = 0.\end{aligned}$$

Therefore, $\operatorname{tr} M^{k}$ are Casimirs for the algebra of elements of *any* matrix M corresponding to a path that starts and terminates at the same cusp. □

In Section 4, we address the question of the the symplectic leaves dimension in some examples.

3.6. *Reduction to the SL_2 Decorated Character Variety*

Let us select the same cusped lamination as in [17]; then, for every arc α in the lamination, we associate a matrix $M_{\alpha} \in SL_k$. This gives $6g - 6 + 3s + 2n$ matrices in SL_k.

It is easy to prove that the following character,

$$\begin{aligned} \operatorname{tr}_K :\ & SL_k(\mathbb{C}) \to \mathbb{C} \\ & M \mapsto \operatorname{tr}(MK), \end{aligned} \qquad \text{where} \quad K = \begin{pmatrix} 0 & \dots & 0 & 0 \\ \dots & \dots & \dots & \dots \\ 0 & \dots & 0 & 0 \\ -1 & 0 & 0 & 0 \end{pmatrix},$$

is well defined on $M^{k}_{g,s,n}$. Recall that $\operatorname{tr}_K(M) = -m_{1k}$.

Define the λ-length of the arc α by $\operatorname{tr}_K(M_\alpha)$ and introduce the following operation:

$$\overset{12}{\operatorname{tr}}_K(\overset{1}{M}_1\overset{2}{M}_2) := \overset{12}{\operatorname{tr}}(\overset{1}{M}_1\overset{1}{K}\overset{2}{M}_2\overset{2}{K}).$$

By taking $\overset{12}{\operatorname{tr}}_K$ in all relations (2.16), (2.17), (2.18) and (2.19), we obtain the λ-lengths algebra on the SL_2 decorated character variety found in [17].

4. Examples of Algebras of Monodromy Data

4.1. Case of Only One Monodromy Datum

In the case when $2g-2+s+n=1$, we have only one monodromy datum. We have two different situations: the first is when M comes back to the same cusp; the second is when it connects different cusps. In the former case, the following lemma holds true:

Lemma 4.1 *For a general-position monodromy datum subject to algebra (3.36), the maximal dimension of the symplectic leaves is $k(k-1)$. The only Casimirs in this case are* $\operatorname{tr}[M^p]$ *with $p=1,\dots,k$.*

Proof The proof is based on the following observation. Let the classical M have a diagonal form $m_{i,j}=\delta_{i,j}\lambda_i$, with all λ_i distinct and non-zero. The Poisson brackets are then non-zero only inside the pairs $(m_{i,j},m_{j,i})$, with $1\le i<j\le k$, for which we have

$$\{m_{i,j},m_{j,i}\}=\lambda_j(\lambda_j-\lambda_i),$$

so these brackets are non-degenerate. The minimal symplectic leaf dimension of the corresponding leaf is thus k^2-k, and it is simultaneously the maximal possible symplectic leaf dimension, as the traces $\operatorname{tr} M^p$, by Theorem 3.3, are k algebraically independent Casimirs of the algebra (3.36). □

We next address the problem of Casimirs for the case when the monodromy datum corresponds to an arc that connects different cusps–algebra (3.37):

Lemma 4.2 [16, 25] *The central elements of the algebra (3.37) in the case of non-restricted matrices M are ratios $M^{\mathrm{UL}}_d/M^{\mathrm{LR}}_{k-d}$ of upper-left and lower-right minors of the respective dimensions $d\times d$ and $(k-d)\times(k-d)$ for $d=1,\dots,k$.*

We now derive analogues of Lemmas 4.1 and 4.2 in the presence of the constraints (3.22).

Lemma 4.3 *For a monodromy datum M subject to the algebra (3.36) with restriction (3.22) imposed, the maximal dimension of symplectic leaves is $k(k-1)/2-[k/2]$, and the k Casimirs are* $\operatorname{tr}[M^p]$*, $p=1,\dots,k$, as in the non-restricted case, plus $[k/2]$ Casimirs defined by*

$$C_i := m_{i,k+1-i}/m_{k+1-i,i},\quad i=1,\dots,[k/2]. \tag{4.1}$$

Proof Traces are Casimirs for the general matrix M and they remain Casimirs for any Poisson reduction. Anti-diagonal elements of M have homogeneous Poisson relations with all other elements:

$$\{m_{i,k+1-i},m_{k,l}\}=m_{i,k+1-i}m_{k,l}[-\delta_{k+1-i,k}+\delta_{k+1-i,l}+\delta_{i,l}-\delta_{i,k}],$$

so the ratios (4.1) have zero Poisson brackets with all $m_{k,l}$. It remains to prove that the highest symplectic leaf dimension matches the number of already-found Casimirs. In order to prove it, take the reduced matrix M in the form in which the diagonal elements $m_{i,i}$ with $1 \le i \le [k/2]$ and all anti-diagonal elements $m_{i,k+1-i}$, $i = 1, \dots, k$, are non-zero and are not algebraically related. Then it is a direct calculation to check that the commutation relations are closed inside quadruples $(m_{i,j}, m_{k+1-i,j}, m_{j,i}, m_{k+1-j,i})$ with $1 \le j < i \le [k/2]$, doubles $(m_{(k+1)/2,j}, m_{j,(k+1)/2})$ with $1 \le j < (k+1)/2$, triples $(m_{i,i}, m_{k+1-i,i}, m_{i,k+1-i})$ and singles $m_{(k+1)/2,(k+1)/2}$. (Of course, doubles and singles occur only for odd n.) It is then a straightforward calculation to show that quadruples and doubles have full symplectic leaf dimension whereas triples and singles have zero symplectic leaf dimensions, so all their elements correspond to Casimirs. But the total number of elements in triples and singles (if any) is exactly $k + [k/2]$, that is, the number of Casimirs listed above. □

Lemma 4.4 [16, 25] *The central elements of the algebra (3.37) in the case of matrices M with restrictions (3.22) are*

$$\hat{C}_d = \frac{M_d^{\mathrm{UL}} \prod_{i=1}^{d} [m_{i,k+1-i} m_{k+1-i,i}]}{M_{k-d}^{\mathrm{UL}}}, \quad d = 0, \dots, \left[\frac{k-1}{2}\right]. \tag{4.2}$$

(See Figure 4.) In this formula, both minors are upper left, of sizes $d \times d$ and $(k-d) \times (k-d)$. We have $[(k+1)/2]$ such Casimirs and the maximal symplectic leaf dimension is $k(k+1)/2 - [(k+1)/2]$.

Proof As illustrated in Figure 4, all matrix elements have homogeneous commutation relations with any minor M_d^{UL}; for the ratio $M_d^{\mathrm{UL}}/M_{k-d}^{\mathrm{UL}}$, the coefficients are $+1, 0, -1$, depending on the region which this matrix element belongs to. They are depicted in Figure 4(a). Next, all matrix elements have homogeneous commutation relations with any element $m_{i,k+1-i}$ on the main anti-diagonal. For the product $\prod_{i=1}^{d} [m_{i,k+1-i} m_{k+1-i,i}]$ of these elements, the corresponding coefficients are depicted

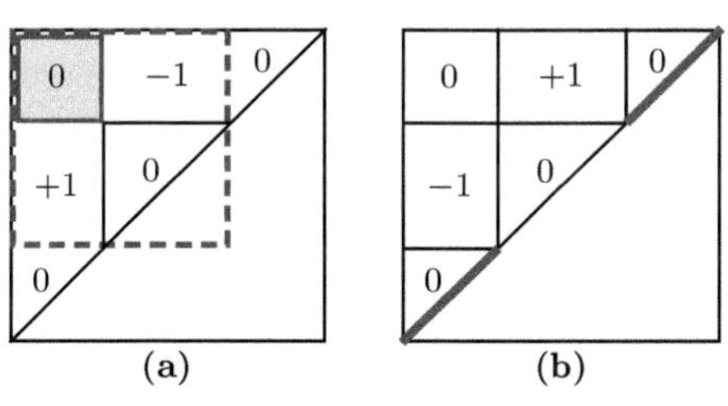

Figure 4 Constructing Casimirs for the restricted matrix M subject to Poisson algebra (3.37). Numbers in the corresponding rectangles or triangles indicate the sign of homogeneous commutation relations between elements in the corresponding region and (a) the ratios of minors $M_d^{\mathrm{UL}}/M_{k-d}^{\mathrm{UL}}$ $(d < k-d)$ and (b) the products $\prod_{i=1}^{d} [m_{i,k+1-i} m_{k+1-i,i}]$ of elements on the anti-diagonal. We see that these signs are complementary in all regions.

in Figure 4(b). We see that the two patterns are complementary, so the product in the right-hand side of (4.2) commutes with all matrix elements.

If we again assume that only anti-diagonal elements and upper half of diagonal elements of M are non-zero, then, as in the proof of Lemma 4.3, we can split the whole set of elements into quadruples, doubles, triples and singles; as in the above proof, all quadruples and doubles will then have the full Poisson dimension but, in contrast to the proof of Lemma 4.3, the symplectic leaf dimension of triples will be 2, not 0 (and it obviously remains 0 for singles). Thus, we are losing exactly $2[k/2]$ central elements, as compared to the previous case, and the maximal Poisson dimension in this case is $k(k+1)/2-[(k+1)/2]$. □

4.2. Monodromy Algebras for $\Sigma_{0,s+1,1}$

We first address the case of monodromy data for a disc bounded by a hole with a single bordered cusp and with s holes in the interior (Figure 5(a)). The basis of monodromy data is constituted by s $(k\times k)$-matrices $M_{2j,2j-1}$, $j=1,\dots,s$, where we order linearly all $2s$ ends of cycles corresponding to these monodromy data; the matrix $M_{2j,2j-1}$ corresponds to the path that starts and terminates at the bordered cusp and circumnavigates the jth hole inside the disc going clockwise. The quantum and Poisson algebras of elements of these matrices are described by relations (2.16), (2.19), (2.20) and (3.33), (3.36).

We introduce also the boundary monodromy datum

$$M_B := M_{21}M_{43}\cdots M_{2s-2,2s-3}M_{2s,2s-1}. \tag{4.3}$$

We begin with the lemma describing a non-restricted case.

Lemma 4.5 *The maximum Poisson dimension of leaves of the $SL_k(\mathbb{C})$-algebra of monodromy data for $\Sigma_{0,s+1,1}$ is $sk(k-1)$. The sk central elements are* $\mathrm{tr}\,[M_{2j,2j-1}]^p$, $j=1,\dots,s$, $p=1,\dots,k$.

Proof That the traces of powers of $M_{2j,2j-1}$ are central was proved in Theorem 3.3. Here we prove that the general dimension of a symplectic leaf is $sk(k-1)$. For this, we again evaluate the rank of the Poisson bivector at a specific point in the phase space. It is convenient to take a point at which all monodromy data are diagonal and all their

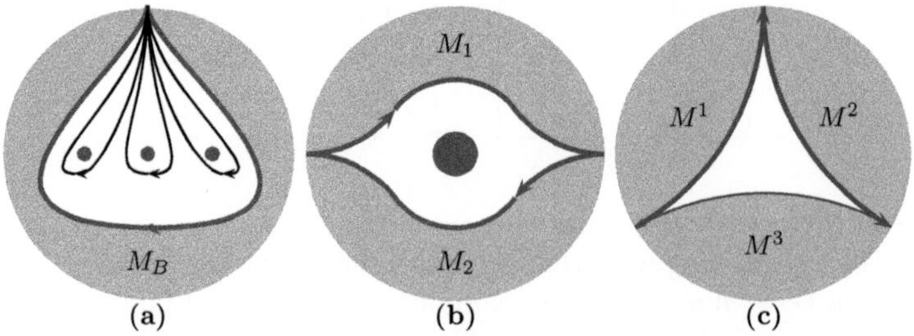

Figure 5 Three cases of monodromy data: $\Sigma_{0,s+1,1}$ (a), $\Sigma_{0,2,2}$ (b) and $\Sigma_{0,1,3}$ (c).

diagonal elements are distinct and non-zero, $m_{2j,2j-1_{p,l}} = \delta_{p,l}\lambda_p^{(j)}$, with all $\lambda_p^{(j)} \neq 0$ and such that $\lambda_{p_1}^{(j_1)} = \lambda_{p_2}^{(j_2)}$ only for $p_1 = p_2$ and $j_1 = j_2$. It is easy to check that non-zero entries in the Poisson bivector correspond to the Poisson brackets between $m_{2j,2j-1_{p,l}}$ and $m_{2j,2j-1_{l,p}}$ with $p \neq l$ and these brackets, evaluated at the chosen point, are

$$\{m_{2j,2j-1_{p,l}}, m_{2j,2j-1_{l,p}}\} = (\lambda_p^{(j)} - \lambda_l^{(j)})[\lambda_p^{(j)}\theta(p-l) + \lambda_l^{(j)}\theta(l-p)], \tag{4.4}$$

$$\{m_{2j,2j-1_{p,l}}, m_{2j,2j-1_{l,p}}\} = (\lambda_p^{(j)} - \lambda_l^{(j)})(\lambda_p^{(i)} - \lambda_l^{(i)})\theta(l-p), \quad j < i. \tag{4.5}$$

So, the Poisson bivector at this point has mostly vanishing entries except $2s$ non-vanishing relations for every fixed pair (p,l), with $p \neq l$. We can organize rows and columns by ordering them by the pairs (p,l), starting from $(1,2)$ and ending with $(k-1,k)$. In this way, the Poisson bivector becomes block diagonal with $2s \times 2s$ blocks of non-zero determinant, provided all $\lambda_p^{(j)}$ are non-zero and distinct. There are $k(k-1)/2$ such blocks, which proves that the rank is $sk(k-1)$. □

We now consider the actual situation with the Lagrangian restriction (3.22) imposed on the monodromy datum M_B (4.3).

Theorem 4.6 *The maximal symplectic leaf dimension of the $SL_k(\mathbb{C})$-algebra of monodromy data for $\Sigma_{0,s+1,1}$ with the restriction (3.22) imposed on the only boundary monodromy datum M_B (4.3) is $sk(k-1) - k(k-1)/2 - [k/2]$. Besides the standard sk central elements that are $\operatorname{tr} M_{2j,2j-1}^p$, $j = 1, \dots, s$, $p = 1, \dots, k$, we have $[k/2]$ central elements having the form (4.1) for the matrix M_B, that is, $C_i = m_{B_{i,k+1-i}}/m_{B_{k+1-i,i}}$.*

Proof We first prove that $C_i = [m_B]_{i,k+1-i}/[m_B]_{k+1-i,i}$ are central. Every element $[m_B]_{i,k+1-i}$ has homogeneous Poisson relations with every matrix element of every matrix $M_{2j,2j-1}$. In Poisson relations, we can identify M_B with M_{41}, and every $M_{2j,2j-1}$—with M_{32}. Actual Poisson brackets coincide with those inside the same matrix M_{41}:

$$\{m_{41_{i,k+1-i}}, m_{32_{p,l}}\} = m_{41_{i,k+1-i}} m_{32_{p,l}} [-\delta_{k+1-i,p} + \delta_{k+1-i,l} + \delta_{i,l} - \delta_{i,p}],$$

so the ratio (4.1) remains central in this case as well.

To address the problem of actual symplectic leaf dimension, it is technically more convenient to remove the last monodromy datum $M_{2s,2s-1}$ from the basis and add M_B to it. Then, all remaining matrices are independent and we can take the restriction on M_B into account explicitly. We again evaluate the Poisson bivector at the point in which all $M_{2j,2j-1}$ with $1 \leq j < s$ are diagonal, and M_B has non-zero anti-diagonal and the upper half-diagonal. We let $S[m_{i,j}]$ denote the orbit of the matrix element $m_{i,j}$ under the action of the symmetry group generated by $S_1[m_{i,j}] = m_{k+1-i,j}$ and $S_2[m_{i,j}] = m_{j,i}$. For a generic (i,j), such an orbit comprises eight elements for $M_{2j,2j-1}$ and four elements for M_B, because of the reduction. It is a cumbersome calculation, omitted, to demonstrate that the Poisson algebra is non-degenerate for the sets $\cup_{j=1}^{s-1} S[m_{2j,2j-1_{k,l}}] \cup S[m_{B_{k,l}}]$ with $1 \leq l < k \leq [(k+1)/2]$ and is highly degenerate for the sets $\cup_{j=1}^{s-1} S[m_{2j,2j-1_{i,i}}] \cup S[m_{B_{i,i}}]$ with $i = 1,\ldots,[k/2]$: the symplectic leaf dimension of every such set comprising $4s-1$ elements is $2(s-1)$. Finally, the set $\cup_{j=1}^{s-1} S[m_{2j,2j-1_{(k+1)/2,(k+1)/2}}] \cup S[m_{B_{(k+1)/2,(k+1)/2}}]$ has zero dimension. So, the total Poisson codimension for even k is $[k/2](2s+1) = sk + k/2$ and, for odd k, it is $[k/2](2s+1) + s = sk + (k-1)/2$, that is, we have $sk + [k/2]$ independent Casimirs, as expected. □

4.2.1. Braid-group action in $\Sigma_{0,s+1,1}$

Let us denote $M_{2r,2r-1}$ by $M_{(r)}$ for brevity. The braid group for $\pi_1(\Sigma_{0,s+1,1})$ is generated by the standard operators B_j:

$$B_j : \left\{ M_{(j-1)} \to M_{(j-1)} M_{(j)} M_{(j-1)}^{-1};\ M_{(j)} \to M_{(j-1)};\ M_{(k)} \to M_{(k)},\ k \neq j, j-1 \right\}. \tag{4.6}$$

Lemma 4.7 *The braid-group action (4.6) preserves the quantum commutation relations for monodromy data. In the semi-classical limit, it also obviously preserves the set of central elements* ($\operatorname{tr} M_{(r)}^k$).

Proof We have to verify the preservation of quantum commutation relations. Checking this for $M_{(k)}$ with $k \neq j, j-1$ is simple: such $M_{(k)}$ have commutation relations of the same form (2.16) with $M_{(j)}$, $M_{(j-1)}$ and with all elements of the multiplicative non-abelian group generated by the matrices $M_{(j)}$ and $M_{(j-1)}$. The new matrices $M'_{(j-1)} = M_{(j-1)} M_{(j)} M_{(j-1)}^{-1}$, and $M'_{(j)} = M_{(j-1)}$, must satisfy the same commutation relations as $M_{(j-1)}$ and $M_{(j)}$. For commutation relations of $M'_{(j)}$ with itself, it trivially holds, so we have to check two non-trivial relations. The first one is

$$\begin{aligned}
&\overset{1}{M}'_{(j-1)} R_{12} \overset{2}{M}'_{(j)} R_{12}^{-1} = \big(\overset{1}{M}_{(j-1)} \overset{1}{M}_{(j)} \overset{1}{M}{}_{(j-1)}^{-1}\big) R_{12} \overset{2}{M}_{(j-1)} R_{12}^{-1} \\
&= \overset{1}{M}_{(j-1)} \overset{1}{M}_{(j)} \underline{\overset{1}{M}{}_{(j-1)}^{-1} R_{12} \overset{2}{M}_{(j-1)}} R_{12}^{-1} \\
&= \overset{1}{M}_{(j-1)} \underline{\overset{1}{M}_{(j)} R_{12}^{-\mathrm{T}} \overset{2}{M}_{(j-1)}} R_{12}^{\mathrm{T}} \overset{1}{M}{}_{(j-1)}^{-1} R_{12} R_{12}^{-1} = \underline{\overset{1}{M}_{(j-1)} R_{12}^{-\mathrm{T}} \overset{2}{M}_{(j-1)} R_{12}^{\mathrm{T}}} \overset{1}{M}_{(j)} R_{12}^{-\mathrm{T}} R_{12}^{\mathrm{T}} \overset{1}{M}{}_{(j-1)}^{-1} \\
&= R_{12} \overset{2}{M}_{(j-1)} R_{12}^{-1} \overset{1}{M}_{(j-1)} \overset{1}{M}_{(j)} \overset{1}{M}{}_{(j-1)}^{-1} = R_{12} \overset{2}{M}'_{(j)} R_{12}^{-1} \overset{1}{M}'_{(j-1)}.
\end{aligned}$$

And the second relation is

$$\begin{aligned}
&R_{12}^{\mathrm{T}} \overset{1}{M}'_{(j-1)} R_{12}^{-\mathrm{T}} \overset{2}{M}'_{(j-1)} = R_{12}^{\mathrm{T}} \overset{1}{M}_{(j-1)} \overset{1}{M}_{(j)} \underline{\overset{1}{M}{}_{(j-1)}^{-1} R_{12}^{-\mathrm{T}} \overset{2}{M}_{(j-1)}} \overset{2}{M}_{(j)} \overset{2}{M}{}_{(j-1)}^{-1} \\
&= R_{12}^{\mathrm{T}} \overset{1}{M}_{(j-1)} \overset{1}{M}_{(j)} R_{12}^{-\mathrm{T}} \overset{2}{M}_{(j-1)} R_{12}^{-1} \underline{\overset{1}{M}{}_{(j-1)}^{-1} R_{12} \overset{2}{M}_{(j)}} \overset{2}{M}{}_{(j-1)}^{-1} \\
&= R_{12}^{\mathrm{T}} \overset{1}{M}_{(j-1)} \underline{\overset{1}{M}_{(j)} R_{12}^{-\mathrm{T}} \overset{2}{M}_{(j-1)}} (R_{12}^{-1} R_{12}) \overset{2}{M}_{(j)} R_{12}^{-1} \overset{1}{M}{}_{(j-1)}^{-1} R_{12} \overset{2}{M}{}_{(j-1)}^{-1} \\
&= \big(\underline{R_{12}^{\mathrm{T}} \overset{1}{M}_{(j-1)} R_{12}^{-\mathrm{T}} \overset{2}{M}_{(j-1)}}\big) \big(\underline{R_{12}^{\mathrm{T}} \overset{1}{M}_{(j)} R_{12}^{-\mathrm{T}} \overset{2}{M}_{(j)}}\big) R_{12}^{-1} \underline{\overset{1}{M}{}_{(j-1)}^{-1} R_{12} \overset{2}{M}{}_{(j-1)}^{-1}}
\end{aligned}$$

$$
\begin{aligned}
&= \overset{2}{M}_{(j-1)} \underline{R_{12}^{-1} \overset{1}{M}_{(j-1)} R_{12} \overset{2}{M}_{(j)}} R_{12}^{-1} \overset{1}{M}_{(j)} (R_{12} R_{12}^{-1}) R_{12}^{-T} \overset{2}{M}{}^{-1}_{(j-1)} R_{12}^{T} \overset{1}{M}{}^{-1}_{(j-1)} R_{12} \\
&= \overset{2}{M}_{(j-1)} \overset{2}{M}_{(j)} R_{12}^{-1} \overset{1}{M}_{(j-1)} (R_{12} R_{12}^{-1}) \underline{\overset{1}{M}_{(j)} R_{12}^{-T} \overset{2}{M}{}^{-1}_{(j-1)}} R_{12}^{T} \overset{1}{M}{}^{-1}_{(j-1)} R_{12} \\
&= \overset{2}{M}_{(j-1)} \overset{2}{M}_{(j)} R_{12}^{-1} \underline{\overset{1}{M}_{(j-1)} R_{12}^{-T} \overset{2}{M}{}^{-1}_{(j-1)} R_{12}^{T}} \overset{1}{M}_{(j)} (R_{12}^{-T} R_{12}^{T}) \overset{1}{M}{}^{-1}_{(j-1)} R_{12} \\
&= \overset{2}{M}_{(j-1)} \overset{2}{M}_{(j)} (R_{12}^{-1} R_{12}) \overset{2}{M}{}^{-1}_{(j-1)} R_{12}^{-1} \overset{1}{M}_{(j-1)} \overset{1}{M}_{(j)} \overset{1}{M}{}^{-1}_{(j-1)} R_{12} = \overset{2}{M'}_{(j-1)} R_{12}^{-1} \overset{1}{M'}_{(j-1)} R_{12}.
\end{aligned}
$$

□

4.2.2. IHX-relations

We now probe the algebra of monodromy data corresponding to intersecting arcs. Our basic example is the case where we have a single intersection of arcs in the case of $\Sigma_{0,s+1,1}$. We let $M_{(k)}$ denote, as in the preceding subsection, the monodromy datum with endpoints $(2k, 2k-1)$; all monodromy data start and terminate at the same bordered cusp. Let us consider the products of monodromy data $M_{(1)}M_{(2)}$ and $M_{(2)}M_{(3)}$ and take their product $\big(\overset{1}{M}_{(1)}\overset{1}{M}_{(2)}\big) R_{12}^{-T} \big(\overset{2}{M}_{(2)}\overset{2}{M}_{(3)}\big)$. For this product, we have

$$
\begin{aligned}
&\overset{1}{M}_{(1)} \underline{\overset{1}{M}_{(2)} R_{12}^{-T} \overset{2}{M}_{(2)}} \overset{2}{M}_{(3)} = \underline{\overset{1}{M}_{(1)} R_{12} \overset{2}{M}_{(2)} R_{12}^{-1}} \overset{1}{M}_{(2)} R_{12}^{-T} \overset{2}{M}_{(3)} \\
&= R_{12} \overset{2}{M}_{(2)} R_{12}^{-1} \overset{1}{M}_{(1)} \overset{1}{M}_{(2)} \underline{R_{12}^{-T}} \overset{2}{M}_{(3)} = R_{12} \overset{2}{M}_{(2)} R_{12}^{-1} \overset{1}{M}_{(1)} \overset{1}{M}_{(2)} \big(q R_{12} - (q^{3/2} - q^{-1/2}) P_{12}\big) \overset{2}{M}_{(3)} \\
&= q R_{12} \overset{2}{M}_{(2)} R_{12}^{-1} \underline{\overset{1}{M}_{(1)} \overset{1}{M}_{(2)} R_{12} \overset{2}{M}_{(3)}} - (q^{3/2} - q^{-1/2}) R_{12} \overset{2}{M}_{(2)} R_{12}^{-1} \overset{1}{M}_{(1)} \overset{1}{M}_{(2)} \underline{P_{12} \overset{2}{M}_{(3)}} \\
&= q R_{12} \big(\overset{2}{M}_{(2)} \overset{2}{M}_{(3)}\big) R_{12}^{-1} \big(\overset{1}{M}_{(1)} \overset{1}{M}_{(2)}\big) R_{12} - (q^{3/2} - q^{-1/2}) R_{12} \big(\overset{2}{M}_{(2)}\big) R_{12}^{-1} \big(\overset{1}{M}_{(1)} \overset{1}{M}_{(2)} \overset{1}{M}_{(3)}\big) P_{12}.
\end{aligned}
\tag{4.7}
$$

In this calculation, we use two identities: the first one is

$$
q^{1/2} R_{12} - q^{-1/2} R_{12}^{-T} = (q - q^{-1}) P_{12},
$$

where $P_{12} = \sum_{i,j} \overset{1}{e}_{i,j} \otimes \overset{2}{e}_{j,i}$ is the standard classical permutation matrix. For this matrix, we have that

$$
\overset{1}{M} P_{12} = P_{12} \overset{2}{M} \quad \text{and} \quad \overset{2}{M} P_{12} = P_{12} \overset{1}{M}
$$

for any (irrespectively classical or quantum) matrix M. Formula (4.7) is similar to the three-term IHX-relation that is common in models of directed intersecting paths in knot theory. (This (local) relation is an unrooted version of the Jacobi identity in the theory of finite type (or Vassiliev) invariants of knots, links and 3-manifolds.) In the right-hand side of (4.7), we have two terms: one is the term in which the original monodromy data enter in opposite order, and the other is the term containing new matrices: $M_{(2)}$ and $M_{(1)}M_{(2)}M_{(3)}$

corresponding to *nonintersecting* paths; these new constitutive monodromy data enjoy commutation relations (2.17). Disregarding the R-matrix structures, we schematically depict the IHX-relations in the following form (the over-/undercrossing indicates which monodromy datum stands to the right):

4.3. *Monodromy Algebras for* $\Sigma_{0,2,2}$

We consider the example of monodromy data for an 'eye'—the disc with the hole inside and with two bordered cusps on the outer boundary (Figure 5(b)). This case is of particular interest as it is a generalization of the modromy data associated to the Dubrovin connection in the theory of Frobenius manifolds [21]. In this case, we have two monodromy data, M_1 and M_2, both subject to restriction (3.22). We then have the following statement about symplectic leaves of this algebra.

Lemma 4.8 *For* $\Sigma_{0,2,2}$ *with two cusps on one hole, the algebra of restricted monodromy data* M_1 *and* M_2 *has maximal symplectic leaf dimension* $k(k-1)$. *The* $2k$ *Casimirs are* $\mathrm{tr}\,[[M_1M_2]^p]$, $p=1,\dots,k$, *and the ratios* $m_{1_{i,k+1-i}}/m_{2_{k+1-i,i}}$, $i=1,\dots,k$.

Proof Anti-diagonal elements of both restricted monodromy matrices have homogeneous commutation relations with all other elements; it is easy to check that ratios $m_{1_{i,k+1-i}}/m_{2_{k+1-i,i}}$ Poisson commute with all elements. Besides that, we have that, the product M_2M_1 is a monodromy datum corresponding to circumnavigating the central hole, so traces of all powers of this product are central elements. (Because of the trace property, it is irrelevant whether we take the monodromy datum to be M_2M_1 or M_1M_2.) We then again consider Poisson bivectors over special matrices M_1 and M_2 with only halves of main diagonal and both anti-diagonals non-zero; we again split elements into octuplets (orbits of $S[m_{1_{i,j}}]\cup S[m_{2_{i,j}}]$ with $1\le j<i\le[k/2]$), quadruplets (orbits of $S[m_{1_{\frac{k+1}{2},j}}]\cup S[m_{2_{\frac{k+1}{2},j}}]$ with $1\le j<[(k+1)/2]$), sextets (orbits of $S[m_{1_{i,i}}]\cup S[m_{2_{i,i}}]$ with $1\le i\le[k/2]$) and doublets $\{m_{1_{\frac{k+1}{2},\frac{k+1}{2}}}, m_{2_{\frac{k+1}{2},\frac{k+1}{2}}}\}$. Octuplets and quadruplets are non-degenerate, sextets have dimension 2 and doublets have dimension 0, so the total codimension is $4[k/2]$ for even k, and $4[k/2]+2$ for odd k; it is easy to see that it is $2k$ in both cases, as expected. □

Remark 4.1 All calculations of Casimirs in Sec.4 were done for GL_k, i.e., for non fixed determinants. This corresponds to the case of r-matrix without the term proportional to $I\otimes I$. Because if we do not have this term, determinants are NOT Casimirs; they however always have homogeneous Poisson brackets with all other variables; when we normalise $GL_k\to SL_k$ by setting $s_{ij}\to s_{ij}/[\det S]^{1/k}$, the additional term $(1/k)I\otimes I$ appears in the r-matrix, and simultaneously all determinants become true Casimirs.

4.4. *Monodromy Algebra for* $\Sigma_{0,1,3}$

The last example pertains to another elementary building block of monodromy data: the ideal triangle $\Sigma_{0,1,3}$ (Figure 5(c)) that, for $k = 2$, corresponds to the modromy data of the Airy equation. In this case, we have two monodromy matrices M^1 and M^2 (here, M^2 follows M^1), both having the reduced (upper-anti-triangular) form—we denote them with an upper index to distinguish them from the other examples. Plus, we have to take into account that their product, M^2M^1, has itself lower-anti-triangular form. This imposes $k(k-1)/2$ restrictions on entries of M^1 and M^2; for the general-position situation, it is not difficult to see that we can express all non-anti-diagonal entries of, say, matrix M^2 in terms of entries of M^1 and the anti-diagonal elements of M^2 and, using relations (2.8), we obtain that, whereas entries of the matrix M^1 enjoy commutation relations (3.37), the Poisson relations for entries of M^1 and anti-diagonal entries of M^2

$$\{m^1_{i,j}, m^2_{r,k+1-r}\} = m^1_{i,j} m^2_{r,k+1-r} \delta_{i,k+1-r}, \; i+j \le k+1, \; r = 1, \dots, k, \tag{4.8}$$

are homogeneous, and all elements of M^1 belonging to the same row commute in the same way with all $m^2_{k,k+1-r}$. All $m^2_{k,k+1-r}$ mutually commute. We then have the following statement about Casimirs of the algebra of the set of elements $\{m^1_{i,j}, \; i+j \le k+1\} \cup \{m^2_{k,k+1-r}, \; r = 1, \dots, k\}$.

Lemma 4.9 *For $\Sigma_{0,1,3}$ with three bordered cusps on a disc (one outer hole), the algebra of restricted monodromy data M^1 and $m^2_{r,k+1-r}$ has maximal symplectic leaf dimension $\frac{k(k+1)}{2} + k - \left[\frac{k}{2}\right]$. The $\left[\frac{k}{2}\right]$ Casimirs are*

$$C_d = \frac{[M^1]^{\mathrm{UL}}_d [M^1]^{\mathrm{UL}}_{k-d}}{\prod_{i=1}^{d} [m^1_{i,k+1-i}]^2 \prod_{i=d+1}^{k-d} [m^1_{i,k+1-i}]} \cdot \frac{\prod_{i=1}^{d} [m^2_{k+1-i,i}]^2}{\prod_{i=1}^{d} [m^2_{i,k+1-i}]^2}, \quad d = 1, \dots, \left[\frac{k}{2}\right],$$

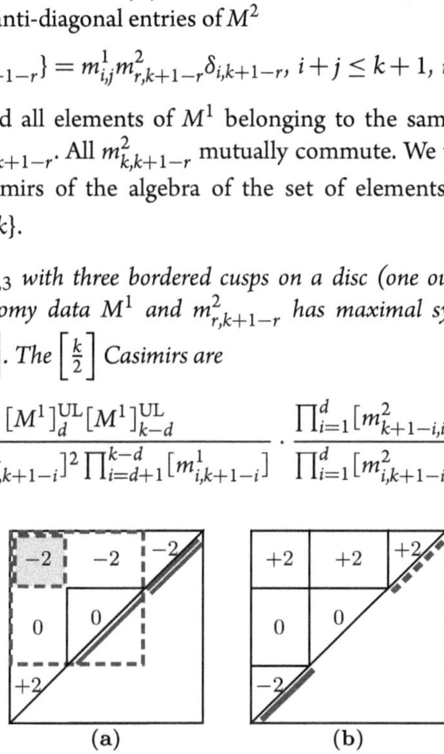

Figure 6 Constructing Casimirs for the restricted matrix M^1 subject to Poisson algebra (3.37) and the anti-diagonal entries of the matrix M^2. Numbers in the corresponding rectangles or triangles indicate the sign of homogeneous commutation relations between elements of M^1 in the corresponding region and (a) the products of minors $[M^1]^{\mathrm{UL}}_d [M^1]^{\mathrm{UL}}_{k-d}$ $(d \le k-d)$ divided by the special products of anti-diagonal elements $\prod_{i=1}^{d} [m^1_{i,k+1-i}]^2 \prod_{i=d+1}^{k-d} [m^1_{i,k+1-i}]$; (b) by the ratios of products $\prod_{i=1}^{d} [[m^2_{k+1-i,i}]^2 [m^2_{i,k+1-i}]^{-2}]$ of elements on the anti-diagonal of the matrix M^2. We see that these signs are complementary in all regions. All anti-diagonal entries of M^2 mutually commute and commute with the above ratios of elements of M^1 because every term contains equal number of elements from the same row of the matrix M^1 in the numerator and the denominator.

where $[M^i]^{\mathrm{UL}}_d$, $i = 1,2$ *denote the upper-left minors of size d of the matrix* M^i *and the product from* $d+1$ *to* $k-d$ *is equal to 1 when* $k = 2$.

Proof The fact that C_d are Casimirs can be verified directly (see Figure 6). We then again consider a Poisson bivector over the pattern in which non-zero elements are $m^1_{i,i}$ with $i \leq [(k+1)/2]$, $m^1_{i,k+1-i}$, and $m^2_{i,k+1-i}$. For quadruples of the matrix M^1, the Poisson brackets are non-degenerate, whereas the dimension of quintuplets $\{m^1_{i,i}, m^1_{i,k+1-i}, m^1_{k+1-i,i}, m^2_{i,k+1-i}, m^2_{k+1-i,i}\}$ for $i = 1,\dots,[k/2]$ is 4 (so each quintuplet adds one Casimir) and the dimension of the doublet $\{m^1_{\left[\frac{k+1}{2}\right],\left[\frac{k+1}{2}\right]}, m^2_{\left[\frac{k+1}{2}\right],\left[\frac{k+1}{2}\right]}\}$ is 2; the total codimension therefore matches the above number of Casimirs. □

5. The Extended Riemann–Hilbert Correspondence

In this section, we conjecture how to define a suitably decorated moduli space $\mathcal{Z}^k_{irr}(\Sigma_{g,s})$ of irregular connections ∇ on a holomorphic rank k vector bundle $E \to \Sigma_{g,s}$ in such a way that the Riemann–Hilbert correspondence

$$RH : \mathcal{Z}^k_{irr}(\Sigma_{g,s}) \to \mathcal{M}^k_{g,s,n}$$

is a Poisson isomorphism.

Given an irregular connection ∇ on a holomorphic rank k vector bundle $E \to \Sigma_{g,s}$, choosing a coordinate z for $\Sigma_{g,s}$ amounts to giving a linear system of differential equations $\nabla_{\frac{\partial}{\partial z}}$ with s poles $a_1,\dots,a_s$ of Poincaré rank $r_1,\dots,r_s$ or, in other words, a meromorphic matrix-valued differential

$$d - A(z)dz$$

with fixed multiplicities $r_1+1,\dots,r_s+1$ at $a_1,\dots,a_s$.

For the sake of simplicity, we restrict to the non-ramified case where $A(z)$ is diagonalizable at each a_p. Then, Krichever proved that the space $\mathcal{E}_{irr}(r_1,\dots,r_s)$ of all possible meromorphic matrix-valued differentials of this form modulo the SL_k action is $(2g+s+r_1+\dots+r_s-2)(k^2-1)$-dimensional and is foliated in symplectic leaves by fixing the exponents, with the symplectic form

$$\omega = -\frac{1}{2}\sum_{t=1}^{kg} \mathrm{res}_{\gamma_t} \mathrm{Tr}\left(Y^{-1}\delta A \wedge \delta Y\right) - \frac{1}{2}\sum_{p=1}^{s} \mathrm{res}_{a_p} \mathrm{Tr}\left(Y_p^{-1}\delta A \wedge \delta Y_p\right),$$

where Y_p is the formal local solution of $dY = A(z)Ydz$ at a_p, and $\gamma_1,\dots,\gamma_{gk}$ are the simple zeroes of the holomorphic sections of the vector bundle $E \to \Sigma_{g,s}$ [33]. This symplectic form induces a Poisson structure on $\mathcal{Z}^k_{irr}(\Sigma_{g,s}) := \mathcal{E}_{irr}(r_1,\dots,r_s) \times \mathbb{C}^{(r_1+\dots+r_s)(k-1)}$, where $\mathbb{C}^{(r_1+\dots+r_s)(k-1)}$ is the space of decorations, or, in other words, a choice of growth rates of the absolute value of the formal solutions Y_p modulo polynomial growth [41]. Following the ideas by Gaiotto, Moore and Neitzke [27], we impose $n = 2(r_1+\dots+r_s)$, so that $\dim(\mathcal{Z}^k_{irr}(\Sigma_{g,s})) = \dim(\mathcal{M}^k_{g,s,n})$. We end this paper with the following

Conjecture *The Riemann–Hilbert correspondence*

$$RH : \mathcal{Z}^k_{irr}(\Sigma_{g,s}) \to \mathcal{M}^k_{g,s,n}$$

is a Poisson isomorphism.

We have tested this conjecture in the case of $g = 0, k = 2$ and connections with only one irregular singularity of Poincaré rank 3. In this case, we have the Jimbo–Miwa linear system associated to the PII equation and $n = 6$. The decorated character variety $\mathcal{M}^2_{0,1,6}$ has dimension 9 with one Casimir. As explained in [18], the isomonodromicity condition means that we need to restrict to a two-dimensional subalgebra in $\mathcal{M}^2_{0,1,6}$ defined by the set of functions that Poisson commute with the frozen cluster variables corresponding to arcs connecting pairs of bordered cusps. On the left-hand side of the Riemann–Hilbert correspondence, the space $\mathcal{Z}^2_{irr}(\Sigma_{0,1})$ has also dimension 9 and, by imposing the isomonodromicity condition (where t is the PII independent variable),

$$\frac{\partial A}{\partial t} - \frac{\partial B}{\partial z} = [B, A]$$

one obtains a restriction to a two-dimensional space [35], which we denote $\tilde{\mathcal{Z}}^2_{irr}(\Sigma_{0,1})$.

Remark 5.1 The space $\tilde{\mathcal{Z}}^2_{irr}(\Sigma_{0,1})$ is the de Rham side of the Riemann–Hilbert correspondence. Recently [42], a complete description of the two-dimensional (family of) holomorphic symplectic moduli spaces of rank 2 Higgs bundles over $\mathbb{P}^1$ having a unique pole of order 4 as singularity, and a regular leading-order term was obtained. This moduli (for a fixed choice of parameters) can be interpreted as the Dolbeault counterpart $\mathcal{M}^2_{D,irr}$ of $\tilde{\mathcal{Z}}^2_{irr}(\Sigma_{0,1})$.

Remark 5.2 A general extended Riemann–Hilbert correspondence for related objects (a moduli space of stable unramified irregular singular parabolic connections on smooth projective curves and a set $\widetilde{\mathcal{R}(g,k,s)}$ of generalized monodromy data coming from topological monodromies, formal monodromies and Stokes data) was proposed by M. Inaba and M. Saito in [31]. They proved that the moduli space of generalized monodromy data is a non-singular affine scheme $\mathcal{R}(g,k,s)$ given by the categorical quotient $\mathcal{R}(g,k,s) = \widetilde{\mathcal{R}(g,k,s)}//G$ for a natural action of a reductive group G. An immediate comparison of dimensions for $\mathcal{M}^k_{g,s,n}$ and $\widetilde{\mathcal{R}(g,k,s)}$ shows a good correspondence. For example, in the case of $\mathcal{M}^2_{0,1,6}$ (whose dimension is 9), the dimension of $\widetilde{\mathcal{R}(0,2,1)}$ is 8 (9 minus one Casimir) and $\dim \mathcal{M}^2_{0,1,6} = \dim \mathcal{R}(0,2,1) = 2$.

6. Acknowledgements

The authors wish to dedicate this paper to Nigel Hitchin, whose beautiful mathematics has inspired many of our papers. We thank A. Alekseev, A. Glutsuk, M. Gualtieri, A. Shapiro, M. Shapiro, N. Nekrasov, B. Pym and P. Severa for helpful discussions. The research of V. R. was partially supported by the Russian Foundation for Basic Research under grant RFBR-15-01-05990 and 18-01-00460. He is thankful to MPIM (Bonn) and SISSA (Trieste), where a part of this work was done, for invitation and excellent working conditions. The work of L. Ch. was partially financially supported by the Russian Foundation for Basic Research under grant RFBR-18-01-00460.

REFERENCES

[1] Allegretti, D., Laminations from the symplectic double. arXiv:1410.3035v1 (2014).

[2] Alekseev, A., Kosmann-Schwarzbach, Y., Meinrenken, E. Quasi-Poisson manifolds. *Canad. J. Math.* **54**, no. 1: 3–29 (2002).

[3] Alekseev, A., Malkin, A., Meinrenken, E. Lie group valued momentum maps. *J. Differential Geom.* **48**, no. 3: 445–95 (1988).

[4] Alekseev, A. Yu., Grosse, H., Schomerus, V. Combinatorial quantization of the Hamiltonian Chern–Simons theory. I. *Comm. Math. Phys.* **172**, no. 2: 317–58 (1995).

[5] Audin, M. Gauge theory and integrable systems. In Hurtubise J., Lalonde F., Sabidussi G. (eds) *Gauge Theory and Symplectic Geometry, NATO Adv. Sci. Inst. Ser. C Math. Phys. Sci.* **488**. Springer, Dordrecht, pp. 69–112 (1997).

[6] Ben-Zvi, D., Brochier, A., Jordan, D. Integrating quantum groups over surfaces. arXiv:1501.04652v3 (2016).

[7] Ben-Zvi, D., Brochier, A., Jordan, D. Quantum character varieties and braided module categories. arXiv:1606.04769v2 (2016).

[8] Berenstein, A., Zelevinsky, A. Quantum cluster algebras. Adv. Math. 195, no. 2: 405–55 (2005).

[9] Boalch, P. Stokes matrices, Poisson Lie groups and Frobenius manifolds. *Invent. Math.* **146**, no. 3: 479–506 (2001).

[10] Boalch, P. Geometry and braiding of Stokes data; fission and wild character varieties. *Ann. Math. (2),* **179**, no. 1: 301–65 (2014).

[11] Boalch, P., Wild character varieties, meromorphic Hitchin systems and Dynkin diagrams. arXiv:1703.1037v1 (2017).

[12] Buffenoir, E., Roche, Ph. Two-dimensional lattice gauge theory based on a quantum group. *Comm. Math. Phys.* **170**, no. 3: 669–98 (1995).

[13] Buffenoir, E., Roche, Ph. Link invariants and combinatorial quantization of Hamiltonian Chern–Simons theory, *Comm. Math. Phys.* **181**, no. 2: 331–65 (1996).

[14] Chekhov, L., Fock, V. Quantum mapping class group, pentagon relation, and geodesics. Proc. Steklov Math. Inst. **226**: 149–63 (1999).

[15] Chekhov, L., Mazzocco, M. Shear coordinates on the versal unfolding of the D_4 singularity. *J. Phys. A. Math. Gen.* **43**, no. 44: 1–13 (2010).

[16] Chekhov, L., Mazzocco, M. On a Poisson space of bilinear forms with a Poisson–Lie action. arXiv:1404.0988 (2014). *Uspekhi Mat. Nauk* **72** (2017), no. 6(438):, 139–190; translation in *Russian Math. Surveys* **72** (2017), no. 6, 1109–1156.

[17] Chekhov, L., Mazzocco, M. Colliding holes in Riemann surfaces and quantum cluster algebras. *Nonlinearity* **31**, no. 1: 54–107 (2018).

[18] Chekhov, L., Mazzocco, M. Rubtsov, V. Painlevé monodromy datum manifolds, decorated character varieties and cluster algebras. *Intl. Math. Res. Not.* **2017**, no. 24: 7639–91 (2017).

[19] Chekhov, L. O., Penner, R. C. Introduction to quantum Thurston theory. Russ. Math. Surv. **58**, no. 6: 1141–83 (2003).

[20] Chekhov, L., M. Shapiro, M. Teichmüller spaces of Riemann surfaces with orbifold points of arbitrary order and cluster variables. Intl. Math. Res. Notices **2014**, no. 10: 2746–72 (2014).

[21] Dubrovin, B. Painlevé transcendents in two-dimensional topological field theory. In *The Painlevé Property One Century Later*, CRM Series in Mathematical Physics, Springer, New York, pp. 287–412 (1999).

[22] Dubrovin, B. A., Mazzocco, M. Monodromy of certain Painlevé-VI transcendents and reflection groups. *Invent. Math.* **141**, no. 1: 55–147 (2000).

[23] Fock, V. V., Goncharov, A. B. Moduli spaces of local systems and higher Teichmüller theory. Publ. Math. Inst. Hautes Études Sci. **103**, no. 1: 1–211, (2006).

[24] Fock, V. V., Goncharov, A. B. Cluster $\mathcal{X}$-varieties, amalgamation, and Poisson–Lie groups. In *Algebraic Geometry and Number Theory*, Progr. Math., **253**, Birkhäuser, Boston, pp. 27–68 (2006).

[25] Fock, A., Marshakov, A. A note on quantum groups and relativistic Toda theory. *Nucl. Phys. B.* **56**, no. 3: 208–14 (1997).

[26] Fock, V., Rosly, A. Moduli space of flat connections as a Poisson manifold. *Intl. J. Mod. Phys. B* **11**, no. 26–27: 3195–6 (1997).

[27] Gaiotto, D., Moore, G. W., Neitzke, A. Wall-crossing, Hitchin systems, and the WKB approximation, *Adv. Math.* **234**: 239–403 (2013).

[28] Goldman, W. M. Invariant functions on Lie groups and Hamiltonian flows of surface group representations. *Invent. Math.* 85, no. 2: 263–302 (1986).

[29] Goldman, W. M. The complex-symplectic geometry of $SL(2,\mathbb{C})$ character varieties over surfaces. In Dani, S. S. and Prasad, G., eds, *Algebraic groups and arithmetic.* Tata Inst. Fund. Res., Mumbai, (2004) pp. 375–407 (2004).

[30] Gualtieri, M., Li, S., Pym, B. The Stokes groupoids. *J. Reine Angew. Math.*, http://doi.org/10.1515/crelle-2015-0057 (2015).

[31] Inaba, M., Saito, M.-H. Moduli of unramified irregular singular parabolic connections on a smooth projective curve. *Kyoto J. Math.* **53**, no. 2: 433–82 (2012).

[32] Korotkin, D., Samtleben, H. Quantization of coset space σ-models coupled to two-dimensional gravity. *Comm. Math. Phys.* **190**, no. 2: 411–57 (1997).

[33] Krichever, I. Isomonodromy datum equations on algebraic curves, canonical transformations and Whitham equations. Dedicated to Yuri I. Manin on the occasion of his 65th birthday. *Mosc. Math. J.* **2**, no. 4: 717–52 (2002).

[34] Harnad, J. Dual isomonodromic deformations and momentum maps to loop algebras. *Commun. Math. Phys.*, **166**, no. 2: 337–65 (1994).

[35] Harnad, J., Routhier, M. R-matrix construction of electromagnetic models for the Painlevé transcendents, *J. Math. Phys.* **36**, no. 9: 4863–81 (1995).

[36] Hitchin, N. Frobenius manifolds (with notes by David Calderbank). In Hurtubise J., Lalonde F., Sabidussi G., eds, *Gauge Theory and Symplectic Geometry, NATO Adv. Sci. Inst. Ser. C Math. Phys. Sci.*, **488**, Springer, Dordrecht, pp. 69–112 (1997).

[37] Mazzocco, M. Painlevé sixth equation as isomonodromic deformations equation of an irregular system. In Kuznetsov, V. B., ed., *The Kowalevski property (Leeds, 2000). CRM Proc. Lecture Notes*, **32**. Amer. Math. Soc., Providence, RI, 219–38 (2002).

[38] Musiker, G., Schiffler, R., Williams, L. Bases for cluster algebras from surfaces. Comp. Math, **149**, no. 2: 217–63 (2013).

[39] Musiker, G., Williams, L. Matrix formulas and skein relations for cluster algebras from surfaces. *Intl. Math. Res. Notices* **2013**, no. 13: 2891–944 (2013).

[40] Roche, Ph., Szenes, A. Trace functionals on non-commutative deformations of moduli spaces of flat connections. *Adv. Math.* **168**, no. 2: 133–92 (2002).

[41] Shende, V., Treumann, D., Williams, H., Zaslow, E. Cluster varieties from Legendrian knots. arXiv:1512.08942v1 (2015).
[42] Ivanics, P., Stipsicz, A. Szabo, Sz. Two-dimensional moduli spaces of rank 2 Higgs bundles over $\mathbb{CP}^1$ with one irregular singular point. arXiv:1604.08503v1 (2016). *J. Geom. Phys.* **130** (2018), 184–212.

4

The Deformed Hermitian–Yang–Mills Equation in Geometry and Physics

TRISTAN C. COLLINS
DAN XIE
SHING-TUNG YAU

Department of Mathematics, Harvard University, 1 Oxford St., Cambridge, MA 02138, USA
Email: tcollins@math.harvard.edu, dxie@cmsa.fas.harvard.edu, yau@math.harvard.edu

To Nigel Hitchin, with admiration, on the occasion of his seventieth birthday.

ABSTRACT. We provide an introduction to the mathematics and physics of the deformed Hermitian–Yang–Mills equation, a fully non-linear geometric PDE on Kähler manifolds which plays an important role in mirror symmetry. We discuss the physical origin of the equation, and some recent progress towards its solution. In dimension 3, we prove a new Chern number inequality and discuss the relationship with algebraic stability conditions.

1. The Deformed Hermitian–Yang–Mills Equation and Mirror Symmetry

It was discovered a long time ago that there are five perturbatively well-defined ten-dimensional superstring theories: type IIA, type IIB, type I, heterotic SO(32) and heterotic $E_8 \times E_8$ string theory; see [4] for an introduction to string theory. To get a realistic particle physics model in four dimensions, one needs to study compactifications of string theory on compact 6-(real)-dimensional manifolds. It turns out that Calabi–Yau three-dimensional manifolds [7] play a crucial role in studying supersymmetric compactifications, for which many computations are under control. By choosing different Calabi–Yau geometries and different ten-dimensional string theories, one gets a huge number of string theory vacua in various dimensions. These theories were originally thought to be independent, with no obvious relations between them. One of the main discoveries of the second string revolution

Collins, T. C., Xie, D., Yau, S.-T., *The Deformed Hermitian–Yang–Mills Equation in Geometry and Physics.* In: *Geometry and Physics: A Festschrift in Honour of Nigel Hitchin,* Jørgen Ellegaard Andersen, Andrew Dancer, Oscar García-Prada (Eds): Oxford University Press (2018). © Oxford University Press. DOI: 10.1093/oso/9780198802006.003.0004

in the mid-nineties was that these string vacua are, in fact, not independent at all, and most of them are related through various kinds of string dualities. A fundamental example of this was the discovery of pairs of Calabi–Yau manifolds $(X, \hat{X})$ for which IIA string theory on X (resp. $\hat{X}$) is equivalent to IIB string theory on $\hat{X}$ (resp. X) [17]. This duality came to be called mirror symmetry. Mirror symmetry has generated a huge amount of interest among physicists and mathematicians, in part due to its surprisingly successful prediction of enumerative curve counts inside Calabi–Yau manifolds [6]. One basic feature of mirror symmetry is that it exchanges the complexified Kahler moduli space and complex structure moduli space of the mirror pairs. Mirror symmetry often maps one hard quantum problem to a simpler classical geometric problem; for example, the very complicated counting of curves on X is reduced to simpler computation of period integrals on $\hat{X}$.

A second major discovery in the second string revolution was the existence of various kinds of extended objects, besides the fundamental string, which are used to define string theory. One of most important classes of these extended objects is the class of D-branes. These new discoveries provided new insights into the understanding of mirror symmetry. Using T-duality and D-branes, Strominger, Yau and Zaslow described a geometric picture of mirror symmetry which is now called the SYZ picture [29].

The first ingredient of the SYZ picture of mirror symmetry is the so-called T-duality symmetry of string theory [4]. T-duality relates different string theories compactified on circles. Consider a string theory $\mathcal{T}$ compactified on a circle S_A with radius R. T-duality predicts that it should be equivalent to a different string theory $\mathcal{T}'$ compactified on a circle S_B with radius $\frac{1}{R}$. The typical example is type IIA string theory and type IIB string theory which, when compactified on a circle, are related by T-duality.

The second ingredient of the SYZ picture of mirror symmetry is D-branes. The name D-brane is derived from the fact that the world-sheet string theory description of these objects has Dirichlet ('D') boundary conditions on the world volume of the brane. We often denote a D-brane as a Dp brane, where p denotes the number of spatial dimensions of the brane world volume, and the full space-time dimension of a Dp brane is $p + 1$. D-branes are extended objects carrying Ramond–Ramond (RR) charges. Not all D-branes are physically realistic. The realistic D-branes are minimizers of some energy functional, and are usually referred to as BPS branes. The BPS branes of type II string theory on a Calabi–Yau manifold X have been classified into two kinds in the large volume/ large complex structure limit: one type of D-brane is a special Lagrangian submanifold of X and the other type is a complex submanifold. A D-brane has a quantum moduli space which is related to the geometry it probes in an interesting way; for example, the quantum moduli space of D0 brane probing a Calabi–Yau manifold X is nothing but X itself.

Let us consider how T-duality acts on D-branes; T-duality maps a D-brane wrapping on circle S_A to a D-brane sitting on a point of the dual circle S_B (and vice versa). Combining T-duality and the existence of D-branes, SYZ proposed the following geometric picture of mirror symmetry: consider a three-dimensional Calabi–Yau manifold X which has a T_3 fibration. If we compactify type IIB string theory on X and apply T-duality to every fibre, we should get a type IIA string theory on the mirror manifold $\hat{X}$. BPS D-branes of the type IIB string are required to be special Lagrangians, while BPS D-branes of the type IIA string are required to be complex submanifolds. If we wrap a D3-brane on a T3 fibre, we get a D0-brane on the mirror manifold after T-duality. The quantum moduli space of the original

D3-brane should be equal to the moduli space of a D0-brane, which is then equal to the mirror manifold $\hat{X}$. Therefore, $\hat{X}$ arises as the dual torus fibration, and its geometry can be understood from the D-brane moduli space associated with the special Lagrangian torus fibre of X.

Conjecture 1.1 (Stominger–Yau–Zaslow, [29]) *Let X and $\hat{X}$ be a mirror pair of CY manifolds. Near the large volume/ large complex structure limits:*

- *X and $\hat{X}$ admit dual special Lagrangian torus fibrations $\mu : X \to B$ and $\hat{\mu} : \hat{X} \to B$ over the same base B.*
- *There exists a fibrewise Fourier–Mukai transform which maps Lagrangian submanifolds of X to coherent sheaves on $\hat{X}$.*

1.1. The D-Brane Effective Action and the Deformed Hermitian–Yang–Mills equation

D-branes play an important role in SYZ picture of mirror symmetry and homological mirror symmetry, so it is interesting to further study their behaviour under mirror symmetry. D-brane dynamics can be studied using the low-energy effective action and, in particular, BPS solutions are described as critical points of this action. The bosonic part of supersymmetric Dirac–Born–Infeld (DBI) action of a single Dp brane has the following form:

$$I_p = I_{DBI} + I_{WZ} = -T_p \int_W d^{p+1}\sigma \sqrt{g_{\mu\nu} + \mathcal{F}_{\mu\nu}} + \mu \int_W C \wedge e^{\mathcal{F}}.$$

Here, $g_{\mu\nu}$ is the pullback of the metric, and $\mathcal{F}_{\mu\nu}$ is the modified 2-form $\mathcal{F}_{\mu\nu} = 2\pi\alpha'$ $(F - B)$, with F being the field strength of a gauge field on the D-brane world volume, and B the pullback of the NS 2-form, often called the B-field; T_p and μ are the brane tension and brane charges, respectively, while C is the formal sum of RR fields $C^{(r)}$:

$$C = \sum_{r=0}^{10} C^{(r)}.$$

This action is invariant under the κ symmetry (a fermionic local symmetry) and space-time supersymmetry, and we can combine both in determining the fraction of unbroken supersymmetry by solving the following equation:

$$(1 - \Gamma)\eta = 0.$$

Here, η is the space-time spinor, and Γ is a Hermitian traceless matrix satisfying

$$\mathrm{tr}\Gamma = 0, \ \Gamma^2 = 1.$$

Let's focus on a Calabi–Yau 3-manifold from now on, and assume we have a nowhere vanishing holomorphic 3-form Ω and a Kahler form J (to be consistent with the physics notation). The solutions of the BPS equation for the DBI action of a Dp brane were derived in [26]. We have

- $p+1=3$: M is special Lagrangian, and the modified field strength $\mathcal{F}_{\mu\nu}=0$. The Lagrangian condition is $J|_M=0$ and the special Lagrangian condition is

$$\operatorname{Im} e^{\sqrt{-1}\theta}\Omega|_M=0.$$

- $p+1=2n$ is even: M is holomorphic, and the modified field strength satisfies the following equation

$$\begin{aligned}&\mathcal{F}^{2,0}=0,\\&\frac{1}{n!}(f^*(J)+\mathcal{F})^n=e^{\sqrt{-1}\theta}\frac{\sqrt{|J+\mathcal{F}|}}{\sqrt{|J|}}vol(M).\end{aligned} \tag{1}$$

Here, $f^*(J)$ is the pullback of Kahler form on cycle M. This equation is called the *deformed Hermitian–Yang–Mills (dHYM) equation*. Notice that there is a α' factor in front of $\mathcal{F}$, so the leading-order term of the second equation is then

$$\mathcal{F}\wedge J^{n-1}=cJ^n,$$

which is simply the Hermitian–Yang–Mills equation. The second equation can also be put in the following form:

$$\operatorname{Im}\left(e^{-\sqrt{-1}\theta}(J+\mathcal{F})^n\right)=0,$$

where J is the Kahler form.

1.2. The Semi-Flat Limit of SYZ Mirror Symmetry

Let's summarize the BPS solution of the DBI action: we either have a special Lagrangian cycle with vanishing gauge field strength, or a holomorphic cycle with connection satisfying the deformed Hermitian–Yang–Mills (dHYM) equation. Mirror symmetry exchanges D-branes wrapping special Lagrangian cycles and D-brane wrapping holomorphic cycles. Thus, under the SYZ mirror picture, we should see the exchange of special Lagrangian branes with flat connections and holomorphic branes with dHYM connections.

The SYZ picture of mirror symmetry uses D-branes wrapping on the whole T^3 fibre of X, and the dual is a brane wrapped on a single point of the dual torus fibre of the mirror manifold $\hat{X}$. If we study BPS D-branes wrapping on a single point of the torus fibre and wrapping on the whole base B, after T-duality the mirror should be a D-brane wrapping on the whole manifold $\hat{X}$.

It is, in general, difficult to study the full moduli space of D-branes, but it is possible to check the above picture by looking at the semi-flat limit of the SYZ torus fibration [24]. We briefly recall the set-up for semi-flat mirror symmetry, but refer the reader to the beautiful paper of Hitchin [19, 20] (see also [23]). Fix an affine manifold D, which we assume is a domain in $\mathbb{R}^n$ (e.g. the fundamental domain of a torus). Let x^i denote coordinates on D, and let y^i be coordinates on TD induced by

$$(y^1,\ldots,y^n)\longmapsto \sum_i y^i\frac{\partial}{\partial x^i}.$$

The bundle TD carries a natural complex structure making the coordinates $z^i = x^i + \sqrt{-1}y^i$ holomorphic. At the same time, the bundle T^*D carries a natural symplectic structure by defining

$$\omega = \sum_i dx^i \wedge dy^i.$$

Let $\pi : TD \to D$, and $\hat{\pi} : T^*D \to D$ be the projections. Let $\phi : D \to \mathbb{R}$ be a smooth strictly convex function solving the Monge–Ampère equation

$$\det\left(\frac{\partial^2\phi}{\partial x_i \partial x_j}\right) = 1.$$

Pulling back ϕ by π to the total space of TD induces a Calabi–Yau metric, and hence a metric on T^*D. By the 2 of out 3 rule for Kähler manifolds, this induces a complex structure on T^*D. We can compactify this picture by taking dual lattices $\Lambda \subset TD$, and $\Lambda^* \subset T^*D$, and passing to the quotient $X := TD/\Lambda$, $\hat{X} := T^*D/\Lambda^*$. In this case, $X, \hat{X}$ are mirror Calabi–Yau manifolds. This is semi-flat mirror symmetry. In local coordinates, the Ricci-flat Kahler metric and Kähler form on X are

$$g = \sum_{i,j}\frac{\partial\phi}{\partial x^i\partial x^j}(dx^i dx^j + dy^i dy^j),$$

$$\omega = \frac{\sqrt{-1}}{2}\sum_{i,j}\frac{\partial\phi}{\partial x^i\partial x^j}dz^i\wedge d\bar{z}^j,$$

$$\Omega = dz^1\wedge\cdots\wedge dz^n.$$

The SYZ mirror $\hat{X}$ is found by T-duality on the torus fibres. Let $\tilde{y}^i$ denote coordinates on T^*D dual to y^i. Define coordinates $\tilde{x}$ by the Legendre transform of ϕ:

$$\frac{\partial\tilde{x}^j}{\partial x^k} = \phi_{jk}.$$

The reader can calculate directly that $\tilde{z}^j = \tilde{x}^j + \sqrt{-1}\tilde{y}^j$ define holomorphic coordinates on $\hat{X}$. In this notation, the geometric data for $\hat{X}$ is

$$\tilde{g} = \sum_{i,j}\phi^{ij}(d\tilde{x}^i d\tilde{x}^j + d\tilde{y}^i d\tilde{y}^j),$$

$$\tilde{\omega} = \frac{\sqrt{-1}}{2}\sum_{i,j}\phi^{ij}d\tilde{z}^i\wedge d\bar{\tilde{z}}^j,$$

$$\tilde{\Omega} = d\tilde{z}^1\wedge\cdots\wedge d\tilde{z}^n.$$

We now consider a section of the fibration $\hat{\pi} : \hat{X} \to D, \sigma := \{x \mapsto \tilde{y}^i(x)\}$. This section will be Lagrangian if

$$\frac{\partial \tilde{y}^i}{\partial x^k} = \frac{\partial \tilde{y}^k}{\partial x^i},$$

which implies that σ can be (locally) written as the graph of a 1-form $df : D \to \hat{X}$. Now we impose the assumption that σ is special Lagrangian, namely

$$\mathrm{Im}\left(e^{-\sqrt{-1}\hat{\theta}}\hat{\Omega}\right)\Big|_{\sigma} = 0$$

for a constant $\hat{\theta}$. It is most convenient to write the graph in terms of the Legendre transformed coordinates. We have

$$\frac{\partial f}{\partial x_j} dx^j = \phi^{j\ell} \frac{\partial f}{\partial x_j} d\tilde{x}^\ell$$

so, in terms of the Legendre transform coordinates, the graph is

$$x \mapsto \left(\tilde{x}^i = \tilde{x}^i(x), \quad \tilde{y}^\ell = \phi^{j\ell}\frac{\partial f}{\partial x_j}\right).$$

It follows that

$$d\tilde{z}_i\Big|_{\sigma} = \left[\phi_{ij} + \sqrt{-1}\left(\phi^{pi}\frac{\partial f}{\partial x_j \partial x_p} - \phi^{pm}\phi_{jmk}\phi^{ki}\frac{\partial f}{\partial x_p}\right)\right] dx^j$$

and thus the special Lagrangian condition is

$$\mathrm{Im}\left[e^{-\sqrt{-1}\hat{\theta}} \det\left(\phi_{ij} + \sqrt{-1}\left(\phi^{pi}\frac{\partial f}{\partial x_j \partial x_p} - \phi^{pm}\phi_{jmk}\phi^{ki}\frac{\partial f}{\partial x_p}\right)\right)\right] = 0.$$

In order to translate this to the mirror manifold X, we will need the Fourier–Mukai transform. Fix a point $x \in D$, and consider the fibre $\hat{T} = \hat{\pi}^{-1}(x) \subset \hat{X}$. This is the dual torus to $T = \pi^{-1}(x) \subset X$. A point $\hat{y} \in \hat{T}$ defines a map $T \mapsto \mathbb{R}/\mathbb{Z}$, by $y \mapsto \hat{y}^j y_j$. This map is induced from integrating the flat connection

$$D_A := d + \sqrt{-1}\tilde{y}^j dy_j$$

on the trivial $\mathbb{C}$ bundle over T. This construction, performed on each fibre, yields a $U(1)$ connection on X, and so a complex line bundle L with connection D_A. The curvature of this bundle is

$$D_A^2 = \sqrt{-1}\sum_{i,j}\frac{\partial \tilde{y}^j}{\partial x_i} dx^i \wedge dy^j.$$

The $(0,2)$ part of the curvature is given by

$$\frac{\partial \tilde{y}^j}{\partial x_i} - \frac{\partial \tilde{y}^i}{\partial x_j}$$

and so the induced bundle L has a holomorphic structure precisely when the section σ is Lagrangian. The complex structure of L is given by the operator

$$\overline{\partial}_A = \overline{\partial} - \frac{1}{2}\tilde{y}^j d\bar{z}^j.$$

Recall that σ is the graph of df. Therefore, a holomorphic frame for L is given by the section $\sigma = e^f$ and, in this frame, the connection is the Chern connection with respect to the metric

$$h = e^{2f}.$$

Let's see what this corresponds to under the Fourier–Mukai transform. The $(1,1)$ component of the curvature of the mirror line bundle L with connection D_A is

$$\begin{aligned} F_{i\bar{j}} dz^i \wedge d\bar{z}^j &= -\frac{1}{2}\left(\frac{\partial \tilde{y}^i}{\partial x_j} + \frac{\partial \tilde{y}^j}{\partial x_i}\right)\sqrt{-1}(dy^i \wedge dx^j - dx^i \wedge dy^j) \\ &= \frac{\partial \tilde{y}^i}{\partial x_j}\sqrt{-1}(x^j \wedge dy^i + dx^i \wedge dy^j), \end{aligned}$$

where, in the last line, we used the Lagrangian condition. Now, using the Legendre transform, we can write

$$\frac{\partial \tilde{y}^i}{\partial x_j} = \left(\phi^{pi}\frac{\partial f}{\partial x_j \partial x_p} - \phi^{pm}\phi_{jmk}\phi^{ki}\frac{\partial f}{\partial x_p}\right)$$

and so the special Lagrangian equation is equivalent to

$$\mathrm{Im}\left(e^{-i\hat{\theta}}(\omega + F_A)^n\right) = 0.$$

Summarizing, we have that the curvature F_A satisfies the following equations:

$$\begin{aligned} F_A^{2,0} &= 0, \\ \mathrm{Im}(\omega + F_A)^n &= \tan(\theta)\,\mathrm{Re}(\omega + F_A)^n, \end{aligned}$$

which is precisely the dHYM equation. This correspondence easily extends to the general setting where σ is equipped with a flat $U(1)$ connection.

2. Analytic Aspects of the dHYM Equation

Let (X,ω) be a compact Kähler manifold, and let $\mathfrak{a} \in H^{1,1}(X,\mathbb{R})$ be a given cohomology class. Often, we will assume that $\mathfrak{a} = c_1(L)$ for some holomorphic line bundle L, but this is only for aesthetic purposes. We do not assume X is Calabi–Yau, as in general BPS D-branes correspond to solutions of the dHYM equation supported on proper submanifolds of a Calabi–Yau. We are interested in the following question.

Question 1 *When does there exist a smooth representative* α *of the fixed class* $\mathfrak{a}$ *so that*

$$(\omega + \sqrt{-1}\alpha)^n = re^{\sqrt{-1}\hat{\theta}}\omega^n, \tag{2}$$

where $e^{\sqrt{-1}\hat{\theta}} \in S^1$ *is a constant, and* $r : X \to \mathbb{R}_{>0}$ *is a smooth function.*

Strictly speaking, comparing the expression (2) with (1), the reader will see that we are considering the dHYM equation for L^{-1}, but this is just a matter of convention. We make a few preliminary observations. First, fix a point $p \in X$, and choose holomorphic coordinates centred at p so that

$$\omega(p) = \frac{\sqrt{-1}}{2}\sum_i dz_i \wedge d\bar{z}_i, \qquad \alpha(p) = \frac{\sqrt{-1}}{2}\sum_i \lambda_i dz_i \wedge d\bar{z}_i.$$

Invariantly, the numbers λ_i are the eigenvalues of the relative endomorphism $\omega^{-1}\alpha$; we will sometimes refer to these as the eigenvalues of α, and we hope that no confusion will result. At p we have

$$\frac{(\omega + \sqrt{-1}\alpha)^n}{\omega^n}(p) = \prod_i (1 + \sqrt{-1}\lambda_i) = r_\omega(\alpha)e^{\sqrt{-1}\Theta_\omega(\alpha)},$$

where

$$r_\omega(\alpha) = \sqrt{\prod_i (1 + \lambda_i^2)}, \qquad \Theta_\omega(\alpha) = \sum_i \arctan(\lambda_i). \tag{3}$$

In this notation the dHYM equation can be written as

$$\Theta_\omega(\alpha) = \hat{\theta} \qquad \mod 2\pi. \tag{4}$$

The constant $e^{\sqrt{-1}\hat{\theta}}$ is determined by cohomology by the requirement

$$\int_X (\omega + \sqrt{-1}\alpha)^n \in \mathbb{R}_{>0}e^{\sqrt{-1}\hat{\theta}}.$$

From this observation, we obtain the first obstruction to existence of solutions to the dHYM equation.

Lemma 2.1 *If there exists a solution to the dHYM equation, then*

$$\int_X (\omega + \sqrt{-1}\alpha)^n \in \mathbb{C}^*.$$

This obstruction is non-trivial in dimensions $n \geq 3$, and we will return to it in the next section. Fix a reference metric $\alpha_0 \in \mathfrak{a}$. By the $\partial\overline{\partial}$-lemma, any representative of $\mathfrak{a}$ can be written as

$$\alpha_\phi := \alpha_0 + \sqrt{-1}\partial\overline{\partial}\phi,$$

where $\phi : X \to \mathbb{R}$. By (4), the dHYM equation is the natural complex geometric analogue of the special Lagrangian graph equation, which we essentially recounted in Section 1. Let us recall this problem explicitly. Let $\mathbb{C}^n = \mathbb{R}^n + \sqrt{-1}\mathbb{R}^n$, which we equip with the standard Calabi–Yau structure

$$\omega = \frac{\sqrt{-1}}{2}\sum_i dz_i \wedge d\bar{z}_i \qquad \Omega = dz_1 \wedge dz_2 \wedge \cdots \wedge dz_n.$$

Let $f : \mathbb{R}^n \to \mathbb{R}$, and consider the graph of the gradient map of $x \mapsto (x, \nabla f(x))$, which we denote by L. We seek f so that L is *special Lagrangian* with respect to the Calabi–Yau structure defined by ω, Ω. That is,

$$\omega|_L = 0 \qquad \Omega|_L = e^{\sqrt{-1}\hat{\theta}} dVol_L$$

for some constant $e^{\sqrt{-1}\hat{\theta}} \in S^1$. A straightforward computation shows that this is equivalent to

$$\sum_{i=1}^n \arctan(\lambda_i) = \hat{\theta} \qquad \mod 2\pi,$$

where λ_i are the eigenvalues of the $D^2 f$. Special Lagrangian manifolds were first introduced by Harvey and Lawson [18] as an example of a calibrated submanifold. In particular, special Lagrangian submanifolds are automatically volume minimizing in their homology class. We refer the reader to [19] for a beautiful introduction to study of special Lagrangian manifolds.

Solutions of the dHYM equation also minimize a certain volume functional. Consider the map

$$\mathfrak{a} \ni \alpha \longmapsto V_\omega(\alpha) := \int_X r_\omega(\alpha)\omega^n,$$

where $r_\omega(\alpha)$ is defined in (3). We have

Proposition 2.2 (Jacob–Yau [21]) *Define $\hat{r} \geq 0$ by*

$$\hat{r} = \left| \int_X (\omega + \sqrt{-1}\alpha)^n \right|.$$

Then we have $V_\omega(\alpha) \geq \hat{r}$. Furthermore, a smooth form α minimizes $V_\omega(\cdot)$ if and only of α solves the dHYM equation. In this case, the minimum value of V_ω is precisely $\hat{r} > 0$.

Note that, since since $\arctan(\cdot) : \mathbb{R} \to \mathbb{R}$ is increasing, $\Theta_\omega(\cdot)$ is an elliptic second-order operator. A consequence of this is

Lemma 2.3 (Jacob–Yau [21]) *Solutions of the dHYM equation are unique, up to addition of a constant.*

Proof Suppose we have functions $\phi_i : X \to \mathbb{R}$ where $i = 1,2$, such that $\alpha_i := \alpha_{\phi_i}$ satisfy

$$\theta_\omega(\alpha_i) = \theta_i$$

for constants θ_i. Then $\phi_1 = \phi_2 + c$ for some constant $c \in \mathbb{R}$. Consider the function $\phi_1 - \phi_2$. Let $p \in X$ be a point where $\phi_1 - \phi_2$ achieves its infimum. Then we have

$$\alpha_1 \geq \alpha_2$$

and hence $\theta_\omega(\alpha_1) \geq \theta_\omega(\alpha_2)$. It follows that $\theta_1 \geq \theta_2$. Swapping $1 \leftrightarrow 2$ we get that $\theta_1 = \theta_2$. Finally, we write

$$0 = \int_0^1 \frac{d}{dt}\Theta_\omega((1-t)\alpha_1 + t\alpha_2) = \left(\int_0^1 L_t^{i\bar{j}} dt\right) \partial_i \partial_{\bar{j}}(\phi_1 - \phi_2)$$

where $L_t^{i\bar{j}}$ is the linearized operator of $\Theta_\omega(\cdot)$ at the point $(1-t)\alpha_1 + t\alpha_2$. Since this is uniformly elliptic, the strong maximum principle implies $\phi_1 - \phi_2$ is constant. □

A slightly more general result is

Lemma 2.4 *Suppose ω is a Kähler form, and $\alpha \in \mathfrak{a}$ has the property that $\mathrm{osc}_X\Theta_\omega(\alpha) < \pi$. Then*

(1) $\int_X (\omega + \sqrt{-1}\alpha)^n \in \mathbb{C}^*$.

(2) *Let $\theta_\alpha \in (-n\frac{\pi}{2}, n\frac{\pi}{2})$ be defined by*

$$\int_X (\omega + \sqrt{-1}\alpha)^n \in \mathbb{R}_{>0} e^{\sqrt{-1}\theta_\alpha} \qquad \theta_\alpha \in [\inf_X \theta_\omega(\alpha), \sup_X \theta_\omega(\alpha)].$$

If α' is another representative of the class $\mathfrak{a}$ with $\mathrm{osc}_X\Theta_\omega(\alpha') < \pi$, then we have $\theta_\alpha = \theta_{\alpha'}$.

Proof The assumption that $\mathrm{osc}_X\theta_\omega(\alpha) < \pi$ implies that

$$\frac{(\omega + \sqrt{-1}\alpha)^n}{\omega^n}$$

lies in a half space, and hence the integral cannot vanish. To prove the second point define the interval

$$I(\alpha) := [\inf_X \Theta_\omega(\alpha), \sup_X \Theta_\omega(\alpha)].$$

Writing $\alpha' = \alpha + \sqrt{-1}\partial\bar{\partial}\phi$ and looking at the maximum and minimum of ϕ we see that $I(\alpha) \cap I(\alpha') \neq \emptyset$. On the other hand, we have points $\theta_\alpha \in I(\alpha)$ and $\theta_{\alpha'} \in I(\alpha')$ with $\theta_\alpha = \hat{\theta} \mod 2\pi = \theta_{\alpha'}$. Since $I(\alpha), I(\alpha')$ have length π, this implies $\theta_\alpha = \theta_{\alpha'}$.□

Definition 2.5 *Supposing that there exists some $\alpha \in \mathfrak{a}$ with $\mathrm{Osc}_X\Theta_\omega(\alpha) < \pi$, we will define $\theta = \theta_\alpha$ as in Lemma 2.4 to be the* lifted angle. *Since this is independent of the choice of α, we will drop the subscript α.*

Remark *We note that the lifted angle is, a priori, not determined by cohomology. We will discuss this issue in the next section.*

Let us now return to the problem of solving the dHYM equation. Jacob–Yau [21] studied the solvability of the dHYM equation via a heat flow method. They considered the flow

$$\frac{d}{dt}\phi = \Theta_\omega(\alpha_\phi) - \theta, \tag{5}$$

where θ is the lifted angle (assuming this is well-defined). They proved

Theorem 2.6 (Jacob–Yau [21]) *Suppose that (X, ω) has non-negative orthogonal bisectional curvature. Let $L \to X$ be an ample line bundle. Let h_0 be a positively curved metric on L. Then for k sufficiently large the heat flow (5) for metrics on L^k with initial data h_0^k exists for all time and converges to a solution of the dHYM equation.*

Remark *The reader can easily check that if $\mathfrak{a}$ is a Kähler class ample then for k sufficiently large the lifted angle of $k\mathfrak{a}$ is well defined.*

Furthermore, in dimension 2, Jacob–Yau showed that the dHYM equation could be rewritten as the complex Monge–Ampère equation. As a result, on complex surfaces they gave necessary and sufficient algebraic conditions for the existence of solutions to the dHYM equation based Yau's solution of the complex Monge–Ampère equation [32] and the Demailly-Păun characterization of the Kähler cone [11]. In general it is desirable to obtain existence results for solutions of dHYM without any assumptions on the curvature of (X, ω). Observe that if a solution α of the dHYM equation exists then for every $1 \leq j \leq n$ we have

$$\theta - \frac{\pi}{2} < \sum_{i \neq j} \arctan(\lambda_i) < \theta + \frac{\pi}{2},$$

where λ_i are the eigenvalues of α. Conversely, we have the following;

Theorem 2.7 (Collins-Jacob–Yau [9]) *Suppose there exists a $(1,1)$ form $\chi \in \mathfrak{a}$ such that*

$$\Theta_\omega(\chi) \in ((n-2)\frac{\pi}{2}, n\frac{\pi}{2}). \tag{6}$$

Let $\theta \in ((n-2)\frac{\pi}{2}, n\frac{\pi}{2})$ be the lifted angle. Suppose that for every $1 \leq j \leq n$ we have

$$\sum_{i \neq j} \arctan(\mu_i) \geq \theta - \frac{\pi}{2}. \tag{7}$$

where μ_i are the eigenvalues of χ. Then there exists a smooth solution of the dHYM equation.

We make a few remarks about the theorem. First of all, the conditions are clearly necessary in order to solve the equation. Second, the assumption that $\theta_\omega(\chi) \in ((n-2)\frac{\pi}{2}, n\frac{\pi}{2})$ is superfluous as soon as the lifted angle θ satisfies

$$\theta \geq (n-2+\frac{2}{n})\frac{\pi}{2}.$$

We remark also that if α is a Kähler form, then for k sufficiently large we can always ensure that $k\alpha$ satisfies (6).

3. Algebraic Aspects of the dHYM Equation

We now turn our attention to the algebraic aspects of the dHYM equation. There are essentially two questions we would like to discuss in this section.

(1) Is it possible to define the lifted angle *algebraically*?

(2) Are the algebraic obstructions to the existence of solutions to the dHYM equation?

In regards to the second point, it is useful to recall the origin of the dHYM equation as the equation of motion for BPS *D*-branes on the B-model. Douglas proposed a notion of Π-stability which he predicted would be related to the existence of BPS D-branes in mirror symmetry [15, 16]. Motivated by these ideas, Bridgeland [5] developed a theory of stability conditions on triangulated categories; we refer the reader to [2] for a nice introduction to these ideas with connections to physics and mirror symmetry. Since the dHYM equation is the geometric equation of motion for a BPS D-brane on the B-model, it is reasonable to expect that the solvability of the equation should be linked with Π-stability, or more generally Bridgeland stability. The study of Bridgeland stability conditions has attracted considerable interest since their introduction. Even a partial recounting of theory of Bridgeland stability conditions, and the many important results in this area, is far beyond the scope of this article. Nevertheless, we will recall briefly the salient features which seem to appear in the study of dHYM; we refer the reader to [25] and the references therein for more on this active area of research.

We will focus specifically on the case of interest to mirror symmetry, so that the triangulated category is $D^bCoh(X)$.

Definition 3.1 *A slicing $\mathcal{P}$ of $D^bCoh(X)$ is a collection of subcategories $\mathcal{P}(\varphi) \subset D^bCoh(X)$ for all $\varphi \in \mathbb{R}$ such that*

(1) *$\mathcal{P}(\varphi)[1] = \mathcal{P}(\varphi + \pi)$ where $[1]$ denotes the 'shift' functor,*

(2) *if $\varphi_1 > \varphi_2$ and $A \in \mathcal{P}(\varphi_1)$, $B \in \mathcal{P}(\varphi_2)$, then $\mathrm{Hom}(A,B) = 0$,*

(3) *every $E \in D^bCoh(X)$ admits a Harder-Narasimhan filtration by objects in $\mathcal{P}(\phi_i)$ for some $1 \leq i \leq m$.*

We refer to [5] for a precise definition of the Harder-Narasimhan property. A Bridgeland stability condition on $D^bCoh(X)$ consists of a slicing together with a *central charge* (see below). For BPS D-branes in the B-model, the relevant central charge is given by

$$D^bCoh(X) \ni E \longmapsto Z_\omega(E) := -\int_X e^{-\sqrt{-1}\omega} ch(E).$$

Often a factor of $\sqrt{Td(X)}$ is also included, but we will take the above choice (see, for example, [1, 3]).

Definition 3.2 *A Bridgeland stability condition on $D^bCoh(X)$ with central charge Z_ω is a slicing $\mathcal{P}$ satisfying the following properties*

(1) *For any non-zero $E \in \mathcal{P}(\varphi)$ we have*

$$Z_\omega(E) \in \mathbb{R}_{>0} e^{\sqrt{-1}\varphi},$$

(2)

$$C := \inf\left\{ \frac{|Z_\omega(E)|}{\|ch(E)\|} : 0 \neq E \in \mathcal{P}(\varphi), \varphi \in \mathbb{R} \right\} > 0$$

where $\|\cdot\|$ is any norm on the finite dimensional vector space $H^{even}(X,\mathbb{R})$.

Given a Bridgeland stability condition we define $\mathcal{A} := \mathcal{P}((0,1])$ which is called the *heart*. An object $A \in \mathcal{A}$ is semi-stable (resp. stable) if, for every surjection $A \twoheadrightarrow B$ we have

$$\varphi(A) \leq (\text{ resp. } <)\ \varphi(B).$$

In order to make aesthetic contact with Bridgeland stability we will consider throughout this section the case when $\alpha = c_1(L)$ for some holomorphic line bundle L. This does not serve any purpose other than to make the formulas slightly more appealing. Furthermore, the dHYM equation with transcendental cohomology class also appears in mirror symmetry as the equation satisfied by 'complexified Kähler forms' [23]. First we note that for representative $\alpha \in c_1(L)$ we have

$$(\omega + \sqrt{-1}\alpha)^n = n!(\sqrt{-1})^n \left[e^{-\sqrt{-1}(\omega+\sqrt{-1}\alpha)} \right]_{top}$$

and hence we have

$$\int_X (\omega + \sqrt{-1}\alpha)^n = n!(\sqrt{-1})^n \int_X e^{-\sqrt{-1}(\omega)} ch(L).$$

We are therefore lead to consider

$$Z_\omega(L) := -\int_X e^{-\sqrt{-1}\omega} ch(L).$$

Note that if L admits a solution of the dHYM equation with $\theta \in ((n-2)\frac{\pi}{2}, n\frac{\pi}{2})$ then $\mathrm{Im}(Z_\omega(L)) > 0$. Define a path $\gamma(t) : [1,\infty) \to \mathbb{C}$ by

$$\gamma(t) := Z_{t\omega}(L) = -\int_X e^{-t\sqrt{-1}\omega} ch(L).$$

If $\gamma(t) \in \mathbb{C}^*$, then we can define

$$\theta(L) := \text{Winding angle } \gamma(t)$$

as t runs from $+\infty$ to 1. In complex dimension 1 we have

$$\gamma(t) = -\int_X (c_1(L) - \sqrt{-1}\omega t) = \sqrt{-1}\int_X (t\omega + \sqrt{-1}c_1(L))$$

and so

$$\theta(L) = \mathrm{Arg}_{p.v.} \int_X (\omega + \sqrt{-1}c_1(L)) + \frac{\pi}{2},$$

where $\mathrm{Arg}_{p.v.}$ denotes the principal value of Arg with values in $(-\pi, \pi]$. In dimension 2 we have

$$\gamma(t) = \frac{1}{2}\int_X t^2\omega^2 - c_1(L)^2 + \sqrt{-1}t\int_X c_1(L) \wedge \omega.$$

If $\gamma(t) = 0$ for some $t \in [1,\infty)$, then we must have

$$\int_X c_1(L) \wedge \omega = 0.$$

But in this case the Hodge index theorem says that $\int_X c_1(L)^2 \leq 0$, and hence $\mathrm{Re}(\gamma(t)) \neq 0$. Thus $\gamma(t)$ lies in $\mathbb{C}^*$ and hence $\theta(L)$ is well defined. Furthermore, we have

$$\theta(L) = \mathrm{Arg}_{p.v.} \int_X (\omega + \sqrt{-1}c_1(L))^2$$

In three dimensions we encounter the first difficulty. We write

$$\gamma(t) = \left(\int_X t^2 \frac{c_1(L) \wedge \omega^2}{2} - \frac{c_1(L)^3}{6}\right) + \sqrt{-1}\left(\int_X t\frac{c_1(L)^2 \wedge \omega}{2} - t^3 \frac{\omega^3}{6}\right).$$

In general, $\gamma(t)$ may pass through $0 \in \mathbb{C}$, and in fact, one can construct examples of such behaviour on the blow up of $\mathbb{P}^3$ in a point. However, assuming we have a solution of the dHYM equation, we can prove that this is not the case.

Proposition 3.3 *Suppose $\alpha \in c_1(L)$ solves $\theta_\omega(\alpha) = \theta$ with $\theta \in (\frac{\pi}{2}, \frac{3\pi}{2})$. Then $\gamma(t) \in \mathbb{C}^*$ for all $t \in [1, \infty)$. This follows from the Chern number inequality*

$$\left(\int_X \omega^3\right)\left(\int_X ch_3(L)\right) < 3\left(\int_X ch_2(L) \wedge \omega\right)\left(\int_X ch_1(L) \wedge \omega^2\right)$$

Proof We will use the dHYM equation pointwise to prove the inequality. Suppose first that $\theta \in (\pi, \frac{3\pi}{2})$. Since

$$\Theta_\omega(\alpha) = \sum_{i=1}^{3} \arctan(\lambda_i) = \theta > \pi$$

we must have that α is a Kähler form. Since $c_1(L)$ admits a solution of the dHYM equation, if $\gamma(t)$ passes through the origin at time T, we must have that $T > 1$. Solving for T we have

$$\left(\int_X T^2 \frac{c_1(L) \wedge \omega^2}{2} - \frac{c_1(L)^3}{6}\right) = 0,$$

and so

$$1 < T^2 = \frac{\int_X c_1(L)^3}{3\int_X c_1(L) \wedge \omega^2}.$$

Plugging this into the equation for $\mathrm{Im}(\gamma(T)) = 0$ we see that we must have

$$\left(\int_X \omega^3\right)\left(\int_X c_1(L)^3\right) = 9\left(\int_X c_1(L)^2 \wedge \omega\right)\left(\int_X c_1 \wedge \omega^2\right)$$

We will show this cannot happen. Fix a point $p \in X$, and left $\lambda = (\lambda_1, \lambda_2, \lambda_3)$ be the eigenvalues of α with respect to ω. We write the dHYM equation as

$$\tan(\theta)\left(\omega^3 - 3\alpha^2 \wedge \omega\right) = 3\alpha \wedge \omega^2 - \alpha^3.$$

Let $\sigma_1, \sigma_2, \sigma_3$ be the symmetric functions of degree 1, 2 and 3, respectively. For example,

$$\sigma_2(\lambda) = \lambda_1\lambda_2 + \lambda_2\lambda_3 + \lambda_1\lambda_3.$$

We have

$$\alpha^3 = \sigma_3(\lambda)\omega^3, \quad \alpha^2 \wedge \omega = \sigma_2(\lambda)\frac{\omega^3}{3}, \quad \alpha \wedge \omega^2 = \sigma_1(\lambda)\frac{\omega^3}{3}$$

and so we can write the dHYM equation as

$$\tan(\theta)(1 - \sigma_2) = \sigma_1 - \sigma_3$$

Since $\lambda_i > 0$ for all i we have

$$\sigma_1 + \tan(\theta)(\sigma_2 - 1) < \sigma_1\sigma_2.$$

Since $\theta \in (\frac{\pi}{2}, 3\frac{\pi}{2})$ we have $1 - \sigma_2 < 0$, and so we obtain

$$\tan(\theta) < \sigma_1.$$

Since θ is constant we integrate both sides with respect to ω^3 to get

$$\tan(\theta)\int_X \omega^3 < 3\int_X \alpha \wedge \omega^2.$$

On the other hand, by definition we have

$$\tan(\theta) = \frac{\int_X \alpha^3 - 3\alpha \wedge \omega^2}{\int_X 3\alpha^2 \wedge \omega - \omega^3}.$$

By the assumption on θ the denominator is positive, and so we can rearrange this inequality to obtain the result. The remaining case, when $\theta \in (\frac{\pi}{2}, \pi]$ is even easier, using only that

$$\tan(\theta) \leq 0 < \sigma_1.$$

We leave the details to the reader. □

With this proposition in hand it is easy to see that $\theta(L)$ is precisely the constant appearing on the right hand side of the dHYM equation, provided a solution exists. The main new difficulty in dimension 3 which is not present in dimension 1 or 2 is to determine the algebraically the lifted angle of solutions to the dHYM equation when $Z_{\omega}(L)$ has

$$\mathrm{Re}(Z_{\omega}(L)) < 0, \mathrm{Im}(Z_{\omega}(L)) > 0.$$

The primary difficult is that solutions to dHYM with phase $\theta \in (-\frac{3\pi}{2}, -\pi] \cup [\pi, 3\frac{\pi}{2})$ are both mapped into this quadrant. One way to distinguish these two cases is to determine whether $\mathrm{Re}(Z_{t\omega}(L))$ is positive or negative when $\mathrm{Im}(Z_{t\omega}(L)) = 0$. This is precisely what the Chern number inequality proved in Proposition 3.3 accomplishes. In arbitrary dimension this problem will be even more complicated as it will require keeping track of the signs of the real and imaginary parts of $Z_{t\omega}(L))$ and any point time where $Z_{t\omega}(L))$ crosses the real or imaginary axes.

We note that conjectural Chern number inequalities involving ch_3 have appeared in the literature on Bridgeland stability conditions [3]. These inequalities play a fundamental role in establishing the existence of stability conditions. We note, however, that a counter example to the conjectural inequality in [3] was found by Schmidt [27]. It would be very interesting to extend these inequalities to higher rank bundles admitting solutions of dHYM. We end by remarking that, in this correspondence between dHYM and Bridgeland stability, the lifted angle $\theta(L)$ is not the same as the slicing angle $\varphi(L)$; instead, the two are related by a constant depending on the dimension of support of L. When L is a line bundle this is nothing but the dimension of X, but similar ideas hold for line bundles supported on proper analytic sets, which appear as torsion sheaves in $D^b Coh(X)$.

We now turn to the problem of finding algebro-geometric obstructions to the existence of solutions to the dHYM equation. Recall that, if we have a solution of the dHYM equation with lifted phase $\theta \in (n-2\frac{\pi}{2}, n\frac{\pi}{2})$, then necessarily there is an element $\chi \in c_1(L)$ such that for each $1 \le j \le n$ we have

$$(n-1)\frac{\pi}{2} > \sum_{i \neq j} \arctan(\mu_i) > \theta - \frac{\pi}{2}.$$

In fact, for every subset $J \subset \{1, 2, \ldots, n\}$ with $\#J = p$ we have

$$(n-p)\frac{\pi}{2} > \sum_{i \notin J} \arctan(\mu_i) > \theta - p\frac{\pi}{2}.$$

Consider the form

$$(\omega + \sqrt{-1}\chi)^{n-1}.$$

Fix a point $p \in X$, and choose coordinates so that

$$\omega(p) = \frac{\sqrt{-1}}{2}\sum_i dz_i \wedge d\bar{z}_i \qquad \chi(p) = \frac{\sqrt{-1}}{2}\sum_i \mu_i dz_i \wedge d\bar{z}_i.$$

we have

$$(\omega+\sqrt{-1}\chi)^{n-1} = \sum_j r_j e^{\sqrt{-1}\sum_{i\neq j}\arctan(\mu_j)}\widehat{dz_j\wedge d\bar{z}_j}$$

where $r_j > 0$, and

$$\widehat{dz_j\wedge d\bar{z}_j} = (\sqrt{-1})^{n-1}dz_1\wedge d\bar{z}_1\cdots\widehat{dz_j\wedge d\bar{z}_j}\cdots dz_n\wedge d\bar{z}_n.$$

Consider the real $(n-1,n-1)$ form given by

$$\mathrm{Im}\left(e^{-\sqrt{-1}(\theta-\frac{\pi}{2})}(\omega+\sqrt{-1}\chi)^{n-1}\right) = \sum_j r_j\mathrm{Im}\left(e^{\sqrt{-1}\left(\sum_{i\neq j}\arctan(\mu_j)-(\theta-\frac{\pi}{2})\right)}\right)\widehat{dz_j\wedge d\bar{z}_j}$$

By assumption we have for each $1\leq j\leq n$

$$0<\sum_{i\neq j}\arctan(\mu_j)-(\theta-\frac{\pi}{2})<\pi$$

and so

$$\mathrm{Im}\left(e^{-\sqrt{-1}(\theta-\frac{\pi}{2})}(\omega+\sqrt{-1}\chi)^{n-1}\right)>0$$

in the sense of $(n-1,n-1)$ forms. In particular, if $V\subset X$ is a irreducible analytic subvariety with $\dim_{\mathbb{C}} V = n-1$, then we must have

$$\mathrm{Im}\left(\int_V e^{-\sqrt{-1}(\theta-\frac{\pi}{2})}(\omega+\sqrt{-1}\chi)^{n-1}\right)>0.$$

Similar inequalities hold for irreducible analytic subvarieties of all dimension, with the same proof.

Proposition 3.4 *Suppose $c_1(L)$ admits a solution of the dHYM equation with $\theta\in\left((n-2)\frac{\pi}{2},n\frac{\pi}{2}\right)$. Then for every irreducible analytic subvariety $V\subset X$ of dimension $1\leq p<n$ we have*

$$\mathrm{Im}\left(\int_V e^{-\sqrt{-1}(\theta-(n-p)\frac{\pi}{2})}(\omega+\sqrt{-1}\chi)^{p}\right)>0. \tag{8}$$

In terms of the central charge we can write this in the following way. Define

$$Z_{\omega,V}(L) = -\int_V e^{-\sqrt{-1}\omega}ch(L)$$

then we must have

$$\mathrm{Im}\left(\frac{Z_{\omega,V}(L)}{Z_{\omega,X}(L)}\right)>0. \tag{9}$$

To relate this to the Bridgeland stability condition we would like to think of inequality (9) as saying that the surjection

$$L \twoheadrightarrow L \otimes \mathcal{O}_V$$

does not destabilize L, where $\mathcal{O}_V$ is the skyscraper sheaf with support on V. Unfortunately this is not quite correct (unless $Td(X) = 1$), since

$$Z_{\omega,V}(L) \neq Z_{\omega,X}(L \otimes \mathcal{O}_V).$$

Finally we note that if L admits a solution of the dHYM equation then by the BPS bound in Proposition 2.2 we have

$$\frac{|Z_{\omega,X}(L)|}{\|ch(L)\|} > 0$$

which is precisely the second condition required in the definition of a Bridgeland stability condition.

Conjecture 3.5 (Collins-Jacob–Yau [9]) *There exists a solution to the dHYM equation in the class* $\mathfrak{a}$ *with lifted angle* $\theta \in (n-2)\frac{\pi}{2}, n\frac{\pi}{2})$ *if and only if* (8) *holds for all proper, irreducible analytic subvarieties* $V \subsetneq X$ *with* $\dim_{\mathbb{C}} V = p$.

We end by noting some evidence for the conjecture. First of all, the conjecture holds in complex dimension 2 [9, 21]. In general, we consider an asymptotic version of the conjecture. Suppose that L is an ample line bundle. We ask the following question

Question 2 *When can we find a metric h on L, such that the induced metric* h^k *on* L^k *solves the deformed Hermtian-Yang–Mills equation for* $k \gg 1$*?*

Replacing $\alpha \mapsto k\alpha$ we see that, up to rescaling, the dHYM equation on L^k is

$$c\alpha^n = n\alpha^{n-1} \wedge \omega + O(k^{-1})$$

where the constant c is determined by

$$c = \frac{n\int_X \alpha^{n-1} \wedge \omega}{\int_X \alpha^n}.$$

This equation has a long history in Kähler geometry. It was discovered independently by Donaldson [12] and Chen [8], and was studied from the analytic point of view by Weinkove [30, 31] and Song-Weinkove [28]. Let us consider the asymptotic version of Conjecture 3.5. First, we observe that

$$ch(L^k) = \sum_{p=1}^{n} k^p \frac{c_1(L)^p}{p!}$$
$$ch(\mathcal{O}_V) = (-1)^{d-1}(d-1)![V] + \textit{higher degree}$$

and therefore

$$\int_X e^{-\sqrt{-1}\omega} ch(L^k \otimes \mathcal{O}_V) = (-1)^{d-1}\frac{k^d}{d}\int_V c_1(L)^d$$
$$- \sqrt{-1}(-1)^{d-1}k^{d-1}\int_V c_1(L)^{d-1} \wedge \omega + \textit{lower order}\,.$$

Note that this agrees with the leading order term in the expansion of $Z_{V,\omega}(L^k)$. We obtain

$$\mathrm{Arg}_{p.v.} Z_{X,\omega}(L^k \otimes \mathcal{O}_V) = \arctan\left(-\frac{1}{k}\frac{d\int_V c_1(L)^{d-1} \wedge \omega}{\int_V c_1(L)^d} + \textit{lower order}\right).$$

The formal limit of Conjecture 3.5 is therefore

Conjecture 3.6 (Lejmi-Székelyhidi [22]) *Let L be an ample line bundle. There exists a Kähler metric in $c_1(L)$ solving the J-equation if and only if*

$$\frac{d\int_V c_1(L)^{d-1} \wedge \omega}{\int_V c_1(L)^d} < \frac{n\int_X c_1(L)^{n-1} \wedge \omega}{\int_X c_1(L)^n}$$

for all irreducible analytic subvarieties $V \subsetneq X$ with $\dim_{\mathbb{C}} V = d$.

In fact, this conjecture was discovered from a very different point of view than the one discuss here. The work of Lejmi-Székelyhidi [22] is based on an extension of K-stability, which plays an important role in the existence of constant scalar curvature Kähler metrics [13, 14]. We have

Theorem 3.7 (Collins-Székelyhidi [10]) *Conjecture 3.6 is true when X is toric.*

Acknowledgements

The authors are grateful to Adam Jacob for some helpful comments on an early draft of this paper.

REFERENCES

[1] D. Arcara, and A. Bertram *Bridgeland-stable moduli spaces for K-trivial surfaces*, J. Eur. Math. Soc. 15 (2013), no. 1, 1–38.

[2] P. Aspinwall *D-branes on Calabi–Yau manifolds*, in *Progress in string theory*, 1–152, World Scientific Publishing, Hackensack, NJ, 2005.

[3] A. Bayer, E. Macri and Y. Toda *Bridgeland stability conditions on threefolds I: Bogomolov-Gieseker type inequalities*, J. Algebraic Geom. 23 (2014), no. 1, 117–63.

[4] K. Becker, M. Becker and J. H. Schwarz *String theory and M-theory: A modern introduction*, Cambridge University Press, 2006.

[5] T. Bridgeland *Stability conditions on triangulated categories,* Ann. Math. (2) 166 (2007), no. 2, 317–45.

[6] P. Candelas, X. De La Ossa, P. S. Green and L. Parkes *A pair of Calabi–Yau manifolds as an exactly soluble superconformal theory,* Nucl. Phys. B 359 (1991), no. 1, 21–74.

[7] P. Candelas, G. T. Horowitz, A. Strominger and E. Witten *Vacuum configurations for superstrings,* Nucl. Phys. B 258 (1996), 243–59.

[8] X.-X. Chen *A new parabolic flow in Kähler manifolds* Comm. Anal. Geom. 12 (2004), no. 4, 837–52.

[9] T. C. Collins, A. Jacob and S.-T. Yau $(1,1)$ *forms with specified Lagrangian phase: A priori estimates and algebraic obstructions,* arXiv:1508.01934.

[10] T. C. Collins and G. Székelyhidi *Convergence of the J-flow on toric manifolds,* J. Differential Geom. 107 (2017), no. 1, 47–81.

[11] J.-P. Demailly and M. Păun *Numerical characterization of the Kähler cone,* Ann. Math. (2) 159 (2004), no. 3, 1247–74

[12] S. K. Donaldson *Moment maps and diffeomorphisms,* Asian J. Math. 3 (1999), no. 1, 1–16.

[13] S. K. Donaldson *Scalar curvature and stability of toric varieties.* J. Differential Geom. **62** (2002), no. 2, 289–349.

[14] S. K. Donaldson *Lower bounds on the Calabi functional,* J. Differential Geom. **70** (2005), no. 3, 453–72.

[15] M. R. Douglas *D-branes on Calabi–Yau manifolds,* European Congress of Mathematics, Vol. II (Barcelona, 2000), 449–66, Progr. Math. 202, Birkhäuser, Basel, 2001.

[16] M. R. Douglas *D-branes, categories and $N = 1$ supersymmetry. Strings, branes, and M-theory,* J. Math. Phys. 42 (2001), no. 7, 2818–43.

[17] B. R. Greene and R. M. Plesser *Duality in Calabi–Yau moduli space,* Nucl. Phys. B 338 (1990), no. 1, 15–37.

[18] R. Harvey and H. B. Lawson *Calibrated geometries,* Acta. Math. 148 (1982), no. 1, 47–157.

[19] N. Hitchin *Lectures on special Lagrangian submanifolds,* Winter School on Mirror Symmetry, Vector Bundles and Lagrangian Submanifolds (Cambridge, MA, 1999), 151Ð182, AMS/IP Stud. Adv. Math. **23**, Amer. Math. Soc., Providence, RI, 2001.

[20] N. Hitchin *The moduli space of special Lagrangian submanifolds.* Dedicated to Enio De Giorgi. Ann. Scuola Norm. Sup. Pisa Cl. Sci. (4) 25 (1997), no. 3–4, 503–15.

[21] A. Jacob and S.-T. Yau *A special Lagrangian type equation for holomorphic line bundles,* Math. Ann. 369 (2017), no. 1–2, 869–98.

[22] M. Lejmi and G. Székelyhidi *The J-flow and stability,* Adv. Math. 274 (2015), 404–31.

[23] N. C. Leung *Mirror symmetry without corrections,* arXiv:0009235 (2001).

[24] N. C. Leung, S.-T. Yau and E. Zaslow *From special Lagrangian to Hermitian–Yang–Mills via Fourier–Mukai,* Adv. Theor. Math. Phys. 4 (2000), no. 6, 1319–41.

[25] E. Macri and B. Schmidt *Lectures on Bridgeland stability,* Proceedings of the CIMPA-CIMAT-ICTP School on Moduli of Curves (Guanajuato, Mexico, 2016), Springer, Cham, 2017.

[26] M. Mariño, R. Minasian G. Moore and A. Strominger *Non-linear instantons from supersymmetric p-branes,* J. High Energy Phys. (2000), no. 1, 005.

[27] B. Schmidt *Counterexample to the generalized Bogomolov-Gieseker inequality for threefolds,* Int. Math. Res. Not. **2017**, (2017) no. 8, 2562–6.

[28] J. Song and B. Weinkove *On the convergence and singularities of the J-flow with applications to the Mabuchi energy,* Comm. Pure Appl. Math. 61 (2008), no. 2, 210–29.

[29] A. Strominger, S.-T. Yau and E. Zaslow *Mirror Symmetry is T-duality,* Nucl. Phys. B 479 (1996), no. 1–2, 243–59.

[30] B. Weinkove *Convergence of the J-flow on Kähler surfaces,* Comm. Anal. Geom. 12 (2004), no. 4, 151–64.

[31] B. Weinkove *On the J-flow in higher dimensions and the lower boundedness of the Mabuchi energy*, J. Differential Geom. 73 (2006), no. 2, 351–8.

[32] S.-T. Yau *On the Ricci curvature of a compact Kähler manifold and the complex Monge–Ampère equation, I*, Comm. Pure Appl. Math. 31 (1978), no. 3, 339–411.

5

Quaternionic Geometry in Dimension 8

DIEGO CONTI
THOMAS BRUUN MADSEN
SIMON SALAMON

Dipartimento di Matematica e Applicazioni, Università di Milano Bicocca, Via Cozzi 55, 20125 Milano, Italy. Email: diego.conti@unimib.it
Department of Mathematics, Aarhus University, Ny Munkegade 118, Bldg 1530, 8000 Aarhus, Denmark. Email: thomas.madsen@math.au.dk
Department of Mathematics, King's College London, Strand, London WC2R 2LS, UK. Email: simon.salamon@kcl.ac.uk

ABSTRACT. We describe the 8-dimensional Wolf spaces as cohomogeneity one SU(3)-manifolds, and discover perturbations of the quaternion-kähler metric on the simply connected 8-manifold $G_2/SO(4)$ that carry a closed fundamental 4-form but are not Einstein.

To Nigel Hitchin on the occasion of his 70th birthday

1. Introduction

Of the 'fundamental geometries' captured by Berger's list of holonomy groups, the quaternionic unitary group stands out in that Riemannian manifolds with holonomy in $Sp(n)\,Sp(1)$ are Einstein but not Ricci-flat, unless locally hyperkähler. Excluding the latter case, the study of these quaternion-kähler manifolds splits into two cases, depending on the sign of the scalar curvature. The negative case is fairly flexible [1, 2, 9, 10], but the situation of positive scalar curvature is extremely rigid. In fact, it is conjectured that a complete positive quaternion-kähler manifold is necessarily one of the symmetric spaces that were first described by Wolf [23]. This rigidity suggests a quest for ways of weakening the holonomy condition. It turns out that dimension 8 harbours a particularly natural type of almost quaternion-kähler manifold.

An appealing way of expressing an almost quaternionic Hermitian structure is to say that our 8-manifold admits a 4-form that is pointwise linearly equivalent to

Conti, D., Bruun Madsen, T., Salamon, S., *Quaternionic Geometry in Dimension 8*. In: *Geometry and Physics: A Festschrift in Honour of Nigel Hitchin*, Jørgen Ellegaard Andersen, Andrew Dancer, Oscar García-Prada (Eds): Oxford University Press (2018). DOI: 10.1093/oso/9780198802006.003.0005

$$\Omega = \tfrac{1}{2}(\omega_1^2 + \omega_2^2 + \omega_3^2), \tag{1.1}$$

where $(\omega_1, \omega_2, \omega_3)$ is the standard hyperkähler triplet on $\mathbb{R}^8 \cong \mathbb{H}^2$:

$$\begin{cases} \omega_1 = dx^{12} + dx^{34} + dx^{56} + dx^{78}, \\ \omega_2 = dx^{13} + dx^{42} + dx^{57} + dx^{86}, \\ \omega_3 = dx^{14} + dx^{23} + dx^{58} + dx^{67}. \end{cases}$$

In these terms, the quaternion-kähler condition then amounts to Ω being parallel for the Levi–Civita connection, $\nabla\Omega = 0$.

Swann [22] observed that it is possible to have Ω non-parallel and closed (and so harmonic), but that closedness of the fundamental 4-form implies quaternion-kähler in dimension at least 12. Using exterior differential systems, Bryant [5] analysed the local existence of the 'harmonic' Sp(2) Sp(1)-structures and showed that solutions exist in abundance, even though the PDE system at first looks overdetermined. In fact, involutivity of the exterior differential system can be deduced by observing that the contraction of Ω with any vector $v \in \mathbb{R}^8$ induces a stable form on the quotient $\mathbb{R}^8/\langle v\rangle$ (cf. [7, 8]). In [21], the third author provided the first compact non-parallel example of such a geometry. Later, many more examples have followed [8, 14] by reducing the internal symmetry group from the quaternionic unitary group to its intersection with SO(6) and SO(7).

These previously known examples of non-parallel harmonic Sp(2) Sp(1)-manifolds all have infinite fundamental group and associated metric of negative scalar curvature. A natural question is whether such structures with positive scalar curvature exist on simply connected manifolds.

Acknowledgements

We all thank Robert Bryant, David Calderbank and Andrew Swann for useful comments. DC was partially supported by FIRB 2012 'Geometria differenziale e teoria geometrica delle funzioni'. TBM gratefully acknowledges financial support from Villum Fonden. The completion of this work was supported by a grant from the Simons Foundation (#488635, Simon Salamon), which will also provide a forum for pursuing topics mentioned in the final section. The interest of all three authors in special holonomy can be traced back to Nigel Hitchin, who recognized long ago the importance of this field.

2. The Wolf Spaces

In Wolf's construction [23], one starts with a compact centreless simple Lie group G with Lie algebra $\mathfrak{g}$ and choice of Cartan subalgebra $\mathfrak{t} \subset \mathfrak{g}$. One then picks a maximal root $\beta \in \mathfrak{t}$ and considers an associated $\mathfrak{sp}(1)$ and its centralizer $\mathfrak{l}_1$ in $\mathfrak{g}$. The Lie algebra $\mathfrak{k} = \mathfrak{sp}(1) \oplus \mathfrak{l}_1$ will model the holonomy algebra of a symmetric space: if we let G be the simply connected compact simple Lie group corresponding to $\mathfrak{g}$, and K the compact subgroup generated by $\mathfrak{k}$, then G/K is a compact symmetric quaternion-kähler manifold, a so-called *Wolf space*, with holonomy K. The associated quaternionic structure on the tangent space is generated by the subgroup $\mathrm{Sp}(1) \subset \mathrm{K}$.

2.1. Quaternionic Projective Plane

The model space for a quaternion-kähler 8-manifold is the quaternionic projective plane

$$\mathbb{HP}(2) = \frac{\mathrm{Sp}(3)}{\mathrm{Sp}(2) \times \mathrm{Sp}(1)}.$$

In Wolf's terms, we can describe this as follows.

Choose the Cartan subalgebra $\mathfrak{h}_{\mathbb{C}}$ of $\mathfrak{sp}(3)_{\mathbb{C}}$ spanned by the three elements

$$H_i = E_{i,i} - E_{i+3,i+3},$$

where the matrix $E_{k,\ell}$ has only non-zero entry, equal to 1, at position (k,ℓ). Then let $L_j \in \mathfrak{h}^*$ be the element satisfying $L_j(H_i) = \delta_{ji}$. The corresponding roots of $\mathfrak{sp}(3)_{\mathbb{C}}$ are the vectors $\pm L_i \pm L_j$. The associated eigenspaces are spanned by

$$\begin{aligned}
&X_{i,j} = E_{i,j} - E_{3+j,3+i},\\
&Y_{i,j} = E_{i,3+j} + E_{j,3+i},\ Z_{i,j} = E_{3+i,j} + E_{3+j,i},\\
&U_i = E_{i,3+i},\ V_i = E_{3+i,i},
\end{aligned}$$

where $i \neq j$ in the first two rows.

A real structure σ is determined by

$$\begin{aligned}
&\sigma(H_i) = -H_i,\ \sigma(X_{i,j}) = -X_{j,i},\\
&\sigma(Y_{i,j}) = -Z_{i,j},\ \sigma(Z_{i,j}) = -Y_{i,j},\\
&\sigma(U_i) = -V_i,\ \sigma(V_i) = -U_i,
\end{aligned}$$

and we can therefore choose a basis of $\mathfrak{sp}(3)$ given by

$$\{\underbrace{iH_k}_{A_k},\overbrace{X_{k,\ell}-X_{\ell,k}}^{P_{k+\ell-2}},\underbrace{i(X_{k,\ell}+X_{\ell,k})}_{P_{k+\ell+1}},\overbrace{Y_{k,\ell}-Z_{k,\ell}}^{Q_{k+\ell-2}},\underbrace{i(Y_{k,\ell}+Z_{k,\ell})}_{Q_{k+\ell+1}},\overbrace{U_k-V_k}^{R_k},\underbrace{i(U_k+V_k)}_{R_{k+3}}\}.$$

In these terms, Wolf's highest root $\mathfrak{sp}(1)$ is given by

$$\mathfrak{sp}(1) = \langle A_1, R_1, R_4 \rangle,$$

and its centralizer is the copy of $\mathfrak{sp}(2)$ described as

$$\mathfrak{l}_1 = \langle A_2, A_3, P_3, P_6, Q_3, Q_6, R_2, R_3, R_5, R_6 \rangle.$$

We now have the direct sum decomposition

$$\mathfrak{sp}(3) = \mathfrak{k} \oplus \mathfrak{p},$$

where $\mathfrak{k} = \mathfrak{sp}(1) \oplus \mathfrak{sp}(2)$ and

$$\mathfrak{p} = \mathfrak{k}^{\perp} = \langle P_1, P_2, P_4, P_5, Q_1, Q_2, Q_4, Q_5 \rangle.$$

The basis

$$\tfrac{1}{4\sqrt{2}}P_1, \tfrac{1}{4\sqrt{2}}P_4, \tfrac{1}{4\sqrt{2}}Q_1, \tfrac{1}{4\sqrt{2}}Q_4, \tfrac{1}{4\sqrt{2}}P_2, \tfrac{1}{4\sqrt{2}}P_5, \tfrac{1}{4\sqrt{2}}Q_2, \tfrac{1}{4\sqrt{2}}Q_5$$

is orthonormal for the Killing metric on the subspace $\mathfrak{p} = (\mathfrak{sp}(1) \oplus \mathfrak{sp}(2))^{\perp}$, and determines an adapted frame for the $\mathrm{Sp}(2)\,\mathrm{Sp}(1)$-structure. Its dual basis is given by

$$f^1 = 4\sqrt{2}p_1, f^2 = 4\sqrt{2}p_4, f^3 = 4\sqrt{2}q_1, f^4 = 4\sqrt{2}q_4, f^5 = 4\sqrt{2}p_2,$$
$$f^6 = 4\sqrt{2}p_5, f^7 = 4\sqrt{2}q_2, f^8 = 4\sqrt{2}q_5,$$

where $p_1, \ldots, q_5$ is the dual basis of $P_1, \ldots, Q_5$.

2.2. *Complex Grassmannian*

Consider next the complex Grassmannian of planes in $\mathbb{C}^4$:

$$\mathrm{Gr}_2(\mathbb{C}^4) = \frac{\mathrm{SU}(4)}{\mathrm{S}(\mathrm{U}(2) \times \mathrm{U}(2))}.$$

In order to describe Wolf's structure on this space, we begin by considering $\mathrm{SU}(4)_{\mathbb{C}} = \mathrm{SL}(4, \mathbb{C})$ with its usual basis:

$$H_i = E_{i,i} - E_{i+1,i+1}, X_1 = E_{1,2}, X_2 = E_{1,3}, \ldots, X_6 = E_{3,4}, Y_1 = E_{2,1}, \ldots, Y_6 = E_{4,3}.$$

From the real structure σ, given by $\sigma(H_i) = -H_i, \sigma(X_i) = -Y_i, \sigma(Y_i) = -X_i$, we see that a basis of $\mathfrak{su}(4)$ can be described as

$$\mathfrak{su}(4) = \langle \underbrace{iH_j}_{A_j}, \underbrace{X_j - Y_j}_{C_j}, \underbrace{i(X_j + Y_j)}_{B_j} \rangle.$$

In these terms, Wolf's highest root $\mathfrak{sp}(1)$ reads

$$\mathfrak{sp}(1) = \langle i(H_1 + H_2 + H_3), C_3, B_3 \rangle$$

and its centralizer is

$$\mathfrak{l}_1 = \langle i(H_1 - H_3), iH_2, C_4, B_4 \rangle.$$

As a result, we have the direct sum decomposition

$$\mathfrak{su}(4) = \mathfrak{k} \oplus \mathfrak{p},$$

where $\mathfrak{k} = \mathfrak{sp}(1) \oplus \mathfrak{k}_1$ and

$$\mathfrak{p} = \mathfrak{k}^{\perp} = \langle C_1, B_1, C_2, B_2, C_5, B_5, C_6, B_6 \rangle.$$

We see that the orthonormal frame

$$\tfrac{1}{4}C_1, \tfrac{1}{4}B_2, \tfrac{1}{4}C_5, \tfrac{1}{4}B_5, \tfrac{1}{4}C_2, \tfrac{1}{4}B_2, \tfrac{1}{4}C_6, \tfrac{1}{4}B_6$$

is adapted to the Sp(2) Sp(1)-structure. Letting $c_1, \ldots, b_6$ denote the dual basis of $C_1, \ldots, B_6$, we then have an Sp(2) Sp(1)-adapted coframe $f^1, \ldots, f^8$ given by $f^1 = 4c_1$, and so forth.

2.3. *The Exceptional Wolf Space*

We finally turn to Wolf's construction of a quaternion-kähler structure on

$$\frac{\mathrm{G}_2}{\mathrm{SO}(4)}.$$

This is slightly more involved, due to the more complicated nature of $\mathfrak{g}_2$.

First, we need to choose a suitable basis of $(\mathfrak{g}_2)_{\mathbb{C}}$. We shall follow [11], where the long roots are given by

$$\alpha_2 = (-\tfrac{3}{2}, \tfrac{\sqrt{3}}{2}) = -\beta_2, \quad \alpha_5 = (\tfrac{3}{2}, \tfrac{\sqrt{3}}{2}) = -\beta_5, \quad \alpha_6 = (0, \sqrt{3}) = -\beta_6$$

and the short roots are

$$\alpha_1 = (1, 0) = -\beta_1, \quad \alpha_3 = (-\tfrac{1}{2}, \tfrac{\sqrt{3}}{2}) = -\beta_3, \quad \alpha_4 = (\tfrac{1}{2}, \tfrac{\sqrt{3}}{2}) = -\beta_4.$$

We shall pick $\beta = (0, \sqrt{3})$ as the highest root.

The real structure σ of $\mathfrak{g}_2^{\mathbb{C}}$ is determined by

$$\sigma(H_i) = -H_i, \quad \sigma(X_i) = -Y_i, \quad \sigma(Y_i) = -X_i$$

and a basis of $\mathfrak{g}_2$ is therefore given by

$$\{A_1 = iH_1, A_2 = iH_2, W_j = X_j - Y_j, Z_j = i(X_j + Y_j) : \ 1 \leqslant j \leqslant 6\}.$$

The highest root $\mathfrak{sp}(1)$ is given by

$$\mathfrak{sp}(1) = \langle A_1 + 2A_2, W_6, Z_6 \rangle$$

and its centralizer in $\mathfrak{g}_2$ is given by

$$\mathfrak{l}_1 = \langle A_1, W_1, Z_1 \rangle.$$

We now have the direct sum decomposition

$$\mathfrak{g}_2 = \mathfrak{k} \oplus \mathfrak{p},$$

where

$$\mathfrak{p} = \mathfrak{k}^{\perp} = \langle W_2, W_3, W_4, W_5, Z_2, Z_3, Z_4, Z_5 \rangle.$$

In particular, the basis

$$W_2, Z_2, -W_5, -Z_5, \tfrac{1}{\sqrt{3}} W_3, \tfrac{1}{\sqrt{3}} Z_3, -\tfrac{1}{\sqrt{3}} W_4, -\tfrac{1}{\sqrt{3}} Z_4,$$

which is orthonormal for the Killing form on the subspace $\mathfrak{p} = \mathfrak{so}(4)^{\perp}$, determines an adapted frame for this Sp(2) Sp(1)-structure. As in the previous cases, we let f^i denote its dual coframe, meaning $f^1 = w_2$, etc.

3. Cohomogeneity One SU(3)-Actions

Each of the eight-dimensional Wolf spaces M admits a cohomogeneity one SU(3)-action, which was studied by Gambioli [12] (see also [19]). We summarize and elaborate on key facts below. In each case, the action comes from the embedding $\mathrm{SU}(3) \subset \mathrm{G}$. In order to explicitly describe the orbits of this SU(3)-action, we choose an element $Z \in \mathfrak{p} \cap \mathfrak{su}(3)^{\perp}$ and write $\gamma(t) = \exp(tZ)$. Then the SU(3)-orbits of $\gamma(t)$ are given by

$$\iota_t \colon \mathrm{SU}(3) \to M, \quad g \mapsto g\gamma(t)\mathrm{K}.$$

As the quotient map

$$\pi : G \to M, \quad g \mapsto g\mathrm{K}$$

has $\ker \pi_{*g} = L_{g*}(\mathfrak{k})$, we can identify $T_{\gamma(t)\mathrm{K}}M$ with $L_{\gamma(t)*}(\mathfrak{p})$, or simply $\mathfrak{p}$ where left translation is then understood. It follows that we can identify ι_{t*} with the map $\mathfrak{su}(3) \to \mathfrak{p}$ given by

$$X \mapsto [\mathrm{Ad}(\gamma(t)^{-1})(X)]_{\mathfrak{p}}.$$

Since Z is orthogonal to $\mathfrak{su}(3) \subset \mathfrak{g}$, it is clear from invariance of the Killing form that $L_{\gamma(t)*}(Z)$ is orthogonal to the SU(3)-orbit of $\gamma(t)$ for all t, and altogether the cohomogeneity one action infinitesimally is described by the mapping

$$\mathfrak{su}(3) \oplus \mathbb{R} \to \mathfrak{p}, \quad X \mapsto [\mathrm{Ad}(\gamma(t)^{-1})(X)]_{\mathfrak{p}}, \frac{\partial}{\partial t} \mapsto Z. \tag{3.1}$$

Table 3.1 The three Wolf space building blocks.

H	$\mathfrak{g}/\mathfrak{h}$	V	G/H
SU(2)	$\mathbb{R}\oplus\mathbb{H}$	Σ^2	S^5
SO(3)	$\odot^2_0\mathbb{R}^3$	$\mathbb{R}^3$	L
U(2)	$[[\Lambda^{1,0}K]]$	$\mathbb{C}^2$	$\mathbb{CP}(2)$

Now, given the adapted quaternion-kähler frame on $T_{eK}M$, we can use (3.1) to pull this back to $\mathfrak{su}(3)\oplus\mathbb{R}$ and thereby get a description of the Wolf space structure that is adapted to the cohomogeneity one setting.

Before doing so, we fix some conventions for $\mathfrak{su}(3)$: in the following, $e^1,\ldots,e^8$ will always denote a basis of $\mathfrak{su}(3)^*$ such that the following structure equations hold:

$$\begin{gathered} de^1=-e^{23}-e^{45}+2e^{67},\quad de^2=e^{13}+e^{46}-e^{57}-\sqrt{3}e^{58},\\ de^3=-e^{12}-e^{47}+\sqrt{3}e^{48}-e^{56},\quad de^4=e^{15}-e^{26}+e^{37}-\sqrt{3}e^{38},\\ de^5=-e^{14}+e^{27}+\sqrt{3}e^{28}+e^{36},\quad de^6=-2e^{17}+e^{24}-e^{35},\\ de^7=2e^{16}-e^{25}-e^{34},\quad de^8=-\sqrt{3}(e^{25}-e^{34}). \end{gathered} \tag{3.2}$$

In terms of matrices, we can express the dual basis $e_1,\ldots,e_8$ as

$$\begin{gathered} e_1=E_{21}-E_{12}, e_2=E_{31}-E_{13}, e_3=E_{32}-E_{23},\\ e_4=-i(E_{23}+E_{32}), e_5=i(E_{13}+E_{31}), e_6=-i(E_{12}+E_{21}),\\ e_7=i(E_{22}-E_{11}), e_8=\tfrac{i}{\sqrt{3}}(2E_{33}-E_{11}-E_{22}). \end{gathered}$$

In the following subsections, we shall show that, from the cohomogeneity one SU(3) point of view, Wolf's quaternion-kähler manifolds arise by combining three basic models that correspond to tubular neighbourhoods $\mathrm{G}\times_{\mathrm{H}}V$ of the relevant singular orbits G/H. These basic building blocks are summarized in Table 3.1, where Σ^2 is the irreducible three-dimensional representation of SU(2), $K=\Lambda^{2,0}$ denotes the one-dimensional representation of U(2) corresponding to the determinant, and $\mathbb{R}^3$ and $\mathbb{C}^2$ are the standard representations of SO(3) and U(2), respectively.

In Table 3.1, L is the symmetric space SU(3)/SO(3) that parametrizes special Lagrangian subspaces of $\mathbb{R}^6\cong\mathbb{C}^3$.

3.1. *Quaternionic Projective Plane*

In order to give a cohomogeneity one description of the quaternionic projective plane, we start by fixing the embedding of SU(3) in Sp(3) given via

$$\mathrm{SU}(3)\subset\left\{\begin{pmatrix} X & 0\\ 0 & (X^{-1})^T\end{pmatrix} : X\in\mathrm{GL}(3,\mathbb{C})\right\}\subset\mathrm{Sp}(3,\mathbb{C}).$$

Correspondingly, we have the following description of $\mathfrak{su}(3)$ at the Lie algebra level:

$$\mathfrak{su}(3)=\langle A_1-A_2,A_2-A_3,P_1,P_2,P_3,P_4,P_5,P_6\rangle.$$

Now choose an element

$$Z \in \mathfrak{p} \cap \mathfrak{su}(3)^{\perp} = \langle Q_1, Q_2, Q_4, Q_5 \rangle.$$

We shall fix $Z = Q_1$. In these terms, the adapted quaternion-kähler frame pulls back as follows.

Lemma 3.1 *Under the mapping* (3.1), *Wolf's coframe pulls back to the dual of* $\mathfrak{su}(3) \oplus \mathbb{R}$ *to give*

$$\begin{aligned} \tilde{e}^1(t) = 4\sqrt{2}\cos(2t)e^6, \tilde{e}^2(t) &= -4\sqrt{2}\cos(2t)e^7, \\ \tilde{e}^3(t) = 4\sqrt{2}dt, \tilde{e}^4 &= \tfrac{4\sqrt{6}}{3}\sin(2t)e^8, \\ \tilde{e}^5(t) = 4\cos(t)(e^2+e^4), \tilde{e}^6(t) &= 4\cos(t)(e^3+e^5), \\ \tilde{e}^7(t) = 4\sin(t)(e^2-e^4), \tilde{e}^8(t) &= 4\sin(t)(e^3-e^5). \end{aligned} \tag{3.3}$$

Proof In order to write things consistently with the structure equations (3.2), we fix on $\mathfrak{su}(3) \subset \mathfrak{sp}(3)$ the basis

$$\begin{aligned} e_1 = A_1 - A_2, e_2 = \tfrac{1}{\sqrt{2}}(P_2 - P_3), e_3 = \tfrac{1}{\sqrt{2}}(P_5 + P_6), e_4 = \tfrac{1}{\sqrt{2}}(P_2 + P_3), \\ e_5 = \tfrac{1}{\sqrt{2}}(P_5 - P_6), e_6 = P_1, e_7 = -P_4, e_8 = \tfrac{1}{\sqrt{3}}(A_1 + A_2 - 2A_3). \end{aligned}$$

Computing the action of $\mathrm{Ad}(\exp(-tQ_1))$ with respect to the bases

$$e_1, \ldots, e_8, \quad P_1, P_4, Q_1, Q_4, P_2, P_5, Q_2, Q_5$$

of $\mathfrak{su}(3)$ and $\mathfrak{p}$, respectively, we find that the map (3.1) is represented by the matrix

$$\begin{pmatrix} 0 & 0 & 0 & 0 & 0 & \cos(2t) & 0 & 0 \\ 0 & 0 & 0 & 0 & 0 & 0 & -\cos(2t) & 0 \\ 0 & 0 & 0 & 0 & 0 & 0 & 0 & 0 \\ 0 & 0 & 0 & 0 & 0 & 0 & 0 & \frac{1}{3}\sqrt{3}\sin(2t) \\ 0 & \frac{1}{2}\cos(t)\sqrt{2} & 0 & \frac{1}{2}\cos(t)\sqrt{2} & 0 & 0 & 0 & 0 \\ 0 & 0 & \frac{1}{2}\cos(t)\sqrt{2} & 0 & \frac{1}{2}\cos(t)\sqrt{2} & 0 & 0 & 0 \\ 0 & \frac{1}{2}\sin(t)\sqrt{2} & 0 & -\frac{1}{2}\sin(t)\sqrt{2} & 0 & 0 & 0 & 0 \\ 0 & 0 & \frac{1}{2}\sin(t)\sqrt{2} & 0 & -\frac{1}{2}\sin(t)\sqrt{2} & 0 & 0 & 0 \end{pmatrix}.$$

Considering its transpose, and recalling that $\frac{\partial}{\partial t}$ maps to Q_1, we see that the adapted coframe $f^1, \ldots, f^8$ pulls back to $\tilde{e}^1(t), \ldots, \tilde{e}^8(t)$ to give the stated result. □

Our computation confirms, more directly, the following result that is indicated in [12].

Proposition 3.2 *The Wolf space* $\mathbb{HP}(2)$ *can be viewed as a cohomogeneity one manifold obtained by glueing together the disc bundles over the singular orbits* $\mathbb{CP}(2) = \mathrm{SU}(3)/\mathrm{U}(2)$ *and* $S^5 = \mathrm{SU}(3)/\mathrm{SU}(2)$. *Each principal orbit is a copy of the exceptional Aloff–Wallach space* $\mathrm{N}^{1,-1} \cong \mathrm{N}^{1,0}$.

Proof The point is to identify the principal and singular stabilizers; the latter appear at $t=0$ and $t=\pi/4$. Since the singular orbits have codimension strictly smaller than 6, both singular stabilizers are connected [16, Corollary 1.9]. In particular, it suffices to work at the Lie algebra level, since connected subgroups of SU(3) are in one-to-one correspondence with subalgebras of $\mathfrak{su}(3)$. It is worthwhile making this more explicit by identifying the Lie algebras of the principal and singular stabilizers.

Regarding the principal orbits, we observe that the coframe (3.3) for generic t annihilates the $\mathfrak{u}(1)$ spanned by e_1. When $t=0$, the coframe is the annihilator of a Lie algebra $\mathfrak{u}(2)$ spanned by the four elements $e_1, e_8, e_2-e_4, e_3-e_5$. Finally, at $t=\pi/4$ the subspace annihilated is the $\mathfrak{su}(2)$ spanned by e_1, e_6, e_7. □

For later reference, let us emphasize that the tangent space of the open set corresponding to the principal orbits at each point decomposes as the U(1)-representation

$$\mathbb{R}^8 \cong 2\mathbb{R} \oplus 2V_1 \oplus V_2, \tag{3.4}$$

where V_k is the irreducible two-dimensional representation on which the principal U(1) acts via matrices of the form

$$\begin{pmatrix} \cos(k\theta) & \sin(k\theta) \\ -\sin(k\theta) & \cos(k\theta) \end{pmatrix};$$

at the infinitesimal level, this follows directly from (3.2).

As a final remark, note that $\mathbb{HP}(2)$ comes with a U(1)-action, generated by the diagonal U(1) in U(3) $\subset$ Sp(3), that commutes with the action of SU(3). Clearly, this circle action generates a Killing vector field. Explicitly, this action is generated by $X=\sqrt{3}e_8+3A_3$ and, in our cohomogeneity one framework, it reads

$$\begin{aligned} \iota_t(g) \mapsto \exp(sX)\iota_t(g) &= g\exp(sX)\gamma(t)K = g\exp(\sqrt{3}se_8)\exp(3sA_3)\gamma(t)\,\mathrm{K} \\ &= g\exp(\sqrt{3}se_8)\gamma(t)\exp(3sA_3)\,\mathrm{K} = \iota_t(g\exp(\sqrt{3}se_8)), \end{aligned}$$

where we have used the fact that A_3 is an element of $\mathfrak{sp}(2) \subset \mathfrak{sp}(2) \oplus \mathfrak{sp}(1)$ that commutes with $Z=Q_1$. Therefore, the Killing vector field X can be identified with the left-invariant vector field $\sqrt{3}e_8$ on SU(3).

3.2. *Complex Grassmannian*

In order to get an explicit description of the cohomogeneity one nature of $\mathrm{Gr}_2(\mathbb{C}^4)$, we fix the copy of SU(3) which comes from the usual embedding

$$\mathrm{SU}(3) \cong \left\{ \left(\begin{smallmatrix} A & \\ & 1 \end{smallmatrix}\right) : A \in \mathrm{SU}(3) \right\} \subset \mathrm{SU}(4).$$

At the level of Lie algebras, this means that we are working with the copy $\mathfrak{su}(3) \subset \mathfrak{su}(4)$ spanned by

$$e_1 = A_1 + A_2, e_2 = \frac{1}{\sqrt{2}}(C_1 + C_4), e_3 = \frac{1}{\sqrt{2}}(B_1 + B_4), e_4 = \frac{1}{\sqrt{2}}(C_1 - C_4),$$
$$e_5 = \frac{1}{\sqrt{2}}(B_1 - B_4), e_6 = C_2, e_7 = -B_2, e_8 = \frac{1}{\sqrt{3}}(A_1 - A_2);$$

this choice of a basis is consistent with (3.2). Proceeding as before, we pick an element

$$Z \in \mathfrak{p} \cap \mathfrak{su}(3)^{\perp} = \langle C_5, B_5, C_6, B_6 \rangle,$$

which, for concreteness, we shall fix to $Z = C_5$. Computations as in the proof of Lemma 3.1 then give the following lemma:

Lemma 3.3 *Under the mapping* (3.1), *Wolf's coframe pulls back to the dual of* $\mathfrak{su}(3) \oplus \mathbb{R}$ *to give*

$$\tilde{e}^1(t) = 2\sqrt{2}\cos(t)(e^2 + e^4), \quad \tilde{e}^2(t) = 2\sqrt{2}\cos(t)(e^3 + e^5), \quad \tilde{e}^3(t) = 4dt,$$
$$\tilde{e}^4(t) = -\tfrac{4\sqrt{3}}{3}\sin(2t)e^8, \quad \tilde{e}^5(t) = 4e^6, \quad \tilde{e}^6(t) = -4e^7,$$
$$\tilde{e}^7(t) = 2\sqrt{2}\sin(t)(-e^2 + e^4), \quad \tilde{e}^8(t) = 2\sqrt{2}\sin(t)(e^3 - e^5).$$

Using these observations, we have the following result.

Proposition 3.4 *The Wolf space* $\mathrm{Gr}_2(\mathbb{C}^4)$ *can be viewed as a cohomogeneity one manifold obtained by glueing together the disc bundles over two copies of the singular orbit* $\mathbb{CP}(2) = \mathrm{SU}(3)/\mathrm{U}(2)$. *Each principal orbit is a copy of the exceptional Aloff–Wallach space* $N^{1,0} \cong N^{1,-1}$. *The form of the metric in the two bundles is the same up to the identification*

$$e_1 \mapsto -e_1,\ e_2 \mapsto e_4,\ e_3 \mapsto -e_5,\ e_4 \mapsto -e_2,\ e_5 \mapsto e_3,\ e_6 \mapsto e_6,\ e_7 \mapsto -e_7,\ e_8 \mapsto e_8.$$

Proof The arguments follow those of Proposition 3.2. For convenience, let us write down the Lie algebras of the stabilizers. For the principal orbits, we have $\mathfrak{u}(1)$, corresponding to e_1. At $t = 0$ we see that the algebra annihilated by the pulled-back coframe is the copy of $\mathfrak{u}(2)$ spanned by $e_1, e_8, e_2 - e_4, e_3 - e_5$. Finally, at $t = \pi/2$ the Lie algebra of the singular stabilizer is spanned by the $\mathfrak{u}(2)$ determined by $e_1, e_8, e_2 + e_4, e_3 + e_5$.

The identification given is the inner automorphism of $\mathfrak{su}(3)$ obtained by conjugating with the matrix $\begin{pmatrix} 0 & i & 0 \\ i & 0 & 0 \\ 0 & 0 & 1 \end{pmatrix}$. It preserves the principal $\mathfrak{u}(1)$ and the quaternion-kähler metric. The second statement follows. □

As for $\mathbb{HP}(2)$, note that the complex Grassmannian comes with a U(1)-action, now generated by the diagonal U(1) in U(3) ⊂ SU(4), commuting with the SU(3)-action. Again, this circle action clearly generates a Killing vector field which in our cohomogeneity one framework can be identified with $e_8 \in \mathfrak{su}(3)$. In this case, $X = A_1 + 2A_2 + 3A_3$ is the sum of $-2\sqrt{3}e_8$ with $3A_1 + 3A_3$, which commutes with $Z = C_5$ and is contained in K.

3.3. *The Exceptional Wolf Space*

The quaternion-kähler structure on the space $G_2/SO(4)$ admits a cohomogeneity one description that comes from the embedding of SU(3) in G_2 as the group generated by the long roots (cf. [18]). At the Lie algebra level, we have that $\mathfrak{su}(3)$ is generated by

$$A_1 + A_2, \tfrac{1}{\sqrt{2}}(W_2 + W_6), \tfrac{1}{\sqrt{2}}(Z_6 - Z_2), \tfrac{1}{\sqrt{2}}(W_2 - W_6),$$
$$-\tfrac{1}{\sqrt{2}}(Z_2 + Z_6), W_5, -Z_5, -\tfrac{1}{\sqrt{3}}(A_1 + 3A_2);$$

this choice of basis is consistent with (3.2). In order to study the orbits of the SU(3)-action, we choose an element

$$Z \in \mathfrak{p} \cap \mathfrak{su}(3)^{\perp} = \langle W_3, W_4, Z_3, Z_4 \rangle.$$

Specifically, we fix $Z = W_3$. Then computations, completely similar to those in the proof of Lemma 3.1, give the following lemma:

Lemma 3.5 *Under the mapping (3.1), Wolf's coframe pulls back to the dual of* $\mathfrak{su}(3) \oplus \mathbb{R}$ *to give*

$$\tilde{e}^1(t) = \tfrac{\sqrt{2}}{2}(\cos(t)^3 - \sin(t)^3)e^2 + \tfrac{\sqrt{2}}{2}(\cos(t)^3 + \sin(t)^3)e^4,$$
$$\tilde{e}^2(t) - \tfrac{\sqrt{2}}{2}(\cos(t)^3 - \sin(t)^3)e^3 - \tfrac{\sqrt{2}}{2}(\cos(t)^3 + \sin(t)^3)e^5,$$
$$\tilde{e}^3(t) = -e^6, \tilde{e}^4(t) = e^7, \tilde{e}^5(t) = \sqrt{3}dt, \tilde{e}^6(t) = -\sin(2t)e^8,$$
$$\tilde{e}^7(t) = -\sqrt{\tfrac{3}{8}}\sin(2t)(\sin(t) - \cos(t))e^2 - \sqrt{\tfrac{3}{8}}\sin(2t)(\sin(t) + \cos(t))e^4,$$
$$\tilde{e}^8(t) = -\sqrt{\tfrac{3}{8}}\sin(2t)(\sin(t) - \cos(t))e^3 - \sqrt{\tfrac{3}{8}}\sin(2t)(\sin(t) + \cos(t))e^5.$$

With the above observations, we have the following result that confirms statements from [12]:

Proposition 3.6 *The exceptional Wolf space* $G_2/SO(4)$ *can be viewed as a cohomogeneity one manifold obtained by glueing together disc bundles over the singular orbits* $\mathbb{CP}(2) = SU(3)/U(2)$ *and* $L = SU(3)/SO(3)$. *Each principal orbit is an exceptional Aloff–Wallach space* $N^{1,0} \cong N^{1,-1}$.

Proof Again, the arguments are like those of Proposition 3.2 but, for convenience, we spell out the Lie algebras of the stabilizers. For the principal orbits, we have $\mathfrak{u}(1)$, corresponding to e_1. At $t = 0$, we see that the algebra annihilated by the pulled-back coframe is the copy $\mathfrak{u}(2)$ spanned by $e_1, e_8, e_2 - e_4, e_3 - e_5$. Finally, at $t = \pi/4$, the Lie algebra of the singular stabilizer is spanned by the copy $\mathfrak{so}(3)$ determined by e_1, e_2, e_3. □

In contrast with the quaternionic projective plane and the complex Grassmannian, $G_2/SO(4)$ clearly does not admit a (global) circle action from a commuting $U(1) \subset G_2$

($\mathrm{SU}(3) \subset \mathrm{G}_2$ is a maximal connected subgroup). However, the open set corresponding to the principal orbits does come with a circle action corresponding to e_8. While X is not a Killing vector field in this case, it turns out to satisfy the generalized condition

$$d(\|X\|^2) \wedge \mathcal{L}_X \Omega = 0, \tag{3.5}$$

as follows by direct computation.

We conclude our cohomogeneity one description of the Wolf spaces with an observation that, in a sense, ties together all three cases.

Proposition 3.7 *The vector bundle* $\mathrm{SU}(3) \times_{\mathrm{U}(2)} \mathbb{C}^2$ *over* $\mathbb{CP}(2)$ *admits three distinct* $\mathrm{SU}(3)$*-invariant quaternion-kähler structures.*

Proof First, note that our analysis of the cohomogeneity one $\mathrm{SU}(3)$-actions shows that the spaces $\mathbb{HP}(2) \setminus S^5$, $\mathrm{Gr}_2(\mathbb{C}^4) \setminus \mathbb{CP}(2)$ (one can choose either copy of $\mathbb{CP}(2)$) and $\mathrm{G}_2/\mathrm{SO}(4) \setminus L$ are all equivalent to the same vector bundle $\mathrm{SU}(3) \times_{\mathrm{U}(2)} \mathbb{C}^2$. The claim then follows since the three quaternion-kähler structures on $\mathbb{HP}(2)$, $\mathrm{Gr}_2(\mathbb{C}^4)$ and $\mathrm{G}_2/\mathrm{SO}(4)$ induce different structures on $\mathrm{SU}(3) \times_{\mathrm{U}(2)} \mathbb{C}^2$, since these Wolf spaces have different holonomy groups and, consequently, different curvature. □

4. Nilpotent Perturbations

Let α be an element of $\Lambda^p(\mathbb{R}^n)^*$, and consider the *(affine) perturbation* by a fixed p-form δ, meaning

$$\beta(t) = \alpha + t\delta, \quad t \in \mathbb{R}.$$

Generally, it is hard to decide whether $\beta(t)$ and α lie in the same $\mathrm{GL}(n,\mathbb{R})$-orbit for all t. However, a useful sufficient criterion can be phrased as follows.

Proposition 4.1 *Let* $A \in \mathfrak{gl}(n,\mathbb{R})$. *If the associated derivation* $\rho(A)$ *satisfies* $\rho(A)^2\alpha = 0$, *then*

$$\beta(t) = \alpha + t\rho(A)\alpha$$

lies in the same $\mathrm{GL}(n,\mathbb{R})$*-orbit as* α *for all* $t \in \mathbb{R}$.

Proof The proof is elementary. We expand $g(t) = \exp(tA)$ to find that

$$g(t)\alpha = \alpha + t\rho(A)\alpha,$$

since the higher-order terms $\frac{t^k}{k!}\rho(A)^k\alpha, k \geqslant 2$, vanish by assumption. So $\beta(t)$ and α lie in the same $\mathrm{GL}(n,\mathbb{R})$-orbit, as claimed. □

Motivated by Proposition 4.1, we would like to characterize the perturbations of α that are parametrized by solutions of

$$\rho(A)^2\alpha = 0. \tag{4.1}$$

Among these solutions, we obviously have elements of the stabilizer $\mathfrak{g}$ of α in $\mathfrak{gl}(n,\mathbb{R})$, but these give rise to trivial perturbations $\beta(t) \equiv \alpha$. In order to eliminate this indeterminacy, we observe the following:

Proposition 4.2 *Let $\alpha \in \Lambda^p(\mathbb{R}^n)^*$. Then every solution A of (4.1) satisfies*

$$\rho(A)\alpha = \rho(N)\alpha,$$

where N is a nilpotent solution of (4.1).

Proof Over the complex numbers we can put A, as an endomorphism of $(\mathbb{C}^n)^*$, into Jordan form. Correspondingly, we obtain a direct sum decomposition $(\mathbb{C}^n)^* = \bigoplus V_i$, where V_i is the generalized eigenspace relative to the eigenvalue λ_i. Denoting by I_{V_i} the matrix corresponding to the projection onto V_i, we have $A = N + \sum \lambda_i I_{V_i}$, where N is nilpotent and real.

In accordance with the above, we can also decompose $\Lambda^p(\mathbb{C}^n)^*$ as a direct sum

$$\bigoplus_{k_1+\cdots+k_m=p} \bigoplus_{j_1<\cdots<j_m} \Lambda^{k_1} V_{j_1} \otimes \cdots \otimes \Lambda^{k_m} V_{j_m}, \tag{4.2}$$

where each summand is closed under $\rho(A)$. Moreover, direct computation shows that

$$\begin{aligned}&(\rho(A) - (\lambda_1 + \lambda_2)I)(\alpha_1 \wedge \alpha_2)\\ &\quad = (\rho(A) - \lambda_1 I)\alpha_1 \wedge \alpha_2 + \alpha_1 \wedge (\rho(A) - \lambda_2 I)\alpha_2,\end{aligned}$$

giving that each summand in (4.2) is contained in the generalized eigenspace of $k_1\lambda_{j_1} + \cdots + k_m\lambda_{j_m}$ relative to $\rho(A)$. It follows that $\rho(A)^2$ also preserves the decomposition (4.2), and its kernel is contained in

$$\bigoplus_{j_1<\cdots<j_m} \bigoplus_{\substack{k_1+\cdots+k_m=p\\ k_1\lambda_{j_1}+\cdots+k_m\lambda_{j_m}=0}} \Lambda^{k_1} V_{j_1} \otimes \cdots \otimes \Lambda^{k_m} V_{j_m}.$$

On this space, $\rho(A)$ and $\rho(N)$ act in the same way. Therefore, if A is a solution of Equation (4.1), then

$$\rho(N)^2\alpha = 0 \quad \text{and} \quad \rho(N)\alpha = \rho(A)\alpha. \qquad \square$$

Motivated by Proposition 4.2, we shall restrict our attention to nilpotent solutions of (4.1) and refer to these as *nilpotent perturbations.*

While Proposition 4.2 is valid in general, our interest is the case where α is the quaternionic form (1.1). If $e_1^0, \ldots, e_8^0$ denotes a standard basis of $\mathbb{R}^8$, and e_0^i its dual, it turns out that solutions to the perturbation problem are most conveniently expressed in terms of the orthonormal basis $(E^1, \ldots, E^8)$ of $(\mathbb{R}^8)^*$ equal to

$$\left(e_0^8, \tfrac{\sqrt{3}}{2}e_0^2 - \tfrac{1}{2}e_0^6, -\tfrac{\sqrt{3}}{2}e_0^1 - \tfrac{1}{2}e_0^5, -\tfrac{\sqrt{3}}{2}e_0^5 + \tfrac{1}{2}e_0^1, \tfrac{\sqrt{3}}{2}e_0^6 + \tfrac{1}{2}e_0^2, -e_0^4, e_0^3, e_0^7\right). \tag{4.3}$$

In these terms, (1.1) reads

$$\begin{aligned}\Omega = &-E^{1247} + \sqrt{3}E^{1248} - E^{1256} - E^{1346} + E^{1357} + \sqrt{3}E^{1358}\\ &+ 2E^{1458} - E^{1678} - E^{2345} + 2E^{2367} - \sqrt{3}E^{2467} + E^{2468} - E^{2578}\\ &- E^{3478} - \sqrt{3}E^{3567} - E^{3568}.\end{aligned}$$

A significant observation, which may at first not be fully appreciated, is that the stabilizer of E^{123} in $\mathrm{Sp}(2)\,\mathrm{Sp}(1)$ is $\mathrm{SO}(3)$. Computations give the following lemma:

Lemma 4.3 *The stabilizer of* $\langle E^1, E^2, E^3\rangle$ *in* $\mathrm{Sp}(2)\,\mathrm{Sp}(1)$ *is the copy of* $\mathrm{SO}(3)$ *whose Lie algebra is spanned by the elements*

$$\begin{pmatrix} 0 & \frac{\sqrt{3}}{2}j \\ \frac{\sqrt{3}}{2}j & -j \end{pmatrix} - \tfrac{1}{2}R_j, \quad \begin{pmatrix} 0 & \frac{\sqrt{3}}{2}k \\ \frac{\sqrt{3}}{2}k & k \end{pmatrix} - \tfrac{1}{2}R_k, \quad \begin{pmatrix} \frac{3}{2}i & 0 \\ 0 & -\frac{1}{2}i \end{pmatrix} - \tfrac{1}{2}R_i$$

of $\mathfrak{sp}(2) \oplus \mathfrak{sp}(1)$.

Above, R_i denotes right multiplication by i, and so forth.

Prompted by Lemma 4.3, we shall decompose $(\mathbb{R}^8)^*$ as the sum of two irreducible $\mathrm{SO}(3)$-modules

$$S^2 = \langle E^1, E^2, E^3\rangle, \quad S^4 = \langle E^4, \dots, E^8\rangle.$$

Using the dual basis E_i, we then have that the 2-forms

$$\beta^1 := (E_2 \wedge E_3) \lrcorner \Omega, \quad \beta^2 := (E_3 \wedge E_1) \lrcorner \Omega, \quad \beta^3 := (E_1 \wedge E_2) \lrcorner \Omega \tag{4.4}$$

define an $\mathrm{SO}(3)$-equivariant linear map $S^2 \to \Lambda^2 S^4$.

In these terms, the following result describes nilpotent perturbations in the quaternionic setting.

Theorem 4.4 *Up to the action of* $\mathrm{Sp}(2)\,\mathrm{Sp}(1)$, *nilpotent solutions of*

$$\rho(A)^2\Omega = 0$$

are parametrized by linear maps $v\colon S^2 \to S^4$ *such that*

$$v^2 \wedge v^3 \wedge \beta^1 + v^3 \wedge v^1 \wedge \beta^2 + v^1 \wedge v^2 \wedge \beta^3 = 0.$$

Explicitly, v *corresponds to the endomorphism* $A = \sum_{i=1}^{3} v^i \otimes E_i$ *of* $(\mathbb{R}^8)^*$.

In terms of forms, the notation $v^i \otimes E_i$ above represents the endomorphism

$$\gamma \mapsto v^i \wedge (E_i \lrcorner \gamma).$$

Remark 4.5 *As a corollary of Theorem 4.4, nilpotent solutions of* $\rho(A)^2\Omega = 0$ *actually satisfy* $A^2 = 0$. *Note that* ρ *is not an algebra homomorphism, so* $A^2 = 0$ *does not imply* $\rho(A)^2 = 0$.

The three-dimensional subspace $\langle E_1, E_2, E_3\rangle$ of $\mathbb{R}^8$, i.e. the annihilator of S^4, is uniquely determined up to the Sp(2) Sp(1)-action. Subspaces of $\mathbb{R}^8$ in its Sp(2) Sp(1)-orbit can be characterized by the angle between quaternionic lines. Indeed, consider the Sp(2) Sp(1)-invariant function

$$Q\colon \mathbb{HP}^1 \times \mathbb{HP}^1 \to \mathbb{R}, \quad Q([v],[w]) = \max_{J\in \mathrm{Sp}(1)} \frac{\langle v, Jw\rangle^2}{|v|^2\,|w|^2}.$$

In an affine chart, we can express Q as

$$Q([1:p],[1:q]) = \frac{\left|1+\bar{p}q\right|^2}{\left|1+\bar{p}q\right|^2 + \left|q-p\right|^2}.$$

Lemma 4.6 *Let $v, w \in \mathbb{R}^8$ be two orthogonal non-zero vectors such that*

$$((v\wedge w)\lrcorner\Omega)^3 = 0.$$

Then, $Q([v],[w]) = \frac{1}{4}$.

Proof Using the action of the quaternionic unitary group, we can assume that $w = ae_8^0$, and $v = be_1^0 + ce_5^0$. Then, a straightforward computation shows that $b = \pm\sqrt{3}c$, giving the asserted result. □

Lemma 4.7 *Let $w_1, \ldots, w_k$ be orthonormal vectors in $\mathbb{R}^8$ such that*

$$((w_i\wedge w_j)\lrcorner\Omega)^3 = 0.$$

Then, $k \leqslant 3$ and, if $k = 3$, up to the action of Sp(2) Sp(1), *we may assume that $w_i = E_i$ for $i = 1, 2, 3$.*

Proof By Lemma 4.6, we have that

$$Q([w_i],[w_j]) = \frac{1}{4}, \quad i \neq j. \tag{4.5}$$

If $k > 2$, the points $[w_i]$ in $\mathbb{HP}^1$ satisfy (4.5) and we can assume, up to Sp(2) Sp(1)-action, that

$$[w_1] = [1:0], \quad [w_2] = [1:p], \quad [w_3] = [1:q].$$

Then

$$\left|p\right|^2 = 3 = \left|q\right|^2, \quad \left|p-q\right|^2 = 3\left|1+\bar{p}q\right|^2;$$

which only has the solution $p = -q$. It is clearly not possible to add a fourth element $[w_4]$ so that (4.5) is satisfied.

Assuming then that $k = 3$, we can use the action of Sp(2) Sp(1), as in the proof of Lemma 4.6, to obtain $w_1 = e_8^0$. This leaves us with an SO(4) symmetry that can be used to obtain $w_2 \in \langle e_1^0, e_2^0, e_3^0, e_4^0, e_6^0\rangle$. The stabilizer in SO(4) of e_6^0 is U(2) and, up to this U(2)-action, we can assume $w_2 = E_2$. The condition $p = -q$ together with orthogonality implies that w_3 is in the span of $\sqrt{3}e_1^0 + e_5^0$ and $\sqrt{3}e_3^0 + e_7^0$. The stabilizer of E_2 in U(2), isomorphic to U(1), acts non-trivially on this two-dimensional space and so allows us to set $w_3 = E_3$. □

Proof of Theorem 4.4 Up to change of basis, nilpotent matrices are classified over the reals by partitions with weight 8, giving 22 possibilities that can be encoded in terms of Young diagrams. For example, the diagram

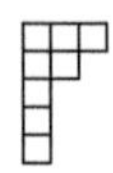

describes the endomorphisms of $(\mathbb{R}^8)^*$ with Jordan blocks of size $3,2,1,1,1$, which, with respect to some basis $\{w^1,\dots,w^8\}$, satisfy $w^3 \mapsto w^2$, $w^2 \mapsto w^1$, $w^5 \mapsto w^4$, with the other vectors mapped to zero. For each diagram Γ, we can fix a representative endomorphism A_Γ and compute the space

$$K_\Gamma = \left\{\alpha \in \Lambda^4(\mathbb{R}^8)^* : \rho(A_\Gamma)^2\alpha = 0\right\}.$$

The equation $\rho(A)^2\Omega = 0$ has a solution with diagram Γ if $\rho(A_\Gamma)^2\alpha = 0$ for some α in the orbit of Ω, and this requires that, for each non-zero $v \wedge w$ in $\Lambda^2\mathbb{R}^8$, the map

$$K_\Gamma \to \Lambda^4(\mathbb{R}^8)^*, \quad \alpha \mapsto ((v \wedge w) \lrcorner \alpha)^2$$

is not identically zero. Computations show that this rules out all cases except

$\Gamma_1 =$ 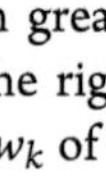 $\Gamma_2 =$ 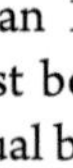 $\Gamma_3 =$ 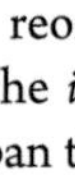 $\Gamma_4 =$ 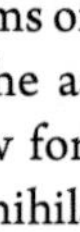 $\Gamma_5 =$ 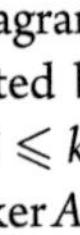 $\Gamma_6 =$ 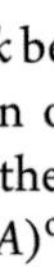

The last diagram corresponds to $A = 0$, and Γ_5 corresponds to A being any rank 1 nilpotent matrix. In either case, the statement of the theorem holds.

Now let A be a solution of $\rho(A)^2\Omega = 0$. In terms of its diagram, let k be the number of rows of length greater than 1, and reorder the associated basis in order that w^i corresponds to the rightmost box in the ith row for $1 \leqslant i \leqslant k$. In other words, the elements $w_1,\dots,w_k$ of the dual basis span the annihilator $(\ker A + \operatorname{im} A)^o$. Then, each $(w_i \wedge w_j) \lrcorner \Omega$ is degenerate, meaning that $((w_i \wedge w_j) \lrcorner \Omega)^3 = 0$. In the case of Γ_4, this holds because

$$2(Aw^1) \wedge (Aw^2) \wedge (w_1 \lrcorner w_2 \lrcorner \Omega) = \rho(A)^2\Omega = 0.$$

For $\Gamma_1, \Gamma_2, \Gamma_3$, the hypothesis implies degeneracy as it forces $w_i \lrcorner w_j \lrcorner w_\ell \lrcorner \Omega$ to be zero.

Without loss of generality, we can assume that the covectors w^i are orthogonal to $\operatorname{im} A$ and orthonormal. From Lemma 4.7, we conclude that $k \leqslant 3$, so that we can rule out Γ_2, and assume $w_i = E_i$. In the case of Γ_3 and Γ_4, it now suffices to write $A = \sum_{i=1}^3 v^i \otimes E_i$, so that, using (4.4), the vanishing of $\rho(A)^2\Omega$ becomes

$$v^{23}\wedge\beta^1+v^{31}\wedge\beta^2+v^{12}\wedge\beta^3=0,$$

as required.

Finally, in order to rule out the case Γ_1, assume the associated basis has the form $E_1,E_2,E_3,v_1,\dots,v_5$, so that, with obvious notation,

$$A=v^1\otimes E_1+v^2\otimes E_2+v^3\otimes E_3+v^4\otimes v_1.$$

Then,

$$\begin{aligned}0&=\rho(A)^2\Omega\\&=2\sum_{1\leqslant i<j\leqslant 3}v^i\wedge v^j\wedge(E_i\lrcorner E_j\lrcorner\Omega)+v^4\wedge(E_1\lrcorner\Omega-2\sum_{i=1}^{3}v^i\wedge(E_i\lrcorner v_1\lrcorner\Omega)).\end{aligned}\tag{4.6}$$

Wedging with v^4 and using (4.4), we get

$$v^{124}\wedge\beta^3+v^{314}\wedge\beta^2+v^{234}\wedge\beta^1=0.\tag{4.7}$$

Projecting Equation (4.6) onto the space $S^2\otimes\Lambda^3S^4$, we find the condition

$$\begin{aligned}v^4\wedge\Big[&E^2\wedge\beta^3-E^3\wedge\beta^2+2v^1\wedge(v_1\lrcorner(E^2\wedge\beta^3-E^3\wedge\beta^2))\\&+2v^2\wedge(v_1\lrcorner(E^3\wedge\beta^1-E^1\wedge\beta^3))+2v^3\wedge(v_1\lrcorner(E^1\wedge\beta^2-E^2\wedge\beta^1))\Big]=0.\end{aligned}$$

This implies that

$$\begin{aligned}0&=v^4\wedge[\beta^2+2v^1\wedge(v_1\lrcorner\beta^2)-2v^2\wedge(v_1\lrcorner\beta^1)]\quad\text{and}\\0&=v^4\wedge[\beta^3+2v^1\wedge(v_1\lrcorner\beta^3)-2v^3\wedge(v_1\lrcorner\beta^1)].\end{aligned}\tag{4.8}$$

Therefore, $v^{124}\wedge\beta^3=v^{314}\wedge\beta^2=2v^{234}\wedge\beta^1$, and (4.7) implies each term is zero. Then Equation (4.8) gives $v^4\wedge\beta^2,v^4\wedge\beta^3\in\langle v^{234}\rangle$, which is absurd. In conclusion, Γ_1 cannot occur, and the proof is complete. □

4.1. U(1)-Invariant Perturbations

If we impose invariance, Theorem 4.4 can be simplified considerably. Indeed, consider the eight-dimensional representation of U(1) that models the tangent space to the open set formed of our principal orbits (cf. Equation (3.4)). The nilpotent perturbations compatible with this action have a simple description.

Proposition 4.8 *Let $\Omega\in\Lambda^4(\mathbb{R}^8)^*$ be a* U(1)*-invariant 4-form with stabilizer group* Sp(2) Sp(1)*. Then there is an orthonormal basis as in (4.3), such that (for the dual basis)*

$$E_1,E_8\in 2\mathbb{R},\quad 2V_1=\langle E_2,E_3\rangle\oplus\langle E_4,E_5\rangle,\quad V_2=\langle E_6,E_7\rangle.$$

The space of U(1)*-invariant nilpotent perturbations is generated by*

$$E^8 \wedge (E_1 \lrcorner \Omega). \tag{4.9}$$

Proof By construction, the 4-form is fixed by U(1) so that this group lies inside Sp(2) Sp(1). We can then find an adapted basis as in (4.3), such that U(1) ⊂ SO(3) ⊂ Sp(2) Sp(1), where the middle subgroup SO(3) preserves the splitting

$$\langle E^1, E^2, E^3\rangle \oplus \langle E^4, \ldots, E^8\rangle = S^2 \oplus S^4;$$

this is because all subgroups U(1) in Sp(2) Sp(1) are conjugate, so we can assume that U(1) is contained in some conjugate of SO(3). Due to the way this U(1) acts on S^4, we deduce that, for the dual basis, one has

$$\langle E_1, E_2, E_3\rangle = \mathbb{R} \oplus V_1, \quad \langle E_4, \ldots, E_8\rangle = \mathbb{R} \oplus V_1 \oplus V_2.$$

Now, by making a change of basis if necessary, we can assume that

$$E_1 \in \mathbb{R}, \quad V_1 = \langle E_2, E_3\rangle,$$

corresponding to U(1) stabilizing E^1 in SO(3), whose precise form can be recovered from Lemma 4.3. Computing its action on S^4, we find

$$E_8 \in \mathbb{R}, \quad V_1 = \langle E_4, E_5\rangle, \quad V_2 = \langle E_6, E_7\rangle.$$

Since on nilpotent matrices the map $A \mapsto \rho(A)\Omega$ is injective, the latter is invariant if and only if A is invariant. This means that the space of invariant nilpotent perturbations is given by

$$v^1 \in \langle E^8\rangle, \quad v^2, v^3 \in \langle E^4, E^5\rangle,$$

where

$$\beta^1 \wedge v^{23} + \beta^2 \wedge v^{31} + \beta^3 \wedge v^{12} = 0.$$

It follows that $v^2 \wedge v^3 = 0$, and by invariance this means that $v^2 = 0 = v^3$. In conclusion, the space of U(1)-invariant perturbations is generated by (4.9), as required. □

5. New Closed Sp(2)Sp(1)-Structures

We are now ready to produce explicit examples of closed Sp(2) Sp(1)-structures. Since the corresponding exterior differential system is effectively underdetermined, it is not surprising that, at least locally, it is possible to obtain such examples by deforming the quaternion-kähler metric on a Wolf space M. In fact, it follows from results of the first author [7] that, if one considers the induced structure on a real analytic hypersurface $N \subset M$

(in the language of [8], an SO(4)-structure with a closed 4-form β), then one can extend it to obtain a closed Sp(2) Sp(1)-structure in a neighbourhood of N.

It is not difficult to see that there is more flexibility than that arising from local diffeomorphisms. In our cohomogeneity one setting, this indeterminacy can be seen by parametrizing invariant forms in the GL(8, $\mathbb{R}$)-orbit of the quaternion-kähler 4-form. These depend on eleven functions, because, relatively to (3.4), the centralizer of U(1) in GL(8, $\mathbb{R}$) has dimension 14 and intersects Sp(2) Sp(1) in a three-dimensional torus. Explicit computations show that closedness of the form corresponds to seven equations, leaving four undetermined functions, while equivariant diffeomorphisms only depend on one function.

While the discussion above emphasizes local flexibility, the method of Section 4 proves to be a particularly useful approach for obtaining examples that are both explicit and global. As it turns out, each Wolf space has a family of closed nilpotent perturbations determined by the vector field X, corresponding to e_8. The perturbed metric happens to be genuinely different from the original only when this vector field is not Killing.

Lemma 5.1 *On each of the three Wolf spaces,* SU(3)*-invariant closed nilpotent perturbations of the quaternion-kähler structure* Ω_{qK} *have the form*

$$\widetilde{\Omega} = \Omega_{qK} + dh \wedge (e_8 \lrcorner \Omega),$$

where h is any smooth SU(3)*-invariant function.*

Proof An SU(3)-invariant perturbation is defined by an SU(3)-invariant section of End(T(G/K)). On the complement of the singular orbits,

$$\mathrm{SU}(3)/\mathrm{U}(1) \times (0, T),$$

Proposition 4.8 implies that the perturbation must be induced by a t-dependent nilpotent endomorphism of $\langle dt, e^8 \rangle$. Concretely, the perturbation must be of the form

$$\Omega_{qK} \pm (\lambda e^8 + \mu dt) \wedge \left(\left(\mu e_8 - \lambda \frac{\partial}{\partial t} \right) \lrcorner \Omega \right).$$

Insisting that the perturbed 4-form is closed forces λ to vanish, since for all three quaternion-kähler metrics the restriction of $e^8 \wedge (\frac{\partial}{\partial t} \lrcorner \Omega)$ to principal orbits is not closed. On the other hand, perturbations of the form $f(t)dt \otimes e_8$ preserve closedness, because $\mathcal{L}_{e_8} \Omega \wedge dt$ is zero in each case.

Having resolved the problem on the complement of singular orbits, we need to address the conditions that ensure that our solution will extend. Recall that there are three basic models to consider, which are summarized in Table 3.1. Let us first consider the vector bundle $\mathrm{SU}(3) \times_{\mathrm{U}(2)} \mathbb{C}^2$, where $\mathbb{C}^2$ is the standard representation of U(2), corresponding to a tubular neighbourhood of each singular orbit $\mathbb{CP}(2)$. Away from the zero section, e_8 defines an invariant vector field. Since $e_8 \in \mathfrak{u}(2)$, this vector field is vertical, i.e. it is a U(2)-invariant vector field on $\mathbb{C}^2$. In appropriate real coordinates (x, y, z, w), we can write

$$e_8 = \tfrac{2\sqrt{3}}{3}\Big(x\frac{\partial}{\partial y} - y\frac{\partial}{\partial x} + z\frac{\partial}{\partial w} - w\frac{\partial}{\partial z}\Big), \qquad tdt = xdx + ydy + zdz + wdw,$$

where the factor in front is due to the period of this U(2)-invariant vector field.

This shows that $f(t)dt \otimes e_8$ extends smoothly if and only if $t \mapsto f(t)/t$ is smooth and even, or equivalently, $t \mapsto f(t)$ is smooth and odd. At tubular neighbourhoods of the other singular orbits, corresponding to the vector bundles $\mathrm{SU}(3) \times_{\mathrm{SU}(2)} \Sigma^2$ and $\mathrm{SU}(3) \times_{\mathrm{SO}(3)} \mathbb{R}^3$, respectively, e_8 defines an invariant direction in the isotropy representation. This means that $f(t)dt \otimes e_8$ is smooth if $f(t)dt$ is. Consequently, $t \mapsto f(t)$ must again be a smooth odd function. In summary, f extends to a smooth function on $\mathbb{R}$ that satisfies

$$f(t) = -f(-t), \quad f(T-t) = -f(T+t).$$

Any primitive h of $f(t)dt$ then satisfies $h(t) = h(-t)$ and $h(T-t) = h(T+t)$ and therefore defines a global SU(3)-invariant function on G/K, as required. □

5.1. *Perturbing with a Killing Vector Field*

When X is Killing, we have rigidity in the sense that nilpotent perturbation just results in different, but SU(3)-equivalent, ways of expressing Wolf's quaternion-kähler structure:

Proposition 5.2 *Applying* SU(3)*-invariant nilpotent perturbations to the quaternion-kähler structure on* $\mathbb{HP}(2)$ *and* $\mathrm{Gr}_2(\mathbb{C}^4)$ *leaves the structures unchanged up to* SU(3)*-equivariant isometry.*

Proof It follows by Lemma 5.1 that each nilpotent perturbation can be associated with an SU(3)-invariant function h and consequently an invariant vector field $-h(t)e_8$. Away from the singular orbits, its flow has the form

$$\phi^s\colon\ \mathrm{SU}(3) \times (0,T) \to \mathrm{SU}(3) \times (0,T), \quad (g,t) \mapsto (g\exp(-sh(t)e_8), t).$$

Then, ϕ^1 is an equivariant diffeomorphism whose differential at (e,t) is given by

$$(v,0) \mapsto (\mathrm{Ad}(\exp(h(t)e_8))v, 0), \quad \frac{\partial}{\partial t} \mapsto -h'(t)e_8 + \frac{\partial}{\partial t}.$$

Since the adjoint action of e_8 preserves the quaternion-kähler metric, we obtain the same metric up to the isometry that corresponds to replacing e^8 by $e^8 + h'(t)dt$, as required. □

5.2. *Perturbing the Exceptional Wolf Space*

As discussed, X is not Killing in the case of $\mathrm{G}_2/\mathrm{SO}(4)$, but does satisfy the condition (3.5). In a sense, this is exactly what is needed to obtain non-trivial perturbation results.

Theorem 5.3 *The exceptional Wolf space* $G_2/SO(4)$ *admits* $SU(3)$*-invariant non-Einstein positive harmonic* $Sp(2)\,Sp(1)$*-structures. The 4-form determining each such structure belongs to the same cohomology class as the quaternion-kähler 4-form.*

Proof By Lemma 5.1, any $SU(3)$-invariant function h defines a closed perturbation

$$\Omega + dh \wedge (e_8 \lrcorner \Omega).$$

In order to verify that we get non-Einstein examples, we compute the Ricci tensor, which equals

$$\begin{pmatrix}
8-\frac{1}{3}\tan(2t)^2h'(t)^2 & 0 & 0 & 0 & 0 & 0 & -\frac{1}{6}\frac{\sqrt{3}h'(t)^2(3+\cos(4t))}{\cos(2t)^2} & -\frac{1}{3}\tan(2t)h''(t)-4h'(t) \\
0 & 8-\frac{1}{3}\tan(2t)^2h'(t)^2 & 0 & 0 & 0 & 0 & -\frac{1}{3}\tan(2t)h''(t)-4h'(t) & \frac{1}{6}\frac{\sqrt{3}h'(t)^2(3+\cos(4t))}{\cos(2t)^2} \\
0 & 0 & 8 & 0 & 0 & 0 & 0 & 0 \\
0 & 0 & 0 & 8 & 0 & 0 & 0 & 0 \\
0 & 0 & 0 & 0 & 8-\frac{4}{3}\tan(2t)^2h'(t)^2 & \frac{4}{3}\tan(2t)^2\sqrt{3}h'(t) & 0 & 0 \\
0 & 0 & 0 & 0 & \frac{4}{3}\tan(2t)^2\sqrt{3}h'(t) & 8 & 0 & 0 \\
-\frac{1}{6}\frac{\sqrt{3}h'(t)^2(3+\cos(4t))}{\cos(2t)^2} & -\frac{1}{3}\tan(2t)h''(t)-4h'(t) & 0 & 0 & 0 & 0 & 8+\frac{1}{3}\tan(2t)^2h'(t)^2 & 0 \\
-\frac{1}{3}\tan(2t)h''(t)-4h'(t) & \frac{1}{6}\frac{\sqrt{3}h'(t)^2(3+\cos(4t))}{\cos(2t)^2} & 0 & 0 & 0 & 0 & 0 & 8+\frac{1}{3}\tan(2t)^2h'(t)^2
\end{pmatrix}.$$

Finally, note that, in particular, this tells us that the scalar curvature is

$$s = 64 - \tfrac{4}{3}\tan(2t)^2h'(t)^2, \qquad (5.1)$$

so that we can get $s > 0$, but generally non-constant, by choosing h suitably.

For the final statement, we notice that the cohomology class of a closed 4-form on $G_2/SO(4)$ is determined by its restriction to the singular orbit $\mathbb{CP}(2)$, which is a quaternionic submanifold. Since explicit verification shows that the quaternion-kähler form and perturbed 4-form both restrict to the volume form of $\mathbb{CP}(2)$, we conclude that they belong to the same cohomology class, as required. □

Remark 5.4 *Different choices of the perturbing function h in Theorem 5.3 yield non-isometric metrics. Indeed, let* ϕ *be an isometry between two such metrics. As both metrics have isometry group* $SU(3)$, ϕ *maps* $SU(3)$*-orbits to* $SU(3)$*-orbits. In addition, corresponding orbits must have the same volume; notice that, regardless of h, the volume of a principal orbit* $SU(3)/U(1) \times \{t\}$ *is a constant multiple of* $\sin(2t)^3\cos(2t)^2$*. Therefore, the perturbing functions must coincide up to a constant.*

Remark 5.5 *As* $SU(3)$*-invariant functions on* $G_2/SO(4)$ *are in one-to-one correspondence with smooth even functions of period* $\pi/2$ *on the reals, it is easy to find explicit closed perturbations where h is real-analytic.*

It is clear from Equation (5.1) that on $G_2/SO(4)$ we only have constant scalar curvature in the quaternion-kähler case. If we are willing to remove the singular orbit $\mathbb{CP}(2)$, however, the conclusion changes:

Corollary 5.6 *The vector bundle* $SU(3) \times_{SO(3)} \mathbb{R}^3$ *admits non-Einstein harmonic* $Sp(2)\,Sp(1)$*-structures with constant scalar curvature.*

Proof It follows from the proof of Theorem 5.3 that, by choosing $h(t) = c\log\sin(2t)$, for any $c \in \mathbb{R}$, we obtain an incomplete closed $\mathrm{Sp}(2)\,\mathrm{Sp}(1)$-structure of constant scalar curvature defined on $\mathrm{G}_2/\mathrm{SO}(4) \setminus \mathbb{CP}(2)$. For $c \neq 0$, this structure is non-Einstein. □

6. Relations to Other Special Geometries

In our list of symmetric spaces with a cohomogeneity one $\mathrm{SU}(3)$-action, one is missing, namely the Lie group $\mathrm{SU}(3)$ itself, realized as the coset space $\mathrm{SU}(3)^2/\Delta\,\mathrm{SU}(3)$. The relevant action is by consimilarity [17]

$$\mathrm{SU}(3) \times \mathrm{SU}(3) \to \mathrm{SU}(3)\colon\ (g,h) \mapsto gh\bar{g}^{-1} = ghg^T.$$

This action preserves the parallel $\mathrm{PSU}(3)$-structure, given by

$$\gamma = \tfrac{1}{6}\sum_{i=1}^{8} e^i \wedge de^i \tag{6.1}$$

in terms of our usual basis $e^1,\dots,e^8$ of $\mathfrak{su}(3)^*$.

Computations, similar to those of Section 3, reveal that there are equivariant isomorphisms $\mathbb{HP}(2) \setminus \mathbb{CP}(2) \cong \mathrm{SU}(3) \setminus L$, and $\mathrm{G}_2/\mathrm{SO}(4) \setminus \mathbb{CP}(2) \cong \mathrm{SU}(3) \setminus S^5$. A priori, the latter identification would seem to suggest the possibility of using the techniques of Section 4 to find new harmonic $\mathrm{PSU}(3)$-structures, as studied by Hitchin [15]. In fact, as for the exceptional Wolf space, $\mathrm{SU}(3)$ has a 'hidden' $\mathrm{U}(1)$-action, which has a natural interpretation in terms of the fibres of the equivariant map

$$\mathrm{SU}(3) \ni P \mapsto P\overline{P},$$

which intertwines action by consimilarity and conjugation.

Computations show the following proposition:

Proposition 6.1 *There are no non-trivial* $\mathrm{SU}(3)$*-invariant harmonic nilpotent perturbations of the* $\mathrm{PSU}(3)$*-structure* (6.1) *on* $\mathrm{SU}(3)$.

Our studies are also related to G_2-holonomy metrics. The starting point is the quotient of the quaternionic projective plane by the circle action generated by the Killing vector field X. More specifically, one has the $\mathrm{SU}(3)$-equivariant map

$$\mathbb{HP}(2) \setminus \mathbb{CP}(2) \to S^7 \setminus \mathbb{CP}(2) \cong \Lambda^2_- \mathbb{CP}(2)$$

that appeared in [4] (see also [20]). It is well known that the negative spinor bundle over $\mathbb{CP}(2)$ admits a complete metric with holonomy G_2, the so-called Bryant–Salamon metric [6]. By building on work of [3, 13], the authors have succeeded in identifying the 3-form determining this G_2-structure in terms of X, the 4-form and other quaternionic data. A more complete study will appear in a forthcoming paper.

REFERENCES

[1] D. V. Alekseevskiĭ. Quaternion Riemann spaces with transitive reductive or solvable groups of motions. *Funct. Anal. Appl.*, 4(4):321–2, 1970.

[2] D. V. Alekseevskiĭ. Classification of quaternionic spaces with transitive solvable group of motions. *Izv. Akad. Nauk SSSR, Ser. Mat.*, 39(2):315–62, 1975.

[3] V. Apostolov and S. Salamon. Kähler reduction of metrics with holonomy G_2. *Commun. Math. Phys.*, 246(1): 43–61, 2004.

[4] M. Atiyah and E. Witten. M-theory dynamics on a manifold of G_2 holonomy. *Adv. Theor. Math. Phys.*, 6(1):1–106, 2002.

[5] R. L. Bryant. Closed Sp(2)Sp(1)-structures on 8-manifolds. 2005. unpublished.

[6] R. L. Bryant and S. Salamon. On the construction of some complete metrics with exceptional holonomy. *Duke Math. J.*, 58(3):829–50, 1989.

[7] D. Conti. Embedding into manifolds with torsion. *Math. Z.*, 268(3–4):725–51, 2011.

[8] D. Conti and T. B. Madsen. Harmonic structures and intrinsic torsion. *Transform. Groups*, 20(3):699–723, 2015.

[9] V. Cortés. Alekseevskian spaces. *Differ. Geom. Appl.*, 6(2):129–68, 1996.

[10] V. Cortés, M. Dyckmanns and S. Suhr. 'Special metrics and group actions in geometry' Springer INdAM Series vol. 23 (2017), 81–106.

[11] W. Fulton and J. Harris. *Representation theory*, volume 129 of *Graduate Texts in Mathematics*. Springer-Verlag, New York, 1991.

[12] A. Gambioli. SU(3)-manifolds of cohomogeneity one. *Ann. Global Anal. Geom.*, 34(1):77–100, 2008.

[13] A. Gambioli, Y. Nagatomo and S. Salamon. Special geometries associated to quaternion-kähler 8-manifolds. *J. Geom. Phys.*, 91:146–62, 2015.

[14] D. Giovannini. *Special structures and symplectic geometry.* PhD thesis, Università di Torino, 2003.

[15] N. Hitchin. Stable forms and special metrics. In A. Gray, M. Fernández and J. A. Wolf (eds) *Global differential geometry: the mathematical legacy of Alfred Gray (Bilbao, 2000)*, volume 288 of *Contemp. Math.*, pages 70–89. Amer. Math. Soc., Providence, RI, 2001.

[16] C. A. Hoelscher. Classification of cohomogeneity one manifolds in low dimensions. *Pac. J. Math.*, 246(1):129–85, 2010.

[17] R. A. Horn and C. R. Johnson. *Matrix Algebra*. Cambridge Univ. Press, Cambridge, 1985.

[18] P. Kobak and A. F. Swann. Quaternionic geometry of a nilpotent variety. *Math. Ann*, 297(4):747–64, 1993.

[19] S. Kuroki. On 8-manifolds with SU(3)-actions. *RIMS Kokyuroku*, 1569:81–93, 2007.

[20] R. Miyaoka. Topology of the Bryant-Salamon G_2-manifolds and some Ricci flat manifolds. *RIMS Kokyuroku* 1502:230–37, 2006.

[21] S. Salamon. Almost parallel structures. In A. Gray, M. Fernández and J. A. Wolf (eds) *Global differential geometry: the mathematical legacy of Alfred Gray (Bilbao, 2000)*, volume 288 of *Contemp. Math.*, pages 162–81. Amer. Math. Soc., Providence, RI, 2001.

[22] A. F. Swann. Aspects symplectiques de la géométrie quaternionique. *C. R. Acad. Sci. Paris Sér. I Math.*, 308(7):225–8, 1989.

[23] J. A. Wolf. Complex homogeneous contact manifolds and quaternionic symmetric spaces. *J. Math. Mech.*, 14(6):1033–47, 1965.

6

Boundary Value Problems in Dimensions 7, 4 and 3 Related to Exceptional Holonomy

SIMON DONALDSON

19 July, 2017

Dedicated to Nigel Hitchin, for his seventieth birthday.

The variational point of view on exceptional structures in dimensions 6, 7 and 8 is one of Nigel Hitchin's seminal contributions. One feature of this point of view is that it motivates the study of boundary value problems, for structures with prescribed data on a boundary. In this article, we consider the case of seven dimensions and 'G_2-manifolds'. We will review briefly a general framework and then go on to examine in more detail symmetry reductions to dimensions 4 (in Section 2) and 3 (in Section 3). In the latter case, we encounter an interesting variational problem related to the real Monge–Ampère equation and, in Section 6.4. we describe a generalization of this.

The author is grateful to Claude Le Brun and Lorenzo Foscolo for valuable comments and discussions.

1. The Volume Functional in Seven Dimensions

Let V be a seven-dimensional oriented real vector space. A 3-form $\phi \in \Lambda^3 V^*$ defines a quadratic form on V with values in the real line $\Lambda^7 V^*$ by the assignment

$$v \mapsto (i_v \phi)^2 \wedge \phi.$$

The fixed orientation means that it makes sense to say that this form is positive definite; in that case, we call ϕ a *positive* 3-form. From the definition, a positive 3-form defines a conformal class of Euclidean structures on V, and the ambiguity of scale can be fixed by

Donaldson, S., *Boundary Value Problems in Dimensions 7, 4 and 3 Related to Exceptional Holonomy*. In: *Geometry and Physics: A Festschrift in Honour of Nigel Hitchin*, Jørgen Ellegaard Andersen, Andrew Dancer, Oscar García-Prada (Eds): Oxford University Press (2018). © Oxford University Press. DOI: 10.1093/oso/9780198802006.003.0006

choosing the Euclidean structure so that $|\phi|^2 = 7$, using the standard induced metric on $\Lambda^3 V^*$.

Now let M be an oriented 7-manifold. Applying the above in each tangent space, we have the notion of a positive 3-form $\phi \in \Omega^3(M)$; such a form defines a metric g_ϕ and volume form $\nu_\phi \in \Omega^7(M)$. The first variation of the volume form with respect to a variation $\delta\phi$ in ϕ is given by

$$\delta\nu_\phi = \delta\phi \wedge \Psi$$

where $\Psi = \Psi(\phi)$ is a 4-form determined by ϕ, which can also be expressed as

$$3\Psi(\phi) = *_{g_\phi}\phi.$$

Suppose that M is a closed 7-manifold and that $c \in H^3(M, \mathbf{R})$ is a cohomology class which can be represented by positive 3-forms, so we have a non-empty set $\mathcal{S}_c \subset \Omega^3(M)$ of closed positive forms representing c. Hitchin's idea ([5], [6]) is to consider the total volume

$$\mathrm{Vol}(\phi) = \int_M \nu_\phi \tag{1}$$

as a functional on $\mathcal{S}_c$. The first variation, with respect to a variation $\delta\phi = da$, in ϕ is

$$\int_M \Psi \wedge da = -\int_M d\Psi \wedge a \tag{2}$$

so the Euler–Lagrange equation defining critical points is

$$d\Psi = 0.$$

By a well-known result of Fernández and Gray, the two equations $d\phi = 0, d *_{g_\phi} \phi = 0$ imply that the G_2 structure defined by ϕ is torsion-free or, equivalently, that the metric g_ϕ has holonomy contained in G_2. So, from this point of view, the search for these special structures can be divided into two stages:

- Identify manifolds M and classes $c \in H^3(M)$ such that $\mathcal{S}_c$ is non-empty;
- Study the variational problem for the volume functional on $\mathcal{S}_c$.

The local theory of such critical points, with respect to small variations in c and ϕ, is well understood. Hitchin proved that any critical point is a *local maximum* and, in fact, a strict local maximum modulo diffeomorphisms. The proof is an application of Hodge theory. One of the many interesting and fundamental questions in this area is whether it is a *global maximum* over the whole space $\mathcal{S}_c$. Another standard fact (proved earlier by Bryant), is that critical points are stable with respect to variations in the cohomology class c: a critical point ϕ belonging to a class c can be deformed to a critical point for nearby classes in $H^3(M)$. That is, the moduli space of G_2-structures is locally modelled on $H^3(M)$.

Now we introduce our boundary value problem. Let M be a compact oriented 7-manifold with boundary an oriented 6-manifold $N = \partial M$. There is a similar notion of a positive 3-form ρ on N: this is just the condition that at each point $p \in N$ the form ρ can be extended to a positive form on TM_p. A basic algebraic fact is that such a positive 3-form in six dimensions is equivalent to a reduction to $SL(3, \mathbf{C})$, that is, to an almost-complex structure with a trivialization of the 'canonical line bundle'. Fix a closed positive 3-form ρ on N. We assume that the class $[\rho] \in H^3(N)$ is in the image of the restriction map from $H^3(M)$. Define an *enhancement* of ρ to be an equivalence class of closed 3-forms on M extending ρ, under the equivalence relation $\phi_1 \sim \phi_1 + da$, where a vanishes on N. Thus, the set of enhancements of ρ is an affine space modelled on $H^3(M,N)$. Fix an enhancement $\hat{\rho}$ of ρ and let $\mathcal{S}_{\hat{\rho}}$ be the set of positive forms on M in this equivalence class. Suppose that $\mathcal{S}_{\hat{\rho}}$ is non-empty and consider the volume functional on this set, just as before. For a variation $\delta\phi = da$ with a vanishing on the boundary, the integration by parts (2) is still valid and the critical points are given by solutions of $d\Psi = 0$, just as before. That is, we are studying G_2-structures on M with the given boundary value ρ and in the given enhancement class. So, we have the same two stages as before: identify enhanced boundary values $\hat{\rho}$ such that $\mathcal{S}_{\hat{\rho}}$ is non-empty and then study the variational problem.

We will not enter into a proper discussion of the local theory of this boundary value problem (with respect to small variations in ϕ and) here, but we make two simple observations. For the first, we say that a G_2-structure ϕ on a manifold M with boundary N is a *formal maximum* of the volume functional if, for any $a \in \Omega^2(M)$ whose restriction to the boundary vanishes, we have

$$\frac{d^2}{dt^2}\mathrm{Vol}(\phi + tda) \leq 0$$

at $t = 0$. In other words, the Hessian of the volume functional is non-negative.

Proposition 1 *Suppose that M is the closure of a domain in a closed G_2-manifold M^+ with 3-form ϕ. Let $\hat{\rho}$ be the enhanced boundary value given by the restriction of ϕ to $N = \partial M$ and M. Then, $\phi|_M$ is a formal maximum of the volume functional on $\mathcal{S}_{\hat{\rho}}$.*

Let f be a defining function for $\partial M = f^{-1}(0)$, positive on the interior of M. Then, any 2-form a on M whose restriction to ∂M is zero can be written as $a = b + \eta \wedge df$, where b vanishes in $TM|_{\partial M}$. For small ϵ, let $\chi_\epsilon : M \to \mathbf{R}$ be the composite of f with a standard cut-off function, such that χ_ϵ vanishes when $f \leq \epsilon$, is equal to 1 when $f \geq \epsilon$ and with $|d\chi_\epsilon| \leq C\epsilon^{-1}$. Set $a_\epsilon = \chi_\epsilon a$. Then,

$$d(a_\epsilon) = d\chi_\epsilon \wedge b$$

satisfies a uniform L^∞ bound, independent of ϵ (since b is $O(\epsilon)$ on the support of $d\chi_\epsilon$). It follows that

$$\frac{d^2}{dt^2}\mathrm{Vol}(\phi + tda) = \lim_{\epsilon \to 0} \frac{d^2}{dt^2}\mathrm{Vol}(\phi + tda_\epsilon),$$

and the latter is non-positive since a_ϵ can be extended by zero over the closed manifold M^+ and then we can apply Hitchin's result.

In the other direction, critical points are not always *strict* local maxima, modulo diffeomorphisms. To give an example of this, we define for $v \in \mathbf{R}^7$ with $|v| < 1/2$ the manifold-with-boundary $M_v \subset \mathbf{R}^7$ to be

$$M_v = \overline{B}^7 \setminus \left(v + \frac{1}{2} B^7 \right),$$

where B^7 is the open unit ball. Let ϕ_0 be the standard flat G_2 structure on $\mathbf{R}^7$ and let ρ_v be its restriction to the boundary of M_v. In this case, $H^3(M_v, \partial M_v) = 0$, so there is no extra enhancement data. We can choose diffeomorphisms $F_v : M_0 \to M_v$ such that the restriction to the boundaries pulls back ρ_v to ρ_0. Then, $F_v^*(\phi_0)$ are critical points for the boundary value problem on M_0 which are not all equivalent, by diffeomorphisms of M_0, to ϕ_0.

2. Reduction to Dimension 4

In this section, we consider an interesting reduction of the seven-dimensional theory to four dimensions, as follows. Take $M = X \times \mathbf{R}^3$, where X is an oriented 4-manifold, and consider 3-forms of the shape

$$\phi = \sum_{i=1}^{3} \omega^i d\theta_i - d\theta_1 d\theta_2 d\theta_3, \tag{3}$$

where θ_i are coordinates on $\mathbf{R}^3$, and ω^i are 2-forms on X. The condition that ϕ is a positive 3-form goes over to the condition that (ω^i) form a 'positive triple', by which we mean that at each point they span a maximal positive subspace for the wedge product form on $\Lambda^2 T^*X$. More invariantly, we are considering positive forms ϕ which are preserved by the translation action of the $\mathbf{R}^3$ factor and such that the orbits are 'associative' submanifolds. The condition that ϕ be closed goes over to the condition that the ω^i are closed 2-forms on X, making up a 'hypersympletic' structure. These structures are of considerable interest in four-dimensional geometry (see e.g. [3], [4]).

Given such a triple $\underline{\omega} = (\omega^i)$, we define a volume form χ on X by the following procedure. Let χ_0 be any volume form and define a matrix (q^{ij}) by

$$\chi_0 q^{ij} = \omega^i \wedge \omega^j.$$

Now put

$$\chi = \det(q^{ij})^{1/3} \chi_0.$$

It is clear that this is independent of the choice of χ_0. The seven-dimensional volume form associated to ϕ is $-\chi\, d\theta_1\, d\theta_2\, d\theta_3$. All our constructions will be invariant under the action of $SL(3, \mathbf{R})$ on $\mathbf{R}^3$ so it will sometimes be clearer to introduce a three-dimensional oriented vector space W with fixed volume element and consider our data $\underline{\omega}$ as an element of $W \otimes \Omega^2(X)$. Then, a choice of coordinate system on W gives the description as a triple $(\omega^1, \omega^2, \omega^3)$.

Given a positive triple ω^i, we define a matrix (λ^{ij}) by

$$\omega^i \wedge \omega^j = \lambda^{ij} \chi. \tag{4}$$

Thus, $\det(\lambda^{ij}) = 1$, by the definition of χ. Write (λ_{ij}) for the inverse matrix and set

$$\Theta_i = \sum_{j=1}^{3} \lambda_{ij} \omega^j. \tag{5}$$

The 4-form defined by ϕ is

$$\Psi = \sum_{\text{cyclic}} \Theta_i d\theta^j d\theta^k + \chi,$$

where the notation means that (ijk) runs over the three cyclic permutations of (123). Thus, the condition that a closed triple (ω^i) defines a G_2 structure is $d\Theta_i = 0$, which is to say:

$$\sum_{j=1}^{3} d\lambda_{ij} \wedge \omega^j = 0. \tag{6}$$

These equations are obviously satisfied when the matrix (λ^{ij}) is constant on X and these solutions are the *hyperkähler metrics*. Of course, we can produce these equations (8) from a four-dimensional reduction of Hitchin's variation formulation: the equations are the Euler–Lagrange equation for the functional

$$\text{Vol}(\underline{\omega}) = \int_X \chi \tag{7}$$

on closed positive triples $\underline{\omega}$, with respect to exact variations of compact support.

It is well known, and easy to show directly, that the only solutions of the equations (6) on a compact 4-manifold are hyperkähler and this gives extra motivation for considering the boundary value problem. So let X be a 4-manifold with boundary Y and consider triples $\underline{\mu} = (\mu^1, \mu^2, \mu^3)$ of closed 2-forms on Y which form a basis for $\Lambda^2 T^* Y$ at each point. In our more invariant set-up, $\underline{\mu}$ lies in $W \otimes \Omega^2(Y)$. We define an enhancement $\underline{\mu}$ in the obvious way, so the space of enhancements of a given $\underline{\mu}$ is an affine space modelled on $W \otimes H^2(X, Y)$. Fix an enhancement $\underline{\hat{\mu}}$ and let $\mathcal{S}_{\underline{\hat{\mu}}}$ be the set of closed positive triples on X in the given equivalence class. So, the reduced versions of our stages are, first, determining whether this set is non-empty and, second, studying the variational problem given by the volume functional (7).

Stokes' theorem implies that the integrals

$$Q^{ij} = \int_X \omega^i \wedge \omega^j$$

are independent of the choice of ω^i in a fixed enhancement class $\underline{\hat{\mu}}$. More invariantly, Q is a quadratic form on our vector space W, and $\det Q$ is defined, as a real number, using the fixed volume form on W. This has two simple consequences.

Proposition 2 *If $\mathcal{S}_{\underline{\hat{\mu}}}$ is non-empty, then Q is positive definite and there is an upper bound*

$$\int_X \chi \leq \det Q,$$

for $\underline{\omega} \in \mathcal{S}_{\underline{\hat{\mu}}}$ and $\chi = \chi(\underline{\omega})$. Equality holds if and only if $\underline{\omega}$ is hyperkähler.

To see that Q is positive definite, it suffices, by change of basis, to see that $Q^{11} > 0$. But this clear since $\omega^1 \wedge \omega^1$ is positive pointwise on X. To establish the upper bound, it suffices, by change of basis, to consider the case when $Q^{ij} = \delta^{ij}$. Recall that we write $\omega^i \wedge \omega^j = \lambda^{ij}\chi$, where $\det(\lambda) = 1$. Then, we have the elementary inequality (the arithmetic-geometric mean inequality for the eigenvalues) $\mathrm{Tr}(\lambda) \geq 3$. So

$$3\int_X \chi \leq \int_X \mathrm{Tr}\lambda\, \chi = \sum_i \int_X (\omega^i)^2 = 3.$$

Equality holds if and only if $\lambda^{ij} = \delta^{ij}$, which means that $\underline{\omega}$ is hyperkähler.

The first statement in the proposition gives a potential obstruction to finding a positive triple with the gives enhanced boundary data. Consider, for example, the case when $Y = S^3$, and X is the 4-ball. There is a well-known quadratic 'Chern–Simons' form Q_{CS} on the closed 2-forms on S^3 defined by

$$Q_{CS}(\mu) = \int_{S^3} a \wedge \mu,$$

where a is any 1-form with $da = \mu$. The necessary condition on our boundary data in this case is that μ^i span a three-dimensional positive subspace with respect to this form Q_{CS}.

3. Reduction to Dimension 3

We specialize further, mimicking the Gibbons–Hawking construction of hyperkähler 4-manifolds. Thus, we suppose that the 4-manifold X is the total space of a principal S^1-bundle over a 3-manifold U, with the action generated by a vector field ξ, and consider closed positive triples $\underline{\omega}$ which are invariant under the action. We assume that the action is Hamiltonian for each symplectic structure ω^i, so we have Hamiltonian functions $h^i : X \to \mathbf{R}$ with

$$dh^i = i_\xi \omega^i,$$

and these functions are fixed by the circle action and so descend to U. More invariantly, writing $\underline{\omega} \in \Omega^2(X) \otimes W$, we have $i_\xi \underline{\omega} \in \Omega^1(X) \otimes W$ and this is the derivative of a map

$\underline{h} : X \to W$. The functions h^i are then the components of $\underline{h}$ with respect to a coordinate system $W = \mathbf{R}^3$. The definitions imply that $\underline{h}$ induces a local diffeomorphism from U to W, so for local calculations we can suppose that the base U is a domain in W and the functions h^i can be identified with standard coordinates x^i on W. One finds that the general form of such a triple is given by

$$\omega^i = \alpha \wedge dx^i + \sum_{\text{cyclic}} \sigma^{ij} dx^k dx^l, \tag{8}$$

where (jkl) run over cyclic permutations, $\sigma = (\sigma^{ij})$ is a symmetric and positive definite matrix (a function of the coordinates x^i) and α is a connection 1-form on X. The condition that σ is symmetric is the same as saying that the connection is the obvious one defined by the metric induced by ω^i, with horizontal subspaces the orthogonal complement of ξ.

We will now investigate the reduced G_2-equations in this context. Write F for the curvature of the connection, so $F = d\alpha$ and can be regarded as a 2-form on U. We write

$$F = \sum_{\text{cyclic}} F^i dx^j dx^k.$$

Now, writing ∂_j for partial derivatives,

$$d\omega^i = (F^i + \partial_j \sigma^{ij}) dx^1 dx^2 dx^3,$$

so the condition that we have a closed triple is that

$$F^i = -\partial_j \sigma^{ij}.$$

Now $dF = 0$, which is to say

$$\sum \partial_i F^i = 0,$$

and, at least locally, any closed 2-form specifies a connection, up to gauge equivalence. So, locally, we can eliminate the connection and curvature and closed triples correspond to matrix-valued functions σ^{ij} with

$$\sum_{ij} \partial_i \partial_j \sigma^{ij} = 0. \tag{9}$$

The volume form defined by the triple (8) is

$$\chi = \det(\sigma)^{1/3} \left(\alpha dx^1 dx^2 dx^3\right).$$

This implies that

$$\lambda_{ij} = \sigma_{ij} (\det \sigma)^{1/3},$$

where σ_{ij} denotes the matrix inverse to σ^{ij} as usual. Now the equation (6) is

$$\sum_k \partial_k \lambda_{ij} dx^k \left(\alpha dx^j + \sum_{p,q,r \text{cyclic}} \sigma^{jp} dx^q dx^r \right) = 0.$$

Expanding this out, we get two conditions:

1. $\partial_k \lambda_{ij} = \partial_j \lambda_{ik}$;
2. $\sum_{jk} (\partial_k \lambda_{ij}) \sigma^{jk} = 0$.

The first condition asserts, at least locally, that λ_{ij} is the Hessian of a function, u, say:

$$\lambda_{ij} = \partial_i \partial_j u.$$

The second condition is implied by the first since

$$\partial_k \lambda_{ij} \sigma^{jk} = -\partial_i \det(\lambda),$$

and the determinant of λ is 1 by construction. To sum up, the three-dimensional reduction of the G_2 equations can be written locally as a pair of equations for two functions u, V on a domain in $\mathbf{R}^3$. First, the Monge–Ampère equation

$$\det(\partial_i \partial_j u) = 1 \tag{10}$$

and, second,

$$\sum \partial_i \partial_j (V u^{ij}) = 0, \tag{11}$$

where u^{ij} is the inverse of the Hessian $u_{ij} = \partial_i \partial_j u$. Given a pair u, V satisfying these equations, we set $\sigma^{ij} = V u^{ij}$ and the discussion above shows that all solutions arise in this way (locally). Notice that, given u, the second equation is a linear equation for V and, in fact, is familiar as the linearization of the Monge–Ampère equation at u. Recall that the linearized operator $\Box_u$ can be written in three different ways,

$$\Box_u f = \sum \partial_i \partial_j (f u^{ij}) = \sum \partial_i (\partial_j f\, u^{ij}) = \sum (\partial_i \partial_j f) u^{ij},$$

using the identity $\sum \partial_i u^{ij} = 0$.

Now we want to set up our boundary value problem in this context. We suppose that U is a 3-manifold with boundary Σ and the circle bundle extends to the boundary, so that $Y = \partial X$ is a circle bundle over Σ. (The extension of the circle bundle over U means that it must be a trivial bundle, but we do not have a canonical trivialization.) We want to consider triples μ^i of closed 2-forms on Y, as before, invariant under the circle action and such that the action is 'Hamiltonian', i.e. there are circle-invariant functions h^i_Y on Y with

$$dh^i_Y = i_\xi \mu^i,$$

These functions give a map $\underline{h}_\Sigma : \Sigma \to W = \mathbf{R}^3$ and it follows from the definitions that this is an immersion. Now, we encounter a potential obstruction of a differential topological nature to the existence of an invariant closed positive triple on X with these boundary values: the immersion $\underline{h}_\Sigma$ must extend to an immersion of U in $\mathbf{R}^3$. But let us suppose here for simplicity that $\underline{h}_\Sigma$ is an embedding of Σ as the boundary of a domain in $\mathbf{R}^3$. Then, for any extension of the ρ^i over X, of the kind considered above, the map $\underline{h}$ must be a diffeomorphism from U to this domain. Thus, we can simplify our notation by taking U to be a domain in $\mathbf{R}^3$ with smooth boundary Σ. To avoid complication, we suppose that U is simply connected, so that Σ is diffeomorphic to a 2-sphere. Thus, our PDE problem is to solve the equations (10) and (11) for functions u, V on $U \subset \mathbf{R}^3$ and the remaining task is to identify the boundary conditions on $\Sigma = \partial U$ defined by a triple μ^i. (The assumption that U is simply connected means that the above local analysis of solutions applies globally on U.)

The differential geometric analysis of invariant triples $\underline{\mu} = (\mu^i)$ is complicated by the fact that there is no natural connection on the circle bundle $\pi : Y \to \Sigma$. But the analysis has a simple conclusion which can be expressed in terms of certain distributions, or currents, which we call *layer currents*. In this analysis, it will be important to keep track of the full $SL(3,\mathbf{R})$-invariance of the set-up, so we work in the three-dimensional vector space W with fixed volume element. So, we have an embedding of Σ in W as the boundary of a domain U, and a triple $\underline{\mu}$ is a section of $W \otimes \Lambda^2 T^*Y$.

We define a layer current supported on Σ to be a linear map from functions on W to $\mathbf{R}$ of the form

$$\mathcal{L}_{\theta_1,\theta_2,v}(f) = \int_\Sigma (\nabla_v f)\,\theta_1 + f\,\theta_2, \tag{12}$$

where θ_1, θ_2 are 2-forms on Σ with $\theta_1 > 0$, and v is an outward-pointing normal vector field along Σ—a section of the tangent bundle of W restricted to Σ and which is complementary to the tangent bundle of Σ. Of course, this depends only on the restriction of f to the first formal neighbourhood of Σ; in particular, it is defined for a function f on U which is smooth up to the boundary.

The point is that the same functional $\mathcal{L}$ can be defined by different data (θ_1, θ_2, v). First, it is obvious that, for any positive function g on Σ, the data $(g\theta_1, \theta_2, g^{-1}v)$ define the same current. Second, if w is a tangential vector field on Σ, we have

$$\int_\Sigma (\nabla_w f)\theta_1 = \int_\Sigma \tilde{\theta}_2 f, \tag{13}$$

where $\tilde{\theta}_2 = -d(i_w\theta_1)$. It follows that a given layer current $\mathcal{L}$ of this kind can be represented using *any* normal vector field, for appropriate θ_1, θ_2. Let $\nu_\Sigma = TW/T\Sigma$ be the normal bundle of Σ in W. The fixed volume element on W gives an isomorphism

$$\nu_\Sigma = \Lambda^2 T^* \Sigma. \tag{14}$$

Let $[v]$ denote the image of v in ν_Σ. The product

$$H_{\mathcal{L}} = [v].\mu_2 \in \left(\Lambda^2 T^*\Sigma\right)^{\otimes 2}$$

is independent of the choice of data (v, μ_1, μ_2) used to represent $\mathcal{L}$; we call $H_{\mathcal{L}}$ the *primary invariant* of $\mathcal{L}$. For a function f which vanishes on Σ, the derivative df along Σ is defined as a section of ν_Σ^* and, for such functions, we have

$$\mathcal{L}(f) = \int_\Sigma H.df, \tag{15}$$

where we use the isomorphism $\nu_\Sigma^* = \Lambda^2 T^\Sigma$, and the pairing with H yields a 2-form $H.df$ on Σ.

Now let $\underline{\mu} \in W \otimes \Omega^2(Y)$ be a closed S^1-invariant triple on the circle bundle Y over Σ such that the inclusion $\Sigma \to W$ is the Hamiltonian map for the action. Let y be a point of Y and $\epsilon \in W^*$ be a conormal to Σ at $x = \pi(y)$, i.e. an element of W^* vanishing on $(T\Sigma)_x \subset W$. Then we have a map

$$\epsilon : (\Lambda^2 T^*Y)_y \otimes W \to (\Lambda^2 T^*Y)_y,$$

and it follows from the definitions that $\epsilon(\underline{\mu})$ lies in the image of the pullback map $\pi^* : (\Lambda^2 T^*\Sigma)_x \to (\Lambda^2 T^*Y)_y$. Thus, we have a unique element $h \in (\Lambda^2 T^*\Sigma)_x$ with $\pi^*(h) = \epsilon(\underline{\rho})$. Multiplying ϵ by a factor κ clearly multiplies h by κ so, using again the isomorphism (14), we get a well-defined section $H^{\underline{\mu}}$ of $(\Lambda^2 T^*\Sigma)^{\otimes 2}$, independent of the choice of ϵ. We call $H^{\underline{\mu}}$ the primary invariant of the triple $\underline{\mu}$.

Next choose a normal vector field v along Σ. At a point $y \in Y$, we transpose $\underline{\mu}(y)$ to give the map

$$\underline{\tilde{\mu}} : W^* \to (\Lambda^2 T^*Y)_y = TY \otimes \Lambda^3 T^*Y.$$

The annihilator of $v(\pi(y))$ is a two-dimensional subspace of W^* and it follows from the definitions that the image of this subspace under $\underline{\tilde{\mu}}$ defines a two-dimensional subspace of TY transverse to the S^1-orbit. In other words, the choice of normal vector field v defines a connection on the S^1 bundle $\pi : Y \to \Sigma$: in fact, giving a connection is equivalent to giving a complementary bundle to $T\Sigma \subset W$. Let Φ be the curvature of this connection, a 2-form on Σ, and define a current $\mathcal{L}^{\underline{\mu},v}$ by

$$\mathcal{L}^{\underline{\mu},v}(f) = \int_\Sigma (H^{\underline{\mu}}.[v]^{-1})\nabla_v f + \Phi f. \tag{16}$$

Here, $H_{\underline{\mu}}.[v]^{-1}$ is the 2-form given by the pairing of $[v]^{-1} \in \nu^* = (\Lambda^2 T\Sigma)^{-1}$ with $H^{\underline{\mu}} \in (\Lambda^2 T^*\Sigma)^{\otimes 2}$.

Proposition 3 *The layer current $\mathcal{L}^{\underline{\mu},v}$ is independent of the choice of normal vector field v so can be written as $\mathcal{L}^{\underline{\mu}}$. Two triples $\underline{\mu},\underline{\mu}'$ are equivalent by S^1-equivariant diffeomorphisms if and only if $\mathcal{L}^{\underline{\mu}} = \mathcal{L}^{\underline{\mu}'}$.*

If we change v by multiplication by a positive function, then we do not change the connection and hence we do not change the integral of Φf. The other term in the integrand is also unchanged because the scalings of $[v]^{-1}$ and ∇_v cancel. So, to prove the first statement, it suffices to consider changing v to $v+w$, where w is a tangential vector field on Σ. Using the formula (13), we have to show that the connection changes by the addition of the 1-form $a = i_w\theta_1$. To see this, we work in coordinates at a given point on Σ, taking $v = \partial_1$ and the tangent space of Σ spanned by ∂_2, ∂_3. Write $\theta_1 = Gdx^2dx^3$ at the given point. If α is the connection 1-form on Y defined by v, then it follows from the definitions that, over this point,

$$\mu^1 = Gdx^2dx^3\,, \mu^2 = \alpha \wedge dx^2\,,\; \mu^3 = \alpha \wedge dx^3.$$

If $w = w^2\partial_2 + w_3\partial_3$ at this point, the annihilator of $v+w$ in W^* is spanned by $dx^2 - w^2dx^1, dx^3 - w^3dx^1$ and this maps to the two-dimensional subspace in Λ^2T^*Y spanned by

$$\alpha \wedge dx^2 - w^2Gdx^2dx^3,\; \alpha \wedge dx^3 - w^3Gdx^2dx^3,$$

which corresponds to the two-dimensional subspace in TY spanned by

$$\partial_3 - Gw^2\xi,\; -\partial_2 - Gw^3\xi.$$

This is the annihilator of the 1-form $\alpha + a$ where $a = Gw^2dx^3 - Gw^3dx^2$, which is the contraction $i_w\theta_1$, as required.

The second statement of the proposition follows easily from the fact that, since Σ is simply connected, a connection is determined up to gauge equivalence by its curvature.

So far, we have considered our structures over the surface $\Sigma \subset W$. Now let σ be a matrix-valued function over $U \subset W$, as before, defining a triple $\underline{\omega}$ on X. Then, for any smooth function f on U, we have

$$\int_U \sum \sigma^{ij}\partial_i\partial_j f - \sum (\partial_i\partial_j\sigma^{ij})f = \mathcal{L}_\sigma(f), \tag{17}$$

where $\mathcal{L}_\sigma$ is the layer current supported on Σ defined by

$$\mathcal{L}_\sigma f = \int_\Sigma \sum \sigma^{ij}\partial_j f - \Big(\sum \partial_i\sigma^{ij}\Big)f. \tag{18}$$

(To clarify notation: in (17) we suppress the volume form on W which defines our measure and in (18) the integrand is written as a vector field, which defines a 2-form on Σ by contraction with the three-dimensional volume form as in (14).) Then we have the following proposition:

Proposition 4 *The boundary value of the triple $\underline{\omega}$ corresponding to σ is equivalent to the triple $\underline{\mu}$ on Y if and only if $\mathcal{L}_\sigma = \mathcal{L}^{\underline{\mu}}$.*

To see this, regard the inverse matrix σ_{ij} as a Riemannian metric on U. The orthogonal complement with respect to this metric defines a normal vector field ν_σ over Σ and hence a connection on $Y \to \Sigma$. We know that σ defines a connection on the circle bundle $X \to U$. with curvature given by $F^i = -\partial_j \sigma^{ij}$. The proposition amounts to the fact that the restriction of this connection to $Y \to \Sigma$ is the same as the connection defined by ν_σ, which we leave for the reader to check.

To illustrate the nature of this boundary condition, consider an example where Σ is locally given by the plane $x^1 = 0$ and take ∂_1 as normal vector field. Then $\mathcal{L}^{\underline{\mu}}$ is locally represented by 2-forms

$$\theta_1 = G_1 dx^2 dx^3, \theta_2 = G_2 dx^2 dx^3,$$

where G_i are functions of x^2, x^3. That is, for functions f supported in this region,

$$\mathcal{L}(f) = \int_{x^1=0} \left(G_1 \frac{\partial f}{\partial x^1} + G_2 f \right) dx^2 dx^3.$$

Now, if σ is defined over U, we have, for such functions f,

$$\mathcal{L}_\sigma(f) = \int_{x^1=0} \left(\sigma^{11} \frac{\partial f}{\partial x^1} + \left(\sigma^{12} \frac{\partial f}{\partial x^2} + \sigma^{13} \frac{\partial f}{\partial x^3} \right) - (\partial_i \sigma^{1i}) f) \right) dx^2 dx^3.$$

Integrating by parts, the sum of the second and third terms is

$$- \int_{x^1=0} f(\partial_1 \sigma^{11} + 2\partial_2 \sigma^{12} + 2\partial_3 \sigma^{13})\, dx^2 dx^3.$$

Our boundary conditions are:

- $\sigma^{11} = G_1$;
- $\partial_1 \sigma^{11} + 2\partial_2 \sigma^{12} + 2\partial_3 \sigma^{13} = -G_2$.

Notice that, if f is an affine-linear function, then $\mathcal{L}_\sigma(f)$ vanishes for any σ on U. This is connected to the following identities on the boundary:

- For a circle bundle $Y \to \Sigma$ with Chern class d and any invariant triple $\underline{\mu}$ on Y, the value of functional $\mathcal{L}^{\underline{\mu}}(1) = 2\pi d$ (here, 1 denotes the constant function).
- Suppose $d = 0$, so Y is diffeomorphic to $S^1 \times \Sigma = S^1 \times S^2$ and there is a lift $[\Sigma] \in H_2(Y)$. Then, for any invariant triple $\underline{\rho}$,

$$\mathcal{L}^{\underline{\mu}}(x^i) = \int_{[\Sigma]} \mu^i.$$

Again, we leave the proofs as exercises for the interested reader.

Putting all this together, we can formulate the dimensionally reduced version of our general boundary value problem as follows. The functional (9) clearly reduces to the functional

$$\mathrm{Vol}(\sigma) = \int_U (\det\sigma)^{1/3}. \tag{19}$$

Variational Problem I

Given a (simply connected) domain $U \subset \mathbf{R}^3$ with smooth boundary Σ and a layer current $\mathcal{L}$ on Σ, find the critical points of the volume functional (19) over all $\sigma = (\sigma^{ij})$ on U satisfying:

(A) $\sum \partial_i \partial_j \sigma^{ij} = 0$;

(B) $\mathcal{L}_\sigma = \mathcal{L}$.

Our first question is now whether the set $\mathcal{S}_\mathcal{L}$ of matrix-valued functions σ satisfying (A), (B) above is non-empty. The integral formula gives an immediate constraint on the boundary data: if $\mathcal{S}_\mathcal{L}$ is not empty, then $\mathcal{L}f \geq 0$ for all convex functions f on U (with equality if and only if f is affine linear). We also have

Proposition 5 *If a solution to the Variational Problem I exists, it is an absolute maximum of the volume functional on $\mathcal{S}_\mathcal{L}$.*

This follows immediately from the facts that both conditions (A) and (B) are linear in σ, and the function $(\det\sigma)^{1/3}$ is concave.

Let f be a convex function on U satisfying the Monge–Ampère equation $\det(f_{ij}) = 1$. Then, for any σ, we have, pointwise on U,

$$\det(\sigma)^{1/3} = \left(\det\sigma_{ij}\det(f_{ij})\right)^{1/3} \leq \frac{1}{3}\sum \sigma^{ij} f_{ij}. \tag{20}$$

So, if σ satisfies the conditions (A) and (B) of Variational Problem I, we have, integrating over U and using the definition of $\mathcal{L}_\sigma$,

$$\mathrm{Vol}(\sigma) \leq \frac{1}{3}\int_U \sum \sigma^{ij} f_{ij} = \frac{1}{3}\mathcal{L}_\sigma f. \tag{21}$$

(Our previous bound, in Proposition 2, arises by taking quadratic functions f.) These bounds, furnished by solutions of the Monge–Ampère equation, lead to a dual formulation of the variational problem, which incorporates the boundary conditions in a simple way. Write $MA(U)$ for the set of convex solutions of the Monge–Ampère equation on U, smooth up to the boundary.

Variational Problem II

Given a (simply connected) domain $U \subset \mathbf{R}^3$ with smooth boundary Σ and a layer current $\mathcal{L}$ on Σ, minimize $\mathcal{L}(f)$ over all $f \in MA(U)$.

Proposition 6 *Variational Problems I and II are equivalent in the sense that, for $u \in MA(U)$, we can find a positive function V such that $\sigma^{ij} = Vu^{ij}$ is a solution of Variational Problem I if and only u is a solution of Variational Problem II.*

In one direction, equality holds in (21) if and only if u_{ij} is a multiple of the inverse of σ^{ij}. We know that a solution to Variational Problem I has the form $\sigma_{ij} = Vu^{ij}$, where u satisfies the Monge-Ampére equation, so, taking $f = u$, equality holds in (21), and it follows that u minimizes $\mathcal{L}(f)$ over $MA(U)$.

In the other direction, suppose that $u \in MA(U)$ is an extremum of the functional $\mathcal{L}$ (it will follow from the discussion below that u is in fact a minimum and is unique up to the addition of an affine-linear function). The Euler–Lagrange equation is $\mathcal{L}(W) = 0$ for all solutions W of the linearized equation $\Box_u W$ over U. We can solve the Dirichlet problem for this linearized equation, to find V such that $\Box_u V = 0$ and so that, if $\sigma^{ij} = Vu^{ij}$, the primary invariant of $\mathcal{L}_\sigma$ is equal to that of $\mathcal{L}$. If we chose a conormal ν^* to Σ, this is just saying that $\sum Vu^{ij}\nu_i^*\nu_j^*$ is a prescribed function on Σ, which, for fixed u, is just prescribing V on Σ. Then, it follows from the previous discussion that $\mathcal{L}_\sigma(W) = 0$ for all solutions W of the linearized equation. Since $\mathcal{L}_\sigma$ and $\mathcal{L}$ have the same primary invariant, the difference can be written as

$$(\mathcal{L}_\sigma - \mathcal{L})(f) = \int_\Sigma \Theta f,$$

for a 2-form Θ on Σ. For any function f on Σ, we can solve the Dirichlet problem for $\Box_u$ with boundary value f and so

$$\int_\Sigma \Omega f = 0$$

for all f. This implies that $\Omega = 0$ so $\mathcal{L}_\sigma = \mathcal{L}$ and we have solved Variational Problem I.

Modifying our problem, we can obtain a decisive existence result. Rather than fixing the full boundary data $\underline{\mu}$, we just fix the primary invariant $H^{\underline{\mu}}$. Given a positive $H \in \Gamma(\Lambda^2 T^*\Sigma)^{\otimes 2}$, we write $\mathcal{C}_H$ for the set of σ over U satisfying $\sigma^{ij}_{,ij} = 0$ and with the primary invariant of $\mathcal{L}_\sigma$ equal to H. As above, in terms of a conormal ν^*, this amounts to prescribing $\nu_i^*\nu_j^*\sigma^{ij}$ on the boundary.

Proposition 7 *If U is strictly convex, there is a unique critical point of the volume functional on $\mathcal{C}_H$ and this is an absolute maximum.*

The uniqueness and the fact that a critical point is an absolute maximum follows from concavity, just as before. For the existence, we first solve (invoking [2]) the Dirichlet problem for the Monge–Ampère equation to get a function $u \in MA(U)$ with $u = 0$ on Σ. Now solve the Dirichlet problem for the linearized equation to find a function V with $\Box_u = 0$ in U and such that $Vu^{ij}\nu_i^*\nu_j^*$ is the prescribed function on the boundary and write $\sigma^{ij} = Vu^{ij}$. We claim that this σ is a critical point of the volume functional on $\mathcal{C}_H$. Let τ^{ij} be an infinitesimal variation within $\mathcal{C}_H$. In other words, $\sum \partial_i\partial_j\tau^{ij} = 0$ in U and on the boundary $\sum \tau^{ij}\nu_i^*\nu_j^* = 0$. Then the variation in the volume functional is

$$3\delta\mathrm{Vol} = \int_U \sum u_{ij}\tau^{ij} = \int_\Sigma \sum \tau_j^{ij} u - \sum \tau^{ij}\partial_j u. \tag{22}$$

The first term on the right-hand side of (22) vanishes, since u vanishes on Σ. In the second term, the derivative of u along Σ vanishes, so there is only a contribution from the normal derivative of u, and the integrand is a multiple of $\sum \tau^{ij}\nu_i^*\nu_j^*$, so this also vanishes.

4. Further Remarks

4.1. *Singularities*

It seems unlikely that Variational Problems I and II always have solutions, even given the constraints we have found. To see this, we consider the well-known singular solutions of the Monge–Ampère equation, going back to Pogerolov. With coordinates x^1, x^2, x^3, set $r = \sqrt{(x^1)^2 + (x^2)^2}$ and consider functions u of the form $u = f(x^1)r^{4/3}$. Then one finds that

$$\det(u_{ij}) = \frac{64}{27} f\left(\frac{ff''}{3} - (f')^2\right),$$

so we can find smooth functions f on an interval, say, $(-\epsilon, \epsilon)$, with $f'' > 0$ and such that u satisfies the Monge–Ampère equation. Fix such a function f and let η be the vector field

$$\eta = 2x^1 \frac{\partial}{\partial x^1} - x^2 \frac{\partial}{\partial x^2} - x^3 \frac{\partial}{\partial x^3}.$$

This vector field generates volume-preserving transformations, so $V = \nabla_\eta u$ satisfies the linearized equation $\square_u V = 0$ and, if we define $\sigma^{ij} = Vu^{ij}$, we get a singular solution of our reduced G_2 equations (provided that $V > 0$). Suppose that, near the origin, Σ is given by the coordinate plane $x^1 = 0$ and let our boundary data be given locally by the layer current

$$\int G_1 \partial f \partial x^1{}_1 + G_2 f dx^2 dx^3,$$

as above, for smooth functions $G_i(x^2, x^3)$. As we saw above, the boundary condition is given by

$$\sigma^{11} = G_1\,,\ \partial_1\sigma^{11} + 2\partial_2\sigma^{12} + 2\partial_3\sigma^{13} = -G_2.$$

One can compute that

$$V = (2x^1 f' - \frac{4}{3}f) r^{4/3},$$

$$\sigma^{11} = \frac{16f^2}{9}(2x^1 f' - \frac{4}{3}f),$$

$$\sum_i \partial_i \sigma^{1i} = \frac{16f^2}{9}(2x^1 f' - \frac{4}{3}f)' - \frac{16}{3}ff'.$$

Thus, $\sigma^{11}, \sum_i \partial_i \sigma^{1i}$ are smooth functions of x^1 so we get a singular solution of our boundary value problem with smooth boundary data G_1, G_2.

4.2. *Connection with the Apostolov–Salamon Construction*

In the discussion above, we passed from seven dimensions to three dimensions by first imposing translational symmetry in three variables to get down to four dimensions and then imposing a circle action to pass from four dimensions to three. We can achieve the same end by imposing the circle action first, to get a reduction to six dimensions, and then studying translation-invariant solutions. The material in this subsection was explained to the author by Lorenzo Foscolo.

G_2 structures on a 7-manifold M invariant under a free circle action have been studied by Apostolov and Salamon [1] and others. The quotient space N has an induced $SU(3)$ structure, that is to say, a 2-form ω and a complex 3-form Ω equivalent at each point to the standard structures on $\mathbf{C}^3$ (with complex 3-form $dz_1 dz_2 dz_3$). The G_2 structure on M can be written as

$$\phi = \alpha \wedge \omega + V^{3/4} \mathrm{Re}\, \Omega, \tag{23}$$

where we identify forms on N with their lifts to M and V is a smooth positive function on N. (In fact, $V^{-1/2}$ is the length of generator of the circle action in the metric g_ϕ.) The 1-form α is a connection form on the circle bundle $M \to N$. Now, one finds that

$$*_{g_\phi} \phi = -V^{1/4} \alpha \wedge \mathrm{Im}\, \Omega + \frac{V}{2}\omega^2.$$

Thus, if $F = d\alpha$ is the curvature of the connection, the conditions to be satisfied for a torsion-free G_2-structure are

$$d\omega = 0\,,\; F \wedge \omega + d(V^{3/4} \mathrm{Re}\, \Omega) = 0, \tag{24}$$

and

$$d(V^{1/4} \mathrm{Re}\Omega) = 0\,,\; dV \wedge \omega^2 = 2V^{1/4} F \wedge \mathrm{Im}\, \Omega. \tag{25}$$

Now let W be a three-dimensional real vector space as before, and set $N = U \times W^*$, where U is an open set in W. Take standard coordinates x^i on W and θ_i on W^* so there is a standard symplectic form

$$\omega = \sum dx^i \wedge d\theta_i.$$

Let u be a convex function on U and define complex 1-forms

$$\epsilon_a = d\theta_a + i \sum u_{ab} dx^b.$$

It is well known that these define a complex structure compatible with ω and with holomorphic 3-form $\epsilon_1\epsilon_2\epsilon_3$. If u satisfies the Monge–Ampère equation $\det u_{ij} = 1$, then this is a Calabi–Yau structure. If V is a positive function on U, we modify this by taking

$$\epsilon'_a = V^{-1/4} d\theta_a + iV^{1/4} \sum u_{ab} dx^b,$$

and

$$\Omega = \epsilon'_1 \epsilon'_2 \epsilon'_3.$$

This complex 3-form is also algebraically compatible with ω. One checks that, if u satisfies the Monge–Ampère equation and F is the 2-form

$$F = -\sum_{j,k,l \text{ cyclic}} (\partial_i) V u^{ij} dx^k dx^l,$$

then V, F, Ω, ω satisfy the equations (24),(25). The condition that F is closed, and so arises as the curvature of a connection, is the equation $\Box_u V = 0$.

4.3. A General Class of Equations and LeBrun's Construction

Our Variational Problem I, and its dual formulation, Variational Problem II, have natural extensions. We can clearly replace $\mathbf{R}^3$ by $\mathbf{R}^n$ but, more interestingly, we can consider a class of different functionals. Let W be an n-dimensional real vector space and write $s^2_+(W) \subset s^2(W)$ for the cone of positive definite quadratic forms on W. Let ν be a smooth positive concave function on $s^2_+(W)$ which is homogeneous of degree 1, so $\nu(k\sigma) = k\nu(\sigma)$. This means that ν is not strictly concave but we suppose that the kernel of the Hessian of ν has dimension 1 everywhere, which means that $\log \nu$ is strictly concave. The basic example, which we have discussed in the case $n = 3$, is to take $\nu(\sigma) = (\det \sigma)^{1/n}$. In the general case, we consider functions σ on a simply connected domain $U \subset W$ taking values in $s^2_+(W)$, which we express in terms of a basis as $\sigma = (\sigma^{ij})$. We consider the functional

$$I = \int \nu(\sigma)$$

subject to the constraint $\sum \partial_i \partial_j \sigma^{ij} = 0$. Initially, we consider variations of this integral with respect to compactly supported variations of σ, and later we consider boundary conditions.

We regard $D\nu$ as a map from $s^2_+(W)$ to $s^2(W^*)$. The homogeneity of ν implies that this map is constant on rays in $s^2_+(W)$, so it cannot map onto an open set in $s^2(W^*)$. Let $\hat{L}$ be the Legendre transform of $L = \log \nu$. So, $\hat{L}$ is a concave function on some open subset Ω of $s^2(W^*)$, and $\hat{N} = \hat{L}^{-1}(1)$ is a smooth hypersurface in Ω. The homogeneity of ν implies that the image of $D\nu$ is exactly V and, for $v \in \hat{N}$, the preimage $(D\nu)^{-1}(v)$ is a single ray in $s^2_+(W)$. Set $N = \nu^{-1}(1) \subset s^2_+(W)$. Then the restriction of $D\nu$ gives a diffeomorphism from N to $\hat{N}$ and we write $\psi : \hat{N} \to N$ for the inverse.

Proposition 8 *The integral I is stationary, with respect to compactly supported variations in σ satisfying the constraint $\sum \partial_i \partial_j \sigma^{ij} = 0$, if and only if there is a function u on U such that*

$$\partial_i \partial_j u = (D\nu)(\sigma).$$

In one direction, if there is such a function u and, if τ is a compactly supported variation of σ with $\partial_i \partial_j \tau^{ij} = 0$, then

$$\delta I = \int \langle D\nu(\sigma), \tau \rangle = \int (\partial_i \partial_j u)\tau^{ij} = \int u \partial_i \partial_j \tau^{ij} = 0.$$

The other direction follows easily from the fact that we can generate solutions of $\partial_i \partial_j \tau^{ij} = 0$ from an arbitrary tensor h^{iaj} which is skew-symmetric in i, a via the formula

$$\tau^{ij} = \partial_a h^{iaj} + \partial_a h^{jai}.$$

Now the general local solution of our variational problem is obtained as follows. First, solve the equation for a function u that $u_{ij} = \partial_i \partial_j u$ lies in $\hat{N}$, that is,

$$\hat{L}(u_{ij}) = 1. \tag{26}$$

Now set $\lambda = (\lambda^{ij}) = \psi(u_{ij})$. We know that $\sigma = V\psi(u_{ij})$ for some positive function V and the remaining equation to solve is the linear equation for V

$$\partial_i \partial_j (\lambda^{ij} V) = 0. \tag{27}$$

We compare this with the linearization of the non-linear equation (26) at u. The derivative of $\hat{L}$ at a point $v \in N$ is given by $\psi(v)$, so the linearized equation is

$$\sum \lambda^{ij} \partial_i \partial_j V = 0. \tag{28}$$

In general, this is not the same as the equation (27), but the two equations are adjoint in that the formal adjoint $\diamond^*_u$ of the operator $\diamond_u(V) = \sum \lambda^{ij} \partial_i \partial_j V$ is $\diamond^*(V) = \sum \partial_i \partial_j (\lambda^{ij} V)$.

In the case when $\nu(\sigma) = (\det\sigma)^{1/n}$, the derivative of $\log\nu$ is the map $\sigma \mapsto n^{-1}\sigma^{-1}$ and we recover the previous set-up. In this special case, the linearized equation is self-adjoint, i.e $\Diamond_u^* = \Diamond_u$.

We can now introduce a boundary value problem on a domain $U \subset W = \mathbf{R}^n$ with smooth boundary Σ and with a given layer current $\mathcal{L}$ supported on Σ, extending the definitions from $n = 3$ in the obvious way. If u solves the non-linear equation $\hat{L}(u_{ij}) = 1$ on U, then, for any σ satisfying the boundary conditions and $\sum \partial_i\partial_j\sigma^{ij} = 0$, we have an inequality

$$\int_U \nu(\sigma) \leq \mathcal{L}(u),$$

and we get a dual variational problem as before.

Claude LeBrun pointed out to the author that there are some striking similarities between the variant of the Gibbons–Hawking construction studied in the previous section and another variant introduced by him in [7], constructing Kähler surfaces of zero scalar curvature. While this does not exactly fit into the general framework above, we will outline how it can be treated in a similar fashion.

We consider a triple of forms ω^i with ω^1 a Kähler form and ω^2, ω^3 the real and imaginary parts of a holomorphic 2-form. This means that we restrict attention to matrix-valued functions σ which are diagonal, with $\sigma^{11} = a$ and $\sigma^{22} = \sigma^{33} = b$ for positive functions a, b. The condition $\partial_i\partial_j\sigma^{ij} = 0$ is then

$$a_{11} + b_{22} + b_{33} = 0, \tag{29}$$

(writing $a_{11} = \partial_1\partial_1 a$, etc.). The variational formulation, generating the zero scalar curvature equation, comes from the 'Mabuchi functional', which in this situation is given by

$$I = \int a(\log(a/b) - 1), \tag{30}$$

and the function $\nu(a,b) = a(\log(a/b) - 1)$ is homogeneous of degree 1. The condition that I is stationary with respect to compactly supported variations satisfying the constraint (29) is that there is a function u with

$$u_{11} = \log(a/b), \quad u_{22} + u_{33} = -(a/b).$$

In other words, u satisfies the non-linear equation

$$e^{u_{11}} + u_{22} + u_{33} = 0. \tag{31}$$

Given such a function u, we set $b = V$, and $a = e^{u_{11}}V$, and the equation (29) is the linear equation for V:

$$(e^{u_{11}}V)_{11} + V_{22} + V_{33} = 0. \tag{32}$$

This is again the adjoint of the linearization of the non-linear equation (31). To relate this to LeBrun's set-up, we put $U = u_{11}$, so

$$\begin{aligned}(e^U)_{11} + U_{22} + U_{33} &= 0,\\ (e^U V)_{11} + V_{22} + V_{33} &= 0,\end{aligned}$$

and these are the equations, for functions U, V, obtained by LeBrun.

REFERENCES

[1] V. Apolostolov and S. Salamon*Kähler reduction of metrics with holonomy G_2*. Comm. Math. Phys. 246 (2004), no. 1, 43–61.

[2] L. Caffarelli, L. Nirenberg and J. Spruck *The Dirichlet problem for non-linear second-order equations I. The Monge–Ampère equation.* Comm. Pure Appl. Math. 37 (1984), no. 2, 369–402.

[3] S. Donaldson *Two forms on four-manifolds and elliptic equations.* In: P. A. Griffiths (ed.) Inspired by S. S. Chern, 153–72, Nankai Tracts Math., 11, World Sci. Publ. 2006.

[4] J. Fine and C. Yao *Hypersymplectic four-manifolds, the G_2-Laplacian flow and extension assuming bounded scalar curvature.* arxiv 1704.07620 (2017).

[5] N. Hitchin *The geometry of three-forms in six dimensions.* J. Differential Geom. 55 (2000), no. 3, 547–76.

[6] N. Hitchin *Stable forms and special metrics.* In: A. Gray, M. Fernández and J. A. Wolf (eds). Global differential geometry: the mathematical legacy of Alfred Gray (Bilbao, 2000), 70–89, Contemp. Math., 288, Amer. Math. Soc. 2001.

[7] C. LeBrun *Explicit self-dual metrics on* $\mathbf{CP}^2\sharp\ldots\sharp\mathbf{CP}^2$. J. Differential Geom. 34 (1999) 223–53.

7

A Hitchin Connection for a Large Class of Families of Kähler Structures

JØRGEN ELLEGAARD ANDERSEN
KENNETH RASMUSSEN

Centre for Quantum Geometry of Moduli Spaces, Department of Mathematics, University of Aarhus, DK-8000 Aarhus, Denmark. Email: andersen@imf.au.dk

ABSTRACT. In this paper, we construct a Hitchin connection in a setting, which significantly generalizes the setting covered by the first author in [A5], which in turn was a generalization of the moduli space case covered by Hitchin in his original work on the Hitchin connection [9]. In fact, our construction provides a Hitchin connection which is a partial connection on the space of all compatible complex structures on an arbitrary but fixed prequantizable symplectic manifold, which satisfies a certain Fano-type condition. The subspace of the tangent space to the space of compatible complex structures on which the constructed Hitchin connection is defined is, in fact, of finite codimension, if the symplectic manifold is compact. In a number of examples, including flat symplectic space, symplectic tori and moduli spaces of flat connections for a compact Lie group, we prove that our Hitchin connection is defined in a neighbourhood of the natural families of complex structures compatible with the given symplectic form, which these spaces admits.

Dedicated to Nigel Hitchin at the conference *Hitchin 70*, Celebrating his seventieth birthday.

1. Introduction

In this paper, we construct a Hitchin connection in a setting which generalises earlier work done by the first author in [A5], where the Hitchin connection is constructed under the rather restrictive assumption that the family of complex structures has the so-called rigid property, which was also the case for the moduli space case in which Hitchin constructed

Supported in part by the centre of excellence grant "Centre for quantum geometry of Moduli Spaces" (DNRF95) from the Danish National Research Foundation.

Ellegaard Andersen, J., Rasmussen, K., *A Hitchin Connection for a Large Class of Families of Kähler Structures.* In: *Geometry and Physics: A Festschrift in Honour of Nigel Hitchin*, Jørgen Ellegaard Andersen, Andrew Dancer, Oscar García-Prada (Eds): Oxford University Press (2018).
DOI: 10.1093/oso/9780198802006.003.0007

his connection first [H]. This means that the corresponding deformations of the metric is by the real part of a global holomorphic symmetric tensor, as we recall in details below. In particular, if a given Kähler manifold has no global holomorphic symmetric tensors beside zero, then the approach of [AS] does not apply; hence, the wording rigid for such families—they constitute rather thin slices in the infinite-dimensional space of all complex structures. In this paper, we relax this condition considerably.

In order to describe our generalization, let us briefly introduce the setting. We let (M,ω) be a symplectic manifold. We let $\mathcal{T}$ be a complex manifold parametrizing a holomorphic family $J\colon \mathcal{T} \to C^\infty(M,\mathrm{End}(TM))$ of complex structures, which are all Kähler with respect to ω. For any point $\sigma \in \mathcal{T}$, we will write M_σ when we refer to the complex manifold (M,J_σ).

We will consider the variation of the family J along a real vector field V on $\mathcal{T}$, which we denote $V[J]$. We consider the splitting of $V = V' + V''$ into types on $\mathcal{T}$ and we consider the symmetric bivector field $G(V') = V'[J]\cdot\tilde{\omega}$, where $\tilde{\omega}$ is the bivector field inverse to ω. We think of G as a 1-form on $\mathcal{T}$ with coefficients in bivector fields and, as such, we write $G(V) = G(V')$. Observe that, if g is the corresponding family of Kähler metrics parametrized by $\mathcal{T}$, then we have that

$$V[g] = G(V) + \overline{G(V)}.$$

The assumption that the *family* J is rigid says that $G(V)$ defines a holomorphic section $G(V)_\sigma \in H^0(M_\sigma, S^2(T'M_\sigma))$ at all points $\sigma \in \mathcal{T}$. This is, of course, a very restrictive condition; however, it is satisfied in the setting in which Hitchin initially introduced his connection. This was the case of Teichmüller space parametrizing Kähler structures for the Narasimhan–Atiyah–Bott–Goldman symplectic form [AB, NS1, NS2, G] on the moduli spaces of flat $SU(n)$ connections on a genus g surface [H]. See also the work of Axelrod, Della Pietra and Witten for a physical derivation of this connection using Chern–Simons theory [ADW], and [AS] for a verification that the two connections agree.

In this paper, we weaken the rigid criterion by adding the possibility of varying the bivector field $G(V)$ by adding a term of the form $\bar{\partial}\beta(V)\cdot\tilde{\omega}$ for an arbitrary vector field $\beta(V)_\sigma \in C^\infty(M_\sigma, T'M_\sigma)$.

Definition 1 *We call the family J* weakly restricted *if there exists a 1-form β on $\mathcal{T}$ with values in $C^\infty(M_\sigma, T'M_\sigma)$ at each point $\sigma \in \mathcal{T}$, such that, for all vector fields V along $\mathcal{T}$ and all $\sigma \in \mathcal{T}$, there exist $G_\beta(V)_\sigma \in H^0(M, S^2(T'M_\sigma))$ such that*

$$G_\beta(V)_\sigma \cdot \omega = V'[J]_\sigma + \bar{\partial}\beta(V)_\sigma. \tag{1}$$

The main result in this article is the construction of a Hitchin connection, when we assume the family to be weakly restricted, on top of a couple of further minor topological assumptions.

It is, of course, interesting to investigate when we can solve the weakly restricted criterion. We let $\mathcal{C}_\omega$ be the space of all complex structures on M compatible with the symplectic form ω and let $J \in \mathcal{C}_\omega(M)$. Then we have that

$$T_J\mathcal{C}_\omega = \ker(\bar{\partial}_J\colon \Omega^{0,1}(M,T'M_J)_\omega \to \Omega^{0,2}(M,T'M_J)),$$

where

$$\Omega^{0,1}(M,T'M_J)_\omega = \{V_J \in \Omega^{0,1}(M,T'M_J) \mid \omega(V_J\cdot,J\cdot) = \omega(\cdot,JV_J\cdot)\}, \tag{2}$$

which is the same as stating that V_J is symmetric with respect to the Kähler metric g_J associated to ω and J. Thus, we see that given $V_J \in T_J\mathcal{C}_\omega$, we can solve the weakly restricted condition, e.g. find $\beta(V)$, whenever we have

$$V_J \in H^0(M_J,S^2(T'M_J))\cdot\omega + Im(\bar{\partial}_J\colon C^\infty(M,T'M_J)_\omega \to \Omega^{0,1}(M,T'M_J)_\omega), \tag{3}$$

where

$$C^\infty(M,T'M_J)_\omega = \{X \in C^\infty(M,T'M_J) \mid \bar{\partial}(i_X\omega) = 0\}.$$

Thus, if the map

$$\cdot\omega\colon H^0(M_J,S^2(T'M_J)) \to H^1(M_J,T'M_J)_\omega$$

is surjective, this is always possible. Here, $H^1(M_J,T'M_J)_\omega$ is defined in analogy with (2), namely to be the symmetric part of this cohomology.

A particular simple case, where we can always solve (1), is, of course, if

$$H^1(M_J,T'M_J)_\omega = 0.$$

In general, we can solve the equation (1) if the cohomology class of $V'[J]_\sigma$ is contained in the image of $\cdot\omega$ in (3). Thus, our construction will only provide a partial connection on the space of all complex structures compatible with the symplectic form on ω. If M is compact, we see that this partial connection is defined on a subspace of finite codimension of the tangent space to the space of all complex structures compatible with ω.

Let us now briefly recall the set-up in geometric quantization. Let (M,ω) be a symplectic manifold and assume that (M,ω) admits a prequantum line bundle $(\mathcal{L},\nabla,\langle\cdot,\cdot\rangle)$ [W]. Let $\mathcal{T}$ be a complex manifold parametrizing a holomorphic family of complex structures J making (M,ω,J_σ) Kähler for each $\sigma \in \mathcal{T}$. Now for each $\sigma \in \mathcal{T}$ we consider the quantum space at level $k \in \mathbb{N}$, which is the subspace $H^{(k)}_\sigma$ of the prequantum space $\mathcal{H}^{(k)} = C^\infty(M,\mathcal{L}^k)$ consisting of holomorphic sections

$$H^{(k)}_\sigma = H^0(M_\sigma,\mathcal{L}^k) \subset \mathcal{H}^{(k)}.$$

We will throughout this paper assume that these quantum spaces form a smooth subbundle $H^{(k)}$ of the trivial bundle

$$\hat{\mathcal{H}}^{(k)} = \mathcal{T} \times \mathcal{H}^{(k)}.$$

Now we let ∇^T denote the trivial connection on $\hat{\mathcal{H}}^{(k)}$, and then we consider a connection of the form

$$\boldsymbol{\nabla}_V = \nabla^T_V + u(V), \tag{4}$$

where $u \in \Omega^1(\mathcal{T}, \mathcal{D}(M, \mathcal{L}^k))$ is a 1-form on $\mathcal{T}$ with values in the space of differential operators on sections of $\mathcal{L}^k$. Our goal is to construct a u such that $\boldsymbol{\nabla}$ preserves the quantum spaces $H_\sigma^{(k)}$ inside each fibre of $\hat{\mathcal{H}}^{(k)}$.

Definition 2 (Hitchin connection) *A Hitchin connection in the bundle $\hat{\mathcal{H}}^{(k)}$ is a connection of the form (4), that preserves the subspaces $H_\sigma^{(k)}$ inside each fibre of $\hat{\mathcal{H}}^{(k)}$.*

We prove the following in this paper.

Theorem 1 (Hitchin connection for weakly restricted families) *Let (M,ω) be a symplectic manifold with a prequantum line bundle $\mathcal{L}$. Assume that M has first Chern class of the form $c_1(M,\omega) = n\left[\frac{\omega}{2\pi}\right]$ for some integer $n \in \mathbb{Z}$ and such that $b_1(M) = 0$. Furthermore, let $J\colon \mathcal{T} \to C^\infty(M, \operatorname{End}(TM))$ be a weakly restricted, holomorphic family of Kähler structures on M, parametrized by a complex manifold $\mathcal{T}$, and assume that the family admits a family of Ricci potentials F. Then there exists a Hitchin connection $\boldsymbol{\nabla}$ in the bundle $\hat{H}^{(k)}$ over $\mathcal{T}$, given by the expression*

$$\boldsymbol{\nabla}_V = \nabla_V^{T} + u(V),$$

where

$$u(V) = \frac{1}{2(2k+n)}\big(\Delta_{G_\beta(V)} + 2\nabla_{G_\beta(V)\cdot dF} - i(2k+n)\nabla_{\beta(V)} + 4kV'[F] - 2ikdF\cdot\beta(V) - ik\delta(\beta(V)) + 2k(k+n)\phi(V) + ik\psi(V)\big),$$

and $\phi(V), \psi(V) \in C^\infty(M)$ are smooth functions, satisfying

$$\bar\partial\phi(V) = \omega\cdot\beta(V) \quad \text{and} \quad \bar\partial\psi(V) = \Omega(V),$$

where $\Omega(V) \in \Omega^1(M)$ is given by

$$\Omega(V) = -\delta(G_\beta(V))\cdot\omega + \delta(\bar\partial\beta(V)) - 2dF\cdot G_\beta(V)\cdot\omega + 2\bar\partial\beta(V)\cdot dF + 4i\bar\partial V'[F].$$

Remark 1 *If none of the complex structures admit non-constant holomorphic functions on M, we get that $\psi(V) = 0$.*

Remark 2 *When M is compact, Hodge theory will provide us with a family of Ricci potentials and, of course, there will, in that case, only be constant holomorphic functions globally on M. So, we don't need the assumption, that such a family of Ricci potentials exist, and we automatically get $\psi(V) = 0$ from Remark 1. Thus, in this case we can reduce the assumptions in Theorem 1 to the two cohomological restrictions*

$$b_1(M) = 0 \quad \text{and} \quad c_1(M,\omega) = n\left[\frac{\omega}{2\pi}\right].$$

Remark 3 *We, of course, expect that the Fano-type condition*

$$c_1(M,\omega) = n\left[\frac{\omega}{2\pi}\right]$$

can be removed by doing metaplectic correction as considered in [AGL].

Remark 4 *The condition $b_1(M) = 0$ is only used to ensure that the closed 1-form $\Omega(V)$ is exact, such that $\psi(V)$ exists. In other words, if we already have a $\bar{\partial}$-primitive for $\Omega(V)$ for all V, this assumption can also be ignored. See details in the proof of Proposition 3.*

Remark 5 *When $\beta(V) = 0$, the family is rigid and this new Hitchin connection restricts to the Hitchin connection in [A5].*

Remark 6 *We stress that we do not need that $\mathcal{T}$ is a complex manifold, in fact, we have a complete analogue of Theorem 1, when $\mathcal{T}$ has no complex structure. Please see Theorem 2 in Section 4.2.*

In the case of rigid families, it was proved that the Hitchin connection in this setting is projectively flat by the first author and Gammelgaard in [AG1] generalizing the projective flatness proofs of Hitchin in [H] and Axelrod, Della Pietra and Witten in [ADW] in the original moduli space setting (see also [vGdJ, R1]). The projective flatness is, of course, very important for its relation to quantum Chern–Simons theory, in particular in relation to the projective representations of the mapping class groups which arises. These actually are the same as the Witten–Reshetikhin–Turaev TQFT (WRT-TQFT) [RT1, RT2, T, B1, BHMV1, BHMV2] representations as it follows by the combination of two results. One is by Laszlo [La1], which identifies the Hitchin connection projectively with the TUY-connection, constructed in conformal field theory by Tsuchiya, Ueno and Yamada [TUY]. The second is by the first author and Ueno and identifies the TQFTs coming from conformal fields theory for the affine Lie algebra of $\mathfrak{sl}(m,\mathbb{C})$ and then the WRT-TQFT constructed from $U_q(\mathfrak{sl}(m,\mathbb{C}))$, for $q = \exp(2\pi i/(k+n))$ [AU1, AU2, AU3, AU4] as proposed in [W1]. This has been exploited to prove a number of results about the WRT-TQFT [A2, A3, A4, A5, AH, A6, AHJMMc].

In the case of weakly restricted families, we cannot expect to prove projective flatness in general, due to the no-go theorem of [GM]. It is, however, very natural to ask for the existence of a Hitchin connection given by differential operators and not just by Toeplitz operators as the L^2-induced connection would be. We also expect that the Hitchin connection in some sense minimizes the possible curvature, something which we hope to return to in future work.

In the final two example sections, we illustrate the applicability of our construction. We show that our construction applies to certain open subsets of the entire family of all complex structures on $\mathbb{R}^{2n}$ with the standard symplectic structure and the entire family of all complex structures on $\mathbb{R}^{2n}/\mathbb{Z}^{2n}$ compatible with the standard symplectic structure. Further, our construction also applies to certain open subsets of the entire family of all complex structures on co-adjoint orbits and on the moduli spaces of flat $SU(n)$-connections on a surface of genus $g > 1$, possibly with central holonomy around a point on the surface, with the Narasimhan–Atiyah–Bott–Goldman symplectic structure on it.

The first author would like to take this opportunity to thank Professor Nigel Hitchin for his patient guidance, which has continued ever so steadily since the first day of the first author's D. Phil. studies at Oxford University.

2. Geometric Quantization

In this chapter, we will very briefly recall the mathematical theory of quantization called 'geometric quantization' and explain some of the problems that arise within this quantization procedure. One of the main points for us is the need to choose a polarization, which in our case will be a Kähler structure compatible with the symplectic form on the manifold, which models the classical mechanical system. This choice is not canonical and is auxiliary to the physical theory, and therefore we would suspect that the theory should, in some sense, be independent of this choice. The Hitchin connection aims to relate these different choices.

A quantization scheme is, in the simplest form, a way to pass from classical mechanics to quantum mechanics. That is, to a classical mechanic system, represented by a phase space consisting of a symplectic $2n$ dimensional manifold (M,ω), it assigns a corresponding Hilbert space $\mathcal{H}$ of quantum states, and to a classical observable given by a smooth function $f \in C^{\infty}(M)$ it assigns a self-adjoint operator $Q(f)$ on $\mathcal{H}$.

For the theory to be physically sound, this assignment should be linear, send the constant function 1 to the identity operator and should fulfil the commutation relation

$$[Q(f), Q(g)] = i\hbar Q(\{f,g\}). \tag{5}$$

Last, applying the quantization to $\mathbb{R}^{2n}$ with the standard symplectic form should yield the canonical quantization (see [W]). It has, however, been shown that such a full quantization cannot exit. The approach we will follow to handle this is to weaken the relation (5), so we only require the asymptotic relation

$$[Q(f), Q(g)] = ihQ(\{f,g\}) + O(h^2) \quad \text{as} \quad h \to 0. \tag{6}$$

2.1. *Prequantization*

The first step in geometric quantization is prequantization. Here, we construct a Hilbert space of quantum observables as sections of tensor powers of a so-called *prequantum line bundle* over the phase space (M,ω).

Definition 3 (Prequantum line bundle) *A prequantum line bundle over the symplectic manifold (M,ω) is a triple $(\mathcal{L}, h, \nabla)$ consisting of a line bundle $\mathcal{L}$ over M with a Hermitian metric h and a compatible connection ∇, such that the curvature of ∇ is related to the symplectic form by the relation*

$$F_{\nabla} = -i\omega. \tag{7}$$

We say that a connection is compatible with the Hermitian structure h if we have

$$Xh(s_1,s_2) = h(\nabla_X s_1, s_2) + h(s_1, \nabla_X s_2) \tag{8}$$

for any real vector field X on M and any two sections $s_1, s_2 \in C^\infty(M,\mathcal{L})$, *and we call* (M,ω) prequantizable, *if there exist a prequantum line bundle over it.*

By looking at the real first Chern class $c_1(\mathcal{L})$, we can see that condition (7) is actually a restrictive condition, since $c_1(\mathcal{L}) = \frac{i}{2\pi}[F_\nabla] = [\frac{\omega}{2\pi}]$, which is not the case for all symplectic manifolds. It is, however, true that (M,ω) is prequantizable, precisely when

$$\left[\frac{\omega}{2\pi}\right] \in Im(H^2(M,\mathbb{Z}) \to H^2(M,\mathbb{R})),$$

which is the so-called *prequantum condition.* Now, given a prequantum line bundle over M, we can define the *prequantum space.* Here we remark that a prequantum structure on a line bundle induces a prequantum structure on any tensor power of the bundle, and we will also use the notation h and ∇ for the induced metric and connection.

Definition 4 (Prequantum space) *The prequantum space of level* $k \in \mathbb{N}$ *is the infinite-dimensional vector space of sections of the k'th tensor power of* $\mathcal{L}$

$$\mathcal{H}^k = C^\infty(M,\mathcal{L}^k). \tag{9}$$

More precisely, we actually consider the L^2*-completion with respect to the Hermitian inner product on compactly supported sections, given by*

$$\langle s_1,s_2\rangle = \frac{1}{n!}\int_M h(s_1,s_2)\omega^n. \tag{10}$$

We will not distinguish between $\mathcal{H}^k$ *and the completion in the following.*

Next, we define the prequantum map, sending a function $f \in C^\infty(M)$ to an operator on $\mathcal{H}^k$ by the expression

$$P_k(f) = -\frac{1}{k}\nabla_{X_f} + if, \tag{11}$$

and, with this definition, the prequantum map satisfies

$$[P_k(f),P_k(g)] = \frac{i}{k}P_k\left(\{f,g\}\right), \tag{12}$$

and thus prequantization satisfies almost all the requirements for a quantization scheme. It does, however, fail to reproduce canonical quantization, since the prequantum space gives wave functions that depend independently on position and momentum. This is not in line with canonical quantization, since we have twice as many degrees of freedom as we are supposed to.

2.2. *Kähler Quantization*

To deal with the problem that prequantization produces with a large Hilbert space, we introduce a polarization. We will only consider the case, where (M,ω) admits a compatible complex structure J, making (M,ω,J) Kähler and we denote the Kähler manifold M_J. In this case, we get complex structures on each $\mathcal{L}^k$ given by $\bar{\partial} = \nabla^{(0,1)}$, since ω has type $(1,1)$ with respect to J and the prequantum condition thus ensures that $(F_\nabla)^{(0,2)} = 0$. This means we can choose the quantum space to be the holomorphic sections

$$H_J^{(k)} = H^0(M_J,\mathcal{L}^k) = \left\{ s \in \mathcal{H}^{(k)} \mid \nabla_J^{0,1} s = 0 \right\}. \tag{13}$$

This is a subspace of the prequantum space $\mathcal{H}^k$, and it is finite dimensional, if M is compact. Now, the prequantum operators do not, in general, preserve the subspace $H_J^{(k)}$, but since it is, in fact, a closed subspace, we have a projection map

$$\pi_J^{(k)} : \mathcal{H}^k \to H_J^{(k)},$$

and we just define the quantum operator by taking the prequantum operator and projecting the result back on $H_J^{(k)}$, that is,

$$Q_k(f)_J = \pi_J^{(k)} \circ P_k(f). \tag{14}$$

With this construction, we lose the commutation relation (5), but it can be shown that we still have the relation (6) (see [BMS, Sch1, Sch2, Sch3, Tuyn, KS, MM1, MM2, MM3]).

3. Families of Kähler Structures

In this section, we explore the properties of families of Kähler structures on a symplectic manifold (M,ω). We start out with a smooth family of integrable almost-complex structures, all compatible with ω, giving us a family of Kähler structures on M.

3.1. *Smooth Families of Kähler Structures*

We let $\mathcal{T}$ be a smooth manifold and (M,ω) a symplectic manifold. Then we say that $\mathcal{T}$ smoothly parametrizes a family of almost-complex structures on (M,ω), if there exist a smooth map

$$J\colon \mathcal{T} \to C^\infty(M,\mathrm{End}(TM)) \quad \text{mapping} \quad \sigma \mapsto J_\sigma \tag{15}$$

such that J_σ is an almost-complex structure for each $\sigma \in \mathcal{T}$. We say that J is smooth, when it defines a smooth section of the pullback bundle $\pi_M^*(\mathrm{End}(TM))$, where $\pi_M\colon \mathcal{T} \times M \to M$ is the projection on M.

We will look at the case where all J_σ are integrable and compatible with the symplectic structure, such that (M, ω, J_σ) is Kähler for each $\sigma \in \mathcal{T}$.

A complex structure gives a splitting of the complexified tangent bundle, so now we have a splitting

$$TM_{\mathbb{C}} = T'M_\sigma \oplus T''M_\sigma,$$

for each $\sigma \in \mathcal{T}$, into the holomorphic and anti-holomorphic parts, i.e. the i and $-i$ eigenspaces of J_σ.

We denote the Kähler metric by g, and it is given by

$$g = \omega \cdot J,$$

where the dot denotes contraction of tensors placed next to each other. We remark that, for a vector X, we have the standard notation $i_X\omega = X \cdot \omega$. We will often need contraction of tensors and in more complicated expressions, we cannot always indicate contraction by placing the tensors next to each other. For this, we will use abstract index notation to denote the entries of the tensor and, following the Einstein summation convention, repeated indices are contracted. In the same spirit, we use subscript indices for covariant entries and superscript indices for contravariant entries. The contraction corresponds to a summation in local coordinates, but our indices only name the entries of the tensor and do not represent a choice of local coordinates. Writing the expression from before in abstract index notation, would look like

$$g_{ab} = \omega_{au} J^u_b,$$

and we remark that we try to use the letters a, b, c, d for entries that are not contracted, and u, v, w, x, y for contracted indices.

We will need the inverses of the metric and the symplectic form. These are the symmetric bivector field $\tilde{g}$ and antisymmetric bivector field $\tilde{\omega}$, such that

$$g \cdot \tilde{g} = \tilde{g} \cdot g = \mathrm{Id} \quad \text{and} \quad \omega \cdot \tilde{\omega} = \tilde{\omega} \cdot \omega = \mathrm{Id}.$$

These exist, since the metric and symplectic forms are non-degenerate. We will sometimes use g and $\tilde{g}$ to either lower or raise an index by contraction. It is also useful to record that the relation between g and ω implies that

$$\tilde{g} = -J \cdot \tilde{\omega}.$$

Another construction, which we will need later, is the canonical line bundle of M_J, e.g. the top exterior power of the holomorphic tangent bundle

$$K_J = \bigwedge^m T'M_J^*. \tag{16}$$

The Hermitian structure on $T'M_J$ induced by g again induces a Hermitian structure in the canonical line bundle, which we will also just denote by h.

Now we want to investigate the variation of the family J. More precisely, we will differentiate along a vector field V on $\mathcal{T}$. This derivative is again a map into the endomorphism bundle, which we denote by

$$V[J] : \mathcal{T} \to C^\infty(M, \mathrm{End}(TM)).$$

Differentiating the equality $J^2 = -1$ and using the Leibniz rule, we get

$$JV[J] = -V[J]J,$$

which shows that $V[J]_\sigma$ interchanges types on $TM_\mathbb{C}$, sending i eigenvectors to $-i$ eigenvectors and the other way around. Thus, $V[J]$ decomposes as

$$V[J] = V[J]' + V[J]'', \tag{17}$$

where the two components

$$\begin{aligned} V[J]'_\sigma &\in C^\infty(M, T''M^*_\sigma \otimes T'M_\sigma) \quad \text{and} \\ V[J]''_\sigma &\in C^\infty(M, T'M^*_\sigma \otimes T''M_\sigma) \end{aligned}$$

are each others, conjugates. Now, since contraction in the first entry of ω defines an isomorphism $i_\omega : TM_\mathbb{C} \to TM^*_\mathbb{C}$, we can get any element in

$$C^\infty(M, \mathrm{End}(TM_\mathbb{C})) = C^\infty(M, TM_\mathbb{C} \otimes TM^*_\mathbb{C})$$

by contraction with a bivector field. We let $\tilde{G}(V) : \mathcal{T} \to C^\infty(M, TM_\mathbb{C} \otimes TM_\mathbb{C})$ be such that the equality

$$V[J] = \tilde{G}(V) \cdot \omega \tag{18}$$

holds at each $\sigma \in \mathcal{T}$ and for each vector field V on $\mathcal{T}$. We get another expression for $\tilde{G}(V)$ by differentiating the equality $\tilde{g} = -J \cdot \tilde{\omega}$ along V, namely

$$V[\tilde{g}] = -V[J] \cdot \tilde{\omega} = -\tilde{G}(V). \tag{19}$$

This is again using the Leibniz rule, and that $\tilde{\omega}$ does not depend on σ, so the derivative along any vector field V vanishes. Since $\tilde{g}$ is symmetric, (19) implies that $\tilde{G}(V)$ is also symmetric.

Looking at the types of $V[J]$ and ω, we see that $\tilde{G}(V)$ has no $(1,1)$-part, and so we get a decomposition of $\tilde{G}(V)$ into

$$\tilde{G}(V) = G(V) + \bar{G}(V),$$

where

$$G(V)_\sigma \in C^\infty(M, S^2(T'M_\sigma)) \quad \text{and} \quad \bar{G}(V)_\sigma \in C^\infty(M, S^2(T''M_\sigma)).$$

Observe also that $\bar{G}(V)$ is actually the conjugate of $G(V)$, since $\tilde{G}(V)$ is real and thus its own conjugate. We can also express the variation of the Kähler metric in terms of $\tilde{G}(V)$ by differentiating the compatibility condition $g = \omega \cdot J$, getting

$$\begin{aligned} V[g] &= \omega \cdot V[J] \\ &= \omega \cdot \tilde{G}(V) \cdot \omega \\ &= g \cdot J \cdot \tilde{G}(V) \cdot g \cdot J \\ &= -g \cdot J \cdot (G(V) + \bar{G}(V)) \cdot J \cdot g \\ &= -i^2 g \cdot G(V) \cdot g - (-i)^2 g \cdot \bar{G}(V) \cdot g \\ &= g \cdot \tilde{G}(V) \cdot g. \end{aligned}$$

This also shows that $V[g] \in C^\infty(M, S^2(T'M_\sigma^*) \oplus S^2(T''M_\sigma^*))$. One more relation about variations, which we will need, is the variation of the Levi–Civita connection ∇^g, here superscripted with g to denote that it is the connection related to the metric g. One has that (see e.g. [Besse], Theorem 1.174)

$$2g(V[\nabla^g]_X Y, Z) = \nabla_X(V[g](Y,Z)) + \nabla_Y(V[g])(X,Z) - \nabla_Z(V[g])(X,Y)$$

for vector fields X, Y, Z on M. To give an explicit expression, we use the above result and write it in the index notation as

$$2V[\nabla^g]^c_{ab} = \nabla_a \tilde{G}(V)^{cu} g_{ub} + g_{au} \nabla_b \tilde{G}(V)^{uc} - g_{au} \tilde{g}^{cw} \nabla_w \tilde{G}(V)^{uv} g_{vb}. \tag{20}$$

3.2. The Canonical Line Bundle of a Family

Before we get started with the calculations in this section, we recall that the Ricci form is the skew-symmetric $(1,1)$-form $\rho = J \cdot r$, where r is the Ricci curvature tensor, which is given by

$$r(X,Y) = \mathrm{Tr}(Z \mapsto R(Z,X)Y) \quad \text{or, in index notation,} \quad r_{ab} = R^w_{wab},$$

where R is the curvature tensor of the Levi–Civita connection. We also recall that the Levi–Civita connection and the curvature $R(X,Y)$ preserves types on $TM_{\mathbb{C}}$. The form part of R is J-invariant and is of type $(1,1)$.

We shall now recall the derivation of an expression for the variation of the Ricci form ρ. This will not seem apparent from the beginning, but we will construct a certain line bundle over the product manifold $\mathcal{T} \times M$ and consider the induced connection in this bundle. The Ricci form will appear in an expression for the curvature in some directions on $\mathcal{T} \times M$ and using the Bianchi identity, we will get a very useful relation.

First, we consider the vector bundle $\hat{T}'M \to \mathcal{T} \times M$, where the fibres are just the holomorphic tangent spaces of M, that is, $\hat{T}'M_{(\sigma,p)} = T'_pM_\sigma$. So, the point $\sigma \in \mathcal{T}$ determines the splitting of the bundle $TM \otimes \mathbb{C}$, and the point $p \in M$ chooses the fibre T'_pM_σ of the subbundle $T'M_\sigma$. We use the hat to denote that we consider it as a bundle over the product $\mathcal{T} \times M$; similarly, we will use a hat on the differential $\hat{d}$ and the connection $\hat{\nabla}$ on this bundle.

We notice that the Kähler metric induces a Hermitian structure $\hat{h}$ on $\hat{T}'M$. We construct a connection on $\hat{T}'M$ in two steps. We first notice that along vector fields on $M = \{\sigma\} \times M$, we can use the Levi–Civita connection on the bundle $T'M_\sigma$, which gives us a partial connection on $\hat{T}'M$ compatible with the Hermitian structure.

Now we think of a section $Z \in C^\infty(\mathcal{T} \times M, \hat{T}'M)$ as a smooth family of sections of the holomorphic tangent bundles, and we let V be a vector field on $\mathcal{T} = \mathcal{T} \times \{p\}$. Since each of the holomorphic tangent spaces sits inside the larger complexified tangent bundle $T'M_\sigma \subset TM \otimes \mathbb{C}$, we can think of Z as a smooth family of vector fields in this bundle. Since $TM \otimes \mathbb{C}$ is unchanged along V, we can take the variation of Z along V in $TM \otimes \mathbb{C}$, and then project the result back onto the holomorphic subbundle. This defines a connection along the directions on $\mathcal{T}$, that is,

$$\hat{\nabla}_V Z = \pi^{1,0} V[Z].$$

We still want this connection to be compatible with the Hermitian structure, so we check this by the following calculation. Let V be a vector field on $\mathcal{T}$ and X, Y sections of $\hat{T}'M$. Then we have

$$\begin{aligned} V[\hat{h}(X,Y)] = V[g(X,\bar{Y})] &= V[g](X,\bar{Y}) + g(V[X],\bar{Y}) + g(X,\overline{V[Y]}) \\ &= g(V[X],\bar{Y}) + g(X,\overline{V[Y]}) \\ &= h(\hat{\nabla}_V X, Y) + h(X, \hat{\nabla}_V Y), \end{aligned}$$

since the $(1,1)$-part of $V[g]$ vanishes as shown earlier. This is exactly the condition, that $\hat{\nabla}$ is compatible with the Hermitian structure. In this way, we have constructed a connection on all of $\hat{T}'M$ compatible with the Hermitian structure.

Next, we consider the top exterior power, which, as in the case of $T'M_\sigma$, gives the *canonical line bundle of the family* of complex structures, and we will denote this

$$\hat{K} = \bigwedge^m \hat{T}'M^* \to \mathcal{T} \times M. \tag{21}$$

Just as for the normal canonical line bundle, we get an induced Hermitian structure and a compatible connection on $\hat{K}$.

We will now state the proposition from [AGL] about the curvature of $\hat{\nabla}^K$.

Proposition 1 *Given vector fields X, Y on M and V on $\mathcal{T}$, the curvature of $\hat{\nabla}^K$ is given by*

$$F_{\hat{\nabla}^K}(X,Y) = i\rho(X,Y), \tag{22}$$

$$F_{\hat{\nabla}^K}(V,X) = \frac{i}{2}\delta\tilde{G}(V) \cdot \omega \cdot X. \tag{23}$$

We remark that we can also get an expression for the curvature for two vector fields V, W on $\mathcal{T}$, but we will not need this result here. Now, applying the Bianchi identity and the results of Proposition 1 gives us the desired result about the variation of the Ricci form.

Proposition 2 *The variation of the Ricci form along a vector field V on $\mathcal{T}$ is given by*

$$V[\rho] = \frac{1}{2}d(\delta\tilde{G}(V)\cdot\omega). \tag{24}$$

Proof Let X, Y be commuting vector fields on M, and V a vector field on $\mathcal{T}$. Then the Bianchi identity for $\hat{\nabla}^K$ gives

$$\begin{aligned}
0 &= V[F_{\hat{\nabla}^K}(X,Y)] + X[F_{\hat{\nabla}^K}(Y,V)] + Y[F_{\hat{\nabla}^K}(V,X)] \\
&= iV[\rho(X,Y)] - X\left[\frac{i}{2}\delta\tilde{G}(V)\cdot\omega\cdot Y\right] + Y\left[\frac{i}{2}\delta\tilde{G}(V)\cdot\omega\cdot X\right] \\
&= iV[\rho(X,Y)] - \frac{i}{2}d(\delta\tilde{G}(V)\cdot\omega)(X,Y),
\end{aligned}$$

where the last equality follows by the invariant formula for the exterior derivative, since X and Y were chosen to commute. Now isolating $V[\rho(X,Y)]$ gives the desired equality. □

3.3. *Holomorphic Families of Kähler Structures*

In this and the next section, we will introduce some extra restriction on a family of complex structures. Assume that $\mathcal{T}$ is a complex manifold. Then we can require, that the family J of holomorphic structures defines a holomorphic map from $\mathcal{T}$ to the space of holomorphic structures. This is defined as follows.

Definition 5 *Let $\mathcal{T}$ be a complex manifold and $J\colon \mathcal{T} \to C^\infty(M, \mathrm{End}(TM))$ a family of integrable almost-complex structures on M. Then J is holomorphic if*

$$V'[J] = V[J]' \quad \text{and} \quad V''[J] = V[J]''$$

for any vector field V on $\mathcal{T}$.

Now let us denote the integrable almost-complex structure on $\mathcal{T}$ by I. Then we get an almost-complex structure $\hat{J}$ on the product manifold $\mathcal{T}\times M$, defined by

$$\hat{J}(V\oplus X) = IV \oplus J_\sigma X, \quad \text{for} \quad V\oplus X \in T_{(\sigma,p)}(\mathcal{T}\times M).$$

It can be shown that holomorphicity of the family of complex structure as defined above is equivalent to $\hat{J}$ being an integrable almost-complex structure on $\mathcal{T}\times M$ (see [AG1]). This is shown by using the criterion of the vanishing of the Nijenhuis tensor.

A useful consequence of holomorphicity is that

$$\tilde{G}(V') = V'[J]\cdot\tilde{\omega} = V[J]'\cdot\tilde{\omega} = G(V), \tag{25}$$

and, similarly, $\tilde{G}(V'') = \bar{G}(V)$.

4. The Hitchin Connection

In this section, we give the construction of the Hitchin connection. The ideas of the proofs follow the original construction by the first author in [A5], although some things are done with a slightly different touch, which in large parts is inspired by [AG1], which is joined work by the first author and Gammelgaard.

The theorem and the set-up are stated in the introduction, Theorem 1, and, as in the rigid setting, the proof is done by constructing a u which satisfies the condition in the following lemma.

Lemma 1 *The connection* $\boldsymbol{\nabla} = \nabla^T + u$ *is a Hitchin connection if and only if*

$$\nabla_\sigma^{0,1} u(V)s = \frac{i}{2} V'[J_\sigma] \cdot \nabla_\sigma^{1,0} s, \tag{26}$$

for any holomorphic section $s \in H_\sigma^{(k)}$ *and any smooth vector field* V *on* $\mathcal{T}$.

Proof By assumption, we need to have that

$$\begin{aligned} 0 &= \nabla_\sigma^{0,1}(\boldsymbol{\nabla}_V s) \\ &= \nabla_\sigma^{0,1} V[s] + \nabla_\sigma^{0,1} u(V)s. \end{aligned}$$

Now, by differentiating $\nabla_\sigma^{0,1} s = 0$ along V, we get that

$$\begin{aligned} 0 = V[\nabla_\sigma^{0,1} s] &= V\left[\frac{1}{2}(Id + iJ_\sigma) \cdot \nabla s\right] \\ &= \frac{i}{2} V[J_\sigma] \cdot \nabla_\sigma^{1,0} s + \nabla_\sigma^{0,1} V[s]. \end{aligned}$$

Combining the above expressions, we get equation (26). □

The construction of the connection is carried out through a number of lemmas. We will start in the most general setting and then add the assumptions in the lemmas, when we need them. First, we just assume that we have a symplectic manifold (M,ω) with a prequantum line bundle $\mathcal{L}$, a family of Kähler structures J, and that we have an arbitrary symmetric bivector field

$$G \in C^\infty(M_\sigma, S^2(T'M_\sigma)),$$

and then we get a linear bundle map

$$G \colon T'M_\sigma^* \to T'M_\sigma,$$

given by contracting with G. Now, using this, we get an operator Δ_G on $\mathcal{H}^{(k)}$ given by

$$\begin{aligned} \Delta_G \colon \mathcal{H}^{(k)} = C^\infty(M, \mathcal{L}^k) &\xrightarrow{\nabla_\sigma^{1,0}} C^\infty(M, T'M_\sigma^* \otimes \mathcal{L}^k) \\ &\xrightarrow{G \otimes \mathrm{Id}} C^\infty(M, T'M_\sigma \otimes \mathcal{L}^k) \end{aligned}$$

$$\xrightarrow{\nabla_\sigma^{1,0}\otimes \mathrm{Id}+\mathrm{Id}\otimes\nabla_\sigma^{1,0}} C^\infty(M, T'M_\sigma^*\otimes T'M_\sigma\otimes\mathcal{L}^k)$$
$$\xrightarrow{\mathrm{Tr}} C^\infty(M, T'M_\sigma\otimes\mathcal{L}^k).$$

In abstract tensor notation, we can write this in the short form

$$\Delta_G s = \nabla_{u'} G^{u'v'} \nabla_{v'} s,$$

where the outer connection is the connection in the tensor product $T'M_\sigma \otimes \mathcal{L}^k$, which is given exactly as described above by the Leibniz rule.

Lemma 2 *Let $G \in H^0(M_\sigma, S^2(T'M_\sigma))$ be any holomorphic bivector field on (M, ω_σ); then we have, for any section $s \in H_\sigma^{(k)}$,*

$$\nabla^{(0,1)}\Delta_G s = -2ik\omega\cdot G\cdot\nabla^{(1,0)}s - i\rho\cdot G\cdot\nabla^{(1,0)}s - ik\omega\cdot\delta(G)s. \tag{27}$$

Proof The proof is a calculation that mainly uses the trick of commuting two covariant derivatives to get one term that disappears because of type considerations plus a curvature term. We will write out the proof using abstract tensor notation, which highlights contraction of terms. So, for $s \in H_\sigma^{(k)}$, we get that

$$\begin{aligned}
\nabla^{(0,1)}\Delta_G s &= \nabla_{a''}\nabla_{u'}G^{u'v'}\nabla_{v'}s\\
&= \nabla_{u'}\nabla_{a''}G^{u'v'}\nabla_{v'}s + [\nabla,\nabla]_{a''u'}G^{u'v'}\nabla_{v'}s\\
&= \nabla_{u'}G^{u'v'}\nabla_{a''}\nabla_{v'}s + [\nabla,\nabla]_{a''u'}(G^{u'v'})\nabla_{v'}s + G^{u'v'}[\nabla,\nabla]_{a''u'}\nabla_{v'}s\\
&= \nabla_{u'}G^{u'v'}[\nabla,\nabla]_{a''v'}s + R^{w}_{a''wu'}G^{u'v'}\nabla_{v'}s - ik\omega_{a''u'}G^{u'v'}\nabla_{v'}s\\
&= -ik\nabla_{u'}G^{u'v'}\omega_{a''v'}s - R^{w}_{wa''u'}G^{u'v'}\nabla_{v'}s - ik\omega\cdot G\cdot\nabla s\\
&= -ik\nabla_{u'}(G^{u'v'})\omega_{a''v'}s - ikG^{u'v'}\omega_{a''v'}\nabla_{u'}s - r_{a''u'}G^{u'v'}\nabla_{v'}s - ik\omega\cdot G\cdot\nabla s\\
&= -ik\omega_{a''v'}\delta(G)^{v'}s - ik\omega\cdot G\cdot\nabla s - J_a^x r_{xy}J_{u'}^y G^{u'v'}\nabla_{v'}s - ik\omega\cdot G\cdot\nabla s\\
&= -2ik\omega\cdot G\cdot\nabla s - ik\omega\cdot\delta(G)s - i\rho_{au'}G^{u'v'}\nabla_{v'}s\\
&= -2ik\omega\cdot G\cdot\nabla s - i\rho\cdot G\cdot\nabla s - ik\omega\cdot\delta(G)s.
\end{aligned}$$
□

Now we will start applying more assumptions to continue the construction.

Corollary 1 *Consider the situation as in Lemma 2 and assume that the family admits a family of Ricci potentials F. Then we get*

$$\nabla^{(0,1)}\Delta_G s = -i(2k+n)\omega\cdot G\cdot\nabla^{(1,0)}s + 2(d\bar\partial_\sigma F_\sigma)\cdot G\cdot\nabla^{(1,0)}s - ik\omega\cdot\delta(G)s. \tag{28}$$

Proof The proof follows directly by inserting the expression for the Ricci form given in terms of the family of Ricci potentials F_σ, i.e. $\rho_\sigma = n\omega_\sigma + 2id\bar\partial_\sigma F_\sigma$, in (27). □

To get rid of the second term here, we can use the following lemma.

Lemma 3 *Under the same assumptions as above, we have that*

$$\nabla^{(0,1)}(\nabla_{G\cdot dF}s) = -ik\omega \cdot G \cdot dFs - (d\bar{\partial}_\sigma F_\sigma) \cdot G \cdot \nabla^{(1,0)}s, \tag{29}$$

and thus we get the equality

$$\nabla^{(0,1)}(\Delta_G s + 2\nabla_{G\cdot dF}s) = -i(2k+n)\omega \cdot G \cdot \nabla^{(1,0)}s - ik\omega \cdot \delta(G)s - 2ik\omega \cdot G \cdot dFs. \tag{30}$$

Proof Again the proof is a calculation. For $s \in H_\sigma^{(k)}$, we have that

$$\begin{aligned}
\nabla^{(0,1)}(\nabla_{G\cdot dF}s) &= \nabla_{a''} G^{u'v'} dF_{v'} \nabla_{u'} s \\
&= G^{u'v'} dF_{v'} \nabla_{a''} \nabla_{u'} s + G^{u'v'} \nabla_{a''}(dF_{v'}) \nabla_{u'} s \\
&= G^{u'v'} dF_{v'} [\nabla, \nabla]_{a''u'} s + \nabla_{a''}(dF_{u'}) G^{u'v'} \nabla_{v'} s \\
&= -ik\omega_{a''u'} G^{u'v'} dF_{v'} s + (\bar{\partial} dF) \cdot G \cdot \nabla s \\
&= -ik\omega \cdot G \cdot dFs + (\bar{\partial}\partial F) \cdot G \cdot \nabla s \\
&= -ik\omega \cdot G \cdot dFs - (\partial\bar{\partial} F) \cdot G \cdot \nabla s.
\end{aligned}$$

Now (30) follows by combining equations (28) and (29). □

4.1. The Weakly Restricted Case

Now we impose the weakly restricted condition and continue the construction in the case where the family of complex structures is holomorphic. The following lemma and proposition are the key components.

Lemma 4 *Consider the situation as in Corollary 1, and assume that the family of holomorphic structures J_σ is weakly restricted. Furthermore, assume that the family admits a family of Ricci potentials F.*

Let $G_\beta(V)$ and $\beta(V)$ be bivector and vector fields associated to the family J_σ satisfying (1). Then we have that the 1-form

$$\Omega(V) = -\delta(G_\beta(V)) \cdot \omega + \delta(\bar{\partial}\beta(V)) - 2dF \cdot G_\beta(V) \cdot \omega + 2\bar{\partial}\beta(V) \cdot dF + 4i\bar{\partial}V'[F] \tag{31}$$

is closed and of type (0, 1).

Proof We start with the equation $\rho_\sigma = n\omega + 2id\bar{\partial}_\sigma F_\sigma$. Differentiating this equation along V', we get

$$\begin{aligned}
V'[\rho] &= 2idV'[\bar{\partial}]F + 2id\bar{\partial}V'[F] \\
&= -\ dV'[J] \cdot dF + d(2i\bar{\partial}V'[F]) \\
&= -\ d(G_\beta(V) \cdot \omega) \cdot dF + d\bar{\partial}\beta(V) \cdot dF + d(2i\bar{\partial}V'[F]) \\
&= -\ d(dF \cdot G_\beta(V) \cdot \omega) + d(\bar{\partial}\beta(V) \cdot dF) + d(2i\bar{\partial}V'[F]).
\end{aligned}$$

Now we use Proposition 2, which gives us

$$2V'[\rho] = d\delta(G(V)\cdot\omega) = d\delta(V'[J]) = d\delta(G_\beta(V))\cdot\omega - d\delta(\bar\partial\beta(V)).$$

Inserting this on the left-hand side of the above equation, we get that

$$\begin{aligned} 0 = &- d(\delta(G_\beta(V))\cdot\omega) + d\delta(\bar\partial\beta(V)) - d(2dF\cdot G_\beta(V)\cdot\omega) \\ &+ d(2\bar\partial\beta(V)\cdot dF) + d(4i\bar\partial V'[F]) = d\Omega(V), \end{aligned}$$

which exactly states that $\Omega(V)$ is closed. By checking each term, it is also seen directly to be of type $(0,1)$. □

Proposition 3 *Consider the set-up of Lemma 4, and, furthermore, assume that $H^1(M,\mathbb{R}) = 0$. Then there exists $\psi(V) \in C^\infty(M)$, such that*

$$\begin{aligned} &\delta(G_\beta(V))\cdot\omega + 2dF\cdot G_\beta(V)\cdot\omega \\ &\quad = 4i\bar\partial V'[F] + 2\bar\partial(dF\cdot\beta(V)) + in\omega\cdot\beta(V) + \bar\partial\delta(\beta(V)) - \bar\partial\psi(V). \end{aligned}$$

If none of the complex structures admits non-constant holomorphic functions on M, which is true, for instance, if M is compact, we get that $\psi(V) = 0$.

Proof We know that $\Omega(V)$ is closed and of type $(0,1)$ and, since we have assumed $H^1(M,\mathbb{R}) = 0$, it is exact. Thus, there exists a function $\psi(V) \in C^\infty(M)$, such that $\Omega(V) = \bar\partial\psi(V)$.

Observe that, if none of the complex structures admits non-constant holomorphic functions on M, we get that $\psi(V) = 0$, since the equation $d\psi(V) = \bar\partial\psi(V)$ shows that $\psi(V)$ is anti-holomorphic.

Combining expressions, we get that

$$\delta(G_\beta(V))\cdot\omega + 2dF\cdot G_\beta(V)\cdot\omega = 4i\bar\partial V'[F] + 2dF\cdot\bar\partial\beta(V) + \delta(\bar\partial\beta(V)) - \bar\partial\psi(V).$$

Now we only need to rewrite the term $\delta(\bar\partial\beta(V))$, and this is again done by an application of commuting covariant derivatives, which goes as follows:

$$\begin{aligned} \delta(\bar\partial\beta(V)) &= \nabla_{u'}\nabla_{a''}\beta(V)^{u'} \\ &= [\nabla,\nabla]_{u'a''}\beta(V)^{u'} + \nabla_{a''}\nabla_{u'}\beta(V)^{u'} \\ &= R^{u'}_{u'a''v'}\beta(V)^{v'} + \bar\partial\delta(\beta(V)) \\ &= r_{a''v'}\beta(V)^{v'} + \bar\partial\delta(\beta(V)) \\ &= J^u_{a''}r_{uw}J^w_{v'}\beta(V)^{v'} + \bar\partial\delta(\beta(V)) \\ &= i\rho_{a''v'}\beta(V)^{v'} + \bar\partial\delta(\beta(V)) \\ &= i\rho\cdot\beta(V) + \bar\partial\delta(\beta(V)) \\ &= in\omega\cdot\beta(V) - 2\partial\bar\partial F\cdot\beta(V) + \bar\partial\delta(\beta(V) \\ &= in\omega\cdot\beta(V) + 2\bar\partial dF\cdot\beta(V) + \bar\partial\delta(\beta(V). \end{aligned}$$

Inserting this above and rewriting

$$\bar{\partial} dF \cdot \beta(V) + dF \cdot \bar{\partial}\beta(V)) = \bar{\partial}(dF \cdot \beta(V)),$$

we obtain the desired equation, which completes the proof. □

We can now insert this in the expression from Lemma 3 and we get that

$$\begin{aligned}
&\nabla^{(0,1)}(\Delta_{G_\beta(V)}s + 2\nabla_{G_\beta(V)\cdot dF}s)\\
&\quad = -i(2k+n)\omega \cdot G_\beta(V) \cdot \nabla^{(1,0)}s - ik(\omega \cdot \delta(G_\beta(V))s + 2\omega \cdot G_\beta(V) \cdot dFs)\\
&\quad = i(2k+n)(G_\beta(V) \cdot \omega) \cdot \nabla^{(1,0)}s + ik(\delta(G_\beta(V)) \cdot \omega + 2dF \cdot G_\beta(V) \cdot \omega)s\\
&\quad = i(2k+n)(V'[J] + \bar{\partial}\beta(V)) \cdot \nabla^{(1,0)}s\\
&\qquad + ik(4i\bar{\partial}V'[F] + 2\bar{\partial}(dF \cdot \beta(V)) + in\omega \cdot \beta(V) + \bar{\partial}\delta(\beta(V)) - \bar{\partial}\psi(V))s\\
&\quad = 2(2k+n)\frac{i}{2}V'[J] \cdot \nabla^{(1,0)}s + i(2k+n)\bar{\partial}\beta(V) \cdot \nabla^{(1,0)}s\\
&\qquad - 4k\bar{\partial}V'[F]s + 2ik\bar{\partial}(dF \cdot \beta(V))s - kn\omega \cdot \beta(V)s + ik\bar{\partial}\delta(\beta(V))s - ik\bar{\partial}\psi(V)s.
\end{aligned}$$

Now we only need one last lemma to get rid of the last first-order term on the right side.

Lemma 5 *For any family of vector fields $\beta(V) \in C^\infty(M_\sigma, T'M_\sigma)$ and a holomorphic section $s \in H^k_\sigma$, we have that*

$$\nabla^{(0,1)}\nabla_{\beta(V)}s = \bar{\partial}\beta(V) \cdot \nabla^{(1,0)}s - ik\omega \cdot \beta(V)s. \tag{32}$$

Proof The result follows directly by the following calculation

$$\begin{aligned}
\nabla^{(0,1)}\nabla_{\beta(V)}s &= \bar{\partial}\beta(V) \cdot \nabla^{(1,0)}s + \beta(V)^{u'}\nabla_{a''}\nabla_{u'}s\\
&= \bar{\partial}\beta(V) \cdot \nabla^{(1,0)}s + \beta(V)^{u'}[\nabla, \nabla]_{a''u'}s\\
&= \bar{\partial}\beta(V) \cdot \nabla^{(1,0)}s - \beta(V)^{u'}ik\omega_{a''u'}s\\
&= \bar{\partial}\beta(V) \cdot \nabla^{(1,0)}s - ik\omega \cdot \beta(V)s.
\end{aligned}$$

□

Now, using this lemma, we get that

$$\begin{aligned}
&\nabla^{(0,1)}(\Delta_{G_\beta(V)}s + 2\nabla_{G_\beta(V)\cdot dF}s - i(2k+n)\nabla_{\beta(V)}s)\\
&\quad = 2(2k+n)\frac{i}{2}V'[J] \cdot \nabla^{(1,0)}s - 4k\bar{\partial}V'[F]s\\
&\qquad + 2ik\bar{\partial}(dF \cdot \beta(V))s - 2k(k+n)\omega \cdot \beta(V)s + ik\bar{\partial}\delta(\beta(V))s - ik\bar{\partial}\psi(V)s,
\end{aligned}$$

and now, by moving all the zeroth order terms to the left side, we get the desired result. Here, $\phi(V) \in C^\infty(M)$ is a smooth function, such that $\bar{\partial}\phi(V) = \omega \cdot \beta(V)$, and thus we get that

$$\begin{aligned}
&\nabla^{(0,1)}(\Delta_{G_\beta(V)}s + 2\nabla_{G_\beta(V)\cdot dF}s - i(2k+n)\nabla_{\beta(V)}s \\
&\qquad + 4kV'[F]s - 2ikdF\cdot\beta(V)s - ik\delta(\beta(V))s + 2k(k+n)\phi(V)s + ik\psi(V)s) \\
&\qquad = 2(2k+n)\frac{i}{2}V'[J]\cdot\nabla^{(1,0)}s.
\end{aligned}$$

This completes the proof of Theorem 1, since we see that we get a Hitchin connection, by setting

$$\begin{aligned}
u(V) = &\frac{1}{2(2k+n)}(\Delta_{G_\beta(V)} + 2\nabla_{G_\beta(V)\cdot dF} - i(2k+n)\nabla_{\beta(V)} \\
&+ 4kV'[F] - 2ikdF\cdot\beta(V) - ik\delta(\beta(V)) + 2k(k+n)\phi(V) + ik\psi(V)).
\end{aligned}$$

4.2. *Hitchin Connection for Smooth Families of Complex Structures*

We have, in the above constructions of the Hitchin connection, assumed that the family of complex structures is holomorphic. We can, however, go through the construction without assuming that $\mathcal{T}$ is a complex manifold. We have not used holomorphicity of the family before assuming it to be weakly restricted so, instead of differentiating along V' in Lemma 4, we differentiate along V.

Doing this, we get that the form

$$\begin{aligned}
&-\delta(G_\beta(V))\cdot\omega - 2dF\cdot G_\beta(V)\cdot\omega + \delta(\bar\partial\beta(V)) + 2\bar\partial\beta(V)\cdot dF + 4i\bar\partial V[F] \\
&-\delta(\bar G(V))\cdot\omega - 2dF\cdot\bar G(V)\cdot\omega
\end{aligned}$$

is closed and, hence, exact. It is, however, no longer of type $(0,1)$, but it splits into a $(1,0)$ and a $(0,1)$ part, which come as ∂ and $\bar\partial$ of a function $\tilde\psi(V)\in C^\infty(M)$. Both of the new terms are of type $(1,0)$, so we get similarly as above

$$\bar\partial\tilde\psi(V) = -\delta(G_\beta(V))\cdot\omega + \delta(\bar\partial\beta(V)) - 2dF\cdot G_\beta(V)\cdot\omega + 2\bar\partial\beta(V)\cdot dF + 4i\bar\partial V[F].$$

Arguing as in the proof of Proposition 3, we get that

$$\begin{aligned}
&\delta(G_\beta(V))\cdot\omega + 2dF\cdot G_\beta(V)\cdot\omega \\
&\qquad = 4i\bar\partial V[F] + 2\bar\partial(dF\cdot\beta(V)) + in\omega\cdot\beta(V) + \bar\partial\delta(\beta(V)) - \bar\partial\psi(V).
\end{aligned}$$

Now going through the construction of the Hitchin connection as above, still assuming weakly restricted but without holomorphicity of the family of Kähler structures, we get the following theorem.

Theorem 2 (Hitchin connection for smooth $\mathcal{T}$) *Consider the same set-up as in Theorem 1, except the manifold $\mathcal{T}$ is only assumed to be smooth and the assumption of holomorphicity of the family J is dropped. Then there exists a Hitchin connection $\boldsymbol{\nabla}$ in the bundle $\hat H^{(k)}$ over $\mathcal{T}$, given by the expression*

$$\boldsymbol{\nabla}_V = \nabla^T_V + u(V),$$

where

$$u(V) = \frac{1}{2(2k+n)}(\Delta_{G_\beta(V)} + 2\nabla_{G_\beta(V)\cdot dF} - i(2k+n)\nabla_{\beta(V)}$$

$$+ 4kV[F] - 2ikdF\cdot\beta(V) - ik\delta(\beta(V)) + 2k(k+n)\phi(V) + ik\tilde{\psi}(V)), \quad (33)$$

where $\phi(V)$ *is defined as in Theorem 1, and* $\tilde{\psi}(V) \in C^\infty(M)$ *satisfies*

$$\bar{\partial}\tilde{\psi}(V) = \tilde{\Omega}(V),$$

and $\tilde{\Omega}(V) \in \Omega^1(M)$ *is the closed and hence exact 1-form*

$$\tilde{\Omega}(V) = -\delta(G_\beta(V))\cdot\omega + \delta(\bar{\partial}\beta(V)) - 2dF\cdot G_\beta(V)\cdot\omega + 2\bar{\partial}\beta(V)\cdot dF + 4i\bar{\partial}V[F].$$

5. The No-Go Theorem and Projective Flatness

In this section, we briefly recall the work [GM], in which Ginzburg and Montgomery show a no-go theorem, stating conditions under which no natural projectively flat connection can exist on the vector bundle of quantizations. It turns out that the Hitchin connection constructed in this paper fulfils the conditions, and thus it cannot be projectively flat in general.

To state the theorem, we need to fix notation. We let $\mathcal{H}$ be the group of Hamiltonian symplectomorphisms of M, and $\mathcal{G}$ be the group of diffeomorphisms of the unit circle bundle U of $\mathcal{L}$ which preserve the connection form. Last, we let $\mathcal{G}_0$ be the identity-connected component of $\mathcal{G}$. Let $J_0 \in \mathcal{C}_\omega(M)$ and assume that there exist a $\mathcal{C}^0_\omega(M)$ neighbourhood of J_0, such that $\mathcal{H}^{(k)}|_{\mathcal{C}^0_\omega(M)}$ is a vector bundle over $\mathcal{C}^0_\omega(M)$.

Theorem 3 (Ginzburg and Montgomery) *Assume that there exist a complex structure* J_0 *with stabilizer* G_{J_0} *in* $\mathcal{H}$ *of positive dimension, and that the infinitesimal representation of* G_{J_0} *on* $H^{(k)}_{J_0}$ *is non-trivial. Then there is no projectively flat connection on* $\mathcal{H}^{(k)}|_{\mathcal{C}^0_\omega(M)}$*, which is invariant under the* $\mathcal{G}_0$ *local action.*

We will now consider an example where we can apply our construction for a certain small-enough neighbourhood of a particular J_0 with such a symmetry group.

Let G be a compact simple and simply connected Lie group. We are going to consider a co-adjoint orbit M in $\mathfrak{g}^*$. On M we are going to consider the Kirillov–Kostant symplectic structure (see e.g. [W]). Furthermore, we have the natural G-invariant complex structure J_0 on M coming from the identification

$$M = G^{\mathbb{C}}/P,$$

where P is a parabolic subgroup determined by M. It is well known that (M, J_0) is rigid and that there exists a small-enough neighbourhood $\mathcal{C}^0_\omega(M)$ of $J_0 \in \mathcal{C}_\omega(M)$ such that

$$H^1(M_J, T_J) = 0$$

for all complex structures $J \in \mathcal{C}^0_\omega(M)$.

We now want to determine $\beta(V)_J$ uniquely for all $J \in \mathcal{C}^0_\omega(M)$ and all $V \in T_J\mathcal{C}^0_\omega(M)$ solving

$$V'[J]_J = -\bar{\partial}_J \beta(V)_J.$$

This we can do uniquely by the above vanishing of $H^1(M_J, T_J)$ and if we impose suitable conditions on $\beta(V)_J$. One possible way is to require that $\beta(V)_J$ be orthogonal to all holomorphic vector fields on (M, J), which will be an invariant condition for all symmetries each $J \in \mathcal{C}^0_\omega(M)$ has. We remark for latter use that another way could be to require special evaluation properties of $\beta(V)_J$ at various points on M; this will, however, typically break local symmetry groups at particular J. Further, we can determine a smooth family of Ricci potentials, by picking, for each complex structure $J \in \mathcal{C}^0_\omega(M)$, the unique potential with zero average. Since M is simply connected, there is a unique prequantum line bundle $(\mathcal{L}, \nabla, \langle\cdot,\cdot\rangle)$ with curvature $-i\omega$. Thus, we satisfy all assumptions of Theorem 1 and so we get the following corollary.

Corollary 2 *For the co-adjoint orbit M, we get a Hitchin connection in the bundle $H^{(k)}$ over the subspace $\mathcal{C}^0_\omega(M)$. This connection is invariant under the local action of the group of bundle automorphisms of the prequantum line bundle $(\mathcal{L}, \nabla, \langle\cdot,\cdot\rangle)$ covering the symplectomorphism group of (M, ω).*

We see that this connection therefore satisfies all the requirements of Ginzburg and Montgomery's Theorem 3 above, thus this connection cannot be projectively flat over $\mathcal{C}^0_\omega(M)$. There still remains the interesting questions of how to compute the curvature of this connection and to what extent this connection fails to be projectively flat.

6. Pull-Backs of the Hitchin Connection

Let us consider a symplectic manifold (M, ω), and assume that we have a rigid subfamily $\mathcal{T} \subseteq \mathcal{C}_\omega(M)$ of all the complex structures compatible with ω.

Furthermore, we assume that we have some connected subspace $\mathcal{C}^0_\omega(M)$ on which we can find a map $\Phi\colon \mathcal{C}^0_\omega(M) \to \mathrm{Diff}(M)$ denoted $J \mapsto \Phi_J$ such that, for each $J \in \mathcal{C}^0_\omega(M)$, there exists a $J' \in \mathcal{T}$ with

$$\Phi_J^*(J') = J \quad \text{and} \quad \Phi_{|\mathcal{T}} = \mathrm{Id}.$$

That is, Φ_J gives a biholomorphism from M with the complex structure J to M with the complex structure J' from the rigid family $\mathcal{T}$.

Now, for each J, we can consider the pullback bundle $\Phi_J^*\mathcal{L} \to M$, which is naturally isomorphic to $\mathcal{L}$ itself, since Φ_J is isotopic to the identity for all $J \in \mathcal{C}^0_\omega(M)$.

Choosing a holomorphic isomorphism $\widetilde{\Psi}_J\colon \mathcal{L} \to \Phi_J^*\mathcal{L}$, we get the following commutative diagram:

$$\begin{array}{ccccc} \mathcal{L} & \xrightarrow{\widetilde{\Psi}_J} & \Phi_J^*\mathcal{L} & \xrightarrow{p} & \mathcal{L} \\ \downarrow & & \downarrow & & \downarrow \\ M & \xrightarrow[\mathrm{Id}]{} & M & \xrightarrow[\Phi_J]{} & M, \end{array} \tag{34}$$

where p is the map given canonically in the construction of the pullback bundle. Composing the maps in the top of the diagram, we get an induced endomorphism on $\mathcal{L}$ given by $\Psi_J = p \circ \widetilde{\Psi}_J$. We need to fix Ψ_J uniquely up to the action of the automorphism group of the line bundle, $\mathrm{Aut}(\mathcal{L}) = C^*$. We seek a section Ψ of the bundle

$$\mathcal{L}(M) = \left\{(J, \Psi) \in \mathcal{C}^0_\omega(M) \times \mathrm{Hom}(\mathcal{L}, \mathcal{L}) \mid \Psi\colon (L, J) \to (L, J') \text{ holo. for some } J' \in \mathcal{T}\right\}$$

over $\mathcal{D}(M)$, where

$$\mathcal{D}(M) = \left\{(J, \Phi) \in \mathcal{C}^0_\omega(M) \times \mathrm{Diff}(M) \mid J = \Phi^*(J') \text{ for some } J' \in \mathcal{T}\right\},$$

which, in turn, is a bundle over $\mathcal{C}^0_\omega(M)$. If we have one point $x \in M$, which is fixed for all Φ_J, $J \in \mathcal{C}^0_\omega(M)$, then we can fix the ambiguity by requiring that

$$(\Psi_J)_x = \mathrm{Id}\colon \mathcal{L}_x \to \mathcal{L}_x,$$

and hence get the required section Ψ. Let us now assume we have a map

$$\pi_{\mathcal{T}}\colon \mathcal{C}^0_\omega(M) \to \mathcal{T},$$

which is compatible with some Φ. Then, any section Ψ, as above, will induce an isomorphism of the bundles

$$\begin{array}{ccccc} H^{(k)}_{|\mathcal{T}} & \longleftarrow & \pi^*_{\mathcal{T}}((H^{(k)})_{|\mathcal{T}}) & \xrightarrow{\cong} & H^{(k)}, \\ & \searrow & & \searrow \quad \swarrow & \\ & & \mathcal{T} & \xleftarrow{\pi_{\mathcal{T}}} & \mathcal{C}^0_\omega(M) \end{array} \tag{35}$$

and thus a projectively flat connection on $(H^{(k)})_{|\mathcal{T}}$, which we have by [AG1], induces a projectively flat connection on the pullback bundle, which then gives a projectively flat connection on $H^{(k)}$. Here, we have used that

$$\pi^*((H^{(k)})_{|\mathcal{T}})_J = H^0(M_J, \Phi_J^*\mathcal{L}^k),$$

giving us the isomorphism on each fibre and, since the diagram (35) commutes, an isomorphism on the level of bundles is obtained.

Example One example where we can construct Φ fulfilling the requirements for (M,ω) is the underlying symplectic manifold of $\mathbb{P}^n$, where, of course, ω is the Fubini–Study symplectic form.

On M every complex structure in a small-enough neighbourhood $\mathcal{C}^0_\omega(M)$ of the standard complex structure (J_0 on $\mathbb{P}^n$) on M, which are all biholomorphic to the standard complex structure, so we let $\mathcal{T} = \{J_0\}$, and thus we have $\mathcal{D}(M)_J \neq \emptyset$ for all $J \in \mathcal{C}^0_\omega(M)$.

It is known that $\mathbb{P}^n$ has the property that there exist $n+1$ points $x_0, x_1, \ldots, x_n \in \mathbb{P}^n$ such that any set of lifts of these to $\mathbb{C}^{n+1}$ is a basis. Further, we have, for any such set of $n+1$ points, that there exists a unique $\Phi_0 \in \mathrm{Aut}(\mathbb{P}^n)$ mapping $\Phi_0(y_i) = x_i$. This means that we can, for any $J \in \mathcal{C}^0_\omega(M)$, determine a unique biholomorphism $\Phi_J \colon (M,J) \to (M,J_0)$ such that $\Phi_J(x_i) = x_i$. This way, we can define a section $\Phi \in C^\infty(\mathcal{C}^0_\omega(M), D(M))$.

Note that $\Phi_{J_0} = \mathrm{Id}$; thus, Φ and $\mathcal{T} = \{J_0\}$ fulfil the requirements outlined above. Thus, we get a flat connection in $H^{(k)}$ over the entire space $\mathcal{C}^0_\omega(M)$. This connection does, however, not have the symmetry required by Theorem 3. Furthermore, it does not agree with the connection obtained in Corollary 2, since that connection is not projectively flat; however, that connection does have the symmetry properties.

We expect that a similar construction can be made to work for any co-adjoint orbit, by a similar 'symmetry-breaking' construction.

7. Further Examples

In this section, we give a number of examples where we can solve the weakly restricted criterion for open subsets of the entire family of complex structures on a given symplectic manifold and thus get a Hitchin connection on such open subspaces of all complex structures on the given symplectic manifold.

The first example we consider is $M = \mathbb{R}^{2n}$, with the standard symplectic structure J_0 and $\mathcal{C}^0_\omega(M)$ an open and small-enough neighbourhood of J_0, such that $\mathcal{H}^{(k)}\,|_{\mathcal{C}^0_\omega(M)}$ is a vector bundle over $\mathcal{C}^0_\omega(M)$.

We may also assume that

$$H^1(M_J, T'M_J) = 0$$

for all complex structures $J \in \mathcal{C}^0_\omega(M)$. This means that we have a solution to the weakly restricted criterion, with $G_\beta(V) = 0$ and $\beta(V)_\sigma$ a solution to

$$V'[J]_\sigma = -\bar{\partial}_\sigma \beta(V)_\sigma$$

for all vector fields V and points σ in $\mathcal{C}^0_\omega(M)$. We will need a smooth family of β's, which we can assume exists by choosing a suitable $\mathcal{C}^0_\omega(M)$.

In this case, the functions $\phi(V)$ and $\psi(V)$ in the expression of the Hitchin connection from Theorem 1 can be calculated explicitly by curve integrals (depending, of course, on a choice of base point in M) of the $\bar{\partial}$-exact forms that they are related to by definition.

Now let us consider the symplectic torus $M = \mathbb{R}^{2n}/\mathbb{Z}^{2n}$ with the standard symplectic structure ω. In this case, it is not true that the moduli space of complex structures is locally a point. We consider the usual moduli space of linear complex structures compatible with the standard symplectic structure, which is the moduli space of principal polarized abelian varieties. In fact, the space of all linear complex structures on $\mathbb{R}^{2n}$ compatible with ω can be identified with

$$\mathbb{H} = \{Z \in M_{n,n}(\mathbb{C}) \mid Z = Z^t, \mathrm{Im}(Z) > 0\}.$$

The complex structure corresponding to $Z \in \mathbb{H}$ we denote J_Z. It is easy to check that the map

$$\cdot\omega\colon H^0(M_{J_Z}, S^2(T'M_{J_Z})) \to H^1(M_{J_Z}, T'M_{J_Z})_\omega$$

is surjective for all $Z \in \mathbb{H}$. Recall that, for any $J \in C_\omega(M)$, there exists a unique $Z \in \mathbb{H}$, for which there exists a unique biholomorphism

$$\Phi_J\colon (M,J) \to (M,J_Z),$$

which induces the identity on $H^1(M,\mathbb{Z})$ and which preserves $0 \in M$. We then also have that

$$\cdot\omega\colon H^0(M_J, S^2(T'M_J)) \to H^1(M_J, T'M_J)_\omega \tag{36}$$

is surjective for all complex structures $J \in \mathcal{C}_\omega(M)$ and further gives us a natural projection map

$$\pi : \mathcal{C}_\omega(M) \to \mathbb{H}.$$

We now fix a prequantum line bundle $(\mathcal{L}, \nabla, \langle\cdot,\cdot\rangle)$ over (M,ω). Consider then the bundle of quantum spaces $\tilde{H}^{(k)} \to \mathbb{H}$, with its usual Hitchin connection (see e.g. [9, A1]) and further the pullback

$$\pi^*\tilde{H}^{(k)} \to \mathcal{C}_\omega(M).$$

Since each Φ_J induces the identity on the first homology, we see that $\Phi_J^*\mathcal{L} \cong \mathcal{L}$ and as, further, $\Phi_J(0) = 0$, we can find a section Ψ as discussed above which induces an isomorphism of the quantum bundles:

$$\Psi^* : \pi^*\tilde{H}^{(k)} \to H^{(k)}.$$

We now pull back the Hitchin connection in $\tilde{H}^{(k)}$ to $\pi^*\tilde{H}^{(k)}$ and push it to $H^{(k)}$ by this isomorphism, to get a projectively flat connection. Again by the no-go theorem, Theorem 3, this connection cannot be natural, which is also clear from its construction.

We will now show that the constructions of this paper apply to provide a construction of a natural partial connection in $H^{(k)}$ over $\mathcal{C}_\omega(M)$. Since we have that (36) is surjective, we see that the equation (1) can be solved for all $J \in \mathcal{C}^0_\omega(M)$ and all tangent vectors $V \in T_J\mathcal{C}^0_\omega(M)$. For any choice of solution to this equation, we get a $\bar\partial$-closed form $\omega \cdot \beta_J(V)$. However, the map

$$\omega\cdot: H^0(M_J, T'M_J) \to H^{0,1}(M)$$

is an isomorphism for all $J \in \mathcal{C}_\omega(M)$, so we can uniquely determine $\beta_J(V)$ as a solution to (1), by requiring that $\omega \cdot \beta_J(V) = 0$ in $H^{0,1}(M)$. This, in turn, means that we can indeed find a unique solution to the equation $\bar{\partial}\phi(V) = \omega \cdot \beta_J(V)$ of zero average.

We now consider the linear map

$$[\Omega]_J: T_J\mathcal{C}^0_\omega(M) \to H^{0,1}(M_J).$$

We see that we can apply our Hitchin connection construction along the distribution

$$\ker[\Omega] \subset T\mathcal{C}^0_\omega(M),$$

simply by just choosing the $\psi(V)$ with zero average which solves

$$\bar{\partial}\psi(V) = \Omega(V).$$

We remark that, if $\ker[\Omega]$ is always trivial, then we would indeed get that our construction gives a connection on the entire space of complex structures in this case.

The last example we will consider is the moduli spaces M of flat $SU(n)$-connections on a surface of genus $g > 1$, possibly with central holonomy around a point on the surface. We have the Narasimhan–Goldman–Atiyah–Bott symplectic form ω on (the smooth part of) M [AB]. We further have that the Chern–Simons functional induces the Chern–Simons line bundle $(\mathcal{L}, \nabla, \langle\cdot,\cdot\rangle)$ over (M,ω) [Fr, RSW].

We first consider the usual family of complex structures J parametrized by Teichmüller space $\mathcal{T}$. In this situation, Hitchin has proved [H] that the map

$$\cdot\omega: H^0(M_J, S^2(T'M_J)) \to H^1(M_J, T'M_J)_\omega$$

is surjective, for all $J \in \mathcal{T}$. We define $\mathcal{C}^0_\omega(M)$ to be the maximal connected subspace of all complex structures on M, which is compatible with ω and for which there exist a unique $J' \in \mathcal{T}$ and a unique biholomorphism

$$\Phi_J: (M,J) \to (M,J'),$$

with the property that it varies smoothly with $J \in \mathcal{C}^0_\omega(M)$, and $\Phi_J = \mathrm{Id}$, for all $J \in \mathcal{T}$. But now we see as above that (1) can always be solved and, since there are no holomorphic vector fields on (M,J) for all $J \in \mathcal{C}^0_\omega(M)$, we get a unique $\beta_J(V)$ with the needed properties for all $J \in \mathcal{C}^0_\omega(M)$, and $V \in T_J\mathcal{C}^0_\omega(M)$. But then we have that Theorem 1 applies and we get a Hitchin connection in $H^{(k)}$ over all of $\mathcal{C}^0_\omega(M)$. We have here normalized $\psi(V)$ and $\phi(V)$ by requiring that they have zero average over M.

REFERENCES

[A1] J.E. Andersen, 'Deformation quantization and geometric quantization of abelian moduli spaces', Comm. Math. Phys. 255(3) (2005), 727–45.

[A2] J.E. Andersen, 'Asymptotic faithfulness of the quantum $SU(n)$ representations of the mapping class groups', Ann. Math. 163(3) (2006), 347–68.

[A3] J.E. Andersen, 'The Nielsen–Thurston classification of mapping classes is determined by TQFT', J. Math. Kyoto Univ. 48 2 (2008), 323–38.

[A4] J.E. Andersen, 'Mapping class groups do not have Kazhdan's property (T)', math.QA/0706.2184 (2007).

[A5] J.E. Andersen, 'Hitchin's connection, Toeplitz operators, and symmetry invariant deformation quantization', Quantum Topol. 3 (3–4) (2012), 293–325.

[AGL] J.E. Andersen, N. L. Gammelgaard & M. Lauridsen, 'Hitchin's connection in half-form quantization', Quantum Topol. 3 (3–4) (2012), 327–57.

[AG1] J.E. Andersen & N.L. Gammelgaard, 'The Hitchin–Witten connection and complex quantum Chern–Simons theory', arXiv:1409.1035 (2014).

[AU1] J.E. Andersen & K. Ueno, 'Abelian conformal field theories and determinant bundles', Int. J. Math. 18(8) (2007), 919–93.

[AU2] J.E. Andersen & K. Ueno, 'Constructing modular functors from conformal field theories', J. Knot Theory Ramifications 16 (2) (2007), 127–202.

[AU3] J.E. Andersen & K. Ueno, 'Modular functors are determined by their genus zero data', Quantum Topol. 3 (3–4) (2012), 255–91.

[AU4] J.E. Andersen & K. Ueno, 'Construction of the Witten–Reshetikhin–Turaev TQFT from conformal field theory,' Invent. Math. 201(2) (2015), 519–59.

[AH] J.E. Andersen & B. Himpel, 'The Witten–Reshetikhin–Turaev invariants of finite order mapping tori II', Quantum Topol. 3(3) (2012), 377–421.

[A6] J.E. Andersen, 'The Witten–Reshetikhin–Turaev invariants of finite order mapping tori I', J. Reine Angew. Math. 2013(681) (2013), 1–38.

[AHJMMc] J.E. Andersen, B. Himpel, S. F. Jørgensen, J. Martens & B. McLellan, 'The Witten–Reshetikhin–Turaev invariant for links in finite order mapping tori I, Adv. Math. 304 (2017), 131–78.

[AB] M. Atiyah & R. Bott, 'The Yang–Mills equations over Riemann surfaces', Phil. Trans. R. Soc. Lond. A **308** (1505) (1982), 523–615.

[ADW] S. Axelrod, S. Della Pietra & E. Witten, 'Geometric quantization of Chern–Simons gauge theory', J. Differential Geom. 33 (1991), 787–902.

[Besse] A.L. Besse, '*Einstein Manifolds*', Springer-Verlag, Berlin (1987).

[B1] C. Blanchet, 'Hecke algebras, modular categories and 3-manifolds quantum invariants.', Topology 39 (1) (2000), 193–223.

[BHMV1] C. Blanchet, N. Habegger, G. Masbaum & P. Vogel, 'Three-manifold invariants derived from the Kauffman bracket', Topology 31(4) (1992), 685–99.

[BHMV2] C. Blanchet, N. Habegger, G. Masbaum & P. Vogel, 'Topological quantum field theories derived from the Kauffman bracket', Topology 34(4) (1995), 883–927.

[BMS] M. Bordeman, E. Meinrenken & M. Schlichenmaier, 'Toeplitz quantization of Kähler manifolds and $gl(N), N \to \infty$ limit', Comm. Math. Phys. 165(2) (1994), 281–96.

[Fr] D.S. Freed, 'Classical Chern–Simons Theory, Part 1', Adv. Math. 113 (1995), 237–303.

[vGdJ] B. Van Geemen & A.J. De Jong, 'On Hitchin's connection', J. Amer. Math. Soc. 11(1) (1998), 189–228.

[GM] V. Ginzburg & R. Montgomery, 'Geometric quantization and no-go theorems', Banach Centre Publ. **51**(1) (2000), 69–77.

[G] W.M. Goldman, 'The symplectic nature of fundamental groups of surfaces', Adv. Math. 54 (2) (1984), 200–25.

[H] N. Hitchin, 'Flat connections and geometric quantization', Comm. Math. Phys. 131(2) (1990), 347–80.

[KS] A.V. Karabegov & M. Schlichenmaier, 'Identification of Berezin–Toeplitz deformation quantization', J. Reine Angew. Math. 2001(540) (2001), 49–76.

[La1] Y. Laszlo, 'Hitchin's and WZW connections are the same', J. Differential Geom. 49 (3) (1998), 547–76.

[MM1] X. Ma & G. Marinescu, '*Holomorphic Morse Inequalities and Bergman Kernels*,' Prog. in Math., 254. Birkhäuser Verlag, Basel (2007).

[MM2] X. Ma & G. Marinescu, 'Toeplitz operators on symplectic manifolds', J. Geom. Anal. 18(2) (2008), 565–611.

[MM3] X. Ma & G. Marinescu, 'Berezin–Toeplitz quantization on Kähler manifolds', J. Reine Angew. Math. 2012(662) (2012), 1–56.

[NS1] M.S. Narasimhan & C.S. Seshadri, 'Holomorphic vector bundles on a compact Riemann surface', Math. Ann. 155(1) (1964) 69–80.

[NS2] M.S. Narasimhan & C.S. Seshadri, 'Stable and unitary vector bundles on a compact Riemann surface', Ann. Math. 82 (1965), 540–67.

[R1] T.R. Ramadas, 'Chern–Simons gauge theory and projectively flat vector bundles on M_g', Comm. Math. Phys. 128 (2) (1990), 421–26.

[RSW] T.R. Ramadas, I.M. Singer & J. Weitsman, 'Some comments on Chern–simons gauge theory', Comm. Math. Phys. 126(2) (1989), 409–20.

[RT1] N. Reshetikhin & V. Turaev, 'Ribbon graphs and their invariants derived from quantum groups', Comm. Math. Phys. 127(1) (1990), 1–26.

[RT2] N. Reshetikhin & V. Turaev, 'Invariants of 3-manifolds via link polynomials and quantum groups', Invent. Math. 103(1) (1991), 547–97.

[Sch1] M. Schlichenmaier, 'Zwei Anwendungen algebraisch-geometrischer Methoden in der theoretischen Physik' Dissertation, Universität Mannheim (1996).

[Sch2] M. Schlichenmaier, 'Deformation quantization of compact Kähler manifolds by Berezin–Toeplitz quantization', in M Flato, G. Dito and D. Sternheimer (eds) *Conférence Moshé Flato 1999, Vol. II (Dijon)*, Math. Phys. Stud., 22, Kluwer Acad. Publ., Dordrecht (2000), 289–306.

[Sch3] M. Schlichenmaier, 'Berezin-Toeplitz quantization and Berezin transform', in S Graffi and A. Martinez (eds) *Long time behaviour of classical and quantum systems (Bologna, 1999)*, Ser. Concr. Appl. Math., 1, World Sci. Publishing, River Edge, NJ (2001), 271–87.

[TUY] A. Tsuchiya, K. Ueno & Y. Yamada, 'Conformal field theory on universal family of stable curves with gauge symmetries', in M. Jimbo, T. Miwa and A. Tsuchiya (eds) *Integrable Sys Quantum Field Theory*, Adv. Stud. Pure Math. 19, Academic Press, New York, NY (1989), 459–566.

[T] V. G. Turaev, '*Quantum Invariants of Knots and 3-Manifolds*,' de Gruyter Stud. Math., **18**, Walter de Gruyter & Co., Berlin (1994).

[Tuyn] G.M. Tuynman, Quantization: Towards a comparision between methods', J. Math. Phys. 28(12) (1987), 2829–40.

[W1] E. Witten, 'Quantum field theory and the Jones polynomial', Comm. Math. Phys 121(3) (1989) 351–98.

[W] N.J. Woodhouse, '*Geometric Quantization*,' Oxford University Press, Oxford (1992).

8

The Taub–NUT Ambitoric Structure

PAUL GAUDUCHON

Paul Gauduchon, CMLS, École Polytechnique, CNRS, Université Paris-Saclay, 91128 Palaiseau, France. Email: paul.gauduchon@polytechnique.edu.

En hommage à Nigel Hitchin pour son soixante-dixième anniversaire, avec toute mon amitié et ma profonde admiration.

1. Introduction

The aim of this paper is to pair two apparently unrelated complete Kähler metrics defined on the standard complex plane $\mathbb{C}^2$ to form a complete extremal *ambitoric pair*, as defined in [5]–[6], i.e. a pair of complete conformally related Riemannian metrics on $\mathbb{R}^4$, both Kähler with respect to complex structures inducing opposite orientations, both toric with respect to the same $\mathbb{T}^2$-action and both extremal in the sense of Calabi. This ambitoric pair appears in a natural way in the general context of the *Gibbons–Hawking ansatz* for the construction of four-dimensional hyperkähler structures admitting a free S^1-action, here the well-known *Taub–NUT hyperkähler structure* on $\mathbb{R}^4$, regarded as the *standard deformation* of the standard flat hyperkähler structure, as defined and described in Paragraph 6.2. The resulting ambitoric structure, called the *Taub–NUT ambitoric pair*, is described in Section 5.

The 'positive' partner of the Taub–NUT ambitoric pair is one of the Kähler structures pertaining to the Taub–NUT hyperkähler structure, determined by the choice of a suitable S^1-action; in particular, the metric g_+ is Ricci-flat and self-dual (cf. Remark 15). The chosen Kähler structure, denoted by (g_+, J_+, ω_+) and called the *Ricci-flat Taub–NUT Kähler structure*, turns out to be identified with a complete Kähler metric on $\mathbb{C}^2$ (cf. Proposition 1; the argument presented here is a variant of C. LeBrun's argument in [19]).

The 'negative' partner of the Taub–NUT ambitoric structure is the almost-Kähler structure, which, in general, is naturally associated to any Gibbons–Hawking construction with a fixed S^1-action, as explained in Paragraph 6.3, and which turns out to be *Kähler* in

Date: 10 November, 2017.

Gauduchon, P., *The Taub–NUT Ambitoric Structure*. In: *Geometry and Physics: A Festschrift in Honour of Nigel Hitchin*, Jørgen Ellegaard Andersen, Andrew Dancer, Oscar García-Prada (Eds): Oxford University Press (2018).
 DOI: 10.1093/oso/9780198802006.003.0008

the Taub–NUT context, as shown in Proposition 8. This Kähler structure induces the opposite orientation and is therefore *self-dual*, or *Bochner-flat*. It will be called the *Bochner-flat Taub–NUT Kähler structure*, denoted by (g_-, J_-, ω_-). This too can be identified with a complete Kähler metric on $\mathbb{C}^2$, actually with one of the complete Bochner-flat Kähler metrics constructed by R. Bryant in [10] (cf. Proposition 4).

The toric structures of the Ricci-flat Taub–NUT Kähler structures and the Bochner-flat Taub–NUT Kähler structures are described in Paragraphs 3.2 and 4.2, respectively.

In Section 5, it is shown that the Ricci-flat Taub–NUT Kähler structure (g_+, J_+, ω_+) comes equipped with a *Hamiltonian 2-form* of maximal order 2, in the sense of [4, Definition 2], compatible with the toric structure. It follows that (g_+, J_+, ω_+) is *orthotoric* and that the Taub–NUT ambitoric pair is *regular* and *of parabolic type* (cf. Propositions 5 and 6).

In the last section, Section 6, we review some main features, which are used throughout the paper, of the general Gibbons–Hawking construction.

The Taub–NUT ambitoric pair presented in this paper is closely related to a family of examples of ambitoric pairs appearing in [8]. Further examples, as well as a classification of complete ambitoric 4-orbifolds, are provided in [13].

2. The Taub–NUT Hyperkähler Structure

In this section, we introduce the 1-parameter family of Taub–NUT hyperkähler structures as the standard deformation, as described in Paragraph 6.2, of the standard flat hyperkähler structure on the Euclidean space $\mathbb{R}^4$, in the framework of the general Gibbons–Hawking construction.

2.1. *The Standard Flat Hyperkähler Structure as the Completion of a Gibbons–Hawking Construction*

We denote by $\mathbb{H}$ the field of quaternions identified with the Euclidean space $\mathbb{R}^4$, by writing each quaternion u in the usual form: $u = u_0 + u_1 i + u_2 j + u_3 k$, where u_0, u_1, u_2, u_3 are real numbers, and i, j, k satisfy $i^2 = j^2 = k^2 = -1$, $ijk = -1$. We then set $e_0 = 1$, $e_1 = i$, $e_2 = j$, $e_3 = k$, so that $\{e_0, e_1, e_2, e_3\}$ forms an orthonormal basis of $\mathbb{R}^4$ with respect to the inner product $(u, v) = \mathrm{Re}(u\bar{v})$.

In the sequel, $\mathbb{R}^4 = \mathbb{H}$ will be regarded as a four-dimensional manifold, the above inner product as the (constant) flat Riemannian metric of $\mathbb{R}^4$, denoted by g_0, the e_i as constant vector fields, forming an orthonormal trivialization of the tangent bundle $\mathrm{T}\mathbb{R}^4$ with respect to g_0, each e_i being identified with the 1-form du_i via the metric g_0. Moreover, the *left* multiplication by i, j, k in $\mathbb{H}$ determines three complex structures on $\mathbb{R}^4$, denoted by I_1, I_2, I_3, respectively, whose expression relative to the orthonormal frame $\{e_0, e_1, e_2, e_3\}$ is

$$I_1 e_0 = e_1, I_1 e_2 = e_3, \quad I_2 e_0 = e_2, I_2 e_3 = e_1, \quad I_3 e_0 = e_3, I_3 e_1 = e_2. \tag{2.1}$$

Together with the metric g_0, I_1, I_2, I_3 determine the *standard flat hyperkähler structure* on $\mathbb{R}^4$. The corresponding Kähler forms $\omega_j = g_0(I_j\cdot, \cdot)$ are then

$$\begin{aligned}\omega_1 &= du_0 \wedge du_1 + du_2 \wedge du_3, \ \omega_2 = du_0 \wedge du_2 + du_3 \wedge du_1, \\ \omega_3 &= du_0 \wedge du_3 + du_1 \wedge du_2. \end{aligned} \tag{2.2}$$

These 2-forms generate the space of *self-dual* 2-forms with respect to g_0 and the standard orientation of $\mathbb{R}^4$, which is also the common orientation induced by I_1, I_2, I_3. The whole hyperkähler structure is preserved by the group Sp_1, of quaternions of norm 1, acting by *right* multiplication on $\mathbb{H} = \mathbb{R}^4$, in particular by the induced action

$$\begin{aligned} ue^{it} &= (u_0 \cos t - u_1 \sin t) + (u_1 \cos t + u_0 \sin t)\, i \\ &\quad + (u_2 \cos t + u_3 \sin t)\, j + (u_3 \cos t - u_2 \sin t)\, k \end{aligned} \tag{2.3}$$

of the subgroup S^1 of complex number of norm 1, whose generator is the vector field T defined by

$$T = -u_1\, e_0 + u_0\, e_1 + u_3\, e_2 - u_2\, e_3; \tag{2.4}$$

T is then Killing with respect to g_0, (real) holomorphic with respect to I_1, I_2, I_3 and Hamiltonian with respect to $\omega_1, \omega_2, \omega_3$; more precisely,

$$\begin{aligned} -\iota_T\omega_1 &= u_0 du_o + u_1 du_1 - u_2 du_2 - u_3 du_3 = dx_1, \\ -\iota_T\omega_2 &= u_3 du_0 + u_2 du_1 + u_1 du_2 + u_0 du_3 = dx_2, \\ -\iota_T\omega_3 &= -u_2 du_0 + u_3 du_1 - u_0 du_2 + u_1 du_3 = dx_3, \end{aligned} \tag{2.5}$$

where the momenta x_1, x_2, x_3 are given by

$$x_1 = \frac{1}{2}\left(u_0^2 + u_1^2 - u_2^2 - u_3^2\right), \quad x_2 = u_1 u_2 + u_0 u_3, \quad x_3 = u_1 u_3 - u_0 u_2. \tag{2.6}$$

The map

$$u \mapsto (z_1 := u_0 + iu_1, z_2 := u_2 + iu_3), \tag{2.7}$$

from $\mathbb{H} = \mathbb{R}^4$ to $\mathbb{C}^2$, is clearly a biholomorphism from $(\mathbb{R}^4, I_1)$ to the standard complex plane $\mathbb{C}^2$. This isomorphism is S^1-equivariant with respect to the S^1-action (2.3) on $\mathbb{R}^4$ and the S^1-action on $\mathbb{C}^2$ described by

$$(z_1, z_2)\, e^{it} = (z_1 e^{it}, z_2 e^{-it}). \tag{2.8}$$

Equivalently,

$$dz_1(T) = iz_1, \qquad dz_2(T) = -iz_2. \tag{2.9}$$

Regarded as functions defined on $\mathbb{C}^2$ via this isomorphism, the momenta x_1, x_2, x_3 are given by $x_1 = \frac{|z_1|^2 - |z_2|^2}{2}$, $x_2 + ix_3 = -iz_1 z_2$. These are also determined by the commutation formula in $\mathbb{H}$:

$$u\, i = \frac{2\,(x_1\, i + x_2\, j + x_3\, k)}{r^2}\, u, \tag{2.10}$$

for any u in $\mathbb{H}$, by setting $r := \left(\sum_{j=1}^{3} x_j^2\right)^{\frac{1}{2}}$. Notice that r is related to the Euclidean norm $\rho := |u| = \left(\sum_{i=0}^{3} u_i^2\right)^{\frac{1}{2}}$ in $\mathbb{R}^4$ by

$$\rho^2 = 2r \tag{2.11}$$

and that

$$\rho_1^2 := u_0^2 + u_1^2 = |z_1|^2 = r + x_1, \qquad \rho_2^2 := u_2^2 + u_3^2 = |z_2|^2 = r - x_1. \tag{2.12}$$

Denote by $\pi : \mathbb{R}^4 \to \mathbb{R}^3$ the *momentum map* defined by $\pi(u) = (x_1, x_2, x_3)$. In restriction to $\mathbb{R}^4 \setminus \{0\}$, π is a S^1-principal bundle over $\mathbb{R}^3 \setminus \{0\}$, whose fibres are the orbit of the S^1-action (2.3). This fibration comes equipped with a S^1-invariant connection, whose connection 1-form, η, viewed as a real 1-form on $\mathbb{R}^4 \setminus \{0\}$, is determined by the conditions $\eta(T) = 1$ and $\mathcal{L}_T \eta = 0$, where $\mathcal{L}_T$ denotes the Lie derivative along T; we then have

$$\eta := \frac{\theta}{\rho^2}, \tag{2.13}$$

where $\theta = -u_1 du_0 + u_0 du_1 + u_3 du_2 - u_2 du_3$ is the 1-form on $\mathbb{R}^4$ dual to T with respect to g_0. The curvature $d\eta$ of this connection can be viewed equivalently as a real 2-form defined on $\mathbb{R}^4 \setminus \{0\}$ or on $\mathbb{R}^3 \setminus \{0\}$, via the map π. By a simple computation, using (2.4) and (2.6), we get

$$\begin{aligned} d\eta &= \frac{1}{\rho^2}\left(\bar{\omega} - \frac{2}{\rho^2}\sum_{j=1}^{3} x_j \omega_j\right) \\ &= -\frac{x_1 dx_2 \wedge dx_3 + x_2 dx_3 \wedge dx_1 + x_3 dx_1 \wedge dx_2}{2r^3} = *_3 d\left(\frac{1}{2r}\right), \end{aligned} \tag{2.14}$$

where the (anti-self-dual) 2-form $\bar{\omega}$ is defined by $\bar{\omega} = du_0 \wedge du_1 - du_2 \wedge du_3$, and $*_3$ denotes the Hodge operator on $\mathbb{R}^3$, oriented by $dx_1 \wedge dx_2 \wedge dx_3$. From (2.5) and (2.13), we readily infer that $\{\frac{dx_1}{\rho}, \frac{dx_2}{\rho}, \frac{dx_3}{\rho}, \rho\,\eta = \frac{\theta}{\rho}\}$ is an *orthonormal coframe* on $\mathbb{R}^4 \setminus \{0\}$ with respect to g_0, so that

$$g_0 = \frac{1}{\rho^2}\sum_{j=1}^{3} dx_j \otimes dx_j + \rho^2\, \eta \otimes \eta. \tag{2.15}$$

In terms of the dx_j, η, the pairs (ω_j, I_j) have the following expression:

$$\begin{aligned} &\omega_1 = dx_1 \wedge \eta + \frac{1}{\rho^2} dx_2 \wedge dx_3,\ \omega_2 = dx_2 \wedge \eta + \frac{1}{\rho^2} dx_3 \wedge dx_1, \\ &\omega_3 = dx_3 \wedge \eta + \frac{1}{\rho^2} dx_1 \wedge dx_2, \\ &I_1 dx_1 = \theta,\ I_1 dx_2 = dx_3, \quad I_2 dx_2 = \theta,\ I_2 dx_3 = dx_1, \quad I_3 dx_3 = \theta,\ I_3 dx_1 = dx_2 \end{aligned} \tag{2.16}$$

(here and henceforth, we agree that the action of an almost-complex structure, say, I, on a 1-form α, is defined by $(I\alpha)(X) = -\alpha(IX)$, for any vector field X).

2.2. The Taub–NUT Hyperkähler Structure as a Deformation of the Standard Flat Hyperkähler Structure on $\mathbb{R}^4$

In the expressions (2.15)–(2.16) of the elements of the standard flat hyperkähler structure on $\mathbb{R}^4$, we recognize the well-known *Gibbons–Hawking construction*, attached to the (positive) harmonic function $V(x_1,x_2,x_3) = \frac{1}{2r} = \frac{1}{\rho^2}$, defined on $\mathbb{R}^3 \setminus \{0\}$ (cf. Section 6). More precisely, the standard flat hyperkähler structure on $\mathbb{R}^4$ here appears as the (smooth) *completion* to $\mathbb{R}^4$ of the hyperkähler structure constructed on $\mathbb{R}^4 \setminus \{0\}$ by the *Gibbons–Hawking ansatz*, via the restriction of the momentum map π to $\mathbb{R}^4 \setminus \{0\}$.

Its standard deformation, in the sense of Paragraph 6.2, associated to the harmonic function $V_a := a^2 + \frac{1}{2r}$, is the 1-parameter family of *Taub–NUT hyperkähler structures*. It can be deduced from Remark 17 that, for any $a \geq 0$, the corresponding Taub–NUT hyperkähler structure is also smoothly defined on $\mathbb{R}^4$. This also readily follows from the following expression, deduced from (6.10)—with $V = \frac{1}{2r} = \frac{1}{\rho^2}$—of the Taub–NUT metric g_a, as a metric defined on $\mathbb{R}^4 \setminus \{0\}$:

$$g_a = (1 + a^2\rho^2)\, g_0 - \frac{a^2\,(2 + a^2\rho^2)}{1 + a^2\rho^2}\, \theta \otimes \theta, \tag{2.17}$$

where g_0, ρ^2 and θ are smoothly defined on $\mathbb{R}^4$. On $\mathbb{R}^4$, the Kähler forms, $\omega_1^a, \omega_2^a, \omega_3^a$, of the Taub–NUT hyperkähler structure are given by

$$\omega_1^a = \omega_1 + a^2\, dx_2 \wedge dx_3, \quad \omega_2^a = \omega_2 + a^2\, dx_3 \wedge dx_1, \quad \omega_3^a = \omega_3 + a^2\, dx_1 \wedge dx_2 \tag{2.18}$$

and the complex structures by

$$I_j^a = I_j + \frac{a^2}{1 + a^2\rho^2}\, \theta \otimes (dx_j)^{\sharp_{g_0}} + a^2\, dx_j \otimes T, \tag{2.19}$$

for $j = 1, 2, 3$.

Remark 1 For any $a \geq 0$, the Taub–NUT metric g_a is preserved by the group $G := (\mathrm{Sp}_1 \times \mathrm{S}^1)/\pm 1 \cong \mathrm{U}_2$, where Sp_1 acts by left multiplication, and S^1 by right multiplication on $\mathbb{H} = \mathbb{R}^4$. For each $u \neq 0$ of $\mathbb{H} = \mathbb{R}^4$, it readily follows from (2.10) that the stabilizer of u in G is the S^1-group in G of pairs of the form $\left(\cos t - \sin t\left(\frac{x_1 i + x_2 j + x_3 k}{r}\right), e^{it}\right)$, with $(x_1,x_2,x_3) = \pi(u)$, for any real t. The action of G on $\mathbb{H} = \mathbb{R}^4$ covers the action of the full rotation group SO_3 on $\mathbb{R}^3$ via the momentum map π and hence acts as SO_3 on the space of Kähler forms $\omega_1^a, \omega_2^a, \omega_3^a$ of the Taub–NUT hyperkähler structure.

Remark 2 A well-known alternative formulation of the Taub–NUT metric g_a is obtained by introducing the 1-forms $\sigma_1 := \frac{1}{\rho^2}\left(-u_2\, du_0 - u_3\, du_1 + u_0\, du_2 + u_1\, du_3\right)$, $\sigma_2 := \frac{1}{\rho^2}\left(-u_3\, du_0 + u_2\, du_1 - u_1\, du_2 + u_0\, du_3\right)$, $\sigma_3 = \eta = \frac{1}{\rho^2}\left(-u_1\, du_0 + u_0\, du_1 + u_3\, du_2 - u_2\, du_3\right)$ on $\mathbb{R}^4$, which are invariant by Sp_1, acting by left multiplication on

$\mathbb{H}$ and satisfy $d\sigma_1 = -2\sigma_2 \wedge \sigma_3$, $d\sigma_2 = -2\sigma_3 \wedge \sigma_1$, $d\sigma_3 = -2\sigma_1 \wedge \sigma_2$. The coframe $\{d\rho, \rho\sigma_1, \rho\sigma_2, \rho\sigma_3\}$ on $\mathbb{R}^4 \setminus \{0\}$ is g_0-orthonormal, so that

$$g_0 = d\rho \otimes d\rho + \rho^2 \left(\sigma_1 \otimes \sigma_1 + \sigma_2 \otimes \sigma_2 + \sigma_3 \otimes \sigma_3\right), \tag{2.20}$$

and, since $\sigma_3 = \eta$, it follows from (2.17) that, on $\mathbb{R}^4 \setminus \{0\}$, g_a has the following expression:

$$g_a = (1 + a^2\rho^2)\left(d\rho \otimes d\rho + \rho^2\left(\sigma_1 \otimes \sigma_1 + \sigma_2 \otimes \sigma_2 + \frac{\sigma_3 \otimes \sigma_3}{(1 + a^2\rho^2)^2}\right)\right). \tag{2.21}$$

By choosing as a new radial coordinate the Gibbons–Hawking function $V = \frac{1}{\rho^2}$, g_a can be put in the canonical form

$$g_a = \frac{1}{4}(a_1 a_2 a_3)^2\, dV \otimes dV + \sum_{j=1}^{3} a_j^2\, \sigma_j \otimes \sigma_j \tag{2.22}$$

(cf. Chapter 4 of N. Hitchin's book [18]), with $a_1(V) = a_2(V) = \frac{(V+a^2)^{\frac{1}{2}}}{V} = \frac{V_a^{\frac{1}{2}}}{V}$, $a_3(V) = \frac{1}{(V+a^2)^{\frac{1}{2}}} = \frac{1}{V_a^{\frac{1}{2}}}$. Up to translating V by a constant, this triplet is the unique solution to the differential system (5.4) in [18, Chapter 4] in the case when $a_1 = a_2$, this system, which ensures that g_a is Ricci-flat, being then reduced to

$$a_1' = \frac{1}{2}a_1 a_3(a_3 - 2a_1), \qquad a_3' = -\frac{1}{2}(a_3)^3. \tag{2.23}$$

3. The Ricci-Flat Taub–NUT Kähler Metric

Within the Taub–NUT hyperkähler structure introduced in the previous section, regarded as a family of Kähler metrics on $\mathbb{R}^4$, we focus our attention on the Kähler structure (g_a, I_1^a, ω_1^a), henceforth simply denoted (g_+, J_+, ω_+) (here, and most often in the sequel, the parameter a is omitted in the notation). We then have

$$g_+ = (1 + a^2\rho^2)g_0 - \frac{a^2(2 + a^2\rho^2)}{(1 + a^2\rho^2)}\theta \otimes \theta, \tag{3.1}$$

$$J_+ = I_1 + \frac{a^2}{(1 + a^2\rho^2)}\theta \otimes (dx_1)^{\sharp_{g_0}} + a^2\, dx_1 \otimes T, \tag{3.2}$$

$$\omega_+ = \omega_1 + a^2\, dx_2 \wedge dx_3, \tag{3.3}$$

with the notation introduced in Section 2. Notice that the choice of (g_a, I_1^a, ω_1^a) is determined by the choice of the S^1-action (2.3).

Like g_0, the Taub–NUT metric g_+ is complete on $\mathbb{R}^4$. Being hyperkähler, g_+ is Ricci-flat, hence scalar-flat, hence *anti-self-dual*, meaning that the self-dual Weyl tensor, W^+, relative to the standard orientation of $\mathbb{R}^4$, is identically zero. In the sequel, (g_+, J_+, ω_+) will be referred to as the *Ricci-flat Taub–NUT Kähler structure.*

3.1. The Ricci-Flat Taub–NUT Kähler Metric as a Complete Kähler Metric on $\mathbb{C}^2$

It is well known that the Ricci-flat Taub–NUT Kähler structure is isomorphic to a (non-flat, Ricci-flat, complete) Kähler metric defined on the standard complex plane $\mathbb{C}^2$ (cf. e.g. [9, Chapter 13], [19]. In the present setting, this can be shown as follows:

Proposition 1 *For any $a \geq 0$, the map*

$$\Phi_a^+ : u \mapsto \left(w_1^+ = e^{a^2 x_1(u)} z_1(u),\, w_2^+ = e^{-a^2 x_1(u)} z_2(u)\right), \tag{3.4}$$

from $\mathbb{R}^4 = \mathbb{H}$ to $\mathbb{C}^2$—with $u = u_0 + u_1 i + u_2 j + u_3 k$, $z_1(u) = u_0 + iu_1$, $z_2(u) = u_2 + iu_3$—is an S^1-equivariant biholomorphism from $(\mathbb{R}^4, I_1^a)$ to the standard complex plane $\mathbb{C}^2$, equipped with the S^1-action: $(w_1^+, w_2^+) \cdot e^{it} = (w_1^+ e^{it}, w_2^+ e^{-it})$.

Proof The fact that w_1^+ and w_2^+ are holomorphic with respect to $J_+ = I_a^1$, as functions defined on $(\mathbb{R}^4, J_+)$, is a direct consequence of Proposition 7 in Section 6, since $dz_1(T) = iz_1$ and $dz_2(T) = -iz_2$ (cf. (2.9)). The map Φ_a^+ is then holomorphic from $(\mathbb{R}^4, J_+)$ to the standard complex plane $\mathbb{C}^2$. Moreover, since the momentum x_1 is S^1-invariant, we clearly have $\Phi_a^+(u \cdot e^{it}) = \Phi_a^+((e^{it} z_1, e^{-it} z_2)) = (e^{it} w_1^+, e^{-it} w_2^+)$. From $d\Phi_a^+ = \left(e^{a^2 x_1}(dz_1 + a^2 z_1 dx_1), e^{-a^2 x_1}(dz_2 - a^2 z_2 dx_1)\right)$, we easily infer that, at each point of $\mathbb{R}^4$, $d\Phi_a^+$, regarded as an element of $\mathrm{End}(\mathbb{R}^4)$, is invertible, of determinant equal to $1 + a^2 \rho^2$. Finally, to check that Φ_a^+ is surjective and one to one, we observe that $w_1^+ w_2^+ = z_1 z_2$, so that

$$\begin{aligned} |w_1^+| &= e^{\frac{a^2}{2}\left(|z_1|^2 - \frac{|w_1^+||w_2^+|}{|z_1|^2}\right)} |z_1| = F_{|w_1^+||w_2^+|}(|z_1|), \\ |w_2^+| &= e^{\frac{a^2}{2}\left(|z_2|^2 - \frac{|w_1^+||w_2^+|}{|z_2|^2}\right)} |z_2| = F_{|w_1^+||w_2^+|}(|z_2|), \end{aligned} \tag{3.5}$$

by setting

$$F_A(x) := x\, e^{\frac{a^2}{2}\left(x^2 - \frac{A}{x^2}\right)}, \tag{3.6}$$

for any $A \geq 0$. Since F_A is (strictly) increasing, with $F_A(0) = 0$ and $F_A(+\infty) = +\infty$, for any $A \geq 0$, it can be regarded as a one-to-one map from $[0, +\infty)$ to itself, of inverse F_A^{-1}, and we then have

$$|z_1| = F^{-1}_{|w_1^+||w_2^+|}(|w_1^+|), \quad |z_2| = F^{-1}_{|w_1^+||w_2^+|}(|w_2^+|). \tag{3.7}$$

The pair $(|z_1|, |z_2|)$, hence also the pair (z_1, z_2), are then entirely determined by the pair (w_1^+, w_2^+), for any (w_1^+, w_2^+) in $\mathbb{C}^2$. It follows that Φ_a^+ is a diffeomorphism and hence a biholomorphism from $(\mathbb{R}^4, J_+)$ to $\mathbb{C}^2$. □

Remark 3 From Remarks 14 and 16 in Section 6, we readily infer that a S^1-invariant Kähler potential for the Ricci-flat Taub–NUT Kähler structure $(g_+, J_+, \omega_+) = (g_a, I_1^a, \omega_1)$ on $\mathbb{R}^4$ is the function φ^+ defined by

$$\varphi^+(u) = \frac{\rho^2}{4} + \frac{a^2}{8}\left(\rho_1^4 + \rho_2^4\right), \tag{3.8}$$

for any $u = u_0 + u_1 i + u_2 j + u_3 k = z_1 + z_2 j$ in $\mathbb{R}^4 = \mathbb{H}$, where, we recall, $\rho_1^2 = u_0^2 + u_1^2 = |z_1|^2$, $\rho_2^2 = u_2^2 + u_3^2 = |z_2|^2$, $\rho^2 = \rho_1^2 + \rho_2^2$. Since

$$|w_1^+| = e^{\frac{a^2}{2}\left(\rho_1^2 - \rho_2^2\right)}\rho_1, \qquad |w_2^+| = e^{\frac{a^2}{2}\left(\rho_2^2 - \rho_1^2\right)}\rho_2, \tag{3.9}$$

this fits with [19, Exercise].

3.2. *The Ricci-Flat Taub–NUT Kähler Structure as a Toric Kähler Structure*

While the Taub–NUT metric $g_a = g_+$ is preserved by the action of $G := \left(\mathrm{Sp}_1 \times \mathrm{S}^1\right)/\pm 1$ (cf. Remark 1 above), the Kähler structure (g_+, J_+, ω_+) is only preserved by the action of the subtorus $\mathbb{T}^2$ of G generated by the right *and* the left multiplication of S^1 on $\mathbb{H}$, which makes (g_+, J_+, ω_+) into a toric Kähler structure. For convenience, the torus $\mathbb{T}^2$ will be identified with $\mathrm{S}^1 \times \mathrm{S}^1$ in such a way that its action can be described by

$$(e^{i\phi_1}, e^{i\phi_2}) \cdot u = e^{i\frac{\phi_1+\phi_2}{2}}\, u\, e^{i\frac{\phi_1-\phi_2}{2}}, \tag{3.10}$$

for any $(e^{i\phi_1}, e^{i\phi_2})$ in $\mathbb{T}^2 = \mathrm{S}^1 \times \mathrm{S}^1$ and any u in $\mathbb{H}$ or, by identifying $u = z_1 + z_2 j$ with the pair (z_1, z_2) in $\mathbb{C}^2$, by

$$(e^{i\phi_1}, e^{i\phi_2}) \cdot (z_1, z_2) = \left(e^{i\phi_1} z_1, e^{i\phi_2} z_2\right). \tag{3.11}$$

We denote by T_1 and T_2 the generators of this action, corresponding to the action of the first and the second factor S^1; we then have

$$T_1 = -u_1\, e_0 + u_0\, e_1, \qquad T_2 = -u_3\, e_2 + u_2\, e_3. \tag{3.12}$$

Notice that the torus action is free on the open subset, $\mathbb{R}_0^4$ of $\mathbb{R}^4$ defined by the condition

$$\rho_1^2 = u_o^2 + u_1^2 > 0, \qquad \rho_2^2 = u_2^2 + u_3^2 > 0 \tag{3.13}$$

and that the vector field T is related to the generators T_1, T_2 by $T = T_1 - T_2$. We denote by $\theta_1 = -u_1\, du_0 + u_0\, du_1$ (resp. $\theta_2 = -u_3\, du_2 + u_2\, du_3$) the 1-form dual to T_1 (resp. T_2) with respect to the standard flat metric g_0, so that $\theta = \theta_1 - \theta_2$ and

$$\theta_1(T_1) = \rho_1^2, \qquad \theta_1(T_2) = \theta_2(T_1) = 0, \qquad \theta_2(T_2) = \rho_2^2. \tag{3.14}$$

We thus have

$$\theta_1 = \rho_1^2\, dt_1, \qquad \theta_2 = \rho_2^2\, dt_2, \tag{3.15}$$

where the *angular coordinates* t_1, t_2 are defined by

$$t_1(u) = \arctan \frac{u_1}{u_0} \mod \pi, \qquad t_2(u) = \arctan \frac{u_3}{u_2} \mod \pi, \tag{3.16}$$

and satisfy

$$dt_1(T_1) = dt_2(T_2) = 1, \quad dt_1(T_2) = dt_2(T_1) = 0. \tag{3.17}$$

From (3.3) and (2.12), we easily infer

$$-\iota_{T_1}\omega_+ = d\mu_1^+, \qquad -\iota_{T_2}\omega_+ = d\mu_2^+, \tag{3.18}$$

where the momenta μ_1^+, μ_2^+ are defined by

$$\mu_1^+ = \frac{1}{2}\rho_1^2(1 + a^2\,\rho_2^2), \qquad \mu_2^+ = \frac{1}{2}\rho_2^2(1 + a^2\,\rho_1^2). \tag{3.19}$$

Notice that

$$\omega_+ = d\mu_1^+ \wedge dt_1 + d\mu_2^+ \wedge dt_2. \tag{3.20}$$

Remark 4 The momentum polytope of the Ricci-flat Taub–NUT toric Kähler structure (g_+, J_+, ω_+) is the whole (closed) positive quadrant in $\mathbb{R}^2$, which is the momentum polytope of the standard flat Kähler structure of $\mathbb{C}^2$. In particular, the total volume of g_+ is infinite for any value of $a \geq 0$. This also directly follows from (3.1)–(3.3), by which we easily get the following expression of the volume form, v_{g_+}, of $g_+ = g_a$:

$$v_{g_+} = \frac{1}{2}\omega_+ \wedge \omega_+ = (1 + a^2\,\rho^2)\, v_{g_0}. \tag{3.21}$$

Remark 5 In general, the scalar curvature, Scal_g, of a toric Kähler manifold (M, g, J, ω) of real dimension $n = 2m$ is given by the *Abreu formula*:

$$\mathrm{Scal}_g = -\sum_{i,j=1}^{m} \frac{\partial^2 H_{ij}}{\partial\mu_i \partial\mu_j}, \tag{3.22}$$

and the Ricci form, $\mathrm{Ric}(\omega)$, is given by

$$\mathrm{Ric}(\omega) = -\frac{1}{2}\sum_{j,k=1}^{m}\left(\sum_{i=1}^{m} \frac{\partial^2 H_{ik}}{\partial\mu_i \partial\mu_j}\right) d\mu_j \wedge dt_k, \tag{3.23}$$

on the open set, M_0, where the torus action is free, with the following notation: for any chosen generators $T_1, \ldots, T_m$ of the torus action, $\mu_1, \ldots, \mu_m$ denote the corresponding momenta with respect to ω, and $t_1,\dots,t_m$ the corresponding angular coordinates, determined by the fact that $\{dt_1,\dots,dt_m,d\mu_1,\dots,d\mu_m\}$ is the (algebraic) dual coframe of the frame $\{T_1,\dots,T_m,JT_1,\dots,JT_m\}$, so that $\omega=\sum_{i=1}^m dx_i\wedge dt_i$ on M_0; the H_{ij} are then defined by

$$H_{ij}=g(T_i,T_j), \tag{3.24}$$

which are viewed as functions of the momenta $\mu_1,\dots,\mu_m$, and the matrix $\mathbf{H}=(H_{ij})$ is the inverse of the Hessian matrix $\mathbf{G}=\left(\frac{\partial^2 u}{\partial\mu_i\partial\mu_j}\right)$, where u denotes the *symplectic potential* of the toric Kähler structure (cf. [17], [1], [2], [14]). In the current setting of the toric Ricci-flat Taub–NUT metric, the chosen toric generators are T_1, T_2, defined by (3.12); by using (3.19), we then get

$$\begin{aligned}
H^+_{11}&:=g_+(T_1,T_1)=a^2\rho_1^2\rho_2^2+\frac{\rho_1^2(1+a^2\rho_2^2)}{1+a^2\rho^2}=\mu_1^++\mu_2^++\frac{2\mu_1^+}{\phi}-\frac{\phi}{2a^2}+\frac{1}{2a^2},\\
H^+_{22}&:=g_+(T_2,T_2)=a^2\rho_1^2\rho_2^2+\frac{\rho_2^2(1+a^2\rho_1^2)}{1+a^2\rho^2}=\mu_1^++\mu_2^++\frac{2\mu_2^+}{\phi}-\frac{\phi}{2a^2}+\frac{1}{2a^2},\\
H^+_{12}&:=g_+(T_1,T_2)=\frac{a^2(2+a^2\rho^2)\rho_1^2\rho_2^2}{1+a^2\rho^2}=\mu_1^++\mu_2^++\frac{\mu_1^++\mu_2^+}{\phi}-\frac{\phi}{2a^2}+\frac{1}{2a^2\phi},
\end{aligned} \tag{3.25}$$

by setting $\phi:=1+a^2\rho^2=\left(1+4a^2(\mu_1+\mu_2)+4a^4(\mu_1^+-\mu_2^+)^2\right)^{\frac12}$. Notice that $\det\mathbf{H}=\rho_1^2\rho_2^2=\frac{1+2a^2(\mu_1^++\mu_2^+)-\phi}{2a^2}$. From (3.25), we infer

$$\begin{aligned}
\frac{\partial^2 H^+_{11}}{\partial\mu_1^+\partial\mu_1^+}&=\frac{\partial^2 H^+_{22}}{\partial\mu_2^+\partial\mu_2^+}=-\frac{\partial^2 H^+_{12}}{\partial\mu_1^+\partial\mu_2^+}=-\frac{8a^2}{\phi^5}\Big(1+4a^2(\mu_1^++\mu_2^+)\\
&\quad+4a^4\big((\mu_1^+)^2+(\mu_2^+)^2+4b\mu_1^+\mu_2^+\big)\Big),\\
\frac{\partial^2 H^+_{11}}{\partial\mu_1^+\partial\mu_2^+}&=\frac{\partial^2 H^+_{22}}{\partial\mu_1^+\partial\mu_2^+}=-\frac{\partial^2 H^+_{12}}{\partial\mu_1^+\partial\mu_1^+}=-\frac{\partial^2 H^+_{12}}{\partial\mu_2^+\partial\mu_2^+}=\frac{48a^4\mu_1^+}{\phi^5}\Big(1+2a^2(\mu_1^++\mu_2^+)\Big).
\end{aligned} \tag{3.26}$$

By substituting in (3.22)–(3.23), we readily check that g_+ is Ricci-flat.

4. The Bochner-Flat Taub–NUT Kähler Metric

In this section, for any $a>0$, we consider the almost-Kähler structure associated to the corresponding Taub–NUT hyperkähler structure (as defined in general in Paragraph 6.3), which will be denoted (g_-,J_-,ω_-), the parameter a being understood; g_- is then conformal to g_+ and the orientation induced by J_- is opposite to the standard orientation of $\mathbb{R}^4$. According to Proposition 8, (g_-,J_-,ω_-) is actually *Kähler*. Moreover, since the Kähler structure (g_+, J_+, ω_+) is anti-self-dual (with respect to the orientation induced by J_+), the Kähler structure (g_-, J_-, ω_-) is *self-dual,* (with respect to the orientation induced by J_-) and hence *Bochner-flat,* since, on any Kähler surface, the Bochner tensor coincides with the anti-self-dual Weyl tensor W^-. It will be called the *Bochner-flat Taub–NUT Kähler structure.*

The Bochner-flat Taub–NUT Kähler metric (g_-, J_-, ω_-) is itself a deformation of the almost-Kähler, actually Kähler, structure associated to the standard flat hyperkähler structure on $\mathbb{R}^4$, namely the Kähler structure $(g_0, \bar{I}_1, \bar{\omega})$, where $\bar{I}_1$ is defined by

$$\bar{I}_1 e_0 = e_1, \quad \bar{I}_1 e_1 = -e_0, \quad \bar{I}_1 e_2 = -e_3, \quad \bar{I}_1 e_3 = e_2, \tag{4.1}$$

and $\bar{\omega} = g_0(\bar{I}_1 \cdot, \cdot) = du_0 \wedge du_1 - du_2 \wedge du_3$ is the 2-form appearing in (2.14). According to (6.24)–(6.26)–(6.25), we then have

$$g_- = \frac{g_+}{(1+a^2\rho^2)^2} = \frac{1}{(1+a^2\rho^2)} g_0 - \frac{a^2(2+a^2\rho^2)}{(1+a^2\rho^2)^3} \theta \otimes \theta, \tag{4.2}$$

$$J_- = \bar{I}_1 + \frac{a^2}{2(1+a^2\rho^2)} \theta \otimes (d\rho^2)^{\sharp_{g_0}} + \frac{a^2}{2} d\rho^2 \otimes T, \tag{4.3}$$

$$\omega_- = \frac{1}{2} d\left(\frac{\theta}{1+a^2\rho^2}\right) = \frac{(2+a^2\rho^2)\,\bar{\omega} - 2a^2 \sum_{j=1}^{3} x_j \omega_j}{2(1+a^2\rho^2)^2}. \tag{4.4}$$

Remark 6 It is easily checked that, in terms of the coframe $\{\frac{d\rho}{\rho}, \sigma_1, \sigma_2, \sigma_3\}$ (cf. Remark 2), the complex structure J_- is determined by

$$J_- \sigma_2 = \sigma_1, \qquad J_- \sigma_3 = -(1+a^2\rho^2)\frac{d\rho}{\rho}, \tag{4.5}$$

hence coinciding with the 'red herring' complex structure mentioned in [19].

Remark 7 In contrast with the Ricci-flat Taub–NUT Kähler structure, the Bochner-flat Taub–NUT Kähler structure considered in this section is preserved by the whole group $G = (\mathrm{Sp}_1 \times \mathrm{S}^1)/\{\pm\}$ (cf. Remark 1).

Proposition 2 *For any $a \geq 0$, $\frac{\rho^2}{4}$ is a globally defined G-invariant Kähler potential for the corresponding sef-dual Kähler Taub–NUT structure.*

Proof From (4.3) we infer

$$J_- \theta_a = \bar{I}_1 \theta = -\frac{1}{2} d\rho^2, \tag{4.6}$$

for any $a \geq 0$, where $\theta_a = \frac{\theta}{1+a^2\rho^2}$ is the 1-form dual to T with respect to $g_a = g_+$. It then follows from (4.4) that $\omega_- = \frac{1}{4} dJ_- d\rho^2$, meaning that $\frac{\rho^2}{4}$ is a (globally defined) G-invariant Kähler potential for the Bochner-flat Taub–NUT Kähler structure for *any* value of $a \geq 0$. □

Remark 8 Denote by $v_{g_-} = \frac{1}{2}\omega_- \wedge \omega_-$ the volume form of the Bochner-flat Taub–NUT metric g_- relative to the orientation induced by J_-, and by v_{g_0} the volume form of the flat metric g_0 relative to the opposite orientation, i.e. the orientation induced by the complex structures I_j, so that $v_{g_0} = \frac{1}{2}\omega_j \wedge \omega_j$, for $j = 1,2,3$. Since $\frac{1}{2}\bar{\omega} \wedge \bar{\omega} = -v_{g_0}$ and $\bar{\omega} \wedge \omega_j = 0$, $j = 1,2,3$, we readily infer from (4.4) that $v_{g_-} = -\frac{v_{g_0}}{(1+a^2\rho^2)^3}$. It follows that, for any $a > 0$, the total volume $\mathrm{Vol}(\mathbb{R}^4, g_-)$ of $(\mathbb{R}^4, g_-)$ is *finite*, equal to $\frac{\pi^2}{2a^4}$.

4.1. The Bochner-Flat Taub–NUT Structure as a Bryant Bochner–Kähler Metric

We now show that, like the Ricci-flat Taub–NUT Kähler structure (g_+, J_+, ω_+), the Bochner-flat Taub–NUT Kähler structure (g_-, J_-, ω_-) is isomorphic to a (complete, Bochner-flat) Kähler metric on the standard complex plane $\mathbb{C}^2$. We first observe that the map $u = u_0 + u_1 i + u_2 j + u_3 k \mapsto (z_1 = u_0 + iu_1, \bar{z}_2 = u_2 - iu_3)$ is clearly a biholomorphism from $(\mathbb{R}^4, \bar{I}_1)$ to the standard complex plane $\mathbb{C}^2$. More generally, we have

Proposition 3 *For any $a \geq 0$, the map, Φ_a^-, from $(\mathbb{R}^4, J_-)$ to the standard complex plane $\mathbb{C}^2$, defined by*

$$u \mapsto \left(w_1^- = z_1\, e^{\frac{a^2}{2}\rho^2}, w_2^- = \bar{z}_2\, e^{\frac{a^2}{2}\rho^2}\right), \tag{4.7}$$

is a biholomorphism from $(\mathbb{R}^4, J_-)$ to $\mathbb{C}^2$.

Proof We only have to show that w_1^- and w_2^- are both holomorphic functions on $\mathbb{R}^4$ with respect to the complex structure J_-. The rest of the argument is similar to the one in Proposition 1, only simpler since $|w^-|^2 = \rho^2 e^{a^2\rho^2}$. By using (4.3) and (4.6), as well as $dz_1(\bar{I}_1 \cdot) = i\, dz_1$ and $dz_1(T) = iz_1$, we get

$$\begin{aligned} dw_1^-(J_aX) - i\, dw_1^-(X) &= e^{\frac{a^2}{2}\rho^2}\, [dz_1(J_aX) - i\, dz(X) + \frac{a^2}{2} z_1 d\rho^2(J_aX) - i\frac{a^2}{2} z_1 d\rho^2(X)] \\ &= e^{\frac{a^2}{2}\rho^2} \frac{a^2}{2} [d\rho^2(X) dz_1(T) + \frac{\theta(X)}{(1+a^2\rho^2)} dz_1\big((d\rho^2)^{\sharp_g}\big) - (J_a d\rho^2)(X) \\ &\quad - i z_1 d\rho^2(X)] \\ &= e^{\frac{a^2}{2}\rho^2} \frac{a^2}{2} z_1\, [i\, d\rho^2(X) + \frac{2\theta(X)}{(1+a^2\rho^2)} - 2\theta_a(X) - i\, d\rho^2(X)] = 0, \end{aligned} \tag{4.8}$$

for any vector field X on $\mathbb{R}^4$. Similarly, by using (4.3) and (4.6) again, as well as $d\bar{z}_2(\bar{I}_1 \cdot) = i\, d\bar{z}_2$ and $d\bar{z}_2(T) = i\bar{z}_2$, we get $dw_2^-(J_aX) - i\, dw_2^-(X) = 0$. □

In his celebrated paper [10], R. Bryant discovered a family of complete, Bochner-flat Kähler—or Bochner–Kähler metrics on the standard complex space $\mathbb{C}^m$, for any $m \geq 2$, parametrized by the m-uplets of non-negative real numbers $\mathbf{k} = (k_1, \ldots, k_m)$, and showed that any complete Bochner-flat Kähler metric on $\mathbb{C}^m$ belongs to this family. For any such

m-uplet $\mathbf{k}$, it can be shown (cf. e.g. [11, Appendix C]), that the corresponding Kähler form on $\mathbb{C}^m$ is $\omega_{\mathbf{k}} = \frac{1}{4} dd^c s_{\mathbf{k}}$, where $s_{\mathbf{k}}$ is determined by

$$s_{\mathbf{k}} = \sum_{i=1}^{m} \frac{|w_i|^2}{e^{k_i s_{\mathbf{k}}}}. \tag{4.9}$$

From (4.9) and Proposition 3, we then infer the following identification:

Proposition 4 *Regarded as a Bochner-flat Kähler structure on $\mathbb{C}^2$ via the biholomorphism Φ_a^- given in Proposition 3, the Bochner-flat Taub–NUT Kähler structure (g_-, J_-, ω_-) of parameter a coincides with the Bryant complete Bochner-flat metric on $\mathbb{C}^2$ of parameter $\mathbf{k} = (a^2, a^2)$.*

Proof Denote by s_a the function defined on $\mathbb{C}^2$ by $s_a(w) = \rho^2\big((\Phi_a^-)^{-1}(w)\big)$. Then s_a is the radial function ρ^2 viewed as a function defined on $\mathbb{C}^2$ via the biholomorphism Φ_a^- of Proposition 3 and is therefore determined by

$$s_a(w) = \frac{|w|^2}{e^{a^2 s_a(w)}}. \tag{4.10}$$

Since $\frac{\rho^2}{4}$ is a Kähler potential for the Bochner-flat Taub–NUT Kähler metric on $(\mathbb{R}^4, J_a)$ (cf. Proposition 2), we infer that the latter is identified, via Φ_a^-, with the Kähler metric on the standard complex plane $\mathbb{C}^2$ whose Kähler potential is the function s_a defined by (4.10) and hence with the complete Bryant Bochner-flat metric of parameter $(k_1 = a^2, k_2 = a^2)$. □

4.2. *The Bochner-Flat Taub–NUT Kähler Structure as a Toric Kähler Structure*

Like the Ricci-flat Taub–NUT Kähler structure (g_+, J_+, ω_+), the Bochner-flat Taub–NUT Kähler structure (g_-, J_-, ω_-) is preserved by the action (3.10) of the torus $\mathbb{T}^2 = \mathrm{S}^1 \times \mathrm{S}^1$ on $\mathbb{R}^4 = \mathbb{H}$. By using the expression of ω_- given in (4.4), the momenta, μ_1^- and μ_2^-, of the generators T_1 and T_2 of this action with respect to ω_-, defined by $-\iota_{T_1}\omega_- = d\mu_1^-$ and $-\iota_{T_2}\omega_- = d\mu_2^-$, respectively, are easily computed, to get

$$\mu_1^- = \frac{\rho_1^2}{2(1 + a^2\rho^2)}, \qquad \mu_2^- = \frac{-\rho_2^2}{2(1 + a^2\rho^2)}, \tag{4.11}$$

up to additive constant. Notice that

$$\omega_- = d\mu_1^- \wedge dt_1 + d\mu_2^- \wedge dt_2. \tag{4.12}$$

Remark 9 The momentum polytope, Δ_a, is determined in $\mathbb{R}^2$ by the affine inequations

$$\xi_1 \geq 0, \quad -\xi_2 \geq 0, \quad -2a^2(\xi_1 - \xi_2) + 1 > 0. \tag{4.13}$$

In particular, the closure, $\overline{\Delta}_a$ of Δ_a is *compact* and $\overline{\Delta}_a \setminus \Delta_a$ is the closed segment $F = \{2a^2(\xi_1 - \xi_2) = 1, \xi_1 \geq 0, \xi_2 \geq 0\}$, which has been 'removed' from $\overline{\Delta}_a$. Since the Euclidean area of $\tilde{\Delta}_a$ is equal to $\frac{1}{8a^2}$, we recover the fact that the (symplectic) area of $(\mathbb{R}^4, g_-$ is equal to $\frac{1}{8a^2} \times (2\pi)^2 = \frac{\pi^2}{2a^2}$ (cf. Remark 8).

Remark 10 By using (4.11), we easily check that the matrix $\mathbf{H}^-$ of scalar products of the generators T_1, T_2 with respect to g_- (cf. Remark 5), is given by

$$\begin{aligned} H_{11}^- &:= g_-(T_1, T_1) = \frac{a^2\,\rho_1^2\rho_2^2}{(1+a^2\,\rho^2)^2} + \frac{\rho_1^2(1+a^2\,\rho_2^2)}{(1+a^2\,\rho^2)^3} \\ &= 2\mu_1^- - 8a^2\,(\mu_1^-)^2 + 8a^4\,(\mu_1^-)^2(\mu_1^- - \mu_2^-), \\ H_{12}^- &:= g_-(T_1, T_2) = \frac{a^2(2+a^2\,\rho^2)}{(1+a^2\,\rho^2)^3}\,\rho_1^2\rho_2^2 \\ &= -8a^2\,\mu_1^-\mu_2^- + 8a^4\,\mu_1^-\mu_2^-(\mu_1^- - \mu_2^-), \\ H_{22}^- &:= g_-(T_2, T_2) = \frac{a^2\,\rho_1^2\rho_2^2}{(1+a^2\,\rho^2)^2} + \frac{\rho_2^2(1+a^2\,\rho_1^2)}{(1+a^2\,\rho^2)^3} \\ &= -2\mu_2^- - 8a^2\,(\mu_2^-)^2 + 8a^4\,(\mu_2^-)^2(\mu_1^- - \mu_2^-). \end{aligned} \tag{4.14}$$

Notice that $\det \mathbf{H}^- = \frac{\rho_1^2\rho_2^2}{(1+a^2\,\rho^2)^4} = 4\mu_1^-\mu_2^-\left(1 - 2a^2(\mu_1^- - \mu_2^-)\right)^2$. From (4.14), we infer

$$\begin{aligned} \frac{\partial^2 H_{11}^-}{\partial\mu_1^-\partial\mu_1^-} &= -16a^2(1 - 3a^2\,\mu_1^- + a^2\,\mu_2^-), \\ \frac{\partial^2 H_{12}^-}{\partial\mu_1^-\partial\mu_2^-} &= -8a^2(1 - 2a^2\,\mu_1^- + 2a^2\,\mu_2^-), \\ \frac{\partial^2 H_{22}^-}{\partial\mu_2^-\partial\mu_2^-} &= -16a^2(1 - a^2\,\mu_1^- + 3a^2\,\mu_2^-), \end{aligned} \tag{4.15}$$

$$\frac{\partial^2 H_{11}^-}{\partial\mu_1^-\partial\mu_2^-} = \frac{\partial^2 H_{12}^-}{\partial\mu_2^-\partial\mu_2^-} = -16a^4\,\mu_1^-, \quad \frac{\partial^2 H_{12}^-}{\partial\mu_1^-\partial\mu_1^-} = \frac{\partial^2 H_{22}^-}{\partial\mu_1^-\partial\mu_2^-} = 16a^4\,\mu_2^-. \tag{4.16}$$

Denote by Scal_{g_-} the scalar curvature of g_-, by $\mathrm{Ric}(\omega_-)$ the Ricci form of the Kähler structure (g_-, J_-, ω_-) and by $\widetilde{\mathrm{Ric}}(\omega_-)$ the *normalized Ricci form*, which is defined by $\widetilde{\mathrm{Ric}}(\omega_-) = \mathrm{Ric}(\omega_-) - \frac{\mathrm{Scal}_-}{6}\,\omega_-$. By using (3.22)–(3.23), we easily infer from (4.15)–(4.16) that Scal_- and $\widetilde{\mathrm{Ric}}(\omega_-)$ have the following expressions:

$$\mathrm{Scal}_{g_-} = 48\,a^2\left(1 - 2a^2(\mu_1^- - \mu_2^-)\right) = \frac{48\,a^2}{1+a^2\,\rho^2}, \tag{4.17}$$

and

$$\begin{aligned}\widetilde{\mathrm{Ric}}(\omega_-) &= 4a^2\left(1-4a^2\,\mu_1^-\right)d\mu_1^-\wedge dt_1\\ &\quad+16a^4\,\mu_1^-\,d\mu_2^-\wedge dt_1-16a^4\,\mu_2^-\,d\mu_1^-\wedge dt_2\\ &\quad+4a^2\left(1+4a^2\,\mu_2^-\right)d\mu_2^-\wedge dt_2.\end{aligned}\tag{4.18}$$

From (4.18), we readily check that the *trace*, $\mathrm{tr}_{\omega_-}(\widetilde{\mathrm{Ric}}(\omega_-)) = \Lambda_{\omega_-}\widetilde{\mathrm{Ric}}(\omega_-)$, and the *pfaffian* $\mathrm{pf}_{\omega_-}(\widetilde{\mathrm{Ric}}(\omega_-)) = \frac{\widetilde{\mathrm{Ric}}(\omega_-)\wedge\widetilde{\mathrm{Ric}}(\omega_-)}{\omega_-\wedge\omega_-}$, of $\widetilde{\mathrm{Ric}}(\omega_-)$ relative to ω_- are given by

$$\begin{aligned}\mathrm{tr}_{\omega_-}(\widetilde{\mathrm{Ric}}(\omega_-)) &= \frac{\mathrm{Scal}_{g_-}}{6} = 8a^2\left(1-2a^2\,(\mu_1^- - \mu_2^-)\right),\\ \mathrm{pf}_{\omega_-}(\widetilde{\mathrm{Ric}}(\omega_-)) &= 16a^4\left(1-4a^2\,(\mu_1^- - \mu_2^-)\right).\end{aligned}\tag{4.19}$$

It follows that $\mathrm{tr}_{\omega_-}(\widetilde{\mathrm{Ric}}(\omega_-))$ and $\mathrm{pf}_{\omega_-}(\widetilde{\mathrm{Ric}}(\omega_-)) = 4a^2\,(\mathrm{tr}_{\omega_-}(\widetilde{\mathrm{Ric}}(\omega_-)) - 4a^2)$ are both *Killing potentials*, i.e. momenta of Hamiltonian Killing vector fields. This fits with the fact that the normalized Ricci form of any self-dual more generally of any *weakly self-dual* Kähler surface is a *Hamiltonian 2-form*, here of order 1 since $\mathrm{pf}_{\omega_-}(\widetilde{\mathrm{Ric}}(\omega_-))$ is an affine function of $\mathrm{tr}_{\omega_-}(\widetilde{\mathrm{Ric}}(\omega_-))$ (cf. [3]). In particular, like any Bochner-flat Kähler manifold, the Bochner-flat Taub–NUT Kähler structure is *extremal*, meaning that Scal_{g_-} is a Killing potential.

5. The Taub–NUT Ambitoric Pair

In this section, for any fixed parameter $a > 0$, we consider the *pair* formed of the Ricci-flat Taub–NUT Kähler structure (g_+, J_+, ω_+) and the Bochner-flat Taub–NUT Kähler structure (g_-, J_-, ω_-), defined by (3.1)–(3.2)–(3.3) and (4.2)–(4.3)–(4.4), respectively. Both Riemannian metrics g_+, g_- are complete on $\mathbb{R}^4$ and they sit in the same conformal class; more precisely, by (4.2), we have $g_- = f_a^2\,g_+$, with

$$f_a = \frac{1}{1+a^2\,\rho^2},\tag{5.1}$$

whereas J_+ induces the standard orientation of $\mathbb{R}^4$, and J_- the opposite one; in particular, J_+ and J_- commute and we denote by τ_a the orthogonal involution defined by

$$\tau_a = -J_+J_- = -J_-J_+.\tag{5.2}$$

Notice that *both* ω_+ and ω_- are J_+- and J_--invariant. Both Kähler structures are preserved by the $\mathbb{T}^2$-action on $\mathbb{R}^4$ defined by (3.10), generated by the vector field T_1, T_2 defined by (3.12). This action is free on the open set $\mathbb{R}^4_0$ defined by (3.13) and the momenta of T_1, T_2 relative to ω_+ (resp. ω_-) are μ_1^+, μ_2^+, given by (3.19) (resp. μ_1^-, μ_2^-, given by (4.11)). The pair (g_+, J_+, ω_+), (g_-, J_-, ω_-) of Kähler structures then constitute a 1-parameter family of complete, 1-parameter *ambitoric structure*, as defined in [5]–[6], here called the *Taub–NUT ambitoric structure*, the parameter a being understood. The $\mathbb{T}^2$-action, which is the same for all values of the parameter a, will be referred to as the *standard* $\mathbb{T}^2$-action.

Remark 11 Since $\mathrm{Scal}_{g_+} = 0$ and $\mathrm{Scal}_{g_-} = 48a^2\big(1-2a^2(\mu_1^- - \mu_2^-)\big)$ are both Killing potentials (cf. Remark 10), the Taub–NUT ambitoric pair is extremal, even *Bach-flat*, since g_+ is Ricci-flat.

Remark 12 Up to scaling, the conformal factor in (5.1) is equal to Scal_{g_-}, which is everywhere positive. This fits with the well-known fact that any Einstein Riemannian metric in the conformal class of a (real) four-dimensional Kähler manifold (M,g,J,ω), if any, is necessarily of the form $\mathrm{Scal}_g^{-2}\,g$ on the open set where Scal_g is positive. This metric is actually Einstein if and only if the *Bach tensor* of the conformal class $[g]$ is zero, which implies in particular that the Kähler structure is extremal (cf. e.g. [3, Paragraphs 1.5, 1.6]).

From (3.2)–(4.3), we infer that the involution τ_a has the following expression:

$$\begin{aligned}\tau_a = \tau_0 &+ \frac{2a^2}{1+a^2\rho^2}\big(a^2\rho_2^2\theta_1 + (1+a^2\rho_1^2)\theta_2\big)\otimes T_1\\ &- \frac{2a^2}{1+a^2\rho^2}\big((1+a^2\rho_2^2)\theta_1 + a^2\rho_1^2\theta_2\big)\otimes T_2,\end{aligned} \tag{5.3}$$

by setting $\tau_0 = -I_1\bar{I}_1 = -\bar{I}_1 I_1$, so that $\tau_0(e_0) = e_0, \tau_0(e_1) = e_1, \tau_0(e_2) = -e_2, \tau_0(e_3) = -e_3$. It follows that

$$\tau_a(d\rho_1^2) = \tau_0(d\rho_1^2) = d\rho_1^2, \qquad \tau_a(d\rho_2^2) = \tau_0(d\rho_2^2) = -d\rho_2^2. \tag{5.4}$$

In particular,

$$\tau_a(df_a^{-1}) = a^2\,\tau_a(d\rho^2) = 2a^2\,dx_1, \tag{5.5}$$

where, we recall, x_1 denotes the momentum of the Killing vector field T with respect to ω_+. We then have

Proposition 5 *For any $a > 0$, the 2-form ψ defined on $\mathbb{R}^4$ by*

$$\psi = a^2 x_1\,\omega_+ + \frac{1}{2}f_a^{-3}\,\omega_- = \frac{a^2}{2}\left(\rho_1^2 - \rho_3^2\right)\omega_+ + \frac{1}{2}\,(1+a^2\rho^2)^3\,\omega_- \tag{5.6}$$

is a Hamiltonian 2-form with respect to the Kähler structure (g_+,J_+,ω_+). It is of maximal order 2 on the dense open set $\mathbb{R}^4_0$ and hence induces an infinitesimal torus action which coincides with the one induced by the standard $\mathbb{T}^2$-action. In particular, the Ricci-flat Taub–NUT Kähler structure (g_+,J_+,ω_+) on $\mathbb{R}^4$ is orthotoric.

Proof In general, for any two conformal Riemannian metrics g, $\tilde{g} = f^{-2}g$ and any 2-form φ, we have

$$\nabla^{\tilde{g}}_X \varphi = \nabla^g_X \varphi + 2f^{-1}\, df(X)\, \varphi + f^{-1} df \wedge \iota_X \varphi - X^{\flat_g} \wedge \iota_{f^{-1} df^{\sharp_g}} \varphi, \tag{5.7}$$

for any vector field X, where $\nabla^{\tilde{g}}$, ∇^g denote the corresponding Levi–Civita covariant derivatives, $X^{\flat_g}$, denotes the 1-form dual to X, and $df^{\sharp_g}$, denotes the dual vector field to df, with respect to g. Since $\nabla^{g_-}\omega_- = 0$, and $g_+ = f_a^{-2} g_-$, we then get

$$\begin{aligned} \nabla^{g_+}_X (f_a^{-3}\omega_-) &= -f_a^{-3} \left(X^{\flat_{g_-}} \wedge \iota_{f_a^{-1} df_a^{\sharp_{g_-}}} \omega_- + \iota_X\big(f_a^{-1} df_a \wedge \omega_-\big)\right) \\ &= 2\big(X^{\flat_{g_+}} \wedge J_- df_a^{-1}\big)_-, \end{aligned} \tag{5.8}$$

where $\big(X^{\flat_{g_+}} \wedge J_- df_a^{-1}\big)_-$ denotes the anti-self-dual part of the 2-form $X^{\flat_{g_+}} \wedge J_- df_a^{-1}$; this being equal to the trace-free, J_+-invariant part of $X^{\flat_{g_+}} \wedge J_- df_a^{-1}$, (5.8) can be rewritten as

$$\nabla^{g_+}_X (f_a^{-3}\omega_-) = X^{\flat_{g_+}} \wedge J_- df_a^{-1} - J_+ X^{\flat_{g_+}} \wedge \tau_a(df_a^{-1}) - \tau_a(df_a^{-1})(X)\, \omega_+. \tag{5.9}$$

From (5.9) and (5.5), we then infer

$$\nabla^{g_+}_X \psi = X^{\flat_{g_+}} \wedge a^2 J_+ dx_1 - J_+ X^{\flat_{g_+}} \wedge a^2\, dx_1. \tag{5.10}$$

Since ω_+ and ω_- are J_+-invariant, ψ is J-invariant as well and, by (5.10), is a *Hamiltonian 2-form* with respect to the Kähler structure (g_+, J_+, ω_+), in the sense of [4, Definition 2]. According to [4, Proposition 3], the *trace*, $\mathrm{tr}_{\omega_+} \psi := 2\frac{\psi \wedge \omega_+}{\omega_+ \wedge \omega_+}$, and the *pfaffian*, $\mathrm{pf}_{\omega_+} \psi := \frac{\psi \wedge \psi}{\omega_+ \wedge \omega_+}$, of any Hamiltonian 2-form ψ with respect to the Kähler form ω_+ are Poisson commuting momenta of Hamiltonian Killing vector fields. In the current case, we have $\mathrm{tr}_{\omega_+} \psi = 2a^2 x_1 = 2a^2(\mu_1^+ - \mu_2^+)$, which is a momentum of $K_1 := 2a^2(T_1 - T_2)$, and $\mathrm{pf}_{\omega_+} \psi = a^2 x_1^2 - \frac{1}{4}(1 + a^2\rho^2)^2 = -\frac{1}{4} - a^2(\mu_1^+ + \mu_2^+)$, which is a momentum of $K_2 := -a^2(T_1 + T_2)$ (notice that $\omega_+ \wedge \omega_- = 0$ and that $\omega_- \wedge \omega_- = -(1 + a^2\rho^2)^{-4}\omega_+ \wedge \omega_+$). For any $a > 0$, the vector fields K_1, K_2 are independent on $\mathbb{R}^4_0$—ψ is therefore a *Hamiltonian 2-form of order* 2 on this open set and evidently generate the same infinitesimal torus action on $\mathbb{R}^4$ as the standard $\mathbb{T}^2$-action. Finally, any Kähler manifold admitting a Hamiltonian 2-form of maximal order on a dense open set is orthotoric in the sense of Definition 6 in [4] or Definition 4 in [5]. □

Proposition 6 *The Taub–NUT ambitoric structure is regular and of parabolic type.*

Proof In general, an ambitoric structure, (g_+, J_+, ω_+), (g_-, J_-, ω_-), with $g_- = f^2 g_+$ and $\tau = -J_+ \circ J_- = -J_- \circ J_+$, is called *regular* if $df \wedge \tau(df) \neq 0$ on a dense open set. In view of (5.4), this condition is satisfied for the Taub–NUT ambitoric structure, since $df_a \wedge \tau_a(df_a) = -2a^4\, d\rho_1^2 \wedge d\rho_2^2 \neq 0$ on $\mathbb{R}^4_0$. Denote by x, y the *roots* of the Hamiltonian 2-form ψ, i.e. the eigenvalues of the corresponding Hermitian operator relative to (g_+, J_+, ω_+), determined by $\mathrm{tr}_{\omega_+} \psi = x + y$ and $\mathrm{pf}_{\omega_+}(\psi) = xy$. We then have

$$x = \frac{1}{2} + a^2\rho_1^2, \qquad y = -\frac{1}{2} - a^2\rho_2^2. \tag{5.11}$$

In particular, it follows from (5.4) that $dx = a^2\, d\rho_1^2$ and $dy = -a^2\, d\rho_2^2$ are orthogonal with respect to the conformal class $[g_+] = [g_-]$. Moreover, if $\bar{g} := f_a g_+ = f_a^{-1} g_- = g_0 - \frac{a^2(2+a^2\rho^2)}{(1+a^2\rho^2)^2}\, \theta \otimes \theta$ denotes the *barycenter metric* of the ambitoric structure, we have

$$|dx|^2_{\bar{g}} = 2a^2(2x-1) =: A(x), \qquad |dy|^2_{\bar{g}} = -2a^2(2y+1) =: B(y). \tag{5.12}$$

By setting $d\tau_1 = -\frac{dt_1 + dt_2}{4a^2}$ and $d\tau_2 = \frac{dt_1 - dt_2}{4a^2}$ on $\mathbb{R}^4_0$, so that $d\tau_1(K_1) = d\tau_2(K_2) = 0$, $d\tau_1(K_2) = \frac{1}{2}$, $d\tau_2(K_1) = 1$, the elements of the Taub–NUT ambitoric structure have then the following expressions:

$$\begin{aligned}
\bar{g} &= \frac{dx^{\otimes 2}}{A(x)} + \frac{dy^{\otimes 2}}{B(y)} + A(x)\frac{(2y\, d\tau_1 + d\tau_2)^{\otimes 2}}{(x-y)^2} + B(y)\frac{(2x\, d\tau_1 + d\tau_2)^{\otimes 2}}{(x-y)^2}\\
g_+ &= (x-y)\,\bar{g}, \qquad g_- = \frac{1}{(x-y)}\,\bar{g},\\
\omega_+ &= dx \wedge (2y\, d\tau_1 + d\tau_2) + dy \wedge (2x\, d\tau_1 + d\tau_2),\\
\omega_- &= \frac{dx \wedge (2y\, d\tau_1 + d\tau_2) - dy \wedge (2x\, d\tau_1 + d\tau_2)}{(x-y)^2},\\
J_+ dx &= J_- dx = \frac{A(x)}{(x-y)}(2y\, d\tau_1 + d\tau_2), \quad J_+ dy = -J_- dy = \frac{B(y)}{(x-y)}(2x\, d\tau_1 + d\tau_2),\\
\psi &= x\, dx \wedge (2y\, d\tau_1 + d\tau_2) + y\, dy \wedge (2x\, d\tau_1 + d\tau_2),
\end{aligned} \tag{5.13}$$

where the pair (x, y) lives in $[\frac{1}{2}, +\infty) \times (-\infty, -\frac{1}{2}]$ and the functions A, B are given by

$$A(t) = 2a^2(2t-1), \qquad B(t) = -2a^2(2t+1). \tag{5.14}$$

In [5, Theorem 3], a general ansatz is presented for regular ambitoric structures on the dense open set $\mathcal{U}$ where the $\mathbb{T}^2$-action is free, depending on the choice of a non-zero element $q = q(t)$ in the space, $\mathcal{P}$, of polynomials of degree ≤ 2 in the variable t; the ambitoric structure is then called *hyperbolic*, *parabolic* or *elliptic*, according to whether the discriminant of q is positive, zero or negative, respectively. In the formalism developed in [5], it has been found convenient to identify the Lie algebra, $\mathfrak{t}$, of the torus $\mathbb{T}^2$ with the orthogonal space, $\mathcal{P}_{0,q}$, to q in $\mathcal{P}$, with respect to the discriminant, viewed as a quadratic form of signature $(2, 1)$ in $\mathcal{P}$. The ansatz presented in [5, Theorem 3] involves $\mathbb{T}^2$-invariant canonical coordinates, x, y, such that dx, dy are orthogonal and $|dx|^2_{\bar{g}} = A(x)$, $|dy|^2_{\bar{g}} = B(y)$ for some positive functions A, B of one variable, and three elements, $d\tau_0, d\tau_1, d\tau_2$, of the dual $\mathfrak{t}^*$, regarded as closed 1-forms on $\mathcal{U}$, defined as follows: for any ξ_p in $\mathfrak{t}$, identified with p in $\mathcal{P}_{0,q}$, $p(t) = p_0 t^2 + 2p_1 t + p_2$, $d\tau_i(\xi_p) = p_i$, $i = 0, 1, 2$; if $q(t) = q_0 t^2 + 2q_1 t + q_2$, the $d\tau_i$'s are then related by $2q_1\, d\tau_1 + q_0\, d\tau_2 + q_2\, d\tau_0 = 0$.

In view of [5, Theorem 3], the above formulas in (5.13) constitute the ansatz for a regular ambitoric structure, where $q \equiv 1$ — the Taub–NUT ambitoric structure is then of parabolic type—the canonical coordinates are the roots x, y of the Hamiltonian 2-form ψ, $A(t) = 2a^2(2t - 1)$ and $B(t) = -2a^2(2t + 1)$, and where the elements of $\mathfrak{t}$ identified with the vector fields K_1, K_2 correspond to the polynomials 1 and t, respectively, in $\mathcal{P}_{0,q}$. □

Remark 13 In general, the scalar curvatures Scal_{g_+} and Scal_{g_-}, of the positive and the negative partner of an ambitoric structure with $q \equiv 1$ are given by $s_{g_+} = -\frac{A''(x)+B''(y)}{x-y}$ and $\mathrm{Scal}_{g_-} = -(x-y)\big(A''(x) + B''(y)\big) + 6\big(A'(x) - B'(y)\big) - \frac{12}{x-y}\big(A(x) + B(y)\big)$, respectively (cf. [5, Paragraph 5.1]). In the current situation, where A, B are given by (5.14), we readily retrieve $\mathrm{Scal}_{g_+} = 0$ and $\mathrm{Scal}_{g_-} = \frac{48a^2}{x-y} = \frac{48a^2}{1+a^2\rho^2}$.

6. The Gibbons–Hawking Ansatz

6.1. *The General Gibbons–Hawking Construction*

In general, the *Gibbons–Hawking ansatz* provides a local description of all hyperkähler 4-manifolds admitting a free S^1-action preserving the whole hyperkähler structure, in terms of a positive harmonic function, V, defined on some connected open subset, $\mathcal{U}$, of the standard three-dimensional Euclidean space $\mathbb{R}^3$ (cf. [16], [18, Chapter 4]). We then have $d(*_3 dV) = \sum_{j=1}^{3} \frac{\partial^2 V}{\partial x_j \partial x_j} dx_1 \wedge dx_2 \wedge dx_3 = 0$, where x_1, x_2, x_3 denote the standard coordinates of $\mathbb{R}^3$, and $*_3$ the Hodge operator of $\mathbb{R}^3$ determined by its standard flat metric and its standard orientation (so that $*_3 dx_1 = dx_2 \wedge dx_3$, etc.) If the de Rham class of the (closed) 2-form $\frac{*_3 dV}{2\pi}$ belongs to the image of $H^2(\mathcal{U}, \mathbb{Z})$ in $H^2(\mathcal{U}, \mathbb{R})$, then $*_3 dV$ can be realized as the curvature form of some S^1-principal bundle, $M_0 \xrightarrow{\pi} \mathcal{U}$, equipped with a (real) connection 1-form η, such that $\pi^*(*_3 dV) = -d\eta$. Then, the Riemannian metric, g, defined on M_0 by

$$g = \frac{1}{V}\eta \otimes \eta + V \sum_{j=1} dx_j \otimes dx_j, \tag{6.1}$$

—here and henceforth, dx_j stands for $\pi^* dx_j$—is the Riemannian metric of a hyperkähler structure, whose Kähler forms are given by

$$\omega_1 = dx_1 \wedge \eta + V\, dx_2 \wedge dx_3,\ \omega_2 = dx_2 \wedge \eta + V\, dx_3 \wedge dx_1,\ \omega_3 = dx_3 \wedge \eta + V\, dx_1 \wedge dx_2, \tag{6.2}$$

and the corresponding complex structures by

$$I_1 = V\, dx_1 \otimes T - \eta \otimes \mathrm{grad}_g x_1 + V\big(dx_2 \otimes \mathrm{grad}_g x_2 - dx_3 \otimes \mathrm{grad}_g x_2\big), \tag{6.3}$$

and similarly for I_2, I_3. The natural S^1-action on M_0 preserves the connection 1-form η and hence the whole hyperkähler structure. The generator, T, of this action is determined by

$$\eta(T) = 1, \qquad \iota_T d\eta = 0. \tag{6.4}$$

The 1-form dual to T with respect to g is then

$$\theta := \frac{1}{V}\eta, \tag{6.5}$$

and the momentum of T relative to ω_J is x_j, i.e. $\iota_T\omega_j = -dx_j$, for $j = 1,2,3$.

Remark 14 For $j = 1,2,3$, it is easily shown that a S^1-invariant (smooth) real function, f, regarded as a function on $\mathbb{R}^3$, is I_j*-pluriharmonic*—meaning that $dI_j df = 0$—if and only if

$$\frac{\partial f}{\partial x_j} = cV, \qquad \sum_{k=1}^{3} \frac{\partial^2 f}{\partial x_k \partial x_k} = 0, \tag{6.6}$$

for some constant c, and that a S^1-invariant (smooth) real function, $\varphi^{(j)}$, is a *Kähler potential* for the Kähler structure (g, I_j, ω_j)—meaning that $\omega_j = dI_j d\varphi^{(j)}$—if and only if $\varphi^{(j)}$ is the sum of a I_j-pluriharmonic function and a function $\varphi_0^{(j)}$ satisfying

$$\frac{\partial \varphi_0^{(j)}}{\partial x_j} = x_j V, \qquad \sum_{k=1}^{3} \frac{\partial^2 \varphi_0^{(j)}}{\partial x_k \partial x_k} = 2V. \tag{6.7}$$

Remark 15 In general, the Ricci tensor of the Riemannian metric g of any hyperkähler structure (g, I_1, I_1, I_3) is identically zero. In (real) dimension 4, the metric g is then *anti-self–dual*, meaning that its self-dual Weyl tensor W_g^+ is identically zero, since g is Kähler, of zero scalar curvature. The curvature, R_g, of g is then reduced to the anti-self-dual Weyl tensor W_g^-. In the Gibbons–Hawking setting, the curvature $R_g = W_g^-$, viewed as a (symmetric, traceless) endomorphism of the vector bundle Λ^- of anti-self-dual 2-forms, can be written as $R_g(\bar\omega_i) = \sum_{j=1}^{3} \Omega_{ij}\bar\omega_j$, where the 2-forms $\bar\omega_1, \bar\omega_2, \bar\omega_3$, defined by

$$\begin{aligned} &\bar\omega_1 = dx_1 \wedge \eta - V\, dx_2 \wedge dx_3, \ \bar\omega_2 = dx_2 \wedge \eta - V\, dx_3 \wedge dx_1, \\ &\bar\omega_3 = dx_3 \wedge \eta - V\, dx_1 \wedge dx_2, \end{aligned} \tag{6.8}$$

form a (normalized) orthonormal frame of Λ^-, and the Ω_{ij} are given by

$$\Omega_{ij} = \frac{\sum_{k=1}^{3} V_k^2}{V^3}\delta_{ij} - \frac{3V_iV_j}{V^3} + \frac{V_{ij}}{V^2}, \tag{6.9}$$

by setting $V_k = \frac{\partial V}{\partial x_k}$ and $V_{ij} = \frac{\partial^2 V}{\partial x_i \partial x_j}$. Since the metric g is Einstein, by [12, Proposition 5], W_g^- is everywhere non-zero or identically zero, and this happens if and only if V is constant or of the form $V = \frac{B}{r_{x^0}}$, for some positive real number B, where r_{x^0} denotes the Euclidean distance to some point x^0 of $\mathbb{R}^3$ (cf. [15, Section 6]).

6.2. The Standard Deformation of a Gibbons–Hawking Metric

If, in the previous construction, we replace V by $V_a := a^2 + V$, for any non-negative real number a, we obtain a 1-parameter family of T-invariant hyperkähler structures, all defined on M_0—called the *standard deformation* of the previous one—whose metric, g_a, has the following expression:

$$g_a := \frac{1}{V_a}\eta \otimes \eta + V_a \sum_{j=1}^{3} dx_j \otimes dx_j = \frac{(a^2+V)}{V} g - \frac{a^2(a^2+2V)}{V^2(a^2+V)}\eta \otimes \eta. \tag{6.10}$$

From (6.2), we readily infer that the Kähler forms, $\omega_1^a, \omega_2^a, \omega_3^a$, of the new hyperkähler structure are given by

$$\omega_1^a = \omega_1 + a^2\, dx_2 \wedge dx_3, \quad \omega_2^a = \omega_2 + a^2\, dx_3 \wedge dx_1, \quad \omega_3^a = \omega_3 + a^2\, dx_1 \wedge dx_2. \tag{6.11}$$

In particular, the x_1, x_2, x_3 are still momenta for the T-action with respect to $\omega_1^a, \omega_2^a, \omega_3^a$, respectively, for any value of the parameter a, as the x_j are T-invariant. Notice that, for $j = 1,2,3$, the gradients, $(dx_j)^{\sharp_g}$ and $(dx_j)^{\sharp_{g_a}}$, of the momentum x_j with respect to g and g_a, respectively, are related to each other by

$$(dx_j)^{\sharp_{g_a}} = \frac{V}{(a^2+V)}(dx_j)^{\sharp_g}, \tag{6.12}$$

whereas the 1-form, θ_a, dual to T with respect to g_a, is given by

$$\theta_a = \frac{V}{a^2+V}\theta \tag{6.13}$$

(an easy consequence of (6.10) in both cases). From (6.3) and (6.12), we then easily deduce that the corresponding complex structures, I_1^a, I_2^a, I_3^a, are given by

$$I_j^a = I_1 + \frac{a^2}{(a^2+V)}\eta \otimes (dx_j)^{\sharp_g} + a^2\, dx_j \otimes T, \quad j = 1,2,3. \tag{6.14}$$

Lemma 1 *Fix $j = 1,2,3$ and let z be any holomorphic function with respect to the complex structure I_j. Then, for any $a \geq 0$, the complex 1-form*

$$\Psi_j := dz - i a^2 dz(T)\, dx_j \tag{6.15}$$

is of type $(1,0)$ with respect to the complex structure I_j^a.

Proof From (6.14), we readily infer

$$dx_j(I_j^a X) = -(I_j^a dx_j)(X) = -\frac{\eta(X)}{(a^2+V)}, \tag{6.16}$$

as well as

$$\begin{aligned} dz(I_j^a X) &= i\,dz(X) \\ &\quad + \frac{a^2}{(a^2+V)}\,\eta(X)\,dz\big((dx_j)^{\sharp_g}\big) + a^2\,dx_j(X)\,dz(T) \\ &= i\,dz(X) - i\,\frac{a^2\,\eta(X)}{(a^2+V)}\,dz(T) + a^2\,dx_j(X)\,dz(T), \end{aligned} \tag{6.17}$$

by using

$$dz\big((dx_j)^{\sharp_g}\big) = -dz(I_j T) = -i dz(T), \tag{6.18}$$

as z is I_j-holomorphic. From (6.16)–(6.17), we then infer

$$\big(dz - i a^2 dz(T)\,dx_j\big)(I_j^a X) = i\big(dz - i a^2\,dz(T)\,dx_j\big)(X), \tag{6.19}$$

meaning that Ψ_j is of type $(1,0)$ with respect to I_j^a. □

As a direct corollary, we get

Proposition 7 *Fix $j = 1,2,3$ and let z be any holomorphic function with respect to I_j. Then $\zeta^+ := z\,e^{a^2 x_j}$ (resp. $\zeta^- := z\,e^{-a^2 x_j}$) is holomorphic with respect to I_j^a if and only if $dz(T) = +iz$ (resp. $dz(T) = -iz$).*

Proof The function $\zeta^\pm$ is I_j^a-holomorphic if and only if $d\zeta^\pm(I_j^a X) = i\,d\zeta^\pm(X)$, for any (real) vector field X. From (3.6), we readily get

$$d\zeta^\pm(I_j^a X) - i\,d\zeta^\pm(X) = a^2\,e^{\pm a^2 x_j}\,dx_j\big(X + i I_j^a X\big)\,(dz(T) \mp iz). \tag{6.20}$$

Since dx_j is real and non-zero, the right-hand side of (6.20) vanishes if and only if $dz(T) = \pm iz$. □

Remark 16 It easily follows from Remark 14 that, for $j = 1,2,3$, if $\varphi_0^{(j)}$ is the Kähler potential for the Kähler structure (g, I_j, ω_j) determined by (6.7), then a T-invariant Kähler potential, $\varphi_a^{(j)}$, for the Kähler structure (g_a, I_k^a, ω_k^a) is given by

$$\varphi_a^{(j)} = \varphi_0^{(j)} + \frac{a^2}{4}\,(r^2 + x_j^2). \tag{6.21}$$

Remark 17 It can be shown that the deformed Gibbons–Hawking hyperkähler structure of parameter $a \geq 0$, hence associated to the harmonic function $a^2 + V$ on $\mathcal{U}$, can be regarded as a hyperkähler reduction of the hyperkähler product $M \times \mathbb{H}$, where M is the Gibbons–Hawking structure associated to V, and $\mathbb{H}$ is equipped with its standard flat hyperkähler structure (cf. Section 2), for the $\mathbb{R}$ action on $M \times \mathbb{H}$ described by

$$(x,u)\cdot s = (\Phi_{as}^T(x), u + s), \tag{6.22}$$

for any x in M, any u in $\mathbb{H}$, any s in $\mathbb{R}$, where Φ_{as}^T denotes the flow of T as the time as (see [15, Section 5] for details).

6.3. The Associated Almost-Kähler Structure

To any Gibbons–Hawking hyperkähler structure (g, I_j, ω_j, T), $j = 1,2,3$, determined by some positive harmonic function V defined on some open subset $\mathcal{U}$ of $\mathbb{R}^3$ and a Killing vector field T as in Section 6, is associated a T-invariant *almost-Kähler structure* (g_K, J, ω_K) defined as follows on the open set where $dV \neq 0$ (cf. e.g. [15, Section 7]):

$$g_K = \frac{\left(\sum_{j=1}^{3} V_j^2\right)^{\frac{1}{2}}}{2V^2}\, g, \qquad \omega_K = \frac{1}{2}\, d\theta = -\frac{1}{2V^2}\sum_{j=1}^{3} V_j \bar{\omega}_j,$$

$$\begin{aligned} J = \frac{2V^2}{\left(\sum_{j=1}^{3} V_j^2\right)^{\frac{1}{2}}}\, \nabla^g T = \frac{V}{\left(\sum_{j=1}^{3} V_j^2\right)^{\frac{1}{2}}} \Bigg[\theta \otimes \sum_{j=1}^{3} V_j dx_j - \sum_{j=1}^{3} V_j dx_j \otimes T \\ + V_1 \left(dx_2 \otimes dx_3^{\sharp_g} - dx_3 \otimes dx_2^{\sharp}\right) + V_2 \left(dx_3 \otimes dx_1^{\sharp_g} - dx_1 \otimes dx_3^{\sharp}\right) \\ + V_3 \left(dx_1 \otimes dx_2^{\sharp_g} - dx_2 \otimes dx_1^{\sharp}\right) \Bigg], \end{aligned} \tag{6.23}$$

where ∇^g denotes the Levi–Civita covariant derivative of g, and the 2-forms $\{\bar{\omega}_1, \bar{\omega}_2, \bar{\omega}_3\}$ have been defined in (6.8). In particular, $d\theta$ is *anti-self-dual* with respect to the orientation induced by the complex structures I_1, I_2, I_3, meaning that J induces the opposite orientation. Notice that the almost-Kähler structure (g_K, J, ω_K) depends on the choice of the vector field T and that T also preserves (g_K, J, ω_K).

For any $a \geq$, we denote by $(g_{a,K}, J_a, \omega_{a,K} = g_{a,K}(J_a \cdot, \cdot))$ the almost-Kähler structure associated to the deformed Gibbons–Hawking hyperkähler structure of parameter a. We then have

Lemma 2 *For any $a \geq 0$, the almost-Kähler structure $(g_{a,K}, J_a, \omega_{a,K})$ is related to (g_K, J, ω_K) by*

$$g_{a,K} = \frac{V}{a^2 + V}\, g_K - \frac{a^2(a^2 + 2V)\left(\sum_{j=1}^{3} V_j^2\right)^{\frac{1}{2}}}{2(a^2 + V)^3}\, \theta \otimes \theta, \tag{6.24}$$

$$J_a = J - a^2 J\theta \otimes T - \frac{a^2 V}{a^2 + V}\, \theta \otimes JT, \tag{6.25}$$

$$\omega_{a,K} = \frac{V}{a^2 + V}\, \omega_K - a^2 \frac{\left(\sum_{j=1}^{3} V_j^2\right)^{\frac{1}{2}}}{2(a^2 + V)^2}\, \theta \wedge J\theta, \tag{6.26}$$

where $(dx_j)^\sharp$ denotes the vector field dual to dx_j with respect to g.

Proof The expression (6.24) of $g_{a,K}$ readily follows from (6.23), applied to the Gibbons–Hawking hyperkähler structure (g_a, I_j^a, ω_j^a), and (6.10). By its very definition, $\omega_{a,K} = \frac{1}{2} d\theta_a$, where θ_a is given by (6.13); we then get (6.26). For any $a \geq 0$, the complex structure J_a is determined by (6.23), by simply replacing the $(dx_j)^{\sharp_g}$ by $(dx_j)^{\sharp_{g_a}}$, while keeping T and the V_j unchanged; by using (6.12), we then obtain (6.25). □

We now show that the Taub–NUT hyperkähler structures are essentially the only Gibbons–Hawking hyperkähler structures for which the associated almost-Kähler structure is Kähler. More precisely, we have

Proposition 8 *The almost-Kähler structure (g_K, J, ω_K) is Kähler—equivalently, J is integrable—if and only if V is either a (non-constant) affine function of x_1, x_2, x_3 or of the form*

$$V = A + \frac{B}{r_{x^0}}, \tag{6.27}$$

for some real numbers $A, B \neq 0$.

Proof If the metric g is flat, i.e. if $V = \frac{B}{r_{x^0}}$ (cf. Remark 15), J, as an almost-complex structure on $\mathbb{R}^4 \setminus \{0\}$, is equal to $\bar{I}_1$, defined by (4.1), which is clearly integrable. We can then assume that $R_g = W_g^-$ is everywhere non-zero. Since the metric g is Einstein, we have $\delta^g W_g^- = 0$, and it then follows from [12, Proposition 5] (cf. also [7, Proposition 3]), that J is integrable if and only if W_g^- is degenerate, as a (trace-free, symmetric) endomorphism of $\Lambda^- M$, i.e. has only two distinct eigenvalues, and the one-dimensional eigen-subbundle of $\Lambda^- M$ is generated by ω_K and hence by $\sum_{j=1}^3 V_j \bar{\omega}_j$. In view of the expression (6.9) of W_g^-, this amounts to demanding that

$$V_{ij} = \lambda \left(V_i V_j - \frac{\sum_{\ell=1}^3 V_\ell^2}{3} \delta_{ij} \right), \tag{6.28}$$

for some function $\lambda = \lambda(x_1, x_2, x_3)$. Then, either $\lambda \equiv 0$, meaning that V is an affine function of x_1, x_2, x_3, and we then easily infer from (6.23) that J is integrable, or $\lambda \neq 0$ on a dense open set $\mathcal{U}$. On $\mathcal{U}$ we have

$$\begin{aligned} V_{ijk} = \lambda^2 &\left(2V_i V_j V_k - \frac{\sum_{\ell=1}^3 V_\ell^2}{3} (V_i \delta_{jk} + V_j \delta_{ki} + V_k \delta_{ij}) \right) \\ &+ \lambda_k V_i V_j - \frac{\sum_{\ell=1}^3 V_\ell^2}{3} \left(\lambda_k + \frac{\lambda^2}{3} V_k \right) \delta_{ij}, \end{aligned} \tag{6.29}$$

for any $i, j, k = 1, 2, 3$, by setting $\lambda_k = \frac{\partial \lambda}{\partial x_k}$. Since $\sum_{\ell=1}^3 V_\ell^2 \neq 0$, it is easy to check that the right-hand side of (6.29) is symmetric in i, j, k if and only if $\lambda_i = \frac{-\lambda^2}{3} V_i$, $i = 1, 2, 3$, if and only if $\lambda = \frac{3}{V - A}$, for some real constant A; (6.28) can then be rewritten as $V_{ij} = \frac{3 V_i V_j}{V - A} - \frac{\sum_{\ell=1}^3 V_\ell^2}{V - A} \delta_{ij}$, or else $f_{ij} = \frac{|df|^2}{2f} \delta_{ij}$, by setting $f = \frac{1}{(V - A)^2}$. It follows that $\frac{|df|^2}{f}$ is constant, equal, say, to $\frac{1}{B^2}$ for some $B > 0$, and that $f = \frac{1}{B^2} \sum_{j=1}^3 (x_j^2 - x_j^0)^2$, for real constants x_j^0, $j = 1, 2, 3$, so that $V = A + \frac{B}{r_{x^0}}$, with $x^0 = (x_1^0, x_2^0, x_3^0)$. □

Acknowledgements

I warmly thank Vestislav Apostolov, Hugues Auvray and David Calderbank for their encouragements and valuable discussions.

....

REFERENCES

[1] M. Abreu, *Kähler geometry of toric varieties and extremal metrics*, Internat. J. Math. **9** (1998), no. 6, 641–51.

[2] M. Abreu, *Kähler geometry of toric manifolds in symplectic coordinates.* In Y Eliashberg, B. A Khesin and F. Lalonde (eds) *Symplectic and contact topology: interactions and perspectives (Toronto/Montreal, 2001)*, Fields Inst. Commun., **35**, Amer. Math. Soc., Providence, RI (2003), 1–24.

[3] V. Apostolov, D. M. J. Calderbank and P. Gauduchon, *The geometry of weakly self-dual Kähler surfaces*, Compos. Math. **135** (2003), no. 3, 279–322.

[4] V. Apostolov, D. M. J. Calderbank and P. Gauduchon, *Hamiltonian 2-forms in Kähler geometry, I General theory*, J. Differential Geom. **73** (2006), no. 3, 359–412.

[5] V. Apostolov, D. M. J. Calderbank and P. Gauduchon, *Ambitoric geometry I: Einstein metrics and extremal ambikähler structures*, J. Reine Angew. Math. **2016** (2016), no. 721, 109–47.

[6] V. Apostolov, D. M. J. Calderbank and P. Gauduchon, *Ambitoric geometry II: Extremal toric surfaces and Einstein 4-orbifolds*, Ann. Sci. Éc. Norm. Supér. (4) **48** (2015), no. 5, 1075–112.

[7] V. Apostolov and P. Gauduchon, *The Riemannain Goldberg–Sachs theorem*, Int. J. Math. **8** (1997), no. 4, 421–39.

[8] V. Apostolov and Y. Rollin, *ALE scalar-flat Kähler metrics on non-compact weighted projective spaces*, Math. Ann. **367** (2017), no. 3–4, 1685–1726.

[9] A. L. Besse, *Einstein manifolds*, Ergeb. Math., 3. Folge, Band 10, Springer-Verlag Berlin–Heidelberg (1987).

[10] R. L. Bryant, *Bochner–Kähler metrics*, J. Amer. Math. Soc. **14** (2001), no. 3, 623–715.

[11] L. David and P. Gauduchon, *The Bochner-flat geometry of weighted projective spaces*, CRM Proc. Lecture Notes, **40**, Amer. Math. Soc., Providence, RI (2006), 109–56.

[12] A. Derdziński, *Self-dual Kähler manifolds and Einstein manifolds of dimension four*, Compos. Math. **49** (1983), no. 3, 405–33.

[13] K. Dixon, *Regular ambitoric 4-manifolds: From Riemannian Kerr to a complete classification*, arXiv:1604.03156 (2016).

[14] S. K. Donaldson, *Kähler geometry on toric manifolds, and some other manifolds with large symmetry.* In L. Ji (ed.) *Handbook of geometric analysis, No. 1*, Adv. Lect. Math., **7**, Int. Press, Somerville, MA 2008, 29–76.

[15] P. Gauduchon, *Autour de l'ansatz de Gibbons–Hawking* (unpublished notes).

[16] G. W. Gibbons and S. Hawking, *Gravitational multi-instantons*, Phys. Lett. B **78** (1978), no. 4, 430–2.

[17] V. Guillemin, *Kähler metrics on toric varietes*, J. Differential Geom. **40** (1994), no. 2, 285–309.

[18] N. Hitchin, *Monopoles, minimal surfaces and algebraic curves*, Sém. Math. Sup., **105**, Presses l'Univ. Montréal, Montreál (1987).

[19] C. LeBrun, *Complete Ricci-flat Kähler metrics on $\mathbb{C}^n$ need not be flat.* In E. Bedford, J. P. D'Angelo, R. E. Greene and S. G. Krantz, eds, *Several Complex Variables and Complex Geometry*, Proc. Symp. Pure Math. **52**, Part 2. American Mathematical Society, Providence (1991) 297–304.

9

Mirror Symmetry with Branes by Equivariant Verlinde Formulas

TAMÁS HAUSEL[1]
ANTON MELLIT[2]
DU PEI[3]

[1] *IST Austria*
Email: tamas.hausel@ist.ac.at
[2] *University of Vienna*
Email: anton.mellit@univie.ac.at
[3] *QGM, Aarhus University*
Walter Burke Institute for Theoretical Physics, Caltech
Email: pei@caltech.edu

18 December 2017

ABSTRACT. We find an agreement of equivariant indices of semi-classical homomorphisms between pairwise mirror branes in the GL_2-Higgs moduli space on a Riemann surface. On one side we have the components of the Lagrangian brane of $U(1,1)$-Higgs bundles, whose mirror was proposed by Nigel Hitchin to be certain even exterior powers of the hyperholomorphic Dirac bundle on the SL_2-Higgs moduli space. The agreement arises from a mysterious functional equation. This gives strong computational evidence for Hitchin's proposal.

1. Introduction

In the conference 'Hitchin 70: Differential Geometry and Quantization' in Aarhus in September 2016, the first and third authors gave talks on recent results on the understanding of equivariant Verlinde formulas on Higgs moduli spaces. Here, we explain how such techniques can be used to compute equivariant indices of semi-classical homomorphisms between branes on Higgs moduli spaces and how, in turn, this could be used to give non-trivial computational evidence for Hitchin's mirror symmetry proposals in [Hi2, §7].

Hausel, T., Mellit A., Pei, D., *Mirror Symmetry with Branes by Equivariant Verlinde Formulas.* In: *Geometry and Physics: A Festschrift in Honour of Nigel Hitchin,* Jørgen Ellegaard Andersen, Andrew Dancer, Oscar García-Prada (Eds): Oxford University Press (2018). DOI: 10.1093/oso/9780198802006.003.0009

Hitchin introduced the moduli space of Higgs bundles on a Riemann surface [Hi1] in 1987. In 2006, the work of Kapustin and Witten [KW] proposed an understanding of the geometric Langlands programme using the S-duality of four-dimensional supersymmetric Yang–Mills theory. One of the main statements is that S-duality induces mirror symmetry between the moduli space $\mathcal{M}_{\mathrm{DR}}(\mathrm{G})$ of flat G-connections on C and $\mathcal{M}_{\mathrm{DR}}(\mathrm{G}^L)$ for the Langlands dual group G^L. The precise statement is still missing, but we have more understanding [DP1] in the semi-classical limit where the mirror symmetry should reduce to a relative Fourier–Mukai type equivalence between $D(\mathcal{M}(\mathrm{G}))$ and $D(\mathcal{M}(\mathrm{G}^L))$, where $\mathcal{M}(\mathrm{G})$ denotes certain moduli space of G-Higgs bundles on C. Even this duality is not completely understood, as we lack the description of the Fourier–Mukai transform on the most singular fibres of the Hitchin map. Consequently, we have few global results proving aspects of this duality.

One global computational result was achieved in [HT] in 2002. There, as a consequence of mirror symmetry, a conjecture was proposed for the agreement of certain Hodge numbers of $\mathcal{M}(\mathrm{SL}_n)$ and $\mathcal{M}(\mathrm{PGL}_n)$, which were proved there for $n = 2, 3$. This conjecture for every n has just recently been settled in [GWZ], using p-adic integration.

Here, we are studying another global approach where deeper properties of the conjectured mirror symmetry could be computationally verified. We set $\mathrm{G} = \mathrm{GL}_2$, and C a smooth complex projective curve of genus $g > 1$. The components of the $\mathcal{M}(\mathrm{U}(1,1))$ Higgs moduli space inside the $\mathcal{M}(\mathrm{GL}_2)$ moduli space can be labeled by certain characteristic classes, as explained in [Hi2]. There are g such components and we will denote these Lagrangians by

$$L_0, \ldots, L_{g-1} \subset \mathcal{M}(\mathrm{GL}_2).$$

To be more precise, L_i is the locus of Higgs bundles of the form

$$M_1 \oplus M_2 \xrightarrow{\Phi} M_1 K_C \oplus M_2 K_C,$$

where M_j are line bundles with $\deg(M_1) = g - 1 - i$, $\deg(M_2) = i - g + 1$ and Φ is off-diagonal. For $i = 0, \ldots, g-2$, the Lagrangians are located in the stable locus $\mathcal{M}(\mathrm{GL}_2)^s$ and so are smooth, and L_i is isomorphic to an affine bundle over the $2i$th symmetric product of C times the Jacobian of degree $g - 1 - i$ line bundles on C. The last component, L_{g-1}, is special as it contains strictly semi-stable points as well.

As explained in [Hi2, §7], L_i is a BAA brane whose mirror should be a BBB brane, i.e. a hyperholomorphic sheaf on the mirror $\mathcal{M}(\mathrm{GL}_2)$ (cf. also [BS] and [GW, §6]). Its support will be $\mathcal{M}(\mathrm{SL}_2) \subset \mathcal{M}(\mathrm{GL}_2)$ and, roughly speaking, it is constructed from a universal bundle $\mathbf{E}$ on $\mathcal{M}(\mathrm{SL}_2) \times C$ by taking $\mathbf{V} := R^1\pi_*(\mathbf{E})$ by the projection π to the first factor. Then, Hitchin proposes that the brane L_i should be mirror to $\Lambda^{2i}\mathbf{V}$. This was checked to be correct on a generic fibre $\chi_a := \chi^{-1}(a)$ of the Hitchin map in the sense that $\mathcal{O}_{L_i}|_{\chi_a}$ is Fourier–Mukai dual to $\Lambda^{2i}\mathbf{V}|_{\chi_a}$. Additionally, the fact that $\Lambda^{2i}\mathbf{V}$ carries a natural hyperholomorphic structure gave more credence to the proposal that $\Lambda^{2i}\mathbf{V}$ should be the mirror of L_i. However, it is unclear how to extend the relative Fourier–Mukai transform to the most singular fibres of the Hitchin map, in particular to the nilpotent cone χ_0, and so we did not have direct checks of the correctness to the proposal that L_i is indeed relative Fourier–Mukai dual to $\Lambda^{2i}\mathbf{V}$.

There have recently been several works computing equivariant Verlinde formulas, on the moduli space of Higgs bundles. Let $\mathcal{L}$ be the determinant line bundle on $\mathcal{M}(\mathrm{G})$, where G, is a simply connected simple group, such as $\mathrm{G} = \mathrm{SL}_n$. The natural

$$\mathbb{T} := \mathbb{C}^\times$$

action on $\mathcal{M}(\mathrm{G})$ by scaling the Higgs field lifts to a $\mathbb{T}$-equivariant structure on $\mathcal{L}$. Although $H^*(\mathcal{M}(\mathrm{G}); \mathcal{L}^k)$ is infinite-dimensional, the induced $\mathbb{T}$-action will have finite-dimensional weight spaces and non-trivial only for non-positive weights:

$$H^*(\mathcal{M}(\mathrm{G}); \mathcal{L}^k) = \bigoplus_{j=\in\mathbb{Z}_{\geq 0}} H^*(\mathcal{M}(\mathrm{G}); \mathcal{L}^k)^{-j}.$$

We define for the $\mathbb{T}$-equivariant line bundle $\mathcal{L}^k$ (and, similarly, later for a $\mathbb{T}$-equivariant sheaf) the equivariant index as

$$\chi_{\mathbb{T}}\left(\mathcal{M}(\mathrm{G}); \mathcal{L}^k\right) = \sum_{i,j} (-1)^i \dim(H^i(\mathcal{M}(\mathrm{G}), \mathcal{L}^k)^j) t^{-j} \in \mathbb{Z}[[t]]. \tag{1.1}$$

In a series of papers [GP, GPYY, AGP] following insights from studying path integrals in certain supersymmetric quantum field theories and TQFT techniques, precise formulas have been achieved for (1.1). Equivalent formulas have been also found in [HL]. There is also a more combinatorial understanding of the results for $\mathrm{G} = \mathrm{SL}_n$ as residue formulas in a TQFT framework in [HSz].

In this paper, we compute similar characters but only on $\mathcal{M}(\mathrm{GL}_2)$ and, eventually, $\mathcal{M}(\mathrm{SL}_2)$. Namely, we will consider the (generically) vector bundles $\Lambda^{2i}\mathbf{V}$ mentioned above, with support in $\mathcal{M}(\mathrm{SL}_2)$ together with a $\mathbb{T}$-equivariant structure. We will also consider the analogues of the determinant bundle $\mathcal{L}$ on $\mathcal{M}(\mathrm{GL}_2)$ with a $\mathbb{T}$-equivariant structure and, in particular, the $\mathbb{T}$-equivariant coherent sheaves $\mathcal{L}^2\mathcal{O}_{L_i}$ on $\mathcal{M}(\mathrm{GL}_2)$ (for the detailed constructions, see Section 2). With the notation $\mathcal{L}_i := \mathcal{L}^2\mathcal{O}_{L_i}$ for $0 \leq i < g-1$ and $\Lambda_j := \Lambda^{2j}\mathbf{V}$, our main result is the following:

Theorem 1.2 *For* $0 \leq i,j < g-1$ *we have*

$$\chi_{\mathbb{T}}\left(\mathcal{M}(\mathrm{GL}_2); \mathcal{L}_j \otimes \Lambda_i\right) = \chi_{\mathbb{T}}\left(\mathcal{M}(\mathrm{GL}_2); \mathcal{L}_i \otimes \Lambda_j\right) \in \mathbb{Z}[[t]]. \tag{1.2}$$

In Corollary 6.3, we will explain how to extend this symmetry for the case of $i = g-1$.

We can interpret the left-hand side of (1.2) as the equivariant index of derived homomorphisms from $(\Lambda^{2i}\mathbf{V})^\vee \cong \Lambda^{2i}\mathbf{V}$ to $\mathcal{L}^2\mathcal{O}_{L_j}$, and the right-hand side as the equivariant index of derived homomorphisms from $\mathcal{L}^2\mathcal{O}_{L_i}^\vee \cong \mathcal{L}^2\mathcal{O}_{L_i}$ to $\Lambda^{2j}\mathbf{V}$. For more details on this see Section 7. This way, our Theorem 1.2 provides computational evidence for Hitchin's proposal in [Hi2, §7].

Interestingly, most of the ingredients used here have already appeared in the first author's PhD dissertation [Ha1], which was written under the supervision of Nigel Hitchin. In a sense, our computation could have been done almost 20 years ago. The present paper thus outlines the importance of ideas from physics, in particular the work of Kapustin and Witten [KW] in 2006 and further mathematical insights by Hitchin [Hi2, §7] in 2013.

The plan of the paper is as follows. In Section 2 we introduce the $\mathbb{C}^\times$-equivariant coherent sheaves $\mathcal{L}_i$ and Λ_i on $\mathcal{M}(\mathrm{GL}_2)$. In Section 3, we compute the equivariant index $\chi_{\mathbb{T}}(\mathcal{M}(\mathrm{GL}_2); \mathcal{L}_i \otimes \Lambda_j)$ for $i < g-1$, by reducing the computation via Hirzebruch–Riemann–Roch to an integral on the $2i$th symmetric product of C, which we evaluate using a residue formula due to Zagier. In Section 4, we find a change of variables which will imply our main symmetry observation for $i,j < g-1$. In Section 5, following the technique of Teleman and Woodward [TW], we compute the equivariant index on the whole moduli stack and then, in Section 6, we prove Corollary 6.3. Finally, in Section 7, we sketch how this should be a consequence of mirror symmetry.

Acknowledgements. We would like to thank Sergei Gukov and Paul Harmsen for encouragement and their interest in this paper, and Jørgen E. Andersen, András Szenes, Laura Schaposnik, Nigel Hitchin, Penghui Li, Iordan Ganev and the referee for useful explanations and comments. The research in this paper was supported by an Advanced Grant 'Arithmetic and physics of Higgs moduli spaces' no. 320593 of the European Research Council, the NCCR SwissMAP of the Swiss National Foundation, START-Project Y963-N35 of the Austrian Science Fund (FWF), the centre of excellence grant 'Centre for Quantum Geometry of Moduli Space' from the Danish National Research Foundation (DNRF95), the Walter Burke Institute for Theoretical Physics, and the US Department of Energy, Office of Science, Office of High Energy Physics, under Award Number DE-SC0011632. In particular, the idea of the computation in this paper arose during the 'Retreat on Higgs bundles, real groups, Langlands duality and mirror symmetry' in the Bernoulli centre at EPF Lausanne in January 2016.

2. Background

Let C be a smooth complex projective curve of genus $g > 1$. Let $\mathcal{M}(\mathrm{GL}_2)$ denote the moduli space of semi-stable degree zero rank 2 Higgs bundles on C. We recall [BGG, Hi2, Sch] that the $\mathrm{U}(1,1)$-Higgs bundles in $\mathcal{M}(\mathrm{GL}_2)$ are degree zero rank 2 Higgs bundles

$$E \stackrel{\Phi}{\to} E \otimes K_C \tag{2.1}$$

such that $\Phi \neq 0$ and

$$E \stackrel{-\Phi}{\to} E \otimes K_C$$

is equivalent with (2.1). In other words, there exists an $a : E \to E$ automorphism of E such that

$$\Phi \circ a = -\Phi.$$

When $\Phi \neq 0$, this implies that such a Higgs bundle has the form

$$M_1 \oplus M_2 \xrightarrow{\begin{pmatrix} 0 & \phi_2 \\ \phi_1 & 0 \end{pmatrix}} (M_1 \oplus M_2) \otimes K_C, \tag{2.2}$$

which consists of a direct sum of two line bundles and an off-diagonal Higgs field. We can assume

$$\deg M_2 \leq \deg M_1 = g - 1 - i$$

for some integer $i \leq g-1$. When $i < g-1$, then

$$\phi_1 \in H^0(C; M_1^{-1} M_2 K_C)$$

cannot be 0 because of semi-stability. As

$$\deg(M_1^{-1} M_2 K_C) = 2i,$$

we have that $i \geq 0$.

The locus of Higgs bundles of the form (2.2) with $\deg(M_1) = g-1-i$ for $0 \leq i \leq g-1$ is denoted by

$$L_i \subset \mathcal{M}(\mathrm{GL}_2).$$

With an argument similar to the one below for SL_2, one can show that for $0 \leq i < g-1$ the locus $L_i \subset \mathcal{M}(\mathrm{GL}_2)$ is isomorphic with a total space of a vector bundle over $\mathcal{J}_{\bar{i}} \times C_{2i}$. We denote by $\mathcal{J}_i$ the Jacobian of degree i line bundles on C, by C_j the jth symmetric product of the curve C and $\bar{i} := g-1-i$.

We will also need to define the sheaf $\mathcal{L}$. Recall the tensor product map

$$\tau : \mathcal{M}(\mathrm{SL}_2) \times T^*\mathcal{J}_0 \to \mathcal{M}(\mathrm{GL}_2),$$

which is a Galois cover, with Galois group $\mathcal{J}[2]$ comprising the 2-torsion points on the Jacobian. We choose $\mathcal{L} \in \mathrm{Pic}(\mathcal{M}(\mathrm{GL}_2))$ such that

$$\tau^*(\mathcal{L}) = \mathcal{L}_{\mathrm{SL}_2} \boxtimes \mathcal{O}_{T^*\mathcal{J}_0},$$

where $\mathcal{L}_{\mathrm{SL}_2}$ is the determinant line bundle on $\mathcal{M}(\mathrm{SL}_2)$ constructed below (2.8). $\mathcal{L}$ is not unique, because

$$\tau^* : \mathrm{Pic}(\mathcal{M}(\mathrm{GL}_2)) \to \mathrm{Pic}(\mathcal{M}(\mathrm{SL}_2) \times \mathcal{J}_0)$$

has a kernel isomorphic to

$$\mathrm{Pic}(\mathcal{M}(\mathrm{GL}_2))[2] \cong \mathcal{J}_0[2].$$

However, when restricted to $\mathcal{M}(\mathrm{U}(1,1))$, the ambiguity disappears. For $0 \leq i < g-1$, we will denote

$$\mathcal{L}_i := \mathcal{L}^2 \otimes \mathcal{O}_{L_i}$$

a coherent sheaf on $\mathcal{M}_{\mathrm{GL}_2}$, while, for $i = g-1$, we define

$$\mathcal{L}_{g-1} := (\mathcal{L}^2 \oplus \mathcal{L}^2) \otimes \mathcal{O}_{L_{g-1}}. \tag{2.3}$$

In our computation below, we will restrict to the part

$$L_i' := L_i \cap \mathcal{M}(\mathrm{SL}_2) \subset \mathcal{M}(\mathrm{SL}_2)$$

inside the moduli space $\mathcal{M}(\mathrm{SL}_2)$ of semi-stable rank 2 Higgs bundles with trivial determinant and trace-free Higgs field. In other words, L_i' is a component of the moduli space of $\mathrm{SU}(1,1)$-Higgs bundles in $\mathcal{M}(\mathrm{SL}_2)$. This means to set $M := M_1$ and choose $M_2 = M^{-1}$ in (2.2).

Now we give a more detailed description of $L_i' \subset \mathcal{M}(\mathrm{SL}_2)$ for $0 \leq i \leq g-2$. For similar discussions, see [Ha1, Lemma 6.1.2] and [HT2, (6.1)].

Define maps $\mathcal{J}_{\bar{i}} \to \mathcal{J}_{2i}$ by sending $M \mapsto M^{-2}K$ and the Abel–Jacobi map $C_{2i} \to \mathcal{J}_{2i}$ by sending $D \mapsto L(D)$. Then, we construct the fibred product of these maps:

$$F_i' := C_{2i} \times_{\mathcal{J}_{2i}} \mathcal{J}_{\bar{i}}. \tag{2.4}$$

We have the two projections $\mathrm{pr}_{\mathcal{J}_{\bar{i}}} : F_i' \to \mathcal{J}_{\bar{i}}$ and $\mathrm{pr}_{C_{2i}} : F_i' \to C_{2i}$. By construction, F_i' is isomorphic to the moduli space of complexes $M \xrightarrow{\phi} M^{-1}K_C$, with $\deg(M) = \bar{i}$ and $\phi \in H^0(\Sigma; M^{-2}K_C)$, in other words, of nilpotent SL_2-Higgs bundles

$$M \oplus M^{-1} \xrightarrow{\Phi} MK_C \oplus M^{-1}K_C$$

given by

$$\Phi = \begin{pmatrix} 0 & 0 \\ \phi & 0 \end{pmatrix}.$$

Let

$$\Delta_{2i} \subset C \times C_{2i}$$

denote the universal divisor, and $\mathcal{P}_{\bar{i}}$ a normalized universal bundle on $C \times \mathcal{J}_{\bar{i}}$. By abuse of notation, we will denote their pullbacks by $\mathrm{pr}_{C_{2i}}$ and, respectively, $\mathrm{pr}_{\mathcal{J}_{\bar{i}}}$ to $C \times F_i'$, with the same Δ_{2i} and $\mathcal{P}_{\bar{i}}$. As $i < g-1$, we have

$$\deg(K_C^2(-D)) > 2g-2$$

for $D \in C_{2i}$ and so

$$H^1(C; K_C^2(-D)) = 0.$$

Consequently,

$$E_i := \mathrm{pr}_{C*}\left(K_C^2(-\Delta_{2i})\right)$$

is a vector bundle on F_i'.

E_i parametrizes $SU(1,1) \subset SL_2$-Higgs bundles as follows. First, note that $\mathcal{P}_{\bar{i}}^2 K_C^{-1}(\Delta_{2i})$ restricted to $x \times C$ is a trivial line bundle for every $x \in F_i'$ and thus, by the push-pull formula, there is a line bundle $M_{F_i'}$ on F_i' such that

$$\mathcal{O}(\Delta_{2i}) \cong M_{F_i'} K_C \mathcal{P}_{\bar{i}}^{-2}. \tag{2.5}$$

Consider the rank 2 vector bundle

$$\mathbf{E}_i := \mathcal{P}_{\bar{i}} \oplus \mathcal{P}_{\bar{i}}^{-1} M_{F_i'} \tag{2.6}$$

on $C \times F_i'$ and the Higgs field

$$\Phi_i = \begin{pmatrix} 0 & \phi_i' \\ \phi_i & 0 \end{pmatrix}$$

with

$$\phi_i \in H^0(C \times E_i; \mathcal{P}_i^{-2} M_{F_i'} K_C) = H^0(C \times E_i; \mathcal{O}(\Delta_{2i}))$$

given by the universal divisor Δ_{2i} and

$$\phi_i' \in H^0(C \times E_i; \mathcal{P}_i^2 L_{E_i}^{-1} K_C) = H^0(C \times E_i; K_C^2(-\Delta_{2i}))$$

the tautological section.

Thus, $(\mathbf{E}_i, \Phi_i)$ is a family of stable SL_2-Higgs bundles parametrized by (the total space of) E_i. Thus, we get the embedding

$$\iota_i : E_i \to \mathcal{M}(SL_2)^s \subset \mathcal{M}(SL_2).$$

We denote its image

$$L_i' := \iota_i(E_i) \subset \mathcal{M}(SL_2)$$

and note that it parametrizes all Higgs bundles of the form $M \oplus M^{-1} \xrightarrow{\Phi} MK_C \oplus M^{-1}K_C$ with Higgs field

$$\Phi = \begin{pmatrix} 0 & \phi' \\ \phi & 0 \end{pmatrix}$$

and such that $\deg M = \bar{i}$. This precisely agrees with a component of the SU(1,1)-Higgs moduli space as described in [Sch] (see also [Hi2, §7] and [Hi1, §10]). Thus, L_i' is a connected smooth Lagrangian subvariety in the stable locus of $\mathcal{M}(\mathrm{SL}_2)$. Note that there is one more component of $\mathcal{M}(\mathrm{SU}(1,1))$ inside $\mathcal{M}(\mathrm{SL}_2)$, namely L_{g-1}'. However, it will no longer be contained in the stable locus and will not be smooth.

We are now going to construct the vector bundles Λ_j on L_i' for $j = 0, \ldots, g-1$. Hitchin [Hi2, §7] defines

$$\Lambda_j := \Lambda^{2j}\mathbf{V}$$

locally, where $\mathbf{V}$ is the Dirac bundle on $U \to \mathcal{M}(\mathrm{SL}_2)^s$ for an étale open U. It is constructed as

$$\mathbf{V} := R^1\mathrm{pr}_{C*}(\mathbf{E}_U \stackrel{\Phi_U}{\to} \mathbf{E}_U K_C)$$

where

$$\mathbf{E}_U \stackrel{\Phi_U}{\to} \mathbf{E}_U K_C$$

is a universal SL_2-Higgs bundle on $C \times U$. The universal Higgs bundle does not exist on the whole $C \times \mathcal{M}(\mathrm{SL}_2)^s$ but on an étale open covering, and the obstruction to glue vanishes for the even exterior powers $\Lambda_j = \Lambda^{2j}\mathbf{V}$. We note that $\mathbf{V}$ is a vector bundle on $\mathcal{M}(\mathrm{SL}_2)^s$ as, for stable degree zero Higgs bundles (E, Φ), the zeroth and second hypercohomology of $E \stackrel{\Phi}{\to} EK_C$ vanishes by [Ha2, Corollary 3.5].

Finally, we will extend the construction of $\mathbf{V}$ to the whole of $\mathcal{M}(\mathrm{SL}_2)$ as a complex of coherent sheaves by defining it for an étale open $U \to \mathcal{M}(\mathrm{SL}_2)$ by the formula

$$\mathbf{V} := R\mathrm{pr}_{C*}(\mathbf{E}_U \stackrel{\Phi_U}{\to} \mathbf{E}_U K_C)[1].$$

Then we define

$$\Lambda_j := \Lambda^{2j}\mathbf{V}$$

as a complex of coherent sheaves on the whole $\mathcal{M}(\mathrm{SL}_2)$.

When we restrict this construction of Λ_j to L_i' for $i < g-1$, then we have almost a universal bundle on $C \times L_i'$. Namely, $(\mathbf{E}_i, \Phi_i)$ is a family of stable SL_2-Higgs bundles on $C \times L_i'$. However, $\mathbf{E}_i$ is not itself an SL_2 bundle, as

$$\det(\mathbf{E}_i) = M_{F_i'}.$$

$M_{F_i'}$ is pulled back from L_i' and ultimately from F_i', which we will see has no square root. However, on the even exterior powers, we can just tensor with integer powers of $\det(\mathbf{E}_i)$ to get the restriction of Λ_j to L_i' and so we get

$$\Lambda_j|_{L_i'} \cong M_{F_i'}^{-j}\Lambda^{2j}(R^1\mathrm{pr}_{L_i'*}(\mathbf{E}_i \stackrel{\Phi_i}{\to} \mathbf{E}_i K_C)), \tag{2.7}$$

which gives the desired vector bundle on L_i'.

We will also need $\mathbb{T}$-equivariant structure on Λ_j. To construct this, we note that Hitchin's construction can be done on $\mathbb{T}$-equivariant étale open $U \to \mathcal{M}(\mathrm{SL}_2)^s$, where, from the universal property, one gets a $\mathbb{T}$-equivariant structure on $\mathbf{E}_U \stackrel{\Phi_U}{\to} \mathbf{E}_U K_C$ (cf. [HT1, §4]). They will then glue to yield a $\mathbb{T}$-equivariant structure on Λ_j. The $\mathbb{T}$-action we endow Λ_j is the one twisted by the trivial line bundle with a weight $-j$ $\mathbb{T}$-action on it.

To see the $\mathbb{T}$-equivariant structure on $\Lambda_j|_{L_i'}$ in (2.7), we endow $\mathcal{P}_i$ with the trivial $\mathbb{T}$-action and $M_{F_i'}$ with a weight minus one $\mathbb{T}$-action, which will give the $\mathbb{T}$-equivariant structure on $\mathbf{E}_i$ in (2.6). Then $(\mathbf{E}_i, \Phi_i)$ becomes a $\mathbb{T}$-equivariant Higgs bundle, provided we twist the second term $\mathbf{E}_i \otimes K_C$ with a weight one $\mathbb{T}$-equivariant trivial line bundle. The $\Lambda_j|_{L_i'}$ will then inherit a $\mathbb{T}$-equivariant structure, which we will further twist with a weight $-j$ trivial line bundle.

Finally, we will need to construct the $\mathbb{T}$-equivariant determinant line bundle $\mathcal{L}_{\mathrm{SL}_2}$ on $\mathcal{M}(\mathrm{SL}_2)$ (which, by abuse of notation, we will abbreviate to $\mathcal{L}$). It is again constructed [Qi] by gluing together

$$\Lambda^{2g-2}(R\mathrm{pr}_{C*}(\mathbf{E}_U)[1]) \tag{2.8}$$

from open sets $U \to \mathcal{M}(\mathrm{SL}_2)$ in an étale open covering.

Again on L_i', we can compute it from our family $(\mathbf{E}_i, \Phi_i)$ and get

$$\mathcal{L}|_{L_i'} = M_{F_i'}^{-g+1} \Lambda^{2g-2}(R\mathrm{pr}_{C*}(\mathbf{E}_i)[1]). \tag{2.9}$$

We can now define our other family of $\mathbb{T}$-equivariant sheaves,

$$\mathcal{L}_i' := \iota_{i*}\mathcal{L}^2|_{L_i'}, \tag{2.10}$$

on $\mathcal{M}(\mathrm{SL}_2)$, supported on L_i'. For a variable s, we also denote

$$\Lambda_{s^2} := \bigoplus_{j=0}^{2g-2} s^{2j} \Lambda_j. \tag{2.11}$$

3. Computation for $i < g - 1$

First, we start with a definition:

Definition 3.1 Let X be a semi-projective $\mathbb{T}$-variety (see [HV]) and $\mathcal{F}$ an $\mathbb{T}$-equivariant sheaf on it. Then define

$$\chi_{\mathbb{T}}(X; \mathcal{F}) := \sum_{i,j} (-1)^i \dim(H^i(X; \mathcal{F})^{-j}) t^j,$$

where $H^i(X; \mathcal{F})^{-j}$ is the $-j$-th weight space of the induced $\mathbb{T}$ action on $H^i(X; \mathcal{F})$.

We expect that, for a semi-projective X, the equivariant index satisfies $\chi_{\mathbb{T}}(X;\mathcal{F}) \in \mathbb{Z}((t))$. The following then is one of our main computational tools.

Proposition 3.2 *For $0 \le i < g-1$ and denoting $v = t^{\frac{1}{2}}$, we have*

$$\begin{aligned}
&\chi_{\mathbb{T}}(\mathcal{M}(\mathrm{GL}_2);\mathcal{L}_i \otimes \Lambda_{s^2}) = \chi_{\mathbb{T}}(L_i';\mathcal{L}_i' \otimes \Lambda_{s^2}) \\
&= \frac{1}{2\pi i}\oint_{|z^2-v^2|=\epsilon} z^{4\bar{i}} \frac{\left((1+\frac{s}{zv})(1+svz)\right)^{g-1+\bar{i}}\left((1+\frac{sv}{z})(1+\frac{sz}{v})\right)^{i}}{(1-v^2z^2)^{2\bar{i}+g-1}(1-v^2/z^2)^{2i-g+1}} \\
&\left(4+\frac{\frac{sv}{z}}{1+\frac{sv}{z}}+\frac{svz}{1+svz}+\frac{4v^2z^2}{1-v^2z^2}+\frac{\frac{4v^2}{z^2}}{1-\frac{v^2}{z^2}}-\frac{\frac{s}{zv}}{1+\frac{s}{zv}}-\frac{\frac{sz}{v}}{1+\frac{sz}{v}}\right)^{g}\frac{dz}{z} \in \mathbb{Z}[[t,s]].
\end{aligned} \tag{3.3}$$

Remark 3.4 In the right-hand side of (3.3), v is treated as a fixed complex number close to 1, and ϵ is a sufficiently small positive real number. In particular, the contour $|z^2-v^2| = \epsilon$ will have two components around the two square roots $\pm v$ of t. Thus, the result could also be computed algebraically as the sum of the residues of the differential form when z equals the two square roots of t.

Proof As L_i' is the total space of the $\mathbb{T}$-equivariant vector bundle E_i, where the $\mathbb{T}$-action has weight two, we have that

$$\begin{aligned}
\chi_{\mathbb{T}}(L_i';\mathcal{L}_i' \otimes \Lambda_{s^2}) &= \chi_{\mathbb{T}}(F_i';\mathcal{L}^2 \otimes \Lambda_{s^2} \otimes \mathrm{Sym}_{t^2}(E_i^*)) \\
&= \int_{F_i'} \mathrm{ch}(\mathcal{L}^2)\mathrm{ch}(\Lambda_{s^2})\mathrm{ch}(\mathrm{Sym}_{t^2}(E_i^*))\mathrm{td}(T_{F_i'}).
\end{aligned} \tag{3.5}$$

Recall $[$HT1, §5$]$ that we have

$$c_1(\Delta_{2i}) = 2i \otimes \sigma + \sum_{l=1}^{2g} \xi_l \otimes e_l + \eta \otimes 1 \in H^2(C_{2i} \times C),$$

where σ is generator of $H^2(C)$, and $e_1,\dots,e_{2g}$ are canonical symplectic generators of $H^1(C)$. Thus, $\eta \in H^2(C_{2i})$, and $\xi_1,\dots,\xi_{2g} \in H^1(C_{2i})$. A theorem of Macdonald $[$Mac, (6.3)$]$ asserts that the cohomology ring $H^*(C_m)$ is generated by η and the ξ_j. It is convenient to introduce

$$\theta_j = \xi_j\xi_{j+g}$$

and

$$\theta = \sum_{j=1}^{g} \theta_j \in H^2(C_{2i}).$$

We note the identity

$$\left(\sum_{l=1}^{2g} \xi_l \otimes e_l\right)^2 = -2\theta\sigma \tag{3.6}$$

used below.

By Grothendieck–Riemann–Roch we find that

$$\begin{aligned}
\mathrm{ch}(E_i) &= \mathrm{pr}_{F'_i *}\left(\mathrm{ch}(K_C^2(-\Delta_{2i}))\mathrm{td}(C)\right)\\
&= \mathrm{pr}_{F'_i *}\left(\exp\left((4g-4-2i)\otimes\sigma - \sum_{l=1}^{2g}\xi_l\otimes e_l - \eta\otimes 1\right)(1+(1-g)\sigma)\right)\\
&= \mathrm{pr}_{F'_i *}\left(\left((4g-4-2i-\theta)\exp(-\eta)\otimes\sigma - \sum_{l=1}^{2g}\exp(-\eta)\xi_l\otimes e_l + \exp(-\eta)\otimes 1\right)(1+(1-g)\sigma)\right)\\
&= \mathrm{pr}_{F'_i *}\left(((4g-4-2i-\theta+1-g)\exp(-\eta))\otimes\sigma + \sum_{l=1}^{2g}\exp(-\eta)\xi_l\otimes e_l + \exp(-\eta)\otimes 1\right)\\
&= (3g-3-2i-\theta)\exp(-\eta) \qquad (3.7)\\
&= (2g-3-2i)\exp(-\eta) + \sum_{i=1}^{g}\exp(-\eta-\theta_i).
\end{aligned}$$

Thus, introducing $\zeta = \exp(\eta)$, we have

$$\begin{aligned}
\mathrm{ch}(\mathrm{Sym}_{t^2}E_i^*) &= \frac{1}{(1-t^2\zeta)^{2\bar{i}-1}\prod_{i=1}^{g}(1-t^2\zeta\exp(\theta_i))}\\
&= \frac{1}{(1-t^2\zeta)^{2\bar{i}+g-1}\exp\left(-\frac{t^2\zeta\theta}{1-t^2\zeta}\right)}
\end{aligned} \tag{3.8}$$

as

$$\begin{aligned}
\prod_{i=1}^{g}(1-t^2\zeta\exp(\theta_i)) &= \prod_{i=1}^{g}(1-t^2\zeta(1+\theta_i)) = \prod_{i=1}^{g}(1-t^2\zeta - t^2\zeta\theta_i)\\
&= (1-t^2\zeta)^g\prod_{i=1}^{g}\left(1-\frac{t^2\zeta\theta_i}{1-t^2\zeta}\right) = (1-t^2\zeta)^g\prod_{i=1}^{g}\exp\left(-\frac{t^2\zeta\theta_i}{1-t^2\zeta}\right)\\
&= (1-t^2\zeta)^g\exp\left(-\frac{t^2\zeta\theta}{1-t^2\zeta}\right).
\end{aligned}$$

Similar computation, using the formula for total Chern class [Mac, (14.5)], yields

$$\mathrm{td}(T_{C_{2i}}) = \left(\frac{\eta}{1-1/\zeta}\right)^{2i-g+1}\exp\left(\frac{\theta}{\zeta-1}-\frac{\theta}{\eta}\right). \tag{3.9}$$

To compute the remaining Chern characters in (3.5), we recall that

$$c_1(\mathcal{P}_{\bar{i}}) = \bar{i} \otimes \sigma + \sum_{l=1}^{2g} \tau_l \otimes e_l \in H^2(\mathcal{J}_{2\bar{i}} \times C).$$

Comparing with (2.5), we see that

$$\mathrm{ch}(\mathcal{P}^2_{2\bar{i}}) = \mathrm{ch}(\Delta_i)\mathrm{ch}(K_C^{-1})\mathrm{ch}(M_{F_i'}^{-1}),$$

and the fact that $M_{F_i'}$ is pulled back from F_i' implies that on F_i' we have

$$\tau_l = \xi_l/2$$

and

$$\mathrm{ch}(M_{F_i'}) = \zeta. \tag{3.10}$$

We need now to compute the $\mathbb{T}$-equivariant Chern character of $\mathbf{E}_i$. From (2.6) and the fact that $\mathbb{T}$ acts on the second summand with weight minus one, we get

$$\mathrm{ch}_{\mathbb{T}}(\mathbf{E}_i) = \mathrm{ch}(\mathcal{P}_{\bar{i}}) \ + \ t\,\mathrm{ch}(\mathcal{P}_{\bar{i}}^{-1})\mathrm{ch}(M_{F_i'}) = \exp\left(\bar{i} \otimes \sigma + \sum_{l=1}^{2g} \xi_l/2 \otimes e_l\right)$$
$$+ t\zeta \exp\left(-\bar{i} \otimes \sigma - \sum_{l=1}^{2g} \xi_l/2 \otimes e_l\right).$$

It follows from Grothendieck–Riemann–Roch that

$$\begin{aligned}
\mathrm{ch}_{\mathbb{T}}(\mathbf{V}_i) &= -\mathrm{ch}_{\mathbb{T}}(R\mathrm{pr}_{C*}(\mathbf{E}_i \stackrel{\Phi_i}{\to} \mathbf{E}_i K_C)) = \mathrm{pr}_{C*}\left(\mathrm{ch}_{\mathbb{T}}(\mathbf{E}_i)(-1 + t^{-1}\mathrm{ch}(K_C))\mathrm{td}(C)\right) \\
&= \mathrm{pr}_{C*}\left(\left(\exp\left(\bar{i} \otimes \sigma + \sum_{l=1}^{2g} \xi_l/2 \otimes e_l\right) + t\zeta \exp\left(-\bar{i} \otimes \sigma - \sum_{l=1}^{2g} \xi_l/2 \otimes e_l\right)\right)\right. \\
&\quad \left.(-1 + t^{-1} + (g-1)(1 + t^{-1})\sigma)\right) \\
&= \mathrm{pr}_{C*}\left(\left((1 + (\bar{i} - \theta/4) \otimes \sigma) + t\zeta\left(1 + (-\bar{i} - \theta/4) \otimes \sigma\right)\right)\right. \\
&\quad \left.(-1 + t^{-1} + (g-1)(1 + t^{-1})\sigma)\right) \\
&= \left((\bar{i} - \theta/4) + t\zeta(-\bar{i} - \theta/4)\right)(t^{-1} - 1) + (1 + t\zeta)(g-1)(t^{-1} + 1) \\
&= \bar{i}(t^{-1} - 1)(1 - t\zeta) - \theta/4(t^{-1} - 1)(1 + t\zeta) + (g-1)(t^{-1} + 1)(1 + t\zeta) \\
&= \bar{i}(t^{-1} - 1)(1 - t\zeta) - e^{\theta_i/4}(t^{-1} - 1)(1 + t\zeta) + g(t^{-1} - 1)(1 + t\zeta) \\
&\quad + (g-1)(t^{-1} + 1)(1 + t\zeta).
\end{aligned}$$

Thus, we have

$$\mathrm{ch}_{\mathbb{T}}(\Lambda_s(\mathbf{V}_i)) = \left(\frac{(1+\frac{s}{t})(1+st\zeta)}{(1+s)(1+s\zeta)}\right)^{\bar{i}} \prod_{i=1}^{g} \frac{(1+se^{\theta_i/4})(1+st\zeta\, e^{\theta_i/4})}{(1+\frac{s}{t}e^{\theta_i/4})(1+s\zeta\, e^{\theta_i/4})} \left(\frac{(1+\frac{s}{t})(1+s\zeta)}{(1+s)(1+st\zeta)}\right)^{g}$$
$$((1+s/t)(1+s\zeta)(1+s)(1+st\zeta))^{g-1}$$
$$= \frac{\exp\left(\frac{s\theta/4}{1+s} + \frac{st\zeta\theta/4}{1+st\zeta}\right)}{\exp\left(\frac{\frac{s}{t}\theta/4}{1+\frac{s}{t}} + \frac{s\zeta\theta/4}{1+s\zeta}\right)} ((1+s/t)(1+st\zeta))^{g-1+\bar{i}}((1+s)(1+s\zeta))^{i}. \tag{3.11}$$

Let

$$s' := \frac{s}{\zeta^{1/2}}.$$

Recalling the extra equivariant twist we endowed Λ_i in the paragraphs after (2.7), we have

$$\mathrm{ch}_{\mathbb{T}}(\Lambda_{s^2}) = (\mathrm{ch}_{\mathbb{T}}(\Lambda_{s'}(\mathbf{V}_i)) + \mathrm{ch}_{\mathbb{T}}(\Lambda_{-s'}(\mathbf{V}_i)))/2. \tag{3.12}$$

Finally, we compute

$$\mathrm{ch}_{\mathbb{T}}(R\mathrm{pr}_{C*}(\mathbf{E}_i)[1])$$
$$= \mathrm{pr}_{C*}\left(-\left(\exp\left(\bar{i}\otimes\sigma + \sum_{l=1}^{2g}\xi_l/2\otimes e_l\right) + t\zeta\exp\left(-\bar{i}\otimes\sigma - \sum_{l=1}^{2g}\xi_l/2\otimes e_l\right)\right)\right.$$
$$\left.(1+(1-g)\sigma)\right)$$
$$= \mathrm{pr}_{C*}\left(-\left(1+(\bar{i}-\theta/4)\otimes\sigma\right) - t\zeta\left(1+(-\bar{i}-\theta/4)\otimes\sigma\right)\right)(1+(1-g)\sigma)\Big)$$
$$= \bar{i}(t\zeta-1) + \theta/4(1+t\zeta) + (g-1)(t\zeta+1)$$
$$= \bar{i}(t\zeta-1) + \sum_{i=1}^{g} e^{\theta_i/4}(1+t\zeta) - (t\zeta+1);$$

thus,

$$\mathrm{ch}_{\mathbb{T}}(\det(R\mathrm{pr}_{C*}(\mathbf{E}_i)[1])) = (t\zeta)^{\bar{i}+g-1}e^{\theta/2}$$

and so, by (2.9) and (3.10),

$$\mathrm{ch}_{\mathbb{T}}(\mathcal{L}|_{L_i'}) = t^{\bar{i}}\zeta^{\bar{i}}e^{\theta/2}. \tag{3.13}$$

Recall the following integral formula of Zagier's from [Th, (7.2)]. For any power series $A(x) \in \mathbb{C}[[x]]$ and $B(x) \in \mathbb{C}[[x]]$,

$$\int_{C_n} A(\eta)\exp(B(\eta)\sigma) = \mathrm{Res}_{x=0}\left(\frac{A(x)(1+xB(x))^g}{x^{n+1}}dx\right).$$

We also let

$$z = e^{\frac{x-u}{2}} \in \mathbb{C}[[x,u]],$$

in particular $dx = \frac{2dz}{z}$. Here, u is such that

$$e^{-u} = t.$$

We also introduce

$$v := e^{\frac{-u}{2}}.$$

By recalling that $\mathrm{pr}_{C_{2i}} : F_i' \to C_{2i}$ is a 2^{2g} cover, we will now compute

$$\begin{aligned}
&\int_{F_i'} \mathrm{ch}_{\mathbb{T}}(\mathcal{L}_i')\, \mathrm{ch}_{\mathbb{T}}(\Lambda_{s'}(\mathbf{V}_i))\, \mathrm{ch}_{\mathbb{T}}(\mathrm{Sym}_{t^2} E_i^*)\, \mathrm{td}(T_{F_i'}) \\
&= \int_{F_i'} (t\zeta)^{2\bar{i}} \frac{\exp\left(\theta + \frac{s'\theta/4}{1+s'} + \frac{s't\zeta\theta/4}{1+s't\zeta} + \frac{t^2\zeta\theta}{1-t^2\zeta} + \frac{\theta}{\zeta-1}\right)}{\exp\left(\frac{\frac{s'}{t}\theta/4}{1+\frac{s'}{t}} + \frac{s'\zeta\theta/4}{1+s'\zeta} + \frac{\theta}{\eta}\right)} \frac{\left((1+\frac{s'}{t})(1+s't\zeta)\right)^{g-1+\bar{i}} \left((1+s')(1+s'\zeta)\right)^i \eta^{2i-g+1}}{(1-t^2\zeta)^{2\bar{i}+g-1}(1-1/\zeta)^{2i-g+1}} \\
&= 2^{2g} \mathrm{Res}_{x=0} z^{4\bar{i}} \frac{\left((1+\frac{s}{zv})(1+svz)\right)^{g-1+\bar{i}} \left((1+\frac{sv}{z})(1+\frac{sz}{v})\right)^i x^{-g}}{(1-v^2z^2)^{2\bar{i}+g-1}(1-v^2/z^2)^{2i-g+1}} \\
&\left(1 + x\left(1 + \frac{\frac{sv}{4z}}{1+\frac{sv}{z}} + \frac{svz/4}{1+svz} + \frac{v^2z^2}{1-v^2z^2} + \frac{\frac{v^2}{z^2}}{1-\frac{v^2}{z^2}} - \frac{\frac{s}{4zv}}{1+\frac{s}{zv}} - \frac{\frac{sz}{4v}}{1+\frac{sz}{v}} - \frac{1}{x}\right)\right)^g dx \\
&= \mathrm{Res}_{x=0} z^{4\bar{i}} \frac{\left((1+\frac{s}{zv})(1+svz)\right)^{g-1+\bar{i}} \left((1+\frac{sv}{z})(1+\frac{sz}{v})\right)^i}{(1-v^2z^2)^{2\bar{i}+g-1}(1-v^2/z^2)^{2i-g+1}} \\
&\left(4 + \frac{\frac{sv}{z}}{1+\frac{sv}{z}} + \frac{svz}{1+svz} + \frac{4v^2z^2}{1-v^2z^2} + \frac{\frac{4v^2}{z^2}}{1-\frac{v^2}{z^2}} - \frac{\frac{s}{zv}}{1+\frac{s}{zv}} - \frac{\frac{sz}{v}}{1+\frac{sz}{v}}\right)^g dx \\
&= \frac{1}{2\pi i} \oint_{|z-v|=\epsilon} z^{4\bar{i}} \frac{\left((1+\frac{s}{zv})(1+svz)\right)^{g-1+\bar{i}} \left((1+\frac{sv}{z})(1+\frac{sz}{v})\right)^i}{(1-v^2z^2)^{2\bar{i}+g-1}(1-v^2/z^2)^{2i-g+1}} \\
&\left(4 + \frac{\frac{sv}{z}}{1+\frac{sv}{z}} + \frac{svz}{1+svz} + \frac{4v^2z^2}{1-v^2z^2} + \frac{\frac{4v^2}{z^2}}{1-\frac{v^2}{z^2}} - \frac{\frac{s}{zv}}{1+\frac{s}{zv}} - \frac{\frac{sz}{v}}{1+\frac{sz}{v}}\right)^g \frac{2dz}{z}.
\end{aligned}$$

We notice that

$$\int_{F_i'} \mathrm{ch}_{\mathbb{T}}(\mathcal{L}_i)\, \mathrm{ch}_{\mathbb{T}}(\Lambda_{-s'}(V_i))\, \mathrm{ch}_{\mathbb{T}}(\mathrm{Sym}_{t^2} E_i^*)\, \mathrm{td}(T_{F_i'})$$

will yield the same result with z replaced by $-z$.

Thus, (3.12) implies the proposition. □

4. The Symmetry for $i, j < g - 1$

First, we rewrite the right-hand side of (3.3) in a more manageable form.

Introduce the notations

$$f(z,s,v) = \frac{z^4(1-v^2/z^2)^2(1+\frac{s}{zv})(1+svz)}{(1-v^2z^2)^2(1+\frac{sv}{z})(1+\frac{sz}{v})} \tag{4.1}$$

and

$$h(z,s,v) = z\frac{\partial f}{\partial z}\frac{(1+\frac{s}{zv})(1+svz)(1+\frac{sv}{z})(1+\frac{sz}{v})}{(1-v^2/z^2)(-v^2z^2+1)}.$$

We can compute

$$z\frac{\frac{\partial}{\partial z}f(z,s,v)}{f(z,s,v)} = \left(4+\frac{\frac{sv}{z}}{1+\frac{sv}{z}}+\frac{svz}{1+svz}+\frac{4v^2z^2}{1-v^2z^2}+\frac{\frac{4v^2}{z^2}}{1-\frac{v^2}{z^2}}-\frac{\frac{s}{zv}}{1+\frac{s}{zv}}-\frac{\frac{sz}{v}}{1+\frac{sz}{v}}\right). \tag{4.2}$$

With this notation, for $i < g-1$, we can rewrite the right-hand side of (1.1) as

$$\frac{1}{2\pi i}\oint_{|(z^2-v^2)(1+\frac{s}{zv})(1+svz)|=\epsilon}\frac{h^{g-1}}{f^i}\frac{df}{f},$$

because the integrand has no pole at $z=-s/v$ or $z=-1/vs$. Thus, for $0\le i<g-1$ and $0\le j\le g-1$, (3.3) gives

$$\chi_{\mathbb{T}}(\mathcal{M}(\mathrm{GL}_2);\mathcal{L}_i\otimes\Lambda_j)) = \frac{1}{(2\pi i)^2}\oint_{|s|=\epsilon}\oint_{|f|=\epsilon}\frac{h^{g-1}}{f^is^{2j}}\frac{df}{f}\frac{ds}{s}. \tag{4.3}$$

Proposition 4.4 *For any integers i and j we have*

$$\frac{1}{(2\pi i)^2}\oint_{|s|=\epsilon}\oint_{|f|=\epsilon}\frac{h^{g-1}}{f^is^{2j}}\frac{df}{f}\frac{ds}{s} = \frac{1}{(2\pi i)^2}\oint_{|s|=\epsilon}\oint_{|f|=\epsilon}\frac{h^{g-1}}{f^js^{2i}}\frac{df}{f}\frac{ds}{s}.$$

Proof To show that h is 'symmetric' in f and s^2, we consider

$$w = \left(\frac{(sz+v)(vz+s)}{(szv+1)(sv+z)}\right)^{\frac{1}{2}}.$$

Then, using (4.5), we find

$$f(w,f(z,s,v)^{\frac{1}{2}},v) = s^2$$

and that

$$h(w, f(z,s,v)^{\frac{1}{2}}, v) = h(z,s,v).$$

One can check these functional equations by noting that h and f depend only on z^2, s^2 and zs; thus, the substitutions $z = w$ and $s = f^{\frac{1}{2}}$ can be done polynomially.

Thus, consider $\varphi : \mathbb{C}^2 \to \mathbb{C}^2$ by

$$\varphi(z,s) = (w, f(z,s,v)^{\frac{1}{2}})$$

defined on a dense open subset of the source, by avoiding the poles of f and w, and also choosing compatible branches of the two square roots, so that

$$wf^{\frac{1}{2}} = \frac{z^2(1-v^2/z^2)(vz+s)}{(1-v^2z^2)(sv+z)}. \tag{4.5}$$

Then we see that

$$\varphi^*\left(\frac{df}{f}\right) = \frac{ds^2}{s^2} = \frac{2ds}{s},$$

$$\varphi^*\left(\frac{ds}{s}\right) = \frac{df^{\frac{1}{2}}}{f^{\frac{1}{2}}} = \frac{1}{2}\frac{df}{f},$$

and so

$$\varphi^*\left(\frac{df \wedge ds}{fs}\right) = \frac{ds \wedge df}{sf}.$$

Thus, we can compute

$$\oint_{|s^2|=|f|=\epsilon} \frac{h^{g-1}}{f^i s^{2j}} \frac{df \wedge ds}{fs} = \oint_{|f|=|s^2|=\epsilon} \varphi^*\left(\frac{h^{g-1}}{f^i s^{2j}} \frac{df \wedge ds}{fs}\right) = \oint_{|f|=|s^2|=\epsilon} \frac{h^{g-1}}{s^i f^{2j}} \frac{ds \wedge df}{sf}$$
$$= \oint_{|s^2|=|f|=\epsilon} \frac{h^{g-1}}{s^i f^{2j}} \frac{df \wedge ds}{fs},$$

proving the proposition. □

Remark 4.6 We can observe that

$$f(1/z,s,v) = f(z,s,v)^{-1}$$

and that

$$\frac{h(1/z,s,v)}{f(1/z,s,v)} = \frac{h(z,s,v)}{f(z,s,v)}.$$

Consequently, after the change of variables $z \mapsto 1/z$, we have

$$\begin{aligned}\oint_{|s^2|=|f|=\epsilon} \frac{h^{g-1}}{f^i s^{2j}} \frac{df \wedge ds}{fs} &= \oint_{|s^2|=|1/f|=\epsilon} \frac{h^{g-1}}{f^{2g-2-i} s^{2j}} \frac{-df \wedge ds}{fs} \\ &= \oint_{|s^2|=|f|=\epsilon} \frac{h^{g-1}}{f^{2g-2-i} s^{2j}} \frac{df \wedge ds}{fs}, \end{aligned} \tag{4.7}$$

where we used the residue theorem and the fact that the poles of the integrand are located at the zeros of f and $1/f$. Using Proposition 4.4, this yields

$$\oint_{|s^2|=|f|=\epsilon} \frac{h^{g-1}}{f^i s^{2j}} \frac{df \wedge ds}{fs} = \oint_{|s^2|=|f|=\epsilon} \frac{h^{g-1}}{f^i s^{2(2g-2-j)}} \frac{df \wedge ds}{fs},$$

which is expected in light of the role of s in labelling the various exterior powers of $\mathbf{V}$ as in (2.11).

Remark 4.8 We can also notice that, for $i < 0$,

$$\oint_{|s^2|=|f|=\epsilon} \frac{h^{g-1}}{f^i s^{2j}} \frac{df \wedge ds}{fs} = 0, \tag{4.9}$$

because the integrand has no pole at zeroes of f. From Proposition 4.4 and (4.7), we get that, for $j < 0$ and $j > 2g-2$,

$$\oint_{|s^2|=|f|=\epsilon} \frac{h^{g-1}}{f^i s^{2j}} \frac{df \wedge ds}{fs} = 0. \tag{4.10}$$

For $0 \leq i < g-1$, this follows from the fact that $\mathbf{V}$ is a vector bundle of rank $4g-4$ on $\mathcal{M}(\mathrm{GL}_2)^s$. We will also give an interpretation of this for $i = g-1$ below in Remark 6.2.

Finally, by combining the previous Proposition 4.4 with (4.3), we have now the following result, which is identical to our Theorem 1.2 from the introduction.

Corollary 4.11 *For* $0 \leq i,j < g-1$, *we have*

$$\chi_{\mathbb{T}}(\mathcal{M}(\mathrm{GL}_2); \mathcal{L}_i \otimes \Lambda_j) = \chi_{\mathbb{T}}(\mathcal{M}(\mathrm{GL}_2); \mathcal{L}_j \otimes \Lambda_i)$$

5. Computation over the Moduli Stack

In this section, we will continue the study of the equivariant indices of Λ_j, but now over the *moduli stack* of Higgs bundles. This means that we will also incorporate the contribution of unstable Hitchin pairs. The benefit is that we can now apply the very powerful techniques developed by Teleman and Woodward in [TW].

As this approach has the advantage of being easily generalizable to other Lie groups, we will first take G to be a general connected finite-dimensional reductive group over $\mathbb{C}$, and will only specify it to be $SL(2,\mathbb{C})$ or $GL(2,\mathbb{C})$ at a later stage. We assume $\pi_1(G)$ has no torsion. Let G_0 be a real form of G, whose fundamental group again is assumed to be free. We consider the moduli space of G_0-Higgs bundles, following [GGM]. The Lie algebra of G_0 is denoted as $\mathfrak{g}_0$ and satisfies $\mathfrak{g}_0 \otimes_{\mathbb{R}} \mathbb{C} \simeq \mathfrak{g}$. It admits the Cartan decomposition

$$\mathfrak{g}_0 = \mathfrak{h} \oplus \mathfrak{m}.$$

Here, $\mathfrak{h}$ is the Lie algebra of the maximal compact subgroup $H \subset G_0$ and, as the adjoint action of H preserves the Cartan decomposition, $\mathfrak{m}$ will carry an H-representation. We denote the set of weights as $\mathfrak{R}(\mathfrak{m})$. Over the curve C, a G_0-Higgs bundle is a pair (E,φ), where E is a holomorphic principal $H^{\mathbb{C}}$-bundle over C and

$$\varphi \in H^0(C, E(\mathfrak{m}^{\mathbb{C}}) \otimes K_C),$$

where $E(\mathfrak{m}^{\mathbb{C}}) := E \times_{H^{\mathbb{C}}} \mathfrak{m}^{\mathbb{C}}$ is the $\mathfrak{m}^{\mathbb{C}}$-bundle associated to E.

One can define the moduli space $\mathcal{M}(G_0)$ of semi-stable G_0-Higgs bundles, over which there is again a $\mathbb{T}$-action acting by scaling the Higgs field, and one can consider the equivariant index of various vector bundles over it. To apply the result of [TW], we will first work with the moduli stack of G_0-Higgs bundles, which is defined as follows. Denote $E^*(\mathfrak{h}^{\mathbb{C}})$ the vector bundle associated to the universal $H^{\mathbb{C}}$-bundle over $C \times \mathrm{Bun}_{H^{\mathbb{C}}}$, and π the projection $C \times \mathrm{Bun}_{H^{\mathbb{C}}} \to \mathrm{Bun}_{H^{\mathbb{C}}}$. Then the moduli stack of G_0-Higgs bundles is

$$\mathfrak{M}(G_0) = \underline{\mathrm{Spec}}_{\mathrm{Bun}_{H^{\mathbb{C}}}} \mathrm{Sym}\left(R\pi_*(E^*(\mathfrak{h}^{\mathbb{C}}))[1]\right).$$

Over $\mathfrak{M}(G_0)$, we say a line bundle L has level $k \in H^4(BH^{\mathbb{C}},\mathbb{Z})$ if $c_1(L)$ is the image of k under the injective transgression homomorphism $H^4(BH^{\mathbb{C}},\mathbb{Z}) \to H^2(\mathrm{Bun}_{H^{\mathbb{C}}},\mathbb{Z})$; k defines a quadratic form on $\mathfrak{h}$ and, following [TW], we call a line bundle *admissible* if $k > -h$ has a quadratic form with $h := -\frac{1}{2}\mathrm{Tr}_{\mathfrak{h}}$. For a line bundle L of level k, we will usually refer to it as $\mathfrak{L}^k$, with slightly abusive use of notation, to make the level explicit. When $H^4(BH^{\mathbb{C}},\mathbb{Z}) = \mathbb{Z}$, k is an integer, and $\mathfrak{L}^k$ will be the kth tensor power of the level-one line bundle $\mathfrak{L}$; $\mathbb{T}$ also acts on $\mathfrak{M}(G_0)$ by scaling the Higgs field, and we will consider admissible line bundles that are equivariant, whose equivariant index is what we are after. But, before presenting the formula for this index, we first need to introduce some notations.

Consider the function on the Cartan subalgebra $\mathfrak{t} \subset \mathfrak{h}^{\mathbb{C}}$,

$$D: \quad \xi \mapsto \frac{k+h}{2}(\xi,\xi) - \mathrm{Tr}_{\mathfrak{h}}\left(\mathrm{Li}_2(te^{\xi})\right),$$

and the map from T, the Cartan of $H^{\mathbb{C}}$, to $T^{\vee}$:

$$\chi_t = e^{(k+h)(\cdot,\cdot)} \prod_{\mu \in \mathfrak{R}(\mathfrak{m})} \left(1 - te^{\mu}\right)^{\mu}.$$

We let F_t be the preimage of $e^{2\pi i\rho}$ in the regular part of T. It is obvious that F_t is invariant under the action of the Weyl group W_{H} of H. Combined with the Killing form, D defines an endomorphism on $\mathfrak{t}$ which we denote as $H_t^\dagger$; finally, we define

$$\theta_t := \frac{\prod_\alpha (1-e^\alpha)\prod_{\mu\in\mathfrak{R}(\mathfrak{m})}(1-te^\mu)}{\det H_t^\dagger},$$

where α in the first product in the numerator runs over all roots of $\mathfrak{h}$. Then we have the following theorem.

Theorem 5.1 (Equivariant Verlinde formula for real groups) *The equivariant index of $\mathfrak{L}^k$ over $\mathfrak{M}(\mathrm{G}_0)$ is given by*

$$\chi_{\mathbb{T}}(\mathfrak{M}(\mathrm{G}_0),\mathfrak{L}^k) = \sum_{f\in F_t/W_{\mathrm{H}}} \theta_t(f)^{1-g}. \tag{5.2}$$

Proof From the map $\mathfrak{M}(\mathrm{G}_0)\to \mathrm{Bun}_{\mathrm{H}^{\mathbb{C}}}$, we have

$$\chi_{\mathbb{T}}(\mathfrak{M}(\mathrm{G}_0),\mathfrak{L}^k) = \chi\left(\mathrm{Bun}_{\mathrm{H}^{\mathbb{C}}}, \mathfrak{L}^k\otimes\Lambda_{-t}R\pi_*(E^*(\mathfrak{h}^{\mathbb{C}}))\right).$$

Here, χ on the right-hand side denotes the usual (non-equivariant) index of a K-theory class over $\mathrm{Bun}_{\mathrm{H}^{\mathbb{C}}}$. Also, we have abusively used $\mathfrak{L}^k$ to denote its restriction to $\mathrm{Bun}_{\mathrm{H}^{\mathbb{C}}}$, and used the identity $\mathrm{Sym}_t\left(V[1]\right) = \Lambda_{-t}V$ for arbitrary class K-theory V. Then, the right-hand side of the equation can be handled by Theorem 2.15 of [TW]. The computation is a straightforward modification of that in the proof of Theorem 7 in [AGP], with the adjoint representation now replaced with $\mathfrak{m}$. □

Remark 5.3 In fact, the theorem holds for more general non-compact subgroups $\mathrm{G}_0\subset\mathrm{G}$ that are not real forms of G. For example, it holds when $\mathrm{G}_0=\mathrm{G}$ is the entire group, for which the index formula just becomes the equivariant Verlinde formula for G studied in [GP, GPYY, AGP, HL].

Example 5.4 ($\mathrm{G}_0=\mathrm{SL}(2,\mathbb{R})$) In this case, $\mathrm{H}=\mathrm{U}(1)$, and $\mathfrak{m}^{\mathbb{C}}$ is the two-dimensional representation that decomposes as $\mathbb{C}_{+2}\oplus\mathbb{C}_{-2}$, where the subscripts denote the weights of the U(1)-action. We parametrize H by an angle ϑ in $[0,2\pi)$. Then F_t is the set of solutions to the equation

$$e^{2ki\vartheta}\left(\frac{1-te^{-2i\vartheta}}{1-te^{2i\vartheta}}\right)^2 = 1,$$

and

$$\theta_t = \frac{(1-te^{2i\vartheta})(1-te^{-2i\vartheta})}{\det H_t^\dagger},$$

where $H_t^\dagger$ is the Hessian of

$$D: \quad \vartheta \mapsto -k\vartheta^2 - \left(\mathrm{Li}_2(te^{2i\vartheta}) + \mathrm{Li}_2(te^{-2i\vartheta})\right).$$

Example 5.5 ($\mathrm{G}_0 = \mathrm{U}(1,1)$) This is very similar to the previous case. We now have $\mathrm{H} = \mathrm{U}(1) \times \mathrm{U}(1)$ and $\mathfrak{m}^{\mathbb{C}} = \mathbb{C}_{1,-1} \oplus \mathbb{C}_{-1,1}$, where subscripts again denote weights under $\mathrm{U}(1) \times \mathrm{U}(1)$. We can identify H with diagonal 2×2 matrices $\mathrm{diag}\{e^{i\vartheta_1}, e^{i\vartheta_2}\}$, and it is convenient to consider the combination $\vartheta = (\vartheta_1 - \vartheta_2)/2$, $\vartheta' = (\vartheta_1 + \vartheta_2)/2$, both in $[0, 2\pi)$. The level is given by two integers k and k', corresponding to, respectively, the non-abelian and abelian parts of $\mathrm{U}(1,1)$. Now, F_t is given by the solutions to the following set of equations:

$$e^{2ki\vartheta}\left(\frac{1-te^{-2i\vartheta}}{1-te^{2i\vartheta}}\right)^2 = 1,$$
$$e^{2k'i\vartheta'} = 1$$

and

$$\theta_t = \frac{(1-te^{2i\vartheta})(1-te^{-2i\vartheta})}{\det H_t^\dagger},$$

where $H_t^\dagger$ is the Hessian of

$$D: \quad \vartheta \mapsto -k\vartheta^2 - k'\vartheta'^2 - \left(\mathrm{Li}_2(te^{2i\vartheta}) + \mathrm{Li}_2(te^{-2i\vartheta})\right).$$

It is easy to relate the $\mathrm{U}(1,1)$ case to the previous case of $\mathrm{G}_0 = \mathrm{SL}(2,\mathbb{R})$, as the equations for ϑ are the same, while the new variable ϑ' takes value in $\{0, \frac{\pi}{k}, \frac{2\pi}{k}, \ldots, \frac{(2k-1)\pi}{k}\}$. Using

$$\det H_t^{\dagger,\mathrm{U}(1,1)} = \det H_t^{\dagger,\mathrm{SL}(2,\mathbb{R})} \cdot k',$$

we have

$$\chi_{\mathbb{T}}\left(\mathfrak{M}_{\mathrm{U}(1,1)}, \mathfrak{L}^k\right) = k'^g \cdot \chi_{\mathbb{T}}\left(\mathfrak{M}_{\mathrm{SL}(2,\mathbb{R})}, \mathfrak{L}^k\right).$$

In our study of mirror symmetry between moduli spaces of Higgs bundles, it is also useful to consider indices of exterior powers of $\mathbf{V}$. On the moduli space, $\mathbf{V}$ can only be defined locally, but it exists globally on the stack $\mathfrak{M}(\mathrm{SL}(2,\mathbb{C}))$ as a hypercohomology complex $R\mathrm{pr}_{C*}(\mathbf{E} \xrightarrow{\Phi} \mathbf{E} \otimes K_C)[1]$, which we will continue to denote as $\mathbf{V}$.

There are several moduli stacks we would like to consider, with natural morphisms between them:

$$\begin{array}{ccc} \mathrm{Bun}_{\mathbb{C}^\times} & \rightarrow & \mathrm{Bun}_{\mathrm{SL}(2,\mathbb{C})} \\ \downarrow & & \downarrow \\ \mathfrak{M}(\mathrm{SU}(1,1)) & \rightarrow & \mathfrak{M}(\mathrm{SL}(2,\mathbb{C})). \end{array}$$

The vertical morphisms are closed immersions, while the horizontal ones are not. The morphism at the bottom will also become a closed immersion if $\mathfrak{M}(\mathrm{SU}(1,1))$ is replaced with $\mathfrak{M}'(\mathrm{SU}(1,1)) = \mathfrak{M}(\mathrm{SU}(1,1))/W$, where $W = \mathbb{Z}/2\mathbb{Z}$ is the Weyl group of $\mathrm{SL}(2,\mathbb{C})$. The complex $\mathbf{V}$ can be restricted to $\mathfrak{M}'(\mathrm{SU}(1,1))$ as well as be pulled back to every corner of the above diagram. To avoid clutter, we will continue denoting the resulting complexes on $\mathfrak{M}'(\mathrm{SU}(1,1))$ and $\mathfrak{M}(\mathrm{SU}(1,1))$ as $\mathbf{V}$, while we will denote the pullback to $\mathrm{Bun}_{\mathbb{C}^\times}$ and $\mathrm{Bun}_{\mathrm{SL}(2,\mathbb{C})}$ in a different font as $\mathfrak{V}$. On the other hand, we will use $\mathfrak{L}^k$ for the kth tensor power of the determinant line bundle that also exists on all the five stacks mentioned above.

To study the index of $\mathfrak{V}$ and its exterior powers, it is useful to have the following proposition.

Proposition 5.6 *In the rational equivariant K-theory of* $\mathrm{Bun}_{\mathrm{SL}(2,\mathbb{C})}$, *we have*

$$\mathfrak{V} = (v^{-1} + v)(g-1)\mathbf{E}_x + (v^{-1} - v)\mathbf{E}_C,$$

where $\mathbf{E}_x$ *and* $\mathbf{E}_C$ *are obtained from the rank 2 universal bundle* $\mathbf{E}$ *over* $\mathrm{Bun}_{\mathrm{SL}(2,\mathbb{C})} \times C$ *by, respectively, restricting to a point* $x \in C$ *and slanting with the fundamental class of* C.

Proof As $\mathfrak{V} = v^{-1}R\pi_*(\mathbf{E} \otimes K_C) - vR\pi_*(\mathbf{E})$, the proposition directly follows from two identities in the K-theory of $\mathrm{Bun}_{\mathrm{SL}(2,\mathbb{C})}$,

$$R\pi_*(\mathbf{E}) = (1-g)\mathbf{E}_x + \mathbf{E}_C,$$

and

$$R\pi_*(\mathbf{E} \otimes K_C) = (g-1)\mathbf{E}_x + \mathbf{E}_C. \qquad \square$$

Now we consider the index of $\Lambda_s\mathbf{V}$ on the moduli stack of $\mathrm{SU}(1,1)$-Higgs bundles $\mathfrak{M}(\mathrm{SU}(1,1))$. As the maximal compact subgroup of $\mathrm{SU}(1,1)$ is $\mathrm{U}(1)$, we use the morphism

$$p : \mathfrak{M}(\mathrm{SU}(1,1)) \quad \to \quad \mathrm{Bun}_{\mathbb{C}^\times} = \mathrm{Pic}(C) \times B\mathbb{C}^\times$$

to reduce the index computation over $\mathfrak{M}(\mathrm{SU}(1,1))$ to an index computation over $\mathrm{Bun}_{\mathbb{C}^\times}$. To give the index formula explicitly, we parametrize the complexified maximal compact subgroup $\mathrm{H}^{\mathbb{C}} = \mathbb{C}^\times$ by z, and denote the set of solutions of

$$z^{2k}\left(\frac{1-v^2z^{-2}}{1-v^2z^2}\right)^2\left(\frac{1+sv^{-1}z^{-1}}{1+sv^{-1}z}\right)\left(\frac{1+svz}{1+svz^{-1}}\right) = 1 \tag{5.7}$$

as $F'_{t,s}$. Define

$$\theta'_{t,s} := \frac{1}{\det H'^{+}_{t,s}} \cdot \frac{(1-v^2z^2)(1-v^2z^{-2})}{(1+sv^{-1}z)(1+sv^{-1}z^{-1})(1+svz)(1+svz^{-1})}, \tag{5.8}$$

where $\det H'^{+}_{t,s}$ is given by

$$\left[\left(2k+\frac{4v^2z^2}{1-v^2z^2}+\frac{4v^2z^{-2}}{1-v^2z^{-2}}-\frac{sv^{-1}z}{1+sv^{-1}z}-\frac{sv^{-1}z^{-1}}{1+sv^{-1}z^{-1}}+\frac{svz}{1+svz}+\frac{svz^{-1}}{1+svz^{-1}}\right)\right]. \quad (5.9)$$

Then we have

Theorem 5.10 *For $\mathfrak{L}^k$ a line bundle on $\mathfrak{M}(\mathrm{SU}(1,1))$ with level $k \geq 0$, the $\mathbb{T}$-equivariant index of $\mathfrak{L}^k \otimes \Lambda_s \mathbf{V}$ is given by*

$$\chi_{\mathbb{T}}\left(\mathfrak{M}(\mathrm{SU}(1,1)), \mathfrak{L}^k \otimes \Lambda_s \mathbf{V}\right) = \sum_{f \in F'_{t,s}} \theta'_{t,s}(f)^{1-g}. \quad (5.11)$$

Proof When $s = 0$, the theorem becomes the Verlinde formula for $\mathrm{SU}(1,1)$-Higgs bundles discussed earlier. We now highlight the effect of turning on s.

The first step is to rewrite $\Lambda_s \mathfrak{V}$ on $\mathrm{Bun}_{\mathbb{C}^\times}$ into a form that Theorem 2.15 of [TW] can handle. To achieve this, we will start with rewriting $\Lambda_s \mathfrak{V}$ over $\mathrm{Bun}_{\mathrm{SL}(2,\mathbb{C})}$. Recall first that, from the universal $\mathrm{SL}(2,\mathbb{C})$-bundle over $\mathrm{Bun}_{\mathrm{SL}(2,\mathbb{C})} \times C$ and a $\mathrm{SL}(2,\mathbb{C})$ representation R, one can obtain a vector bundle, denoted as $E(R)$. We obviously have $\mathbf{E} = E(R_F)$, where R_F is the two-dimensional fundamental representation of $\mathrm{SL}(2,\mathbb{C})$. One can obtain classes $E_x(R)$ and $E_C(R)$ in the K-theory of $\mathrm{Bun}_{\mathrm{SL}(2,\mathbb{C})}$ similar to the construction of $\mathbf{E}_x$ and $\mathbf{E}_C$. Using the identity

$$\Lambda_s = \exp\left[-\sum_{p>0} \frac{(-s)^p \psi^p}{p}\right]$$

with ψ^p denoting the pth Adams operation and the relation $\psi^p E_C(R) = E_C(\psi^p R)/p$, one can rewrite

$$\Lambda_s \mathfrak{V} = E_x(\Lambda_s((v+v^{-1})R_F))^{\otimes(g-1)} \otimes \exp\left[-\sum_{p=1}^{\infty} \frac{(-s)^p E_C(\psi^p((v^{-1}-v)R_F))}{p^2}\right]. \quad (5.12)$$

Similarly, on $\mathrm{Bun}_{\mathbb{C}^\times}$, we denote $E(R_n)$ the line bundle associated to the universal line bundle via the weight-n representation over $\mathrm{Bun}_{\mathbb{C}^\times} \times C$. As before, we can construct K-theory classes $E_x(R_n)$ and $E_C(R_n)$ over $\mathrm{Bun}_{\mathbb{C}^\times}$. From (5.12), one can deduce the following identity in the K-theory of $\mathrm{Bun}_{\mathbb{C}^\times}$:

$$\begin{aligned}\Lambda_s \mathfrak{V} = {} & E_x(\Lambda_s((v+v^{-1})(R_1+R_{-1})))^{\otimes(g-1)} \\ & \otimes \exp\left[-\sum_{p=1}^{\infty} \frac{((-s/v)^p - (sv)^p)\, E_C(R_p + R_{-p})}{p^2}\right].\end{aligned} \quad (5.13)$$

Since

$$\mathfrak{M}(\mathrm{SU}(1,1)) = \underline{\mathrm{Spec}}_{\mathrm{Bun}_{\mathbb{C}^\times}} \mathrm{Sym}(R\pi_* E(R_2 + R_{-2})[1]),$$

we have the equality

$$\chi_{\mathbb{T}}\left(\mathfrak{M}(\mathrm{SU}(1,1)), \mathfrak{L}^k \otimes \Lambda_s \mathbf{V}\right) = \chi\left(\mathrm{Bun}_{\mathbb{C}^\times}, \mathfrak{L}^k \otimes \Lambda_s \mathfrak{V} \otimes \Lambda_{-t} R\pi_* E(R_2 + R_{-2})\right),$$

and the theorem follows from a direct computation using Theorem 2.15 of [TW] applied to $\mathrm{Bun}_{\mathbb{C}^\times}$. With the parametrization of the complexified maximal compact subgroup of SU(1, 1) by z, the character of R_n is given by z^n. Then Theorem 2.15 of [TW] tells us that the first factor $E_x(\Lambda_s((v+v^{-1})(R_1+R_{-1}))^{\otimes(g-1)}$ in (5.13) will result in an additional s-dependent factor,

$$\mathrm{Tr}_{R_s}(z),$$

in $\theta'^{1-g}_{s,t}$, with R_s being the virtual representation $\left(\Lambda_s\left((v+v^{-1})(R_1+R_{-1})\right)\right)^{\otimes(g-1)}$. It is easy to see that the character of R_s is given by

$$\left[(1+svz)(1+svz^{-1})(1+sv^{-1}z^{-1})(1+sv^{-1}z)\right]^{g-1}.$$

On the other hand, the second factor in (5.13) will modify both $\theta'_{s=0,t}$ and $F'_{s=0,t}$. The modification to the former is only through $\det H'^{+}_{t,s}$, which is now given by the Hessian of the function

$$D'_s = D'_{s=0} + \sum_{p=1}^{\infty} \frac{((-sv)^p - (-s/v)^p)\,\mathrm{Tr}_{\psi^p(R_1+R_{-1})}(z)}{p^2}.$$

Since $\mathrm{Tr}_{\psi^p(R_1+R_{-1})}(z) = z^p + z^{-p}$, the summation over p in the above can be written as the sum of four terms,

$$\sum_{p=1}^{\infty} \frac{((-sv)^p - (-s/v)^p)\left(z^p + z^{-p}\right)}{p^2} = \mathrm{Li}_2(-svz^{-1}) + \mathrm{Li}_2(-svz) - \mathrm{Li}_2(-sv^{-1}z) - \mathrm{Li}_2(-sv^{-1}z^{-1}).$$

Computing the Hessian of D'_s gives the desired $\det H'^{+}_{t,s}$.

Now we determine the set $F'_{s,t}$. When $s=0$, it is given by the set of solutions to the 'Bethe ansatz equations',

$$z^{2k}\left(\frac{1-v^2z^{-2}}{1-v^2z^2}\right)^2 = 1.$$

When s is turned on, from Theorem 2.15 of [TW], a new factor will be included in the left-hand side of the above equation, given by

$$\exp\left[-\sum_{p=1}^{\infty}\frac{((-sv)^p-(-s/v)^p)(z^p-z^{-p})}{p}\right]=\left(\frac{1+sv^{-1}z^{-1}}{1+sv^{-1}z}\right)\left(\frac{1+svz}{1+svz^{-1}}\right).$$

Taking into account all three modifications proves the theorem. □

In previous sections, we studied the index over the smooth part of the 'real slice' $\mathfrak{M}'(\mathrm{SU}(1,1))\subset\mathfrak{M}(\mathrm{SL}(2,\mathbb{C}))$, for which we have the following.

Theorem 5.14 *The equivariant index of $\mathfrak{L}^k\otimes\Lambda_s\mathbf{V}$ on $\mathfrak{M}'(\mathrm{SU}(1,1))$,*

$$\chi_{\mathbb{T}}\left(\mathfrak{M}'(\mathrm{SU}(1,1)),\mathfrak{L}^k\otimes\Lambda_s\mathfrak{V}\right)=\frac{1}{2}\sum_{f\in F'_{t,s}}\theta'_{t,s}(f)^{1-g},\tag{5.14}$$

is equal to one-half of the equivariant index $\chi_{\mathbb{T}}\left(\mathfrak{M}(\mathrm{SU}(1,1)),\mathfrak{L}^k\otimes\Lambda_s\mathbf{V}\right)$ on $\mathfrak{M}(\mathrm{SU}(1,1))$.

Proof This theorem follows from the facts that $\mathfrak{M}'(\mathrm{SU}(1,1))=\mathfrak{M}(\mathrm{SU}(1,1))/W$, and $\mathfrak{L}^k\otimes\Lambda_s\mathbf{V}$ is invariant under the action of the Weyl group $W=\mathbb{Z}/2\mathbb{Z}$. □

6. Symmetry for $i=g-1$

To compare the formula (5.15) for $k=2$ with the formula (4.3), we rewrite it in the same residue form. First, we note that the left-hand side of (5.7) for $k=2$ equals $f(z,s,v)$ from (4.1). We also see that (5.9) agrees with

$$z\frac{\frac{\partial}{\partial z}f(z,s,v)}{f(z,s,v)},$$

which is computed in (4.2). Finally, we see from (5.8) that

$$\theta'_{t,s}=\frac{f(z,s,v)}{h(z,s,v)}.$$

By recalling that $F'_{t,s}$ denotes the set of solutions of $f=1$, which are all simple zeroes, we can compute

$$\begin{aligned}\sum_{\varphi\in F'_{t,s}}\theta'_{t,s}(\varphi)^{1-g}&=\sum_{a\in F'_{t,s}}\left(\frac{h(a,s,v)}{f(a,s,v)}\right)^{g-1}=\frac{1}{2\pi i}\oint_{|1-f|=\epsilon}\left(\frac{h}{f}\right)^{g-1}\frac{1}{1-f}\frac{-df}{f}\\&=\frac{1}{2\pi i}\oint_{|f|=\epsilon}\left(\frac{h}{f}\right)^{g-1}\frac{1}{1-f}\frac{df}{f}+\frac{1}{2\pi i}\oint_{|z^2v^2-1|=\epsilon}\left(\frac{h}{f}\right)^{g-1}\frac{1}{1-f}\frac{df}{f}\\&=\frac{1}{2\pi i}\oint_{|f|=\epsilon}\sum_{j=0}^{\infty}\left(\frac{h}{f}\right)^{g-1}f^j\frac{df}{f}-\frac{1}{2\pi i}\oint_{|z^2v^2-1|=\epsilon}\sum_{j=1}^{\infty}\left(\frac{h}{f}\right)^{g-1}f^{-j}\frac{df}{f}\end{aligned}$$

$$= \frac{1}{2\pi i}\oint_{|f|=\epsilon}\sum_{j=0}^{\infty}\left(\frac{h}{f}\right)^{g-1} f^{j}\frac{df}{f} + \frac{1}{2\pi i}\oint_{|f|=\epsilon}\sum_{j=1}^{\infty}\left(\frac{h}{f}\right)^{g-1} f^{-j}\frac{df}{f}$$

$$= \frac{1}{2\pi i}\oint_{|f|=\epsilon}\left(\frac{h}{f}\right)^{g-1}\frac{df}{f} + \sum_{i=0}^{g-2}\chi_{\mathbb{T}}(\mathcal{M}(\mathrm{GL}_2);\mathcal{L}_i\otimes\Lambda_{s^2}) + \frac{1}{2\pi i}\oint_{|f|=\epsilon}\sum_{j=1}^{\infty}\left(\frac{h}{f}\right)^{g-1} f^{j}\frac{df}{f}$$

$$= \frac{1}{2\pi i}\oint_{|f|=\epsilon}\left(\frac{h}{f}\right)^{g-1}\frac{df}{f} + 2\sum_{i=0}^{g-2}\chi_{\mathbb{T}}(\mathcal{M}(\mathrm{GL}_2);\mathcal{L}_i\otimes\Lambda_{s^2}).$$

In the third equation, we used the residue theorem and the fact that the poles of the integrand are the zeroes of $1-f, f$ and z^2v^2-1. In the fourth equation we see that the power series expansions converge on the respective contours. In the fifth equation, we used the residue theorem again, noting that the only poles of the integrand are the zeroes of f and z^2v^2-1. In the sixth, we used (4.3), (4.9) and (4.7). Finally, in the seventh equation, we used (4.3) and (4.9) again.

Thus, from (5.14),

$$\chi_{\mathbb{T}}\left(\mathfrak{M}'(\mathrm{SU}(1,1)),\mathfrak{L}^2\otimes\Lambda_s\mathfrak{V}\right) = \frac{1}{2}\frac{1}{2\pi i}\oint_{|f|=\epsilon}\left(\frac{h}{f}\right)^{g-1}\frac{df}{f}$$

$$+\sum_{j=1}^{g-1}\frac{1}{2\pi i}\oint_{|f|=\epsilon}\left(\frac{h}{f}\right)^{g-1} f^{j}\frac{df}{f}. \tag{6.1}$$

Remark 6.2 From the observation in Remark 4.8, we see that, even though **V** and, consequently, Λ_s are no longer vector bundles on $\mathfrak{M}'(\mathrm{SU}(1,1))$, cohomologically, **V** behaves like a rank $4g-4$ vector bundle on $\mathfrak{M}'(\mathrm{SU}(1,1))$. Interestingly, the same observation about the virtual Dirac bundle **V** was the crucial point in the proof of the vanishing of the intersection form on the rank 2 odd degree Higgs moduli space in [Ha1, Ha2].

From (6.1), we also get

$$2\chi_{\mathbb{T}}\left(\mathfrak{M}'(\mathrm{SU}(1,1)),\mathfrak{L}^2\otimes\Lambda_s\mathfrak{V}\right) - 2\sum_{i=0}^{g-2}\chi_{\mathbb{T}}(\mathcal{M}(\mathrm{GL}_2);\mathcal{L}_i\otimes\Lambda_{s^2}) = \frac{1}{2\pi i}\oint_{|f|=\epsilon}\left(\frac{h}{f}\right)^{g-1}\frac{df}{f}.$$

We can now combine Theorem 5.14 and Proposition 4.4 with the computation above and deduce the following:

Corollary 6.3 *For $0\le j<g-1$, we have*

$$2\chi_{\mathbb{T}}\left(\mathfrak{M}'(\mathrm{SU}(1,1)),\mathfrak{L}^2\otimes\Lambda^{2j}\mathfrak{V}\right) - 2\sum_{i=0}^{g-2}\chi_{\mathbb{T}}(\mathcal{M}(\mathrm{GL}_2);\mathcal{L}_i\otimes\Lambda_j)$$

$$= \chi_{\mathbb{T}}(\mathcal{M}(\mathrm{GL}_2);\mathcal{L}_j\otimes\Lambda_{g-1}).$$

Remark 6.4 For a completely symmetrical picture, we would like to see the left-hand side agree with $\chi_{\mathbb{T}}(\mathcal{M}(\mathrm{GL}_2);\mathcal{L}_{g-1}\otimes\Lambda_j)$ with the special convention for $\mathcal{L}_{g-1}$ in (2.3). It seems, however, that the non-semi-stable locus in $\mathfrak{M}'(\mathrm{SU}(1,1))$ will contribute non-trivially to the equivariant index. We expect that a modification of the extension of the Dirac complex $\Lambda_s\mathbf{V}$ to $\mathfrak{M}(\mathrm{SL}_2)$ will solve this problem.

7. Reflection of Mirror Symmetry

If $\mathcal{F}_1$ and $\mathcal{F}_2$ are $\mathbb{T}$-equivariant coherent sheaves on a semi-projective (cf. [HV]) $\mathbb{T}$-variety X, then we can define (cf. [Gu, (2.23),(3.4)]) the *equivariant Euler form* as

$$\begin{aligned}\chi_{\mathbb{T}}(X;\mathcal{F}_1,\mathcal{F}_2)&=\sum_{k,l}\dim(H^k(R\mathcal{H}om(\mathcal{F}_1,\mathcal{F}_2))^l)(-1)^k t^{-l}\\&=\sum_{k,l}\dim(\mathrm{Hom}_{\mathbf{D}_{coh}(X)}(\mathcal{F}_1,\mathcal{F}_2[k])^l)(-1)^k t^{-l}\\&=\sum_{k,l}\dim(\mathrm{Ext}^k(X;\mathcal{F}_1,\mathcal{F}_2)^l)(-1)^k t^{-l}\in\mathbb{C}((t)),\end{aligned}$$

which we can compute by Hirzebruch–Riemann–Roch and localization:

$$\begin{aligned}\chi_{\mathbb{T}}(X;\mathcal{F}_1,\mathcal{F}_2)&=\int_X \mathrm{ch}_{\mathbb{T}}(\mathcal{F}_1)^*\,\mathrm{ch}_{\mathbb{T}}(\mathcal{F}_2)\,\mathrm{td}_{\mathbb{T}}(X)\\&=\int_{X^{\mathbb{T}}}\mathrm{ch}_{\mathbb{T}}(\mathcal{F}_1)^*|_{X^{\mathbb{T}}}\,\mathrm{ch}_{\mathbb{T}}(\mathcal{F}_2|_{X^{\mathbb{T}}})\,\mathrm{ch}_{\mathbb{T}}(\mathrm{Sym}(N^*_{X^{\mathbb{T}}}))\,\mathrm{td}(X^{\mathbb{T}}),\end{aligned}$$

where $N_{X^{\mathbb{T}}}$ is the normal bundle of $X^{\mathbb{T}}$ in X and for

$$a=a_0+a_2+\cdots+a_{2\dim(X)}\in H^{2*}(X).$$

We define the class

$$a^*=\sum(-1)^i a_{2i}\in H^{2*}(X).$$

Corollary 7.1 *When* $0\le i,j<g-1$,

$$\chi_{\mathbb{T}}(\mathcal{M}(\mathrm{GL}_2);\mathcal{L}_i,\Lambda_j)=-t^{g-1}\chi_{\mathbb{T}}(\mathcal{M}(\mathrm{GL}_2);\Lambda_i,\mathcal{L}_j).$$

Proof We expect that

$$\mathcal{L}_i^\vee\cong\mathcal{L}_i[-4g+3]\tag{7.2}$$

and

$$\Lambda_i{}^\vee\cong\Lambda_i[-2g],\tag{7.3}$$

as self-dualities in the derived category $D_{coh}(\mathcal{M}(\mathrm{GL}_2))$. Indeed, (7.3) holds on $\mathcal{M}(\mathrm{GL}_2)^s$ because there it is a push-forward of a vector bundle from $\mathcal{M}(\mathrm{SL}_2)^s$ of codimension $2g$ with a trivial normal bundle and, as [Hi2, §7, Remarks 1, 2] argues, the vector bundle $\Lambda_i|_{\mathcal{M}(\mathrm{SL}_2)^s}$ is self-dual, due to the quaternionic structure on it. On the other hand, (7.2) holds for $i < g-1$, because of [Hu, Corollary 3.40], using that $\mathcal{M}(\mathrm{GL}_2)^s$ is symplectic and by the following

Lemma 7.4 *For $0 \le i < g-1$, we have*

$$K_{L_i} = \mathcal{L}^4|_{L_i} = \mathcal{L}_i^2|_{L_i}. \qquad \square$$

Proof This follows from $c_1(K_{L_i'}|_{F_i'}) = c_1(\mathcal{L}^4|_{F_i'})$. Starting with [Mac, (14.10)] and (3.7), we compute

$$\begin{aligned} c_1(K_{L_i'}|_{F_i'}) &= c_1(K_{F_i'})c_1(E_i^*) = (g-2i-1)\eta + \theta + (3g-3-2i)\eta + \theta \\ &= (4g-4-4i)\eta + 2\theta = 4c_1(\mathcal{L}|_{F_i'}). \end{aligned}$$

from (3.13). $\square$

To also trace how the $\mathbb{T}$-action intertwines these isomorphisms, we compute using [CG, Lemma 5.4.9]:

$$\begin{aligned} \mathrm{ch}_{\mathbb{T}}(\mathcal{L}_i)^*|_{L_i} &= \mathrm{ch}_{\mathbb{T}}\left(\sum_j (-1)^j \Lambda_j N_{L_i}^*\right)^* \mathrm{ch}_{\mathbb{T}}(\mathcal{L}^{-2}|_{L_i}) \\ &= \mathrm{ch}_{\mathbb{T}}\left(\sum_j (-1)^j \Lambda_j N_{L_i}\right) \mathrm{ch}_{\mathbb{T}}(\mathcal{L}^{-2}|_{L_i}) \\ &= \frac{(-1)^{4g-3} e^{c_1^{\mathbb{T}}(N_{L_i})}}{\mathrm{ch}_{\mathbb{T}}(\mathrm{Sym} N_{L_i}^*)} \mathrm{ch}_{\mathbb{T}}(\mathcal{L}^{-2}|_{L_i}). \end{aligned}$$

Consequently,

$$\begin{aligned} \mathrm{ch}_{\mathbb{T}}(\mathcal{L}_i)^*|_{F_i'} &= \frac{(-1)^{4g-3} e^{c_1^{\mathbb{T}}(N_{L_i}|_{F_i'})}}{\mathrm{ch}_{\mathbb{T}}(\mathrm{Sym} N_{L_i}^*)} \mathrm{ch}_{\mathbb{T}}(\mathcal{L}^{-2}|_{F_i'}) = \frac{(-1)^{4g-3} e^{c_1^{\mathbb{T}}(E_i^*) + c_1^{\mathbb{T}}(T_{F_i}^*) + c_1^{\mathbb{T}}(\mathcal{L}^{-2}|_{F_i'})}}{\mathrm{ch}_{\mathbb{T}}(\mathrm{Sym} N_{L_i}^*)} \\ &= \frac{(-1)^{4g-3} e^{(g-1+2\bar{i})(\eta-u)+\theta+(g-1-2i)\eta+\theta+(g+2i)u-2\bar{i}(\eta-u)-\theta}}{\mathrm{ch}_{\mathbb{T}}(\mathrm{Sym} N_{L_i}^*)} \\ &= \frac{-\mathrm{ch}_{\mathbb{T}}(\mathcal{L}^2|_{F_i'})t^{-1}}{\mathrm{ch}_{\mathbb{T}}(\mathrm{Sym} N_{L_i}^*)} = -\mathrm{ch}_{\mathbb{T}}(\mathcal{L}_i|_{F_i'})t^{-1}. \end{aligned} \tag{7.5}$$

Here, we used

$$c_1^{\mathbb{T}}(E_i^*) = (g-1+2\bar{i})(\eta-u)+\theta$$

from (3.7) and the facts that $\mathrm{rank}(E_i^*) = g - 1 + 2\bar{i}$ and the $\mathbb{T}$-action has weight minus one on E_i^*. We also used

$$c_1^{\mathbb{T}}(T_{F_i}^*) = (g - 1 - 2i)\eta + \theta + (g + 2i)u$$

from [Mac, (14.10)] and $\mathrm{rank}(T_{F_i}^*) = g + 2i$ and the $\mathbb{T}$-action has weight one on $T_{F_i}^*$. Finally,

$$c_1^{\mathbb{T}}(\mathcal{L}^{-2}|_{F_i'}) = -2\bar{i}(\eta - u) - \theta,$$

from (3.13) . Similarly,

$$\mathrm{ch}(\Lambda_j)^*|_{F_i'} = \frac{(-1)^{2g} t^{-g} \mathrm{ch}_{\mathbb{T}}(\Lambda^{j*}|_{F_i'})}{\mathrm{ch}_{\mathbb{T}}(\mathrm{Sym}N_{F_i'}^{F_i})} = \frac{t^{-g} \mathrm{ch}_{\mathbb{T}}(\Lambda_j|_{F_i'})}{\mathrm{ch}_{\mathbb{T}}(\mathrm{Sym}N_{F_i'}^{F_i})},$$

because, from (3.11) and (3.12), we can deduce

$$\mathrm{ch}_{\mathbb{T}}(\Lambda^{j*}|_{F_i'}) = \mathrm{ch}_{\mathbb{T}}(\Lambda_j|_{F_i'}).$$

Thus, we can compute from (7.5)

$$\chi_{\mathbb{T}}(\mathcal{M}(\mathrm{GL}_2); \mathcal{L}_i, \Lambda_j) = \int_{F_i'} \mathrm{ch}_{\mathbb{T}}(\mathcal{L}_i)^*|_{F_i'} \mathrm{ch}_{\mathbb{T}}(\Lambda_j|_{F_i'}) \mathrm{ch}_{\mathbb{T}}(\mathrm{Sym}N_{F_i'}^*) \mathrm{td}(F_i')$$

$$= \int_{F_i'} -t^{-1} \mathrm{ch}(\mathcal{L}^2|_{F_i'}) \mathrm{ch}_{\mathbb{T}}(\Lambda_j|_{F_i'}) \mathrm{ch}_{\mathbb{T}}(\mathrm{Sym}E_i^*) \mathrm{td}(F_i') = -t^{-1} \chi_{\mathbb{T}}(\mathcal{M}(\mathrm{GL}_2); \mathcal{L}_i \otimes \Lambda_j).$$

On the other side, we have

$$\chi_{\mathbb{T}}(\mathcal{M}(\mathrm{GL}_2); \Lambda_i, \mathcal{L}_j) = \int_{F_j'} \mathrm{ch}_{\mathbb{T}}(\Lambda_i)^*|_{F_i'} \mathrm{ch}_{\mathbb{T}}(\mathcal{L}_j|_{F_j'}) \mathrm{ch}_{\mathbb{T}}(\mathrm{Sym}N_{F_i'}^*) \mathrm{td}(F_i')$$

$$= \int_{F_i'} t^{-g} \mathrm{ch}_{\mathbb{T}}(\Lambda_i|_{F_j'}) \mathrm{ch}_{\mathbb{T}}(\mathrm{Sym}N_{F_i'}^*) \mathrm{td}(F_i') = t^{-g} \chi_{\mathbb{T}}(\mathcal{M}(\mathrm{GL}_2); \Lambda_i \otimes \mathcal{L}_j),$$

and the result follows. □

Remark 7.6 Based on the Fourier–Mukai transform on the generic fibres of the Hitchin map, we expect that (the semi-classical limit of the) mirror of $\mathcal{L}_i$ will be the shifted coherent sheaf $\Lambda_i[-g]$ and that of Λ_j the shifted coherent sheaf $\mathcal{L}_j[-3g + 3]$. This explains the sign in Corollary 7.1.

Remark 7.7 The extra t^{g-1} factor in Corollary 7.1 seems to be necessary, and the reason for its appearance could be the fact that the canonical bundle $K_{\mathcal{M}(\mathrm{SL}_2)}$ is trivial with a weight $3g - 3$ equivariant structure and therefore has no square root for g even.

REFERENCES

[AGP] Andersen, J. E., Gukov, S. and Pei, D.: The Verlinde formula for Higgs bundles. arXiv:1608.01761 (2017).

[BS] Baraglia, D. and Schaposnik, L.: Real structures on moduli spaces of Higgs bundles. *Adv. Theor. Math. Phys.* 20 (2016), no. 3, 525–51.

[BGG] Bradlow, S. B., García-Prada, O. and Gothen, P.B.: Surface group representations and $U(p,q)$-Higgs bundles. *J. Differ. Geom.* 64, (2003), no. 1, 111–70.

[CG] Chriss, N. and Ginzburg, V. *Representation theory and complex geometry.* Birkhäuser Boston, Inc., Boston, MA (1997).

[DP1] Donagi, R. and Pantev, T.: Langlands duality for Hitchin systems. *Invent. Math.* 189 (2012), no. 3, 653–735.

[GGM] García-Prada, O., Gothen, P. B. and Mundet i Riera, I: The Hitchin–Kobayashi correspondence, Higgs pairs and surface group representations. arXiv:0909.4487 (2009).

[GWZ] Gröchenig, M., Wyss, D. and Ziegler, P.: Topological mirror symmetry for moduli spaces of Higgs bundles via p-adic integration. arXiv:1707.06417 (2017).

[Gu] Gukov, S.: Quantization via mirror symmetry. *Jpn. J. Math.* 6 (2011), no. 2, 65–119.

[GP] Gukov, S. and Pei, D.: Equivariant Verlinde formula from fivebranes and vortices. arXiv:1501.01310 (2015).

[GPYY] Gukov, S., Pei, D., Yan, W. and Ye, K.: Equivariant Verlinde algebra from superconformal index and Argyres–Seiberg duality. arXiv:1605.06528 (2016).

[GW] Gaiotto, D. and Witten, E.: S-Duality of boundary conditions in N=4 super Yang–Mills theory. *Adv. Theor. Math. Phys.* 13 (2009), no. 3, 721–896.

[Ha1] Hausel, T.: Geometry of the moduli space of Higgs bundles. PhD thesis, University of Cambridge. arXiv:math.AG/0107040 (1998).

[Ha2] Hausel., T.: Vanishing of intersection numbers on the moduli space of Higgs bundles. *Adv. Theor. Math. Phys.* **2** (1998), no. 5, 1011–1040.

[HL] Halpern-Leistner, D.: The equivariant Verlinde formula on the moduli of Higgs bundles. arXiv:1608.01754 (2016).

[HV] Hausel, T. and Villegas, F. R.: Cohomology of large semiprojective hyperkähler varieties. *Astérisque* 2015 (2015), no. 370, 113–56.

[HT1] Hausel, T. and Thaddeus, M.: Generators for the cohomology ring of the moduli space of rank 2 Higgs bundles. *Proc. London Math. Soc.* 88 (2004), no. 3, 632–58.

[HT2] Hausel, T. and Thaddeus, M.: Relations in the cohomology ring of the moduli space of rank 2 Higgs bundles, *J. Am. Math. Soc.*, 16 (2003), no. 2, 303–29.

[HT] Hausel, T. and Thaddeus, M.: Mirror symmetry, Langlands duality, and the Hitchin system. *Invent. Math.* 153 (2003), no. 1, 197–229.

[HSz] Hausel, T. and Szenes, A.: Equivariant Verlinde algebra for Higgs bundles. *In preparation.*

[Hi1] Hitchin, N.: The self-duality equations on a Riemann surface. *Proc. London Math. Soc. (3)* **55** (1987), no. 1, 59–126.

[Hi2] Hitchin, N.: Higgs bundles and characteristic classes. In Ballmann, W., Blohmann, C., Faltings, G., Teichner, P., and Zagier, D., eds., *Arbeitstagung Bonn 2013; In Memory of Friedrich Hirzebruch*, Progress in Mathematics, Vol. 319, Birkhäuser, Boston (2016), 247–64.

[Hu] Huybrechts, D.: *Fourier–Mukai transforms in algebraic geometry.* Oxford Mathematical Monographs. Oxford University Press, Oxford (2006).

[KW] Kapustin, K. and Witten. E.: Electric-magnetic duality and the geometric Langlands program. *Commun. Number Theory Phys.* 1 (2007), no. 1, 1–236.

[Mac] Macdonald, I. G.: Symmetric products of an algebraic curve. *Topology* 1 (1962) no. 4, 319–43.

[Qi] Quillen, D.: Determinants of Cauchy–Riemann operators on Riemann surfaces. Funktsional. Anal. i Prilozhen. 19 (1985), no. 1, 31–4.

[Sch] Schaposnik, L. P.: Spectral data for $U(m,m)$-Higgs bundles. *Int. Math. Res. Not.* **2015**(2015) no. 11, 3486–98.

[Th] Thaddeus, M.: Stable pairs, linear systems and the Verlinde formula. *Invent. Math.* 117 (1994), no. 1, 317–53.

[TW] Teleman, C. and Woodward, C. T.: The index formula for the moduli of G-bundles on a curve. *Ann. Math.* **170** (2009), no. 2, 495–527.

10

Deformation Theory of Lie Bialgebra Properads

SERGEI MERKULOV[1]
THOMAS WILLWACHER[2]

[1] *Sergei Merkulov: Mathematics Research Unit, Luxembourg University, Maison du Nombre, 6 Avenue de la Fonte, L-4364 Esch-sur-Alzette, Grand Duchy of Luxembourg, Email: sergei.merkulov@uni.lu*
[2] *Thomas Willwacher: Department of Mathematics, ETH Zurich, Rämistrasse 101, 8092 Zurich, Switzerland, Email: thomas.willwacher@math.ethz.ch*

ABSTRACT. We compute the homotopy derivations of the properads governing even and odd Lie bialgebras as well as involutive Lie bialgebras. The answer may be expressed in terms of the Kontsevich graph complexes. In particular, this shows that the Grothendieck–Teichmüller group acts faithfully (and essentially transitively) on the completions of the properads governing even Lie bialgebras and involutive Lie bialgebras, up to homotopy. This shows also that, in contrast to the even case, the properad governing odd Lie bialgebras admits precisely one non-trivial automorphism—the standard rescaling automorphism–and that it has precisely one non-trivial deformation, which we describe explicitly.

1. Introduction

1.1. Deformation Theory of Lie Bialgebras and Graph Complexes

An *even* Lie bialgebra is a vector space which carries both a Lie algebra and a Lie coalgebra structure of the same $\mathbb{Z}_2$ parity, satisfying a certain compatibility relation. Lie bialgebras were introduced by Drinfeld in [D1] in the context of the theory of Yang–Baxter equations. They have since seen numerous application, in particular in the theory of Hopf algebra deformations of universal enveloping algebras (cf. [ES] and references therein).

If the composition of the cobracket and the bracket of a Lie bialgebra is zero, the Lie bialgebra is called *involutive*. This additional condition is satisfied in many interesting examples studied in homological algebra, string topology, symplectic field theory, Lagrangian Floer

Merkulov, S., Willwacher, T., *Deformation Theory of Lie Bialgebra Properads.* In: *Geometry and Physics: A Festschrift in Honour of Nigel Hitchin*, Jørgen Ellegaard Andersen, Andrew Dancer, Oscar García-Prada (Eds): Oxford University Press (2018). © Oxford University Press. DOI: 10.1093/oso/9780198802006.003.0010

theory of higher genus and the theory of cohomology groups $H(\mathcal{M}_{g,n})$ of moduli spaces of algebraic curves with labellings of punctures *skew-symmetrized* [C, CFL, MW2, Tu, Sc].

Odd Lie bialgebras are, by definition, the ones in which Lie brackets and cobrackets have opposite $\mathbb{Z}_2$ parities. They have seen applications in Poisson geometry [M1], deformation quantization of Poisson structures [M2] and, most surprisingly, in the theory of cohomology groups $H(\mathcal{M}_{g,n})$ of moduli spaces of algebraic curves with labellings of punctures *symmetrized* [MW2].

We study in this paper the deformation theory of Lie bialgebras in both the even and odd cases. Let $\mathcal{L}ieb_{c,d}$ denote the properad governing Lie bialgebras with Lie brackets in degree $1-c$, and Lie cobrackets in degree $1-d$, so that the case $c+d \in 2\mathbb{Z}$ corresponds to even Lie bialgebras, and the case $c+d \in 2\mathbb{Z}+1$ to odd ones. The involutivity condition is non-trivial only in the even case, and we denote by $\mathcal{L}ieb^{\diamond}_{c,d}$ (with $c+d \in 2\mathbb{Z}$ by default) the properad governing involutive Lie bialgebras.

The main purpose of this note is to finish the study of the homotopy derivations of the properads $\mathcal{L}ieb_{c,d}$ and $\mathcal{L}ieb^{\diamond}_{c,d}$ initiated in [CMW], where a minimal resolution $\mathcal{H}olieb^{\diamond}_{c,d}$ of $\mathcal{L}ieb^{\diamond}_{c,d}$ was constructed; minimal resolutions $\mathcal{H}olieb_{c,d}$ of $\mathcal{L}ieb_{c,d}$ have been found earlier in [Va, Ko3, MaVo] for $c+d \in 2\mathbb{Z}$ and in [M1, M2] for $c+d \in 2\mathbb{Z}+1$. Informally speaking, one question we want to answer is which universal deformations of any (involutive) Lie bialgebra structure one can construct, using only the bracket and cobracket operations.

The answers may be formulated in terms of the cohomology of the oriented graph complex GC^{or}_d, whose elements are linear combinations of isomorphism classes of directed graphs with no closed paths of directed edges. More concretely, in [CMW] the authors describe maps of dg Lie algebras,

$$F\colon \mathsf{GC}^{or}_{c+d+1} \to \mathrm{Der}(\widehat{\mathcal{H}olieb_{c,d}}), \qquad F^{\diamond}\colon \mathsf{GC}^{or}_{c+d+1}[[\hbar]] \to \mathrm{Der}(\widehat{\mathcal{H}olieb^{\diamond}_{c,d}}),$$

where the symbol $\widehat{\ }$ stands for the genus completion, $\hbar$ for a formal variable of homological degree $c+d$, and $\mathrm{Der}(\mathcal{P})$ for the dg Lie algebra of (continuous) derivations of a dg properad $\mathcal{P}$.

Precise definitions of all the properads and the graph complexes, and in particular the dg Lie algebra structures on GC^{or}_{c+d+1}, will be given in §2 below. Explicit formulas for the morphisms F and $F^{\diamond}$ as well for the dg Lie algebra structures in $\mathrm{Der}(\widehat{\mathcal{H}olieb_{c,d}})$ and $\mathrm{Der}(\widehat{\mathcal{H}olieb^{\diamond}_{c,d}})$ are given in §3.

1.2. *Main Theorems*

The main results of the present note are the following two theorems, which are proven in § 4.

1.2.1. Theorem

For any $c, d \in \mathbb{Z}$, the map

$$F\colon \mathsf{GC}^{or}_{c+d+1} \to \mathrm{Der}(\widehat{\mathcal{H}olieb_{c,d}})$$

is a quasi-isomorphism, up to one class in $\mathrm{Der}(\widehat{\mathcal{H}olieb}_{c,d})$*, represented by the series*

$$\sum_{m,n}(m+n-2)$$ 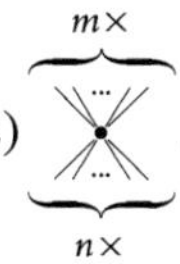 .

1.2.2. Theorem

For any $c, d \in \mathbb{Z}$ *with* $c+d \in 2\mathbb{Z}$*, the map*

$$F^{\diamond} : \mathsf{GC}^{or}_{c+d+1}[[\hbar]] \to \mathrm{Der}(\widehat{\mathcal{H}olieb}^{\diamond}_{c,d})$$

is a quasi-isomorphism, up to classes $T\mathbb{K}[[\hbar]] \in \mathrm{Der}(\widehat{\mathcal{H}olieb}^{\diamond}_{c,d})$*, generated over* $\mathbb{K}[[\hbar]]$ *by the series*

$$T = \sum_{m,n,p}(m+n+2p-2)\hbar^{p}$$ 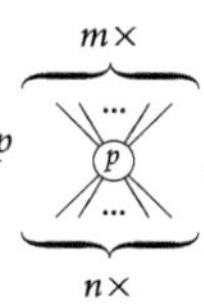 .

1.3. *Some Applications*

The most important cases for applications correspond to (i) the case $c = d = 1$, where one deals with the classical (i.e. the ones in which with all the generators have homological degree zero) Lie bialgebra properads $\mathcal{L}ieb := \mathcal{L}ieb_{1,1}$ and $\mathcal{L}ieb^{\diamond} := \mathcal{L}ieb^{\diamond}_{1,1}$, and the case (ii) $c = 0, d = 1$, where one gets a properad $\mathcal{L}ieb_{odd} := \mathcal{L}ieb_{0,1}$ of odd Lie bialgebras which has the property [M1, M2] that the representations of its minimal resolution $\mathcal{H}olieb_{odd}$ in a graded vector space V are in 1-1 correspondence with formal graded Poisson structures $\pi \in \mathcal{T}^{\geq 1}_{poly}(V)$ on V which vanish at $0 \in V$.

As has been shown in [W1, W2], the cohomology of the oriented graph complexes $H(\mathsf{GC}^{or}_{d+1})$ may be identified with the cohomology of the 'plain' graph complexes[1] GC^{2}_{d} introduced (for $d = 2$) by Kontsevich in [Ko2]. The latter complexes have been studied in [W1], where it was proven, in particular, that there is an isomorphism of Lie algebras,

$$H^{0}(\mathsf{GC}^{2}_{2}) = \mathfrak{grt}_1,$$

where $\mathfrak{grt}_1$ is the Lie algebra of the Grothendieck–Teichmüller group GRT_1 introduced by Drinfeld in [D2]. Then, according to [W1, W2] and, respectively, [CMW], one concludes that

$$H^{0}(\mathsf{GC}^{or}_{3}) = \mathfrak{grt}_1, \text{ and } H^{0}(\mathsf{GC}^{or}_{3}[[\hbar]]) = \mathfrak{grt}_1,$$

and hence obtains (as corollaries to Theorems 1.2.1 and 1.2.2 above) the faithful actions of the group GRT_1 on the completed properads $\widehat{\mathcal{H}olieb}$ and $\widehat{\mathcal{H}olieb}^{\diamond}$ and hence on their

[1] The superscript 2 in the symbol GC^{2}_{d} means that we consider graphs with at least bivalent vertices, while the symbol GC_{d} is reserved traditionally for a complex of graphs with at least trivalent vertices.

representations (the precise meaning of representation of a *completed* properad is given in Section 3.3 below).

It is a folklore conjecture that the cohomology group $H^1(\mathsf{GC}_2^2) = H^1(\mathsf{GC}_3^{or})$ vanishes; if proven, the Main Theorems would imply that the properads $\mathcal{H}olieb$ and $\mathcal{H}olieb^\diamond$ and their completions are *rigid*, i.e. admit no non-trivial deformations.

By contrast, one has $H^0(\mathsf{GC}_2^{or}) = 0$, $H^1(\mathsf{GC}_2^{or}) = \mathbb{K}$ and $H^2(\mathsf{GC}_2^{or}) = \mathbb{K}$. Then Theorem 1.2.1 says that the properad $\mathcal{H}olieb_{odd}$ and its genus completion admits precisely one homotopy non-trivial automorphism—the standard rescaling automorphism—and that its completed version $\widehat{\mathcal{H}olieb}_{odd}$ has precisely one non-trivial deformation, which we describe explicitly in Section 4.3. This unique deformed version of $\widehat{\mathcal{H}olieb}_{odd}$ leads to the notion of *quantizable Poisson structures* [MW3, KMW], the ones whose deformation quantization is a trivial procedure not requiring a choice of an associator as in the case of deformation quantization of ordinary Poisson structures.

1.4. Some Notation

The set $\{1,2,\ldots,n\}$ is abbreviated to $[n]$; its group of automorphisms is denoted by $\mathbb{S}_n$; the trivial one-dimensional representation of $\mathbb{S}_n$ is denoted by $\mathbb{1}_n$, while its one-dimensional sign representation is denoted by sgn_n. The cardinality of a finite set A is denoted by $\#A$.

We work throughout in the category of $\mathbb{Z}$-graded vector spaces over a field $\mathbb{K}$ of characteristic zero. If $V = \oplus_{i\in\mathbb{Z}} V^i$ is a graded vector space, then $V[k]$ stands for the graded vector space with $V[k]^i := V^{i+k}$, and s^k for the associated isomorphism $V \to V[k]$; for $v \in V^i$, we set $|v| := i$. For a pair of graded vector spaces V_1 and V_2, the symbol $\mathrm{Hom}_i(V_1, V_2)$ stands for the space of homogeneous linear maps of degree i, and $\mathrm{Hom}(V_1, V_2) := \bigoplus_{i\in\mathbb{Z}} \mathrm{Hom}_i(V_1, V_2)$; for example, $s^k \in \mathrm{Hom}_{-k}(V, V[k])$.

For a properad $\mathcal{P}$, we denote by $\mathcal{P}\{k\}$ the properad which is uniquely defined by the following property: for any graded vector space V, a representation of $\mathcal{P}\{k\}$ in V is identical to a representation of $\mathcal{P}$ in $V[k]$. The degree-shifted operad of Lie algebras $\mathcal{L}ie\{d\}$ is denoted by $\mathcal{L}ie_{d+1}$ while its minimal resolution by $\mathcal{H}olie_{d+1}$; representations of $\mathcal{L}ie_{d+1}$ are vector spaces equipped with Lie brackets of degree $-d$.

For a right (resp. left) module V over a group G, we denote by V_G (resp. ${}_GV$) the $\mathbb{K}$-vector space of coinvariants: $V/\{g(v) - v \mid v \in V, g \in G\}$ and by V^G (resp. ${}^GV$) the subspace of invariants: $\{\forall g \in G \ : \ g(v) = v,\ v \in V\}$. If G is finite, then these spaces are canonically isomorphic, as $char(\mathbb{K}) = 0$.

For a vector space V and a formal parameter $\hbar$, we denote by $V[[\hbar]]$ the topological vector space of formal power series in $\hbar$ with coefficients in V and, by $\hbar^k V[[\hbar]]$, a vector space of $V[[\hbar]]$ spanned by series of the form $\hbar^k f$ for some $f \in V[[\hbar]]$.

1.5. Remark

A part of this paper contains an extended version of an appendix in the preprint [CMW]; this was removed from the final version of the paper, following the recommendation of a referee.

Acknowledgements

S.M. acknowledges support by the Swedish Vetenskaprådet, grant 2012-5478. T.W. acknowledges partial support by the Swiss National Science Foundation, grant 200021_150012, and by the NCCR SwissMAP of the Swiss National Science Foundation.

2. Properads of Lie Bialgebras and Graph Complexes

2.1. Lie n-Bialgebras

A *Lie n-bialgebra* is a graded vector space V equipped with linear maps

$$\triangle\colon V \to V\wedge V \quad \text{and} \quad [\ ,\ ]: \wedge^2(V[n]) \to V[n]$$

such that the first operation $\triangle$ makes V into a Lie coalgebra, the second operation $[\ ,\ ]$ makes $V[n]$ into a Lie algebra, and the compatibility condition

$$\triangle\,[a,b] = \sum a_1\otimes[a_2,b] + [a,b_1]\otimes b_2 - (-1)^{(|a|+n)(|b|+n)}([b,a_1]\otimes a_2 + b_1\otimes[b_2,a])$$

holds for any $a,b\in V$ with $\triangle\, a =: \sum a_1\otimes a_2$, $\triangle\, b =: \sum b_1\otimes b_2$. The case $n=0$ gives us the ordinary definition of Lie bialgebra [D1]. The case $n=1$ is of interest because minimal resolutions of Lie 1-bialgebras control local Poisson geometry [M1, M2]. For n even, it makes sense to introduce an *involutive Lie n-bialgebra* as a Lie n-bialgebra $(V,[\ ,\ ],\triangle)$ such that the composition map

$$\begin{array}{ccccc} V & \xrightarrow{\triangle} & \Lambda^2 V & \xrightarrow{[\ ,\ ]} & V[-n] \\ a & \longrightarrow & \sum a_1\otimes a_2 & \longrightarrow & [a_1,a_2] \end{array}$$

vanishes (for odd n, this condition is trivial, for symmetry reasons).

2.2. Properads of (Involutive) Lie Bialgebras

Let $\mathcal{L}ieb_n$ (resp. $\mathcal{L}ieb_n^{\diamond}$) denote the properad of (resp. involutive) Lie n-bialgebras. Let us consider their degree-shifted versions,

$$\mathcal{L}ieb_{c,d} = \mathcal{L}ieb_{c+d-2}\{1-c\}, \quad \mathcal{L}ieb^{\diamond}_{c,d} = \mathcal{L}ieb_{c+d-2}\{1-c\}$$

in which the cobracket generator has degree $1-c$, and the bracket generator degree $1-d$. It is worth emphasizing that the symbol $\mathcal{L}ieb^{\diamond}_{c,d}$ tacitly assumes that $c+d\in 2\mathbb{Z}$, i.e. that the numbers c and d have the same parity. Let us describe these properads and their minimal resolutions explicitly.

By definition, $\mathcal{L}ieb_{c,d}$ is a quadratic properad given as the quotient

$$\mathcal{L}ieb_{c,d} := \mathcal{F}ree\langle E\rangle/\langle\mathcal{R}\rangle$$

of the free properad generated by an $\mathbb{S}$-bimodule $E = \{E(m,n)\}_{m,n\geq 1}$ with all $E(m,n) = 0$ except

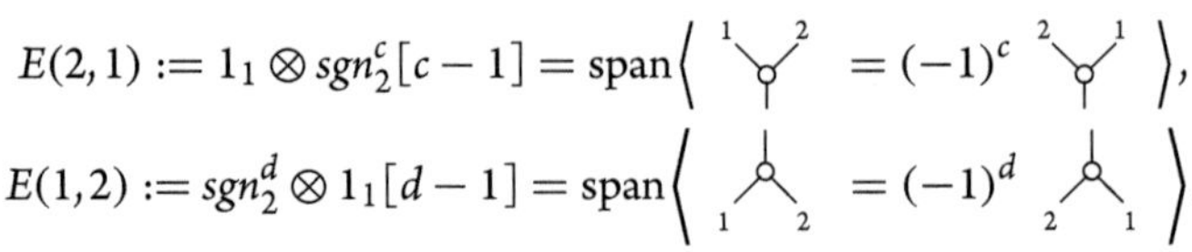

by the ideal generated by the following elements

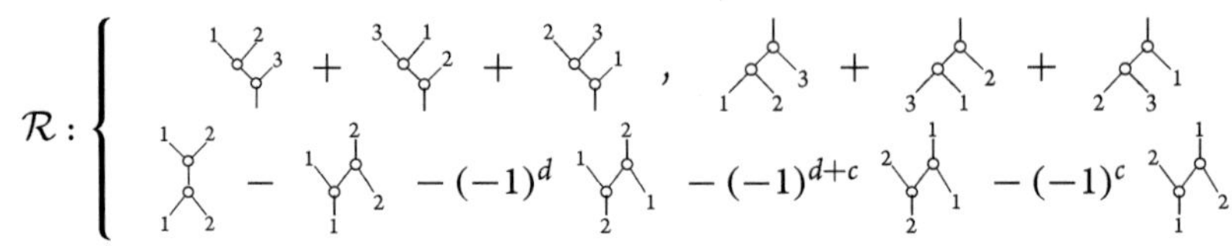 (1)

Similarly, $\mathcal{L}ieb^{\diamond}_{c,d}$ (with $c + d \in 2\mathbb{Z}$ by default) is a quadratic properad $\mathcal{F}ree\langle E\rangle/\langle \mathcal{R}_\diamond\rangle$ generated by the same $\mathbb{S}$-bimodule E modulo the relations

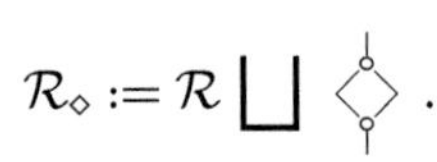

It is clear from the association $\Delta \leftrightarrow$, $[\ ,\ ] \leftrightarrow$ that there is a one-to-one correspondence between representations of $\mathcal{L}ieb_{c,d}$ (resp. $\mathcal{L}ieb^{\diamond}_{c,d}$) in a finite-dimensional space V and (resp. involutive) Lie $(c+d-2)$-bialgebra structures in $V[c-1]$.

The minimal resolution $\mathcal{H}olieb_{c,d}$ of the properad $\mathcal{L}ieb_{c,d}$ was constructed in [Va, Ko3, MaVo] for $d + c \in 2\mathbb{Z}$ and in [M1, M2] for $d + c \in 2\mathbb{Z} + 1$. It is generated by the (skew-)symmetric corollas of degree $1 + c(1-m) + d(1-n)$

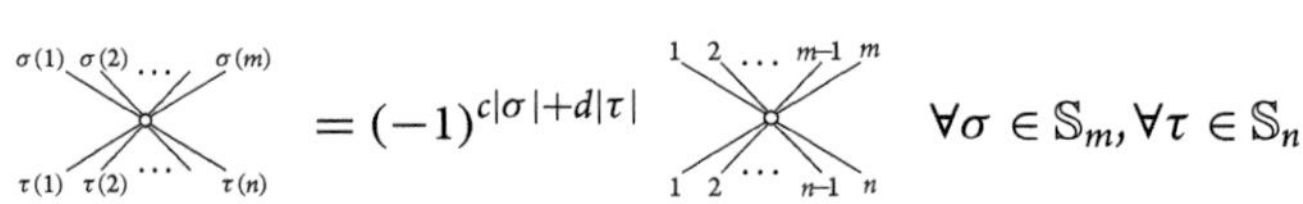 (2)

and has the differential given on the generators by

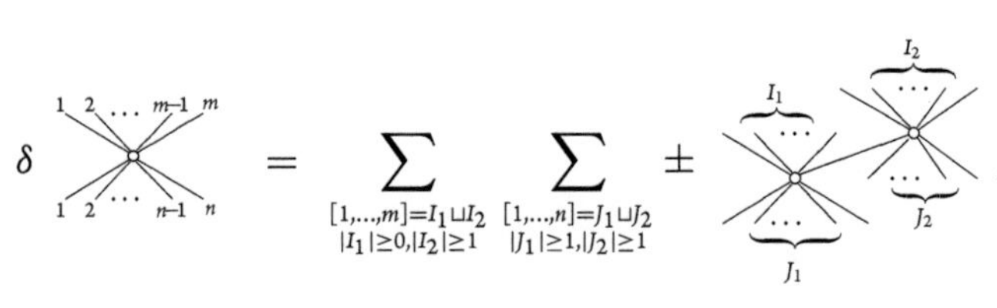 (3)

where the signs on the right-hand side are uniquely fixed for $c + d \in 2\mathbb{Z}$ by the fact that they all equal to $+1$ if c and d are even integers; for $c + d \in 2\mathbb{Z} + 1$, the signs are given explicitly in [M1].

The minimal resolution $\mathcal{H}olieb^{\diamond}_{c,d}$ of the properad $\mathcal{L}ieb^{\diamond}_{c,d}$ was constructed in [CMW]. It is a free properad generated by the following (skew-)symmetric corollas of degree $1 + c(1 - m - a) + d(1 - n - a)$:

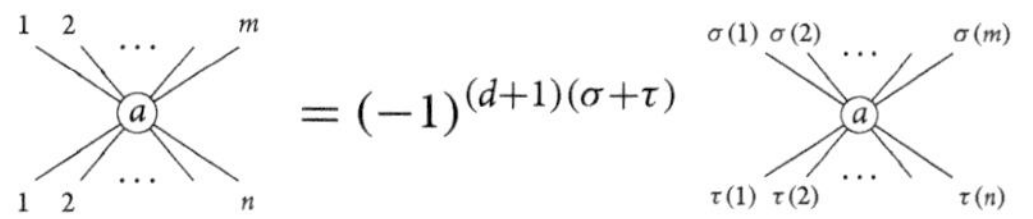

$$= (-1)^{(d+1)(\sigma+\tau)} \qquad \forall \sigma \in \mathbb{S}_m, \forall \tau \in \mathbb{S}_n, \tag{4}$$

where $m + n + a \geq 3$, $m \geq 1$, $n \geq 1$, $a \geq 0$. The differential in $\mathcal{H}olieb^{\diamond}_{d}$ is given on the generators by

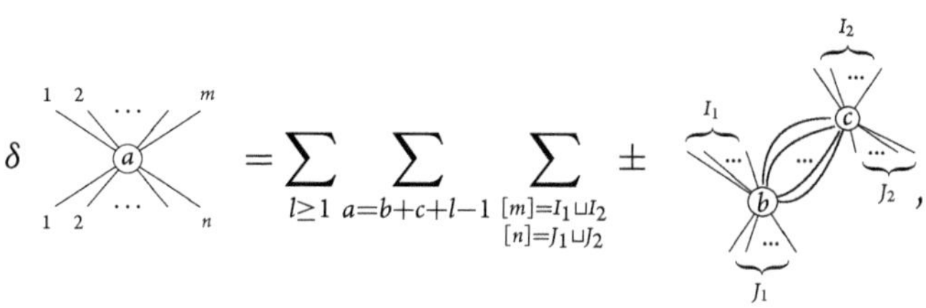

$$\delta(\ldots) = \sum_{l \geq 1} \sum_{a=b+c+l-1} \sum_{\substack{[m]=I_1 \sqcup I_2 \\ [n]=J_1 \sqcup J_2}} \pm \ (\ldots), \tag{5}$$

where the summation parameter l counts the number of internal edges connecting the two vertices on the right-hand side, and the signs are fixed by the fact that they are all equal to $+1$ for c and d odd integers.

Our purpose in this paper is to relate deformation complexes of all the properads considered above to various graph complexes whose cohomology is partially computed, and whose relations with the Grothendieck-Teichmüller Lie algebra are well understood.

2.3. *Complete Variants*

Note that the defining relations for the properads $\mathcal{L}ieb_{c,d}$ and $\mathcal{L}ieb^{\diamond}_{c,d}$ do not mix composition diagrams of different loop orders. It follow that the mentioned properads are all graded by the loop order (here also called genus) of composition diagrams. In particular, fixing the arity, the operations are finite linear combinations (not series) of composites of generators. For some applications, including in particular the integration of derivations to automorphisms, it is more convenient to consider the completed versions by the genus grading $\widehat{\mathcal{L}ieb}_{c,d}$ and $\widehat{\mathcal{L}ieb}^{\diamond}_{c,d}$. Concretely, the operations of fixed arity in the complete versions of our properads are given by infinite series (instead of just linear combinations) of composites of generators.

Similarly, the resolutions $\mathcal{H}olieb_{c,d}$ and $\mathcal{H}olieb^{\diamond}_{c,d}$ receive a grading by the loop order (or genus), and we may also consider the completed versions (with respect to this grading) $\widehat{\mathcal{H}olieb}_{c,d}$ and $\widehat{\mathcal{H}olieb}^{\diamond}_{c,d}$. Here it should be noted that, with respect to the genus grading, the generator (4) must be considered as living in degree a to make this grading consistent with the definition of differential (5).

2.4. Directed Graph Complexes

A *graph* Γ is a one-dimensional *CW* complex whose 0-cells are called *vertices* and whose 1-cells are called *edges*. The set of vertices of Γ is denoted by $V(\Gamma)$, and the set of edges by $E(\Gamma)$. A graph Γ is called *directed* if its edge $e \in E(\Gamma)$ comes equipped with an orientation or, plainly speaking, with a choice of a direction.

Let $G_{n,l}$ be a set of directed graphs Γ with n vertices and l edges such that some bijections $V(\Gamma) \to [n]$ and $E(\Gamma) \to [l]$ are fixed, i.e. every edge and every vertex of Γ has a fixed numerical label. There is a natural right action of the group $\mathbb{S}_n \times \mathbb{S}_l$ on the set $G_{n,l}$, with $\mathbb{S}_n$ acting by relabelling the vertices, and $\mathbb{S}_l$ by relabelling the edges.

For each fixed integer d, a collection of $\mathbb{S}_n$-modules,

$$\mathcal{DG}ra_d = \left\{ \mathcal{DG}ra_d(n) := \prod_{l\geq 0} \mathbb{K}\langle G_{n,l}\rangle \otimes_{\mathbb{S}_l} sgn_l^{\otimes|d-1|}[l(d-1)] \right\}_{n\geq 1},$$

is an operad with respect to the following operadic composition:

$$\begin{array}{rccc} \circ_i: & \mathcal{DG}ra_d(n) \times \mathcal{DG}ra_d(m) & \longrightarrow & \mathcal{DG}ra_d(m+n-1), \ \forall\, i \in [n] \\ & (\Gamma_1, \Gamma_2) & \longrightarrow & \Gamma_1 \circ_i \Gamma_2, \end{array}$$

where $\Gamma_1 \circ_i \Gamma_2$ is defined by substituting the graph Γ_2 into the i-labelled vertex v_i of Γ_1 and taking a sum over reattachments of dangling edges (attached before to v_i) to vertices of Γ_2 in all possible ways.

For any operad $\mathcal{P} = \{\mathcal{P}(n)\}_{n\geq 1}$ in the category of graded vector spaces, the linear map

$$\begin{array}{rccc} [\ ,\ ]: & \mathsf{P} \otimes \mathsf{P} & \longrightarrow & \mathsf{P} \\ & (a \in \mathcal{P}(n), b \in \mathcal{P}(m)) & \longrightarrow & [a,b] := \sum_{i=1}^{n} a \circ_i b - (-1)^{|a||b|} \sum_{i=1}^{m} \\ & & & b \circ_i a \ \in \mathcal{P}(m+n-1) \end{array}$$

makes a graded vector space $\mathsf{P} := \prod_{n\geq 1} \mathcal{P}(n)$ into a Lie algebra [KM]; moreover, these brackets induce a Lie algebra structure on the subspace of invariants $\mathsf{P}^{\mathbb{S}} := \prod_{n\geq 1} \mathcal{P}(n)^{\mathbb{S}_n}$. In particular, the graded vector space

$$\mathsf{dfGC}_d := \prod_{n\geq 1} \mathcal{DG}ra_d(n)^{\mathbb{S}_n}[d(1-n)]$$

is a Lie algebra with respect to the above Lie brackets and, as such, it can be identified with the deformation complex $\mathsf{Def}(\mathcal{L}ie_d \xrightarrow{0} \mathcal{DG}ra_d)$ of a zero morphism. Hence, non-trivial Maurer–Cartan elements of $(\mathsf{fGC}_d, [\ ,\ ])$ give us non-trivial morphisms of operads:

$$f : \mathcal{L}ie_d \longrightarrow \mathcal{DG}ra_d.$$

One such non-trivial morphism f is given explicitly on the generator of $\mathcal{L}ie_d$ by [W1]:

$$f\left(\begin{array}{c}\curlywedge\\ {}_1\ \ {}_2\end{array}\right) = \textcircled{1}\!\to\!\textcircled{2} \; - (-1)^d \; \textcircled{2}\!\to\!\textcircled{1} \; =: \bullet\!\!\to\!\!\bullet \tag{6}$$

Note that elements of dfGC_d can be identified with graphs from $\mathcal{DG}ra_d$ whose vertices' labels are symmetrized (for d even) or skew-symmetrized (for d odd) so that, in pictures, we can forget about labels of vertices and denote them by unlabelled black bullets as in the formula above. Note also that graphs from dfGC_d come equipped with an orientation, *or*, which is a choice of ordering of edges (for d even) or a choice of ordering of vertices (for d odd) up to an even permutation on both cases. Thus, every graph $\Gamma \in \mathsf{dfGC}_d$ has at most two different orientations, or and or^{opp}, and one has the standard relation, $(\Gamma, or) = -(\Gamma, or^{opp})$; as usual, the data (Γ, or) is abbreviates as Γ (with some choice of orientation implicitly assumed). Note that the homological degree of graph Γ from dfGC_d is given by $|\Gamma| = d(\#V(\Gamma) - 1) + (1-d)\#E(\Gamma)$.

The above morphism (6) makes $(\mathsf{dfGC}_d, [\ ,\ ])$ into a *differential* Lie algebra with the differential

$$\delta := [\bullet\!\!\to\!\!\bullet, \ ].$$

This dg Lie algebra contains a dg subalgebra dGC_d spanned by connected graphs with at least bivalent vertices. It was proven in [W1] that

$$H^\bullet(\mathsf{dfGC}_d) = \mathsf{dGC}_d$$

so that there is no loss of generality of working with dGC_d instead of dfGC_d. Moreover, one has an isomorphism of Lie algebras [W1],

$$H^0(\mathsf{dGC}_d) = \mathfrak{grt}_1,$$

where $\mathfrak{grt}_1$ is the Lie algebra of the Grothendieck–Teichmüller group GRT_1 introduced by Drinfeld in the context of deformation quantization of Lie bialgebras. Nowadays, this group plays an important role in many other areas of mathematics (e.g. in the knot theory, in deformation quantization of Poisson manifolds and in the classification theory of solutions of Kashiwara–Vergne problems; see [F] for a review, and many references cited there).

2.4.1. Remark

Often one considers instead of $\mathcal{DG}ra_d$ an operad $\mathcal{G}ra_d = \{\mathcal{G}ra_d(n)\}$ defined by

$$\mathcal{G}ra_d(n) := \prod_{l\geq 0} \mathbb{K}\langle G_{n,l}\rangle \otimes_{\mathbb{S}_l \ltimes (\mathbb{S}_2)^l} sgn_l^{|d|} \otimes sgn_2^{\otimes l|d-1|}[l(d-1)],$$

where the group $(\mathbb{S}_2)^l$ acts on graphs from $G_{n,l}$ by flipping directions of the edges. Then, arguing as above, one arrives at the graph complex

$$\mathsf{fGC}_d := \mathsf{Def}(\mathcal{L}ie_d \xrightarrow{f} \mathcal{G}ra_d)$$

of *undirected* graphs. It contains three important dg Lie subalgebras: (i) $\mathsf{fcGC}_d \subset \mathsf{fGC}_d$, which is spanned by connected graphs; (ii) $\mathsf{GC}_d^2 \subset \mathsf{fcGC}_d$, which is spanned by graphs with at least bivalent vertices; and (iii) $\mathsf{GC}_d \subset \mathsf{GC}_d^2$, which is spanned by graphs with at least trivalent vertices. It was shown in [Ko1, W1] that the cohomology of these subalgebras (which determine completely the cohomology of the full graph complex fGC_d) are related to each other as follows:

$$H^\bullet(\mathsf{fGC}_d) = H^\bullet(\mathsf{GC}_d^2) = H^\bullet(\mathsf{GC}_d) \oplus \bigoplus_{\substack{j\geq 1 \\ j\equiv 2d+1 \mod 4}} \mathbb{K}[d-j],$$

where the summand $\mathbb{K}[d-j]$ is generated by the loop-type graph with j binary vertices. It was proven in [W1] that the complex of directed graphs and the complex of undirected graphs have the same cohomology,

$$H^\bullet(\mathsf{dGC}_d) = H^\bullet(\mathsf{GC}_d^2).$$

In the present context, it is more suitable to work with the directed complex dGC_d rather than with GC_d^2.

2.5. *Oriented Graph Complexes*

A graph Γ from the operad $\mathcal{DG}ra_d$ is called *oriented* if it contains no *wheels*, i.e. directed paths of edges forming a closed circle. The subspace $\mathcal{G}ra_d^{or} \subset \mathcal{DG}ra_d$ spanned by oriented graphs is a suboperad. For example,

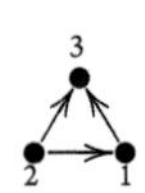 $\in \mathcal{G}ra_d^{or}$ but 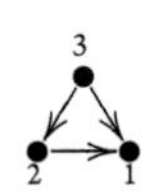 $\notin \mathcal{G}ra_d^{or}(3)$.

The morphism (6) factors through the inclusion $\mathcal{G}ra_d \subset \mathcal{DG}ra_d$ so that one can consider a graph complex

$$\mathsf{fGC}_d^{or} := \mathsf{Def}\left(\mathcal{L}ie_d \xrightarrow{f} \mathcal{G}ra_d^{or}\right)$$

and its subcomplex GC_d^{or} spanned by connected graphs with at least bivalent vertices and with no bivalent vertices of the form $\to\bullet\to$. This subcomplex determines the cohomology of the full graph complex, $H^\bullet(\mathsf{fGC}_d^{or}) = \odot^\bullet(H^\bullet(\mathsf{GC}_d^{or})$. It was proven in [W2] that

$$H^\bullet(\mathsf{GC}_{d+1}^{or}) = H^\bullet(\mathsf{dGC}_d) = H^\bullet(\mathsf{GC}_d^2).$$

In particular, one has a remarkable isomorphism of Lie algebras, $H^0(\mathsf{GC}_3^{or}) = \mathfrak{grt}$. Moreover, $H^i(\mathsf{GC}_3^{or}) = 0$ for $i \leq -2$, and $H^{-1}(\mathsf{GC}_3^{or})$ is a one-dimensional space generated by the graph .

Consider next a Lie algebra $(\mathsf{GC}_3^{or}[[\hbar]], [\ ,\ ]$, where $\mathsf{GC}_3^{or}[[\hbar]]$ is the topological vector space spanned by formal power series in a formal parameter $\hbar$ of homological degree 2, and $[\ ,\ ]$ are the Lie brackets obtained from the standard ones in GC_d^{or} by the continuous extension. It was shown in [CMW] that the formal power series

$$\Phi_\hbar := \sum_{k=1}^{\infty} \hbar^{k-1} \tag{7}$$

k edges

is a Maurer–Cartan element in the Lie algebra $(\mathsf{fGC}_3^{or}[[\hbar]], [\ ,\ ]$ and hence makes the latter into a *differential* Lie algebra with the differential

$$\delta_\hbar = [\Phi_\hbar, \ ].$$

It was proven in [CMW] that $H^0(\mathsf{GC}_3^{or}[[\hbar]], \delta_\hbar) \simeq H^0(\mathsf{GC}_3^{or}, \delta) \simeq \mathfrak{grt}_1$ as Lie algebras. Moreover, $H^i(\mathsf{GC}_3^{or}[[\hbar]], \delta_\hbar) = 0$ for all $i \leq -2$, and $H^{-1}(\mathsf{GC}_3^{or}[[\hbar]], \delta_\hbar)$ is a one-dimensional vector space class generated by the formal power series $\sum_{k=2}^{\infty}(k-1)$ $\hbar^{k-2}$  .

k edges

Sometimes we do not show in our pictures directions of edges of oriented graphs, assuming tacitly that the flow goes from the bottom to the top (as in the case of properads).

3. Deformation Complexes of Properads and Directed Graph Complexes

3.1. *Deformation Complexes of Properads*

For $\mathcal{C}$, a coaugmented co(pr)operad, we will denote by $\Omega(\mathcal{C})$ its cobar construction. Concretely, $\Omega(\mathcal{C}) = \mathcal{F}ree\langle\overline{\mathcal{C}}[-1]\rangle$ as a graded (pr)operad where $\overline{\mathcal{C}}$ is the cokernel of the coaugmentation, and $\mathcal{F}ree\langle\dots\rangle$ denotes the free (pr)operad generated by an $\mathbb{S}$-(bi)module. We will often use complexes of derivations of (pr)operads and deformation complexes of (pr)operad maps. For a map of properads $f : \Omega(\mathcal{C}) \to \mathcal{P}$, we will denote by

$$\mathsf{Def}(\Omega(\mathcal{C}) \xrightarrow{f} \mathcal{P}) \cong \prod_{m,n} \mathrm{Hom}_{\mathbb{S}_m \times \mathbb{S}_n}(\mathcal{C}(m,n), \mathcal{P}(m,n)) \tag{8}$$

the associated convolution complex. It is natural structure of a dg Lie algebra [MV] controlling deformations of the morphism f.

We will also consider the Lie algebra $\mathrm{Der}(\mathcal{P})$ of derivations of the properad $\mathcal{P}$; in fact, we will use a minor variation of the standard definition (given e.g. in [Ta]), as follows. Let $\mathcal{P}^+$ be the free properad generated by $\mathcal{P}$ and one other operation  of arity $(1,1)$ and of cohomological degree $+1$. On $\mathcal{P}^+$ we define a differential δ^+ by setting its value on the new generator by

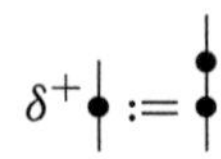

and on any other element $a \in \mathcal{P}(m,n)$ (which we identify pictorially with the (m,n)-corolla whose vertex is decorated with a) by the formula

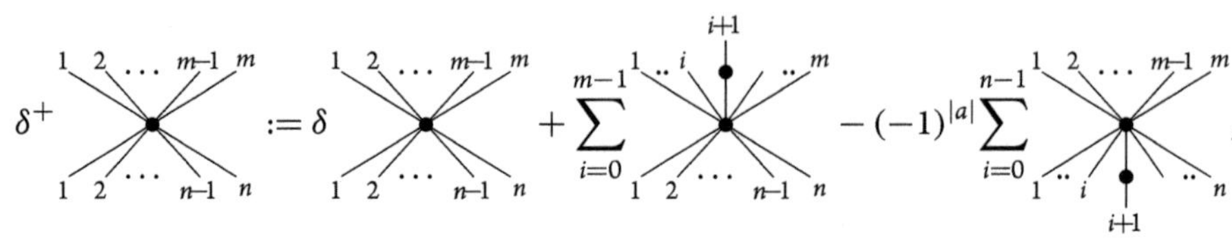

where δ is the original differential in $\mathcal{P}$. The dg properad $(\mathcal{P}^+, \delta^+)$ is uniquely characterized by the property that there is a 1-1 correspondence between representations

$$\rho : \mathcal{P}^+ \longrightarrow \mathcal{E}nd_V$$

of $(\mathcal{P}^+, \delta^+)$ in a dg vector space (V, d), and representations of $\mathcal{P}$ in the same space V but equipped with a deformed differential $d + d'$, where $d' := \rho(\bullet)$. Clearly, any $\mathcal{P}$-algebra is a $\mathcal{P}^+$-algebra by letting d' act trivially, so that we have a properad map $\pi\colon \mathcal{P}^+ \to \mathcal{P}$; we also have a canonical injection $\mathcal{P} \to \mathcal{P}^+$ of *non-differential* properads. Now, slightly abusively, we define $\mathrm{Der}(\mathcal{P})$ as the *complex of derivations of $\mathcal{P}^+$ with values in $\mathcal{P}$*, i.e. as a Lie algebra generated by linear maps

$$D : \mathcal{P}^+ \longrightarrow \mathcal{P}$$

which satisfy the conditions

$$D(a \boxtimes_{1,1} b) = D(a) \boxtimes_{1,1} \pi(b) + (-1)^{|a||D|} \pi(a) \boxtimes_{1,1} D(b), \ \ \forall a, b \in \mathcal{P}^+,$$

where $\boxtimes_{1,1}$ stands [Va] for the properadic composition in $\mathcal{P}$ along any connected oriented 2-vertex graph. The differential d in $\mathrm{Der}(\mathcal{P})$ is given by the formula

$$dD := \delta \circ D - (-1)^{|D|} D \circ \delta^+.$$

Concretely, in all relevant cases, $\mathcal{P} = \Omega(\mathcal{C})$ is the cobar construction of a coaugmented coproperad $\mathcal{C}$. The definition is then made such that $\mathrm{Der}(\mathcal{P})[-1]$ is identified with (8) as a complex. On the other hand, if we were using ordinary derivations, we would have to modify (8) by replacing $\mathcal{C}$ by the cokernel of the coaugmentation $\overline{\mathcal{C}}$ on the right-hand side, thus complicating statements of several results. We assure the reader that this modification is minor and made for technical reasons in the cases we consider, and results about our $\mathrm{Der}(\mathcal{P})$ can be easily transcribed into results about the ordinary derivations if necessary. Note, however, that $\mathrm{Der}(\mathcal{P})$ carries a natural Lie bracket through the commutator.

The deformation complex of a (wheeled) properad $\mathcal{P}$ is, by definition, the dg Lie algebra $\mathrm{Der}(\tilde{\mathcal{P}})$ of derivations of a cofibrant resolution $\tilde{\mathcal{P}} \xrightarrow{\sim} \mathcal{P}$. It may be identified as a complex with the deformation complex of the identity map $\tilde{\mathcal{P}} \to \tilde{\mathcal{P}}$ (which controls deformations of $\mathcal{P}$-algebras) up to a degree shift:

$$\mathrm{Der}(\tilde{\mathcal{P}}) \cong \mathsf{Def}(\tilde{\mathcal{P}} \to \tilde{\mathcal{P}})[1].$$

Note, however, that both $\mathrm{Der}(\mathcal{P})$ and $\mathsf{Def}(\tilde{\mathcal{P}} \to \tilde{\mathcal{P}})$ have natural dg Lie (or $\mathcal{L}ie_\infty$) algebra structures that are *not* preserved by the above map. Furthermore, there is a quasi-isomorphism of dg Lie algebras

$$\mathsf{Def}(\tilde{\mathcal{P}} \to \tilde{\mathcal{P}}) \to \mathsf{Def}(\tilde{\mathcal{P}} \to \mathcal{P}). \tag{9}$$

The zeroth cohomology $H^0(\mathrm{Der}(\tilde{\mathcal{P}}))$ is of particular importance. It is a differential graded Lie algebra whose elements act on the space of $\tilde{\mathcal{P}}$-algebra structures on any vector space. We shall see that, in the examples we are interested in, this dg Lie algebra is very rich, and that it acts non-trivially in general.

Using the explicit structure of the minimal resolutions of the properads $\mathcal{L}ieb_{c,d}$ and $\mathcal{L}ieb^\diamond_{c,d}$ (see Section 2.2 above), we can write down explicit models for the deformation complexes:

$$\begin{aligned}
\mathrm{Der}(\mathcal{H}olieb_{c,d}) &= \prod_{m,n\geq 1} \left(\mathcal{H}olieb_{c,d}(m,n) \otimes sgn_m^{\otimes|c|} \otimes sgn_n^{\otimes|d|}\right)^{\mathbb{S}_m \times \mathbb{S}_m} \\
&\quad [1 + c(1-m) + d(1-n)], \\
\mathrm{Der}(\mathcal{H}olieb^\diamond_{c,d}) &= \prod_{m,n\geq 1} \left(\mathcal{H}olieb^\diamond_{c,d}(m,n) \otimes sgn_m^{\otimes|c|} \otimes sgn_n^{\otimes|d|}\right)^{\mathbb{S}_m \times \mathbb{S}_n} \\
&\quad [[\hbar]][1 + c(1-m) + d(1-n)].
\end{aligned}$$

Here, $\hbar$ is a formal variable of degree $c+d$. Each of the models on the right has a natural combinatorial interpretation as a graph complex. For example, $\mathrm{Der}(\mathcal{H}olieb_{c,d})$ may be interpreted as a complex of directed graphs which have incoming and outgoing legs but have no closed paths of directed edges, for example

$\Gamma =$ 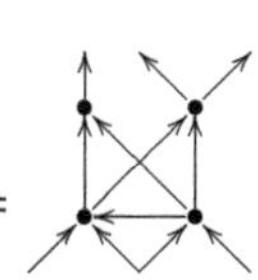 $\in \mathrm{Der}(\mathcal{H}olieb_{c,d}).$

The value of the differential δ on an element $\Gamma \in \mathrm{Der}(\mathcal{H}olieb_{c,d})$ is obtained by splitting vertices of Γ and by attaching new corollas at each single external leg of Γ:

$\delta\Gamma = \delta_{\mathcal{H}olieb_{c,d}}\Gamma \pm \sum$ 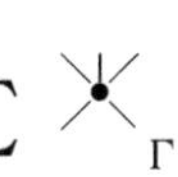 $\pm \sum$ 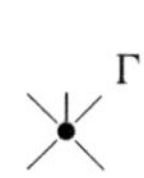 .

Here, $\delta_{\mathcal{H}olieb_{c,d}}$ acts on the vertices of Γ by formula (3).

Similarly, $\mathrm{Der}(\mathcal{H}olieb^{\diamond}_{c,d})$ may be interpreted as a complex of $\hbar$-power series of graphs with weighted vertices, for example,

$$\Gamma =$$ 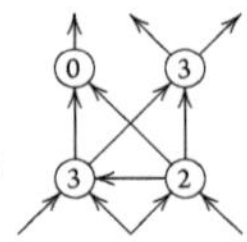 $$\in \mathrm{Der}(\mathcal{H}olieb^{\diamond}_{c,d}).$$

The value of the differential δ on an element $\Gamma \in \mathrm{Der}(\mathcal{H}olieb^{\diamond}_{c,d})$ consists of three summands,

$$\delta\Gamma = \delta_{\mathcal{H}olieb^{\diamond}_{c,d}}\Gamma \pm \sum \hbar^{p+k-1}$$ 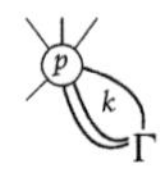 $$\pm \sum \hbar^{p+k-1}$$ 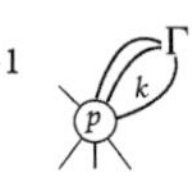 ,

where the first summand comes from the action of the $\mathcal{H}olieb^{\diamond}_{c,d}$-differential on internal vertices of Γ by formula (5) and, in the two terms on the right, one sums over all ways of attaching a new vertex to some subset of the incoming or outgoing legs (k many), and sums over all possible decorations p of the added vertex, with an appropriate power of $\hbar$ as prefactor. Note that the power of $\hbar$ counts the number of loops added to the graph, if we count a vertex decorated by p as contributing p loops.

The Lie bracket is combinatorially obtained by inserting graphs into vertices of another.

The cohomology of all these graph complexes is hard to compute. We may, however, simplify the computation of three of the above four complexes by using formula (9) and equivalently study instead the following much 'smaller' complexes,

$$\mathsf{Def}(\mathcal{H}olieb_{c,d} \to \mathcal{L}ieb_{c,d}) = \prod_{m,n\geq 1} \left(\mathcal{L}ieb_{c,d}(m,n) \otimes sgn_m^{\otimes |c|} \otimes sgn_n^{\otimes |d|}\right)^{\mathbb{S}_m \times \mathbb{S}_m} [c(1-m)+d(1-n)],$$

$$\mathsf{Def}(\mathcal{H}olieb^{\diamond}_{c,d} \to \mathcal{L}ieb^{\diamond}_{c,d}) = \prod_{m,n\geq 1} \left(\mathcal{L}ieb^{\diamond}_{c,d}(m,n) \otimes sgn_m^{\otimes |c|} \otimes sgn_n^{\otimes |d|}\right)^{\mathbb{S}_m \times \mathbb{S}_n} [[\hbar]][c(1-m)+d(1-n)].$$

Note, however, that, in passing from $\mathrm{Der}(\ldots)$ to the (quasi-isomorphic) simpler complexes $\mathsf{Def}(\ldots)$ above, we lose the dg Lie algebra structure, or rather, there is a different Lie algebra structure on the above complexes. The above complexes may again be interpreted as graph complexes. For example, $\mathsf{Def}(\mathcal{H}olieb_{c,d} \to \mathcal{L}ieb_{c,d})$ consists of oriented trivalent graphs with incoming and outgoing legs, modulo the Jacobi and Drinfeld five-term relations. The differential is obtained by attaching a trivalent vertex at one external leg in all possible ways.

3.2. *Complete Variants*

It is more convenient for our purposes to consider (genus-)completed versions of the deformation complexes of the previous subsection. In particular, the genus filtration endows the properads $\widehat{\mathcal{H}olieb}_{c,d}$ and $\widehat{\mathcal{H}olieb}^{\diamond}_{c,d}$ with complete topologies, and we define the complexes of *continuous* derivations $\mathrm{Der}(\widehat{\mathcal{H}olieb}_{c,d})$ and $\mathrm{Der}(\widehat{\mathcal{H}olieb}^{\diamond}_{c,d})$.

More concretely, the difference between the complete and incomplete versions is as follows:

- Elements of $\mathrm{Der}(\mathcal{H}olieb_{c,d})$ of a fixed cohomological degree can be identified with possible infinite series of graphs. However, in each fixed arity, the series must be a finite linear combination. On the other hand, elements of $\mathrm{Der}(\widehat{\mathcal{H}olieb}_{c,d})$ in a fixed degree are simply all series of graphs.
- Similarly, elements of $\mathrm{Der}(\mathcal{H}olieb^{\diamond}_{c,d})$ may be understood as power series in $\hbar$ with coefficients in the series of graphs with the same finiteness condition as before. On the other hand, elements of $\mathrm{Der}(\widehat{\mathcal{H}olieb}^{\diamond}_{c,d})$ are power series in $\hbar$ with coefficients arbitrary series of graphs.

Below, we will always work with the complete versions of our properads and deformation/derivation complexes.

3.3. *A Map from the Graph Complex* GC^{or}_{c+d+1} *to* $\mathrm{Der}(\widehat{\mathcal{H}olieb}_{c,d})$

A derivation D of any free properad $\mathcal{P}$ is uniquely determined by its values, $D(e) \in \mathcal{P}$, on the generators e of $\mathcal{P}$.

There is a natural right action of the dg Lie algebra $\mathsf{fGC}^{or}_{c+d+1}$ on the genus completed dg properad $\widehat{\mathcal{H}olieb}_{c,d}$ by properadic derivations (cf. [CMW]), i.e. there is a canonical morphism of dg Lie algebras,

$$
\begin{array}{cccc} F: & \mathsf{fGC}^{or}_{c+d+1} & \to & \mathrm{Der}(\widehat{\mathcal{H}olieb}_{c,d}) \\ & \Gamma & \to & F(\Gamma) \end{array}, \tag{10}
$$

with values of the derivation $F(\Gamma)$ on the generators of the (genus) completed properad $\widehat{\mathcal{L}ieb}_\infty$ given explicitly by

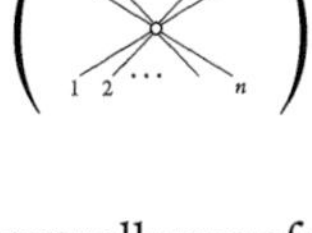

$$\cdot F(\Gamma) = \sum_{\substack{s:[n]\to V(\Gamma) \\ \hat{s}:[m]\to V(\Gamma)}}$$

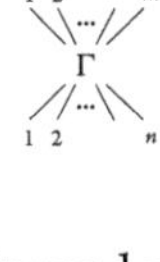

, (11)

where the sum is taken over all ways of attaching the incoming and outgoing legs to the graph Γ, and we set to zero every resulting graph if it contains a vertex with valency < 3 or with at least one incoming or outgoing edge.

Let us first check that the map f has degree zero. If $\Gamma \in \mathsf{fGC}^{or}_{c+d+1}$ has p vertices and l edges, then $|\Gamma| = (c+d+1)(p-1) - (c+d)l$ so that the total degree of the left-hand side in (11) equals

$$\begin{aligned}&(c+d+1)(p-1) - (c+d)l + 1 + c(1-m) + d(1-n)\\ &\quad = (c+d+1)p - (c+d)l - cm - dn.\end{aligned}$$

On the other hand, each summand on the right-hand side gives us a graph in $Holieb_{c,d}$ with p vertices, and each vertex v has $|v|_{out} + \#\hat{s}^{-1}(v)$ output edges and $|v|_{in} + \#s^{-1}(v)$ input edges, where $|v|_{in}$ (resp. $|v|_{out}$) counts the number of outputs (resp. inputs) of v in Γ. The degree of such a graph is equal to

$$\begin{aligned}&\sum_{v\in V(\Gamma)} \left(1 + c(1 - |v|_{out} - \#\hat{s}^{-1}(v)) + d(1 - |v|_{in} - \#s^{-1}(v))\right)\\ &\qquad = p + c(p-l-m) + d(p-l-n) = (c+d+1)p - (c+d)l - cm - dn,\end{aligned}$$

as required. There is an implicit rule of signs in-built into formula (11) which is completely analogous to the one defined in Section 7 of [MaVo].

Consider for a moment $\widehat{\mathcal{H}olieb}_{c,d}$ as a *non-differential* (completed) free properad, and let $\mathrm{Der}_{non-d}(\widehat{\mathcal{H}olieb}_{c,d})$ stand for its Lie algebra of derivations, which can be identified with the vector space

$$\begin{aligned}&\prod_{m,n} \mathrm{Hom}_{\mathbb{S}^{op}_m\times\mathbb{S}_n}\left(\mathbb{K}\langle\mathfrak{C}^m_n\rangle, \widehat{\mathcal{H}olieb}_{c,d}(m,n)\right)\\ &\quad = \prod_{m,n\geq 1}\left(\mathcal{L}ieb_{c,d}(m,n)\otimes sgn_m^{\otimes|c|}\otimes sgn_n^{\otimes|d|}\right)^{\mathbb{S}_m\times\mathbb{S}_m}[c(1-m)+d(1-n)],\end{aligned}$$

as any derivation is uniquely determined by its values on the generators

$\mathfrak{C}^m_n :=$ 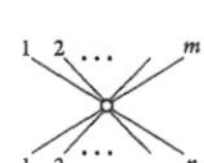 ,

which can be chosen arbitrary. It is immediate from the definition of the Lie algebra structure $[\ ,\ ]$ in GC^{or}_{c+d+1} that the map

$$F : \mathsf{fGC}^{or}_{c+d+1} \longrightarrow \mathrm{Der}_{non-d}(\widehat{\mathcal{L}ieb}_\infty)$$

given by the formula (11) respects the Lie brackets

$$\left(\mathfrak{C}^m_n \cdot F(\Gamma_1)\right)\cdot F(\Gamma_2) - (-1)^{|\Gamma_1||\Gamma_2|}\left(\mathfrak{C}^m_n\cdot F(\Gamma_2)\right)\cdot F(\Gamma_1) = \mathfrak{C}^m_n \cdot F([\Gamma_1,\Gamma_2]),$$

for any $\Gamma_1, \Gamma_2 \in \mathsf{GC}^{or}_{c+d+1}$. This result implies that any Maurer–Cartan element Γ in the Lie algebra $(\mathsf{fGC}^{or}_{c+d+1}, [\ ,\ ])$ gives rise to a continuous differential

$$d_\Gamma : \mathfrak{C}^m_n \longrightarrow \mathfrak{C}^m_n \cdot F(\Gamma)$$

in the properad $\widehat{\mathcal{H}olieb}_{c,d}$. A remarkable (and almost obvious) fact is that the Maurer–Cartan element

$\Gamma =$ 

induces the standard differential (3) in $\widehat{\mathcal{H}olieb}_{c,d}$. This implies that the morphism (10) induces (by changing the right action into a left action via a standard sign factor) a map of *dg* Lie algebras,

$$F\colon \mathsf{GC}_3^{c+d+1} \to \mathrm{Der}(\widehat{\mathcal{H}olieb}_{c,d}), \tag{12}$$

which is proven below to be a quasi-isomorphism (up to one class).

3.3.1. Remark

Interpreting the right-hand side in (12) as a graph complex itself (see Section 3.1), we see that the map F sends a graph $\Gamma \in \mathsf{GC}^{or}_{c+d+1}$ to the series of graphs

$$\sum_{m,n\geq 1} \sum_{\substack{s:[n]\to V(\Gamma)\\ \hat{s}:[m]\to V(\Gamma)}}$$ 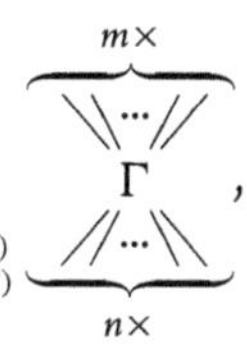 ,

where the second summation symbol has exactly the same meaning as in (11).

3.4. *A Map from the Graph Complex* $\mathsf{GC}^{or}_{c+d+1}[[\hbar]]$ *to* $\mathrm{Der}(\widehat{\mathcal{H}olieb}^{\diamond}_{c,d})$

There is a natural right action of the *non-differential* Lie algebra $(\mathsf{GC}^{or}_{c+d+1}[[\hbar]], [\ ,\ ])$ on the *non-differential* free operad $\mathcal{H}olieb^{\diamond}_{c,d}$ by continuous derivations, i.e. there is a continuous morphism of topological Lie algebras,

$$F^{\diamond} : (\mathsf{fGC}^{or}_{c+d+1}[[\hbar]], [\ ,\ ]) \longrightarrow \mathrm{Der}(\widehat{\mathcal{H}olieb}^{\diamond}_{c,d}).$$

For any monomial $\hbar^k\Gamma \in \mathsf{GC}^{or}_{c+d+1}[[\hbar]]$, the value of the associated derivation $F^{\diamond}(\hbar^k\Gamma)$ on the generators of $\widehat{\mathcal{H}olieb}^{\diamond}_{c,d}$ is given, by definition, by

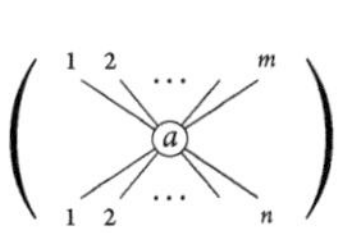 $\cdot F^{(}\hbar^k\Gamma) := \begin{cases} \sum_{\substack{s:[n]\to V(\Gamma)\\ \hat{s}:[m]\to V(\Gamma)}} \sum_{\substack{a=k+\sum_{v\in V(\Gamma)} a_v\\ a_v\geq 0}} \end{cases}$ 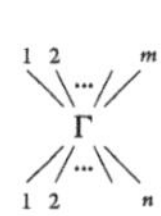 $\begin{cases} \text{if } k\leq a \\ 0 \quad \text{if } k > a, \end{cases}$ (13)

where the first sum is taken over all ways to attach m output legs and n input legs to the vertices of the graph Γ, and the second sum is taken over all ways to decorate the vertices of Γ with non-negative integers $a_1,\ldots,a_{\#V(\Gamma)}$ such they sum to $a-k$; moreover, we set a graph on the right-hand side to zero if there is at least one vertex v with the number n_v of incoming edges equal to zero, or the number m_v of outgoing number equal to zero, or if the condition $n_v+m_v+a_v\geq 3$ is violated. There is an implicit rule of signs in formula (13) which is identical to the one in the subsection above.

It is easy to check that $f^\diamond$ has degree zero and respects Lie brackets. Therefore, it sends any Maurer–Cartan element in $(\mathsf{fGC}_3^{or}[[\hbar]],[\ ,\ ])$ into a differential in the free prop $\widehat{\mathcal{H}olieb}^\diamond_{c,d}$. It is again almost immediate to see that the differential induced by the Maurer–Cartan element (7) is precisely the one given in (5), i.e. the one which makes $\widehat{\mathcal{H}olieb}^\diamond_{c,d}$ into a minimal resolution of $\widehat{\mathcal{L}ieb}^\diamond_{c,d}$. Therefore, we conclude that there is a morphism of *dg* Lie algebras

$$f^\diamond : \left(\mathsf{GC}_3^{or}[[\hbar]],\delta_\hbar=[\Phi_\hbar,\ ]\right)\longrightarrow\left(\mathrm{Der}(\widehat{\mathcal{L}ieb}^\diamond_\infty),d:=[\delta,\ ]\right). \tag{14}$$

We shall prove below that this map is almost a quasi-isomorphism.

3.4.1. Remark

Interpreting $\mathrm{Der}(\widehat{\mathcal{H}olieb}^\diamond_{c,d})$ as a graph complex (see Section 3.1), we can reformulate the map $F^\diamond$ as the one which sends a monomial $\hbar^k\Gamma\in\mathsf{GC}^{or}_{c+d+1}[[\hbar]]$ to the series of graphs

$$\sum_{m,n\geq 1}\ \sum_{\substack{s:[n]\to V(\Gamma)\\ \hat{s}:[m]\to V(\Gamma)}}\ \sum_{\substack{a=k+\sum_{v\in V(\Gamma)}a_v\\ a_v\geq 0}}\ \begin{array}{c}\overbrace{\backslash\backslash\cdots//}^{m\times}\\ \Gamma\\ \underbrace{//\cdots\backslash\backslash}_{n\times}\end{array},$$

where the second and third summation symbols have exactly the same meaning as in (13). We shall use this fact below.

4. Computations of the Cohomology of Deformation Complexes

In this section, we compute the cohomology of several of the deformation complexes, show Theorems 1.2.1 and 1.2.2, and discuss their concrete applications.

4.1. *The Proof of Theorem 1.2.1*

Let us recall the definition of the graph complex $\widehat{\mathsf{GC}}_d^{or}$ from [W2, Section 3.3]. The elements of $\widehat{\mathsf{GC}}_d^{or}$ are $\mathbb{K}$-linear series in directed acyclic graphs with outgoing legs such that all vertices are at least bivalent, and such that there are no bivalent vertices with one incoming and one outgoing edge.[2] We set to zero graphs containing vertices without outgoing edges. Here is an example graph:

The degrees are computed just as for graphs occurring in GC_d^{or}, with the external legs considered to be of degree zero. For the description of the differential, we refer the reader to [W2, Section 3.3].

There is a map $\Psi : \mathsf{GC}_d^{or} \to \widehat{\mathsf{GC}}_d^{or}$ sending a graph Γ to the linear combination

$$\Gamma \mapsto \sum_{j=1}^{\infty} \overbrace{\quad}^{j\times}$$ 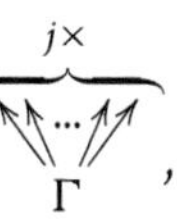 , (15)

where the picture on the right means that one should sum over all ways of connecting j outgoing edges to the graph Γ. Graphs for which there remain vertices with no outgoing edge are identified with 0.

The following proposition has been shown in loc. cit.

4.1.1. Proposition

[Proposition 3 of [W2]] *The map $\Psi : \mathsf{GC}_d^{or} \to \widehat{\mathsf{GC}}_d^{or}$ is a quasi-isomorphism up to the class in $H(\widehat{\mathsf{GC}}_d^{or})$ represented by the graph cocycle*

$$\sum_{j\geq 2} (j-1) \overbrace{\quad}^{j\times}$$  . (16)

There is a map $G : \widehat{\mathsf{GC}}_{c+d+1}^{or} \to \mathrm{Der}(\widehat{\mathcal{H}olieb_{c,d}})$ sending a graph 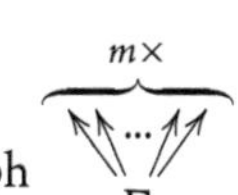 $\in \widehat{\mathsf{GC}}_{c+d+1}^{or}$ to the series

$$G(\Gamma) = \sum_{n}$$ 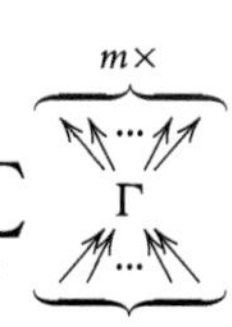 . (17)

[2] The last condition is again not present on [W2], but it does not change the cohomology.

The map $F : \mathsf{GC}^{or}_{c+d+1} \to \mathrm{Der}(\widehat{\mathcal{H}olieb}_{c,d})$ from Theorem 1.2.1 factors through the map G above, i. e. it can be written as the composition

$$\mathsf{GC}^{or}_{c+d+1} \xrightarrow{\Psi} \widehat{\mathsf{GC}}^{or}_{c+d+1} \xrightarrow{G} \mathrm{Der}(\widehat{\mathcal{H}olieb}_{c,d}).$$

In view of Proposition 4.1.1, Theorem 1.2.1 hence follows immediately from the following result.

4.1.2. Proposition

The map $G : \widehat{\mathsf{GC}}^{or}_{c+d+1} \to \mathrm{Der}(\widehat{\mathcal{H}olieb}_{c,d})$ *is a quasi-isomorphism.*

Proof For a graph in $\mathrm{Der}(\widehat{\mathcal{H}olieb}_{c,d})$, we will call its *skeleton* the graph obtained in the following way:

1. Remove all input legs and recursively remove all valence 1 vertices created.
2. Remove valence 2 vertices with one incoming and one outgoing edge and connect the two edges.

An example of a graph and its skeleton the following:

We put a filtration on $\mathrm{Der}(\widehat{\mathcal{H}olieb}_{c,d})$ by the total number of vertices in the skeleton. Let $\mathrm{grDer}(\widehat{\mathcal{H}olieb}_{c,d})$ be the associated graded. Note that, for elements in the image of some graph $\Gamma \in \widehat{\mathsf{GC}}^{or}_{c+d+1}$ under G, the skeleton is just the graph Γ, and hence there is a map of complexes $\widehat{\mathsf{GC}}^{or}_{c+d+1} \to \mathrm{grDer}(\widehat{\mathcal{H}olieb}_{c,d})$, where we consider the left-hand side with zero differential. We claim that the induced map $\widehat{\mathsf{GC}}^{or}_{c+d+1} \to H(\mathrm{grDer}(\widehat{\mathcal{H}olieb}_{c,d}))$ is an isomorphism. From this claim, the proposition follows immediately by a standard spectral sequence argument.

The differential on $\mathrm{grDer}(\widehat{\mathcal{H}olieb}_{c,d})$ does not change the skeleton. Hence, the complex $\mathrm{grDer}(\widehat{\mathcal{H}olieb}_{c,d})$ splits into a direct product of complexes, say $\tilde{C}_\gamma$, one for each skeleton γ:

$$\mathrm{grDer}(\widehat{\mathcal{H}olieb}_{c,d}) = \prod_\gamma \tilde{C}_\gamma.$$

Furthermore, each skeleton represents an automorphism class of graphs, and we may write

$$\tilde{C}_\gamma = C_{\tilde{\gamma}}^{\mathrm{Aut}_\gamma},$$

where $C_{\tilde{\gamma}}$ is an appropriately defined complex for one representative $\tilde{\gamma}$ of the isomorphism class γ, and Aut_γ is the automorphism group associated to the skeleton. In other words, the $\tilde{\gamma}$ now has distinguishable vertices and edges. More concretely, the complex $C_{\tilde{\gamma}}$ is the complex of $\mathbb{K}$-linear series of graphs obtained from $\tilde{\gamma}$ by

1. Adding some bivalent vertices with one input and one output on edges. We call these vertices 'edge vertices'.
2. Attaching input forests at the vertices, such that all vertices are at least trivalent and have at least one input and one output. We call the forest attached to a vertex the forest of that vertex.

An example is the following:

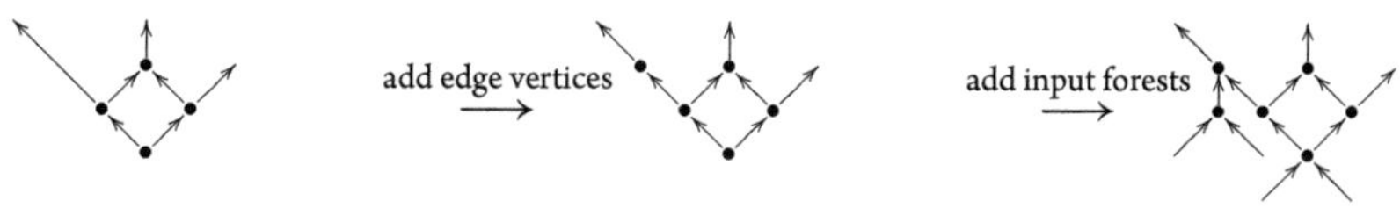

We next put another filtration on $C_{\tilde{\gamma}}$ by the number of edge vertices added in the first step above and consider the associated graded $\mathrm{gr}C_{\tilde{\gamma}}$. Note that the differential of $\mathrm{gr}C_{\tilde{\gamma}}$ acts on each of the forests attached to the vertices separately and hence the complex splits into a (completed) tensor product of complexes, one for each such vertex. Let us call the complex made from the possible forests at the vertex v the forest complex at that vertex. By the same argument showing that the cohomology of a free $\mathcal{L}ie_\infty$ algebra generated by a single generator is two-dimensional, we find that the forest complex at v has either one- or two-dimensional cohomology. If the vertex v has no incoming edge, then the cohomology of the forest complex is one-dimensional, the class being represented by the forests

If the vertex v already has an incoming edge, then there is one additional class obtained by not adding any input forest.

Hence we find that $H(\mathrm{gr}C_{\tilde{\gamma}})$ is spanned by graphs obtained from $\tilde{\gamma}$ as follows:

1. Add some bivalent vertices with one input and one output on edges.
2. For each vertex that is either not at least trivalent or does not have an incoming edge, sum over all ways of attaching incoming legs at that vertex.
3. For at least trivalent vertices with an incoming edge, there is a choice of either not adding anything at that vertex or summing over all ways of attaching incoming legs at that vertex. Let us call the vertices for which the first choice is made 'bald vertices', and the others 'hairy'.

Let us look at the next page in the spectral sequence associated to our filtration on $C_{\tilde{\gamma}}$. The differential creates one edge vertex by either splitting an existing edge vertex or by splitting a skeleton vertex. Again, the complex splits into a product of complexes, one for each edge of $\tilde{\gamma}$. For each such edge, we have to consider three cases separately:

1. Both endpoints in $\tilde{\gamma}$ are hairy.
2. Both endpoints in $\tilde{\gamma}$ are bald.
3. One endpoints is hairy and one is bald.

We leave it to the reader to check that:

1. In the first case, the cohomology is one-dimensional, represented by a single edge without edge vertices.
2. In the third case, the cohomology vanishes.

Since there is necessarily at least one hairy vertex in the graph, the second assertion implies that, if there is a bald vertex as well, the resulting complex is acyclic. Hence, all vertices must be hairy. By the first assertion, the cohomology is one-dimensional for each skeleton. One easily checks that the representative is exactly the image of the skeleton considered as element in $\widehat{\mathsf{GC}}^{or}_{c+d+1}$. Hence, the proposition follows. □

4.1.3. Remark

There is also an alternative way of computing the cohomology of the deformation complex $\mathrm{Der}(\mathcal{H}olieb_{c,d})$ (and of its completion). Namely, by Koszulness of $\mathcal{L}ieb_{c,d}$, this complex is quasi-isomorphic to $\mathsf{Def}(\mathcal{H}olieb_{c,d} \to \mathcal{L}ieb_{c,d})[1]$. It is well known that the prop governing Lie bialgebras $\mathcal{L}iebP$ may be written as

$$\mathcal{L}iebP(n,m) \cong \bigoplus_N \mathcal{L}ieP(n,N) \otimes_{\mathbb{S}_N} \mathcal{L}ieCP(N,m),$$

using the props governing Lie algebras and Lie coalgebras. Interpreting elements of the above prop as linear combinations of directed acyclic graphs, the sub-properad $\mathcal{L}ieb$ may be obtained as that formed by the connected such graphs. It is hence an easy exercise to check that $\mathsf{Def}(\mathcal{H}olieb_{c,d} \to \mathcal{L}ieb_{c,d})[1]$ is identical to the complex $\mathsf{Def}(\mathrm{hoe}_{c+d} \to \mathrm{e}_{c+d})_{\mathrm{conn}}[1]$ from [W1], up to unimportant completion issues. The cohomology of the latter complex has been computed in loc. cit. to be

$$H(\mathsf{GC}_{c+d}) \oplus \bigoplus_{\substack{j\geq 1 \\ j\equiv 2(c+d)+1 \bmod 4}} \mathbb{K}[c+d-j] \oplus \mathbb{K}.$$

Using the main result of [W2], this agrees with the cohomology as computed by Theorem **1.2.1**. Conversely, the above proof of Theorem **1.2.1** together with this remark yields an alternative proof of the main result of [W2].

4.2. *The Proof of Theorem 1.2.2*

Let us next consider Theorem **1.2.2**, whose proof will be a close analogue of that of Theorem **1.2.1** in the previous subsection. There is a natural differential graded Lie algebra structure on $\widehat{\mathsf{GC}}^{or}_{c+d+1}$ such that the map $\Psi : \mathsf{GC}^{or}_{c+d+1} \to \widehat{\mathsf{GC}}^{or}_{c+d+1}$ from the previous section is a map of Lie algebras. The map Ψ extends $\hbar$-linearly to a map of graded Lie algebras

$$\Psi_\hbar : \mathsf{GC}^{or}_{c+d+1}[[\hbar]] \to \widehat{\mathsf{GC}}^{or}_{c+d+1}[[\hbar]].$$

The Maurer–Cartan element $\Phi_\hbar \in \mathsf{GC}^{or}_{c+d+1}[[\hbar]]$ from (7) is sent to a Maurer–Cartan element $\hat{\Phi}_\hbar := \Psi_\hbar(\Phi_\hbar) \in \widehat{\mathsf{GC}}^{or}_{c+d+1}[[\hbar]]$. We endow $\widehat{\mathsf{GC}}^{or}_{c+d+1}[[\hbar]]$ with the differential

$$d_\hbar \Gamma = [\hat{\Phi}_\hbar, \,].$$

In particular, it follows that we have a map of differential graded Lie algebras

$$\Psi_\hbar : (\mathsf{GC}^{or}_{c+d+1}[[\hbar]], d_\hbar) \to (\widehat{\mathsf{GC}}^{or}_{c+d+1}[[\hbar]], d_\hbar).$$

The map $F_\hbar$ from Theorem **1.2.2** factors through $\widehat{\mathsf{GC}}^{or}_{c+d+1}[[\hbar]]$:

$$\mathsf{GC}^{or}_{c+d+1}[[\hbar]] \xrightarrow{\Psi_\hbar} \widehat{\mathsf{GC}}^{or}_{c+d+1}[[\hbar]] \xrightarrow{G_\hbar} \mathrm{Der}(\mathcal{L}ieb^{\diamond}_{c,d}).$$

The second map $G_\hbar : \widehat{\mathsf{GC}}^{or}_{c+d+1}[[\hbar]] \to \mathrm{Der}(\mathcal{H}olieb_{c,d})$ sends $\hbar^N \Gamma$, for $\Gamma \in \widehat{\mathsf{GC}}^{or}_{c+d+1}$ to

$$\sum_{j\geq 1} \sum_{k=0}^{N} \hbar^{N-k}$$ 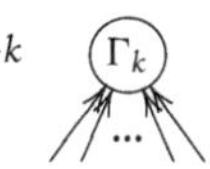 ,

where we again sum over all ways of attaching the incoming legs, setting to zero graphs with vertices without incoming edges. Furthermore, Γ_k is the linear combination of graphs obtained by summing over all ways of assigning weights to the vertices of Γ such that the total weight is k. We have the following two results, from which Theorem **1.2.2** immediately follows.

4.2.1. Proposition

The map $\Psi_\hbar : (\mathsf{GC}^{or}_{c+d+1}[[\hbar]], d_\hbar) \to (\widehat{\mathsf{GC}}^{or}_{c+d+1}[[\hbar]], d_\hbar)$ *is a quasi-isomorphism up to the classes* $T\mathbb{K}[[\hbar]] \subset \widehat{\mathsf{GC}}^{or}_{c+d+1}[[\hbar]]$ *where*

$$T = \sum_{m,p} (m + 2p - 2)$$ 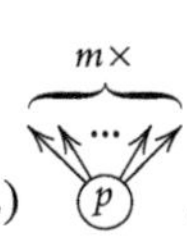 .

Proof sketch Take filtrations on $\mathsf{GC}^{or}_{c+d+1}[[\hbar]]$ and $\widehat{\mathsf{GC}}^{or}_{c+d+1}[[\hbar]]$ by the power of $\hbar$. The differential on the associated graded spaces is the $\hbar$-linear extension of the differentials on GC^{or}_{c+d+1} and $\widehat{\mathsf{GC}}^{or}_{c+d+1}$. Hence, by Proposition **4.1.1**, the map $\Psi_\hbar$ is a quasi-isomorphism on the level of the associated graded spaces, up to the classes above. The result follows by a standard spectral sequence argument, noting that the above element T is indeed $d_\hbar$-closed. □

4.2.2. Proposition

The map $G_\hbar : \widehat{\mathsf{GC}}^{or}_{c+d+1}[[\hbar]] \to \mathrm{Der}(\widehat{\mathcal{H}olieb}^{\diamond}_{c,d})$ *is a quasi-isomorphism.*

Proof Take filtrations on $\widehat{\mathsf{GC}}^{or}_{c+d+1}[[\hbar]]$ and

$$\mathrm{Der}(\widehat{\mathcal{H}olieb}^{\diamond}_{c,d}) \cong \prod_{m,n\geq 1} \left(\widehat{\mathcal{L}ieb}^{\diamond}_{c,d}(m,n)\otimes sgn_m^{\otimes|c|}\otimes sgn_n^{\otimes|d|}\right)^{\mathbb{S}_m\times\mathbb{S}_n}$$
$$[[\hbar]][c(1-m)+d(1-n)]$$

by genus and by powers of $\hbar$. Then we claim that the induced map on the associated graded complexes $\mathrm{gr}G_\hbar : \mathrm{gr}\widehat{\mathsf{GC}}^{or}_{c+d+1}[[\hbar]] \to \mathrm{grDer}(\widehat{\mathcal{H}olieb}^{\diamond}_{c,d})$ is a quasi-isomorphism, thus showing the proposition by a standard spectral sequence argument.

To show the claim, we proceed analogously to the proof of Proposition **4.1.2.** Let us go through the proof again and highlight only the differences. The skeleton of a graph is defined as before, except that one also forgets the weights of all vertices. The complex $\mathrm{grDer}(\widehat{\mathcal{H}olieb}^{\diamond}_{c,d})$ splits into a product of subcomplexes that we again call $\tilde{C}_\gamma$, one for each skeleton γ. Again

$$\tilde{C}_\gamma = C_{\tilde{\gamma}}^{\mathrm{Aut}_\gamma}$$

for some representative $\tilde{\gamma}$ of the isomorphism class γ. Hence, it again suffices to compute the cohomology of $C_{\tilde{\gamma}}$. Graphs contributing are again obtained by adding edge vertices and input forests, except that now all vertices are also assigned an arbitrary weight. Again, we take a filtration on the number of edge vertices, which leaves us with the task of computing the cohomology of a complex associated to one forest attached to a vertex v. We find that representatives of cohomology classes are either:

- Vertex v with any weight and no attached forest. Let us call such a v again 'bald'.
- Vertex v with weight zero and input legs attached in all possible ways. Let us call such a v again 'hairy'.

The differential on the second page of the spectral sequence again adds one edge vertex, which however can have a non-zero weight now and, if it has a non-zero weight, it may be bald. We may introduce another filtration by the number of non-hairy edge vertices. The differential on the associated graded creates one hairy edge vertex. The resulting complex is a tensor product of complexes, one for each edge. The complexes associated to each edge again can have three different types: (i) both endpoints in the skeleton are hairy, (ii) both are bald or (iii) one is hairy, one is bald. Again one checks that, in case (iii), the complex is acyclic and, in case (i), one-dimensional, the cohomology class represented by a single edge. Hence, since at least one vertex must be hairy, all vertices must be. Hence, we recover at this stage the image of $\widehat{\mathsf{GC}}^{or}_{c+d+1}[[\hbar]]$ and are done. □

4.3. *Some Applications*

As remarked in the introduction, the most interesting cases for applications are 'classical' properads:

$$\mathcal{L}ieb := \mathcal{L}ieb_{1,1}, \quad \mathcal{L}ieb^{\diamond} := \mathcal{L}ieb^{\diamond}_{1,1}, \quad \mathcal{L}ieb_{odd} := \mathcal{L}ieb_{0,1}.$$

In the even case, the Lie and co-Lie generators of these Lie bialgebra properads have homological degree zero; in the odd case, the co-Lie generator has degree 1 and the Lie generator degree zero. In all three cases, the associated dg Lie algebras

$$\mathrm{Der}(\widehat{\mathcal{H}olieb}), \quad \mathrm{Der}(\widehat{\mathcal{H}olieb^{\diamond}}), \quad \mathrm{Der}(\widehat{\mathcal{H}olieb_{odd}})$$

are generated by graphs Γ with $\#V(\Gamma) + \#E(\Gamma) \geq 1$ so that all the three are *positively* graded dg Lie algebras with respect to this parameter, and it makes sense to talk about the *groups*

$$\exp\left(\mathrm{Der}^0(\widehat{\mathcal{H}olieb})\right), \quad \exp\left(\mathrm{Der}^0(\widehat{\mathcal{H}olieb^{\diamond}})\right), \quad \exp\left(\mathrm{Der}^0(\widehat{\mathcal{H}olieb_{odd}})\right)$$

which can be identified with, respectively, degree zero subalgebras $\mathrm{Der}^0(\widehat{\mathcal{H}olieb})$, $\mathrm{Der}^0(\widehat{\mathcal{H}olieb^{\diamond}})$, $\mathrm{Der}^0(\widehat{\mathcal{H}olieb_{odd}})$ equipped with the standard Baker–Campbell–Hausdorff multiplication. The subsets of co-cycles

$$Z^0(\widehat{\mathcal{H}olieb}) \subset \mathrm{Der}^0(\widehat{\mathcal{H}olieb}), \quad Z^0(\widehat{\mathcal{H}olieb^{\diamond}}) \subset \mathrm{Der}^0(\widehat{\mathcal{H}olieb^{\diamond}}), \quad Z^0(\widehat{\mathcal{H}olieb}) \subset \mathrm{Der}^0(\widehat{\mathcal{H}olieb_{odd}})$$

are precisely the automorphism groups of the completed properads,

$$\mathrm{Aut}(\widehat{\mathcal{H}olieb}), \quad \mathrm{Aut}(\widehat{\mathcal{H}olieb^{\diamond}}), \quad \mathrm{Aut}(\widehat{\mathcal{H}olieb_{odd}}),$$

with the zero elements corresponding to the identity automorphisms. We say that two automorphisms $f, g \in \mathrm{Aut}(\ldots)$ from the list above are homotopy equivalent, $f \sim g$, if they differ by a coboundary, $f - g = dh$ as elements in the corresponding dg Lie algebra $\mathrm{Der}(\ldots)$, i.e. if f and g define the same cohomology classes in $H^0(\mathrm{Der}(\ldots))$. The subset of automorphisms homotopy equivalent to the identity automorphism (i.e. to zero in $\mathrm{Der}(\ldots)$) is a normal subgroup in Aut(...), and the quotient by this normal subgroup is called the *group of homotopy non-trivial* automorphisms and is denoted, respectively, by

$$\mathbf{Aut}(\widehat{\mathcal{H}olieb}),\quad \mathbf{Aut}(\widehat{\mathcal{H}olieb^{\diamond}}),\quad \mathbf{Aut}(\widehat{\mathcal{H}olieb_{odd}}). \tag{18}$$

Then we have the following three corollaries to the Main Theorems.

4.3.1. Proposition

(i) *The group* $\mathbf{Aut}(\widehat{\mathcal{H}olieb})$ *is equal to the Grothendieck–Teichmüller group* $GRT = GRT_1 \ltimes \mathbb{K}^*$ *with the subgroup* $\mathbb{K}^*$ *acting on* $\widehat{\mathcal{H}olieb}$ *by the following rescaling transformations of the generators:*

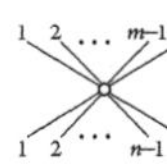 $\longrightarrow \lambda^{m+n-2}$ 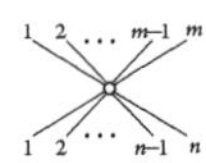 $\forall \lambda \in \mathbb{K}^*. \quad (19)$

(ii) *The group* $\mathbf{Aut}(\widehat{\mathcal{H}olieb^{\diamond}})$ *is equal to* $GRT = GRT_1 \ltimes \mathbb{K}^*$ *with the subgroup* $\mathbb{K}^*$ *acting on* $\widehat{\mathcal{H}olieb}$ *by rescaling transformations,*

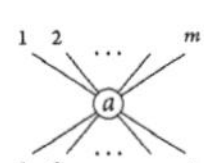 $\longrightarrow \lambda^{m+n+a-2}$ 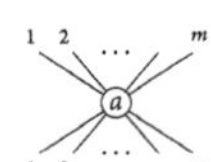 .

(iii) *The group* $\mathbf{Aut}(\widehat{\mathcal{H}olieb_{odd}})$ *is equal to* $\mathbb{K}^*$*, which acts* $\widehat{\mathcal{H}olieb_{odd}}$ *by rescaling transformations of the generators as in (19).*

Proof All three groups in (18) can be identified as sets with, respectively, the zeroth cohomology groups

$$H^0(\mathrm{Der}(\widehat{\mathcal{H}olieb})),\quad H^0(\mathrm{Der}(\widehat{\mathcal{H}olieb^{\diamond}})),\quad H^0(\mathrm{Der}(\widehat{\mathcal{H}olieb_{odd}}))$$

which in turn, by the Main Theorems, can be identified with the zeroth cohomology groups

$$H^0(\mathsf{GC}_3^{or},\delta)\oplus\mathbb{K},\ \ H^0(\mathsf{GC}_3^{or}[[\hbar]],\delta_\hbar)\oplus\mathbb{K},\ \ H^0(\mathsf{GC}_2^{or},\delta)\oplus\mathbb{K},$$

which in turn, according to [W1], [CMW] and, respectively, [W2], are equal as Lie algebras to

$$\mathfrak{grt}_1\oplus\mathbb{K},\quad \mathfrak{grt}_1\oplus\mathbb{K},\quad \text{and, respectively,}\quad \mathbb{K}.$$

This proves all the claims. □

This proposition implies a highly non-trivial action of the Grothendieck–teichmüller group GRT_1 on the completed properads $\widehat{\mathcal{H}olieb}$ and $\widehat{\mathcal{H}olieb^\diamond}$ and hence on their representations. However, one must be careful when talking about representations of the *completed* free properads. We introduce a topology on properads $\widehat{\mathcal{H}olieb}/\widehat{\mathcal{H}olieb}_{odd}$ (resp. on $\widehat{\mathcal{H}olieb^\diamond}$) as the one induced by the genus filtration (resp. by the filtration defined by the parameter 'genus + total a-weight'; cf. [CMW]). If W is a topological vector space, then, by a representation of $\widehat{\mathcal{H}olieb}/\widehat{\mathcal{H}olieb}_{odd}$ (resp. $\widehat{\mathcal{H}olieb^\diamond}$) in W, we mean a *continuous* map of topological properads

$$\rho : \widehat{\mathcal{H}olieb}/\widehat{\mathcal{H}olieb}_{odd} \longrightarrow \mathcal{E}nd_W, \quad \rho^\diamond : \widehat{\mathcal{H}olieb^\diamond} \longrightarrow \mathcal{E}nd_W.$$

Let us describe a sufficiently large class of such representations. Consider an arbitrary dg space V, and let $\hbar$ be a formal parameter of degree zero. The vector space $W := V[[\hbar]]$ can be equipped with the standard $\hbar$-adic topology. Then, maps ρ and $\rho^\diamond$ satisfying on the generators the condition

ρ (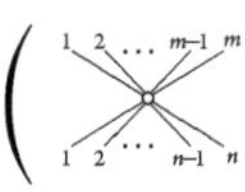) $\in \hbar^{m+n-2}\mathrm{Hom}(V^{\otimes n}, V^{\otimes m}))[[\hbar]],$

$\rho^\diamond$ (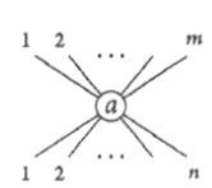) $\in \hbar^{m+n+a-2}\mathrm{Hom}(V^{\otimes n}, V^{\otimes m}))[[\hbar]]$

define a *continuous* representation of the properads $\widehat{\mathcal{H}olieb}/\widehat{\mathcal{H}olieb}_{odd}$ and, respectively, $\widehat{\mathcal{H}olieb^\diamond}$ in the topological space $V[[\hbar]]$.

4.3.2. On the Unique Non-Trivial Deformation of $\widehat{\mathcal{H}olieb}_{odd}$

It was proven in [W2] that the cohomology group $H^1(\mathsf{GC}_2^{or})$ is one-dimensional and is spanned by the following graph:

$\Upsilon_4 :=$ 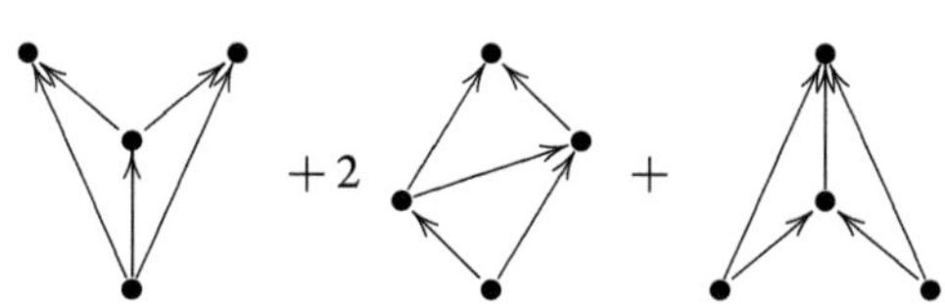 .

Moreover, $H^2(\mathsf{GC}_2^{or}) = \mathbb{K}$ and is spanned by a graph with four vertices. This means that one can construct by induction a new Maurer–Cartan element in the Lie algebra GC_2^{or} (the integer subscript in the summand Υ_n stands for the number of vertices of graphs)

$$\Upsilon_{KS} = \bullet\!\!\longrightarrow\!\!\bullet + \Upsilon_4 + \Upsilon_6 + \Upsilon_8 + \cdots,$$

as all obstructions have more than seven vertices and hence do not hit the unique cohomology class in $H^2(\mathsf{GC}_2^{or})$. Up to gauge equivalence, this new Maurer–Cartan element Υ is the *only* non-trivial deformation of the standard Maurer–Cartan element $\bullet\!\!\longrightarrow\!\!\bullet$. We call this element *Shoikhet* element as it was first found by Boris Shoikhet in [Sh] with a reference to an important contribution by Maxim Kontsevich via an informal communication.

By Theorem 1.2.1, the Maurer–Cartan element Υ_{KS} equips the completed non-differential properad $\widehat{\mathcal{H}olieb}_{odd}$ with a new differential denoted by δ_{KS}. If continuous representations of $\widehat{\mathcal{H}olieb}_{odd}$ equipped with the standard differential δ (originating from $\bullet\!\!\longrightarrow\!\!\bullet$) in a topological vector space $V[[\hbar]]$ can be identified with ordinary formal Poisson structures $\pi \in \mathcal{T}_{poly}(V)[[\hbar]]$, the continuous representations of $\widehat{\mathcal{H}olieb}_{odd}$ equipped with the new differential δ_{KS} give us a notion of *quantizable Poisson structure* $\pi^{quant} \in \mathcal{T}_{poly}(V)[[\hbar]]$ (this notion can be globalized from a vector space V to an arbitrary manifold M). It was proven in [MW3] that, for *finite*-dimensional vector spaces V (or manifolds M), there is a one-to-one correspondence between ordinary Poisson structures and quantizable ones, but the correspondence

$$\left\{\text{Ordinary Poisson structures } \pi \text{ on } M\right\} \stackrel{1:1}{\leftrightarrow} \left\{\text{Quantizable Poisson structures } \pi^{quant} \text{ on } M\right\}$$

is highly non-trivial and depends on a choice of Drinfeld associator. Moreover, quantizable Poisson structures can be deformation quantized in a trivial (and essentially unique) perturbative way [MW3] so that all the subtleties of the deformation quantization are hidden in the above correspondence.

REFERENCES

[CMW] R. Campos, S. Merkulov and T. Willwacher, *The Frobenius properad is Koszul*, Duke Math. J. **165** (2016), no. 1, 2921–89.

[C] M. Chas, *Combinatorial Lie bialgebras of curves on surfaces*, Topology **43** (2004), no. 3, 543–68.

[CFL] K. Cieliebak, K. Fukaya and J. Latschev, *Homological algebra related to surfaces with boundary*, arXiv:1508.02741 (2015).

[CS] M. Chas and D. Sullivan, *Closed string operators in topology leading to Lie bialgebras and higher string algebra*, in O. A. Laudal and R. Piene (eds) The Legacy of Niels Henrik Abel, Springer, Berlin (2004)), pp. 771–84.

[D1] V. Drinfeld, *Hamiltonian structures on Lie groups, Lie bialgebras and the geometric meaning of the classical Yang–Baxter equations*, Soviet Math. Dokl. **27** (1983), no. 1, 68–71.

[D2] V. Drinfeld, *On quasitriangular quasi-Hopf algebras and a group closely connected with Gal($\bar{Q}/Q$)*, Leningrad Math. J. **2** (1991), no. 4, 829–60.

[ES] P. Etingof and O. Schiffmann, Lectures on Quantum Groups, International Press, Somerville (2002).

[F] H. Furusho, *Four groups related to associators*, arXiv:1108.3389 (2011).

[KM] M. Kapranov and Yu. I. Manin, *Modules and Morita theorem for operads*, Amer. J. Math. **123** (2001), no. 5, 811–38.

[KMW] A. Khoroshkin, S. Merkulov and T. Willwacher, *On quantizable odd Lie bialgebras*, Lett. Math. Physics **106**, (2016), no. 9, 1199–215.

[Ko1] M. Kontsevich, *Formal (non)commutative symplectic geometry*, in Gelfand I.M., Corwin L., Lepowsky J. (eds) *The Gelfand Mathematical Seminars, 1990–1992*. Birkhauser, Boston (1993), pp. 173–88.

[Ko2] M. Kontsevich, *Formality Conjecture*, in D. Sternheimer, J. H. Rawnsley and S. Gut (eds.), Deformation Theory and Symplectic Geometry, Kluwer, Dordrecht (1997), pp. 139–56.

[Ko3] M. Kontsevich, a letter to Martin Markl, November 2002.

[MaVo] M. Markl and A.A. Voronov, *PROPped up graph cohomology*, in J. I. Manin and Y. Tschinkel, eds, Algebra, arithmetic, and geometry: in honor of Yu. I. Manin. Vol. II, Progr. Math., 270, Birkhäuser, Boston, MA (2009) pp. 249–81.

[M1] S.A. Merkulov, *Prop profile of Poisson geometry*, Commun. Math. Phys. **262** (2006), no.1, 117–35.

[M2] S.A. Merkulov, *Graph complexes with loops and wheels*, in: 'Algebra, Arithmetic and Geometry - Manin Festschrift' (eds. Yu. Tschinkel and Yu. Zarhin), Progress in Mathematics, Vol. 269, Birkhaüser, Boston (2010), pp. 311–54.

[MV] S.A. Merkulov and B. Vallette, *Deformation theory of representations of prop(erad)s I & II, J. Reine Angew. Math.* **2009**, no. 634, 51–106, and no. 636, 123–174 (2009).

[MW1] S. Merkulov and T. Willwacher, *Grothendieck–Teichmüller and Batalin-Vilkovisky*, Lett. Math. Phys. **104** (2014) No. 5, 625-34.

[MW2] S. Merkulov and T. Willwacher, *Props of ribbon graphs, involutive Lie bialgebras and moduli spaces of curves*, preprint arXiv:1511.07808 (2015) 51pp.

[MW3] S. Merkulov and T. Willwacher, *An explicit two step quantization of Poisson structures and Lie bialgebras*, preprint arXiv:1612.00368 (2016).

[Sc] T. Schedler, *A Hopf algebra quantizing a necklace Lie algebra canonically associated to a quiver*. Intern. Math. Res. Notices **2005** (2005), no. 12, 725–60.

[Sh] B. Shoikhet, *An L_∞ algebra structure on polyvector fields*, preprint arXiv:0805.3363, (2008).

[Ta] D. Tamarkin, *Action of the Grothendieck-Teichmueller group on the operad of Gerstenhaber algebras*, preprint math/0202039 (2002).

[Tu] V.G. Turaev, *Skein quantization of Poisson algebras of loops on surfaces*, Ann. Sci. Ecole Norm. Sup. (4) 24, no. 6, (1991) 635–704.

[Va] B. Vallette, *A Koszul duality for props*, Trans. Amer. Math. Soc., **359** (2007), no. 10, 4865–943.

[W1] T. Willwacher, *M. Kontsevich's graph complex and the Grothendieck–Teichmüller Lie algebra*, Invent. Math. 200 (2015), no. 3, 671–760.

[W2] T. Willwacher, *The oriented graph complexes*, Comm. Math. Phys. 334 (2015), no. 3, 1649–666.

11

Vertex Algebras and 4-Manifold Invariants

MYKOLA DEDUSHENKO[1,3]
SERGEI GUKOV[1,3]
PAVEL PUTROV[2]

[1] *Walter Burke Institute for Theoretical Physics, California Institute of Technology, Pasadena, CA 91125, USA*
[2] *School of Natural Sciences, Institute for Advanced Study, Princeton, NJ 08540, USA*
[3] *Centre for Quantum Geometry of Moduli Space, Department of Mathematics, Aarhus University Aarhus, DK-8000, Denmark*

ABSTRACT. We propose a way of computing 4-manifold invariants, old and new, as chiral correlation functions in half-twisted 2d $\mathcal{N} = (0,2)$ theories that arise from compactification of fivebranes. Such formulation gives a new interpretation of some known statements about Seiberg–Witten invariants, such as the basic class condition, and gives a prediction for structural properties of the multi-monopole invariants and their non-abelian generalizations.

1. Introduction and Motivation

There are at least three parallel tracks that lead to the study of $\overline{Q}_+$-cohomology and chiral correlation functions in a certain class of 2d $\mathcal{N} = (0,2)$ theories.

1.1. *Unorthodox Invariants of Smooth 4-Manifolds*

Searching for new invariants of smooth 4-manifolds, which potentially could go beyond the Seiberg–Witten and Donaldson invariants, it was proposed in [1] to consider a two-dimensional quantum field theory $T[M_4, G]$ as a rather unusual invariant of smooth structures on a 4-manifold M_4. Specifically, $T[M_4, G]$ is a 2d $\mathcal{N} = (0,2)$ superconformal theory that, apart from M_4, also depends on a choice of a root system G and is invariant under the Kirby moves.

Dedushenko, M., Gukov, S., Putrov, P., *Vertex Algebras and 4-Manifold Invariants.* In: *Geometry and Physics: A Festschrift in Honour of Nigel Hitchin*, Jørgen Ellegaard Andersen, Andrew Dancer, Oscar García-Prada (Eds): Oxford University Press (2018). DOI: 10.1093/oso/9780198802006.003.0011

Luckily, conformal field theories in two dimensions exhibit rich mathematical structure which, on the one hand, is rich enough to (potentially) describe the wild world of smooth 4-manifolds and, on the other hand, is rigourous enough to hope for a precise mathematical definition of the invariant $T[M_4, G]$. In fact, for many practical purposes and applications in this paper, a mathematically inclined reader can think of $T[M_4, G]$ as a functor that assigns a vertex operator algebra (VOA) to a smooth 4-manifold M_4 (and a 'gauge' group G). Composing it with other functors that assign various quantities to 2d conformal theories, one can obtain more conventional invariants of smooth 4-manifolds:

$$M_4 \rightsquigarrow T[M_4; G] \rightsquigarrow Z_{T[M_4;G]} = \text{4-manifold invariant.} \tag{1.1}$$

Here, Z can be any invariant of a 2d conformal theory with $\mathcal{N} = (0,2)$ supersymmetry, e.g. its elliptic genus, chiral ring, moduli space of marginal couplings, or central charge. Since 2d theory $T[M_4; G]$ is systematically determined by M_4 and invariant under the Kirby moves, all such invariants lead to various 4-manifold invariants; some are simple and some are quite powerful. In particular, it was conjectured in [1] that chiral ring of the theory $T[M_4; G]$ for $G = SU(2)$ or, equivalently, its $\overline{Q}_+$-cohomology knows about Donaldson invariants of M_4.

One of the main goals in this paper is to present some evidence to this conjecture and to build a bridge between $\text{VOA}_G[M_4]$ and more traditional 4-manifold invariants. Conjecturally, at least for manifolds with $b_2^+ > 1$, one can trade $SU(2)$ Donaldson invariants for Seiberg–Witten invariants defined in a simpler gauge theory, with gauge group $G = U(1)$, which would be too trivial if not for an extra ingredient, the additional spinor fields. We wish to study how these invariants are realized in 2d theory $T[M_4; G]$ with $G = SU(2)$ and $G = U(1)$, respectively. In particular, we shall see that, in 2d realization of Seiberg–Witten invariants, much like in gauge theory on M_4, the non-trivial information about the 4-manifold comes from an extra ingredient, a particular vertex operator $\mathcal{S}(z)$.

The functor that associates a vertex operator algebra to a smooth 4-manifold is a natural generalization of the pioneering work by Nakajima [2], where connection between gauge theory on ALE spaces and Kac–Moody algebras was observed (see also [3] for a visionary exposition). These results alone suffice to determine $\text{VOA}_G[M_4]$ for many negative-definite 4-manifolds by analysing its implications for the Kirby moves; in physics, this might be called 'bootstrapping' $T[M_4]$ with ALE spaces and Kirby moves.

A separate line of development that motivates the study of $\text{VOA}_G[M_4]$ has to do with categorification of quantum group invariants.

1.2. 4d TQFT

In 1994, Crane and Frenkel envisioned a 4d TQFT that categorifies quantum group invariants of knots and 3-manifolds. Their paper [4] was way ahead of its time since, even 20 years later, the proposed 4d TQFT is slowly being built, one brick at a time. Much like any 4d TQFT that obeys Atiyah-Segal axioms, it should assign numerical invariants to closed 4-manifolds and vector spaces to 3-manifolds. Moreover, if the 4d TQFT in question admits two-dimensional topological defects—which we shall call either 'foams' or 'surface operators'—then it should also assign numbers to closed surfaces $D \subset M_4$ and vector spaces to knots and links (cf. Figure 1). A non-trivial requirement is that

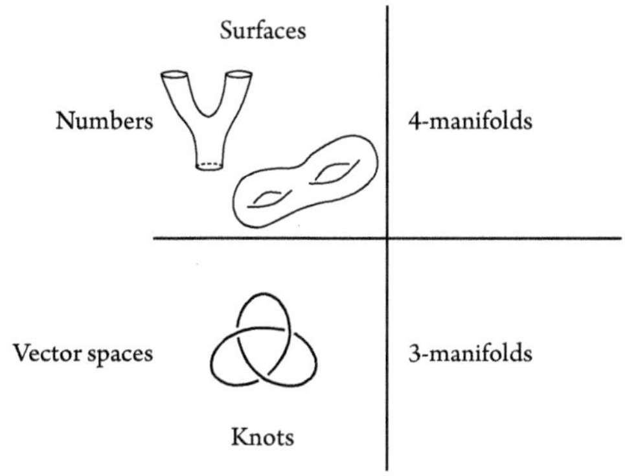

Figure 1 Various corners of a 4d TQFT with topological 'surface operators' or 'foams'. It assigns vector spaces to knots and 3-manifolds, maps between these homological invariants to the corresponding cobordisms, and numerical invariants to closed 4-manifolds and embedded surfaces.

Witten–Reshetikhin–Turaev (WRT) invariants arise as graded Euler characteristics of vector spaces assigned to knots and 3-manifolds.

The first major piece of the desired structure came with the construction of Khovanov–Rozansky homology [5–7] that belongs to the lower left corner in Figure 1. This corner is by far the most developed element of the sought after 2d–4d TQFT on $D \subset M_4$, and even that only for $M_4 = \mathbb{R}^4$ and $D = \mathbb{R} \times K$. Its physical interpretation, proposed in [8], led to many new predictions and connections between various areas, which include knot contact homology [9], gauge theory [10, 11] and algebras of interfaces [12, 13], just to name a few. (A more complete account of these connections can be found e.g. in [13, 14].)

Homological invariants of knots and links were soon generalized to invariants of cobordisms and closed surfaces [15–20]. This generalization, illustrated in the top left corner of Figure 1, corresponds to passing from $D = \mathbb{R} \times K$ to more general surfaces, while keeping $M_4 = \mathbb{R}^4$. Apart from a lone exception [13], the 2d TQFT on a 'foam' D has not been studied in the physics literature.

The situation is roughly reversed as it comes to generalization in a different direction, namely to homological invariants of non-trivial 3-manifolds. It was recently studied in the physics literature [21, 22], while in math the lower right corner of Figure 1 remains a largely unexplored territory. If the homological invariants of 3-manifolds proposed in loc. cit. are functorial (perhaps under certain conditions), then one must be able to replace $\mathbb{R} \times M_3$ by a more general cobordism M_4, finally taking us to the upper right corner of Figure 1:

$$\begin{array}{ccc} \mathbb{R} \times M_3 & & M_4 \\ \cup & \rightsquigarrow & \cup \\ \mathbb{R} \times K & & D. \end{array} \tag{1.2}$$

Now comes the key point. The physical set-up that extends homological invariants of knots and 3-manifolds to their cobordisms is, in fact, precisely what defines the $\overline{Q}_+$-cohomology of 2d $\mathcal{N} = (0,2)$ theory $T[M_4, G]$ or, equivalently, the vertex operator algebra $\text{VOA}_G[M_4]$.

From this perspective, the goal of this paper is to bring the categorification programme closer to its original motivation and move from homological invariants of 3-manifolds with knots to the corresponding invariants of 4-manifolds with embedded surfaces, i.e. from the lower half to the upper half of Figure 1.

1.3. G_2 Perspective

Another motivation for this work comes from M-theory compactifications on 7-manifolds with special holonomy group G_2. Such compactifications lead to $\mathcal{N}=1$ supersymmetric physics in the remaining four space-time dimensions, which is interesting for a number of reasons. If supersymmetry is a part of Nature, it may very well be the geometry of the world we live in. Moreover, 4d $\mathcal{N}=1$ physics is very interesting on its own, exhibiting a wide spectrum of physical phenomena—dynamical SUSY breaking; confinement and other phases; rich landscapes of superconformal points; etc.—many of which offer enormous potential for future discoveries, perhaps through G_2 compactifications.

One way to explore the physics of such compactifications is with the help of extended objects, which, in the context of M-theory, basically limits us to either M2-branes or M5-branes. Since the dimension of M2-brane world volume is barely enough for it to wrap a supersymmetric cycle in a G_2-manifold, fivebranes quickly take the centre stage. In particular, one can try to probe the physics of G_2 compactifications by studying M5-branes supported on coassociative 4-manifolds. In 4d $\mathcal{N}=1$ theory, such objects look like half-BPS 'cosmic strings' that preserve 2d $\mathcal{N}=(0,2)$ supersymmetry on their world sheet. A natural question, then, is: what degrees of freedom does such a string carry?

As the reader might have anticipated by now, the answer is $T[M_4;G]$ for $G=U(N)$, where N is the number of fivebranes. Indeed, a neighbourhood of every coassociative 4-manifold M_4 in the ambient G_2 space looks like a bundle of self-dual 2-forms, $\Lambda^{2,+}(M_4)$,

$$\begin{array}{lccc} \text{space-time:} & \mathbb{R}^4 & \times & \Lambda^{2,+}(M_4) \\ & \cup & & \cup \\ N\,\text{M5-branes:} & \mathbb{R}^2 & \times & M_4 \end{array} \tag{1.3}$$

and the partial topological twist of 6d (2,0) theory on the fivebrane world volume is precisely what defines 2d $\mathcal{N}=(0,2)$ theory $T[M_4;G]$ in two of its remaining space-time directions. These directions can also be 'twisted', so that $\mathbb{R}^2$ can be replaced by a more general world-sheet Σ of a cosmic string, with $\mathbb{R}^4$ accordingly replaced by $T^*\Sigma$. Since the resulting theory on a Riemann surface Σ is not fully topological, it is often called *half-twisted A-model* and the corresponding twist is often called *holomorphic* (as opposed to topological).

The paper is organized as follows. In sections 2 and 3, we discuss 2d $\mathcal{N}=(0,2)$ theory $T[M_4;G]$ and its impurity vertex operators. In particular, we propose a framework in which gauge theoretic 4-manifold invariants (not necessarily abelian!) are given by correlation functions of two vertex operators, $\mathcal{S}(z)$ and $\mathcal{S}_+(z)$, and we find an intriguing relation between these two vertex operators.

One of our main results is the definition of vertex operators $\mathcal{S}(z)$ and $\mathcal{S}_+(z)$ in terms of intersecting fivebranes, which we then compare to the ordinary (winding-momentum) vertex operators V_λ similarly defined via M2-branes ending on fivebranes. Moreover, we show that the former can be expressed as linear combinations of the latter,

$$\mathcal{S}(z) = \sum_{\lambda} \text{SW}(\lambda)\, V_\lambda(z), \tag{1.4}$$

where the expansion coefficients are precisely the Seiberg–Witten invariants of M_4. This relation can be viewed as a UV–IR correspondence relating codimension-2 defects (originating from intersecting M5-branes) and codimension-4 defects (the boundaries of M2-branes) in 6d fivebrane theory.

This framework leads to new predictions, which we verify in section 4 using traditional gauge theory methods in the case of multi-monopole generalization of Seiberg–Witten theory.[1] Mathematically, this theory is interesting in its own right and formulating it on arbitrary 4-manifolds requires extra care. We address these challenges using equivariant techniques extensively used in physics in recent years, starting with [23] and culminating in the formulation of the Nekrasov instanton partition function [24] on 4-manifolds with $U(1) \times U(1)$ symmetry. We use similar methods to carefully define and study multi-monopole invariants of arbitrary 4-manifolds and comment on the structure of the corresponding Floer theory for 3-manifolds. In particular, another new result of this paper is a concrete and explicit expression for the multi-monopole invariants of M_4 as correlators (3.68) of multiple vertex operators $\mathcal{S}(z_i)$ and $\mathcal{S}_+(z_j)$. One corollary is that multi-monopole invariants on M_4 can be expressed in terms of the standard Seiberg–Witten invariants.

Once we reproduce the structure of 2d chiral correlators in the abelian gauge theory, in section 5 we comment on non-abelian generalizations. In particular, as an illustration, we show how correlation functions of $\mathcal{S}(z)$ and $\mathcal{S}_+(z)$ lead to new predictions for invariants of 4-manifolds in a non-abelian gauge theory. It would be interesting to verify these predictions by direct gauge theoretic techniques in a way similar to the analysis of section 4 and explore more general theories. Returning to our original motivation, in section 5 we also discuss the structure of the 4d TQFT illustrated in Figure 1 and propose a simple criterion that can help to identify new opportunities for constructing 4-manifold invariants via 4d $\mathcal{N} = 2$ theories (possibly non-Lagrangian).

Since, most of the time, the choice of G is clear from the context, to reduce clutter, we often omit it and refer to $T[M_4; G]$ simply as $T[M_4]$.

2. Flux Vacua of $T[M_4]$

Many applications of string theory and M-theory, from AdS/CFT to building semi-realistic models of particle physics, involve dimensional reduction (a.k.a. compactification) on a non-trivial manifold in the presence of background fluxes. Generically, such background fluxes break supersymmetry, unless geometric moduli of the compactification manifold obey certain conditions. Sometimes, these conditions can be interpreted as equations for

[1] A mathematically inclined reader may want to skip directly to section 4.

a critical point of a function $\mathcal{W}$ that, in turn, often admits a simple physical interpretation in the low-dimensional effective theory.

The six-dimensional (2,0) theory on the fivebrane world volume, which can be thought of as a younger sister of M-theory [25], also admits flux compactifications. Thanks to non-gravitational physics of such flux vacua, they can serve as simpler examples for a much richer landscape of fluxes in M-theory. In particular, in order to preserve supersymmetry, background values of fields in 6d fivebrane theory must obey certain conditions; equivalently, one can interpret these conditions as constraints on the geometry of the compactification manifold M when fluxes are non-zero. For example, in compactification on a 4-manifold M_4, the relevant condition is the anti-self-duality equation for the 2-form flux F on M_4,

$$F^+ = 0. \tag{2.1}$$

One of the goals in the present section is to describe such SUSY flux vacua in 2d $\mathcal{N} = (0,2)$ theory $T[M_4]$.

2.1. *Theory* $T[M_4]$

Before we incorporate fluxes and vertex operators, our first task is to describe 2d theory $T[M_4]$ itself. Defined as a reduction of 6d (2,0) theory on a 4-manifold M_4 with a partial topological twist along M_4, the resulting 2d theory $T[M_4]$ carries $\mathcal{N} = (0,2)$ supersymmetry. In other words, its right-moving sector has $\mathcal{N} = 2$ supersymmetry and the left-moving sector is basically the vertex operator algebra $\text{VOA}_G[M_4]$.

In general, deriving $T[M_4]$ is a rather non-trivial task, which so far has been achieved only for particular types of 4-manifolds (see e.g. [1] for simply connected M_4 with definite intersection form). Luckily, in the special case of a single fivebrane, i.e. for $G = U(1)$, which we need in this paper, the 2d $\mathcal{N} = (0,2)$ theory $T[M_4]$ can be derived for *any* M_4 using the standard rules of the Kaluza–Klein reduction.[2] In this special case, the 6d fivebrane theory is simply a free theory of a (2,0) tensor multiplet, whose bosonic fields include a self-dual 2-form gauge field B and five real scalars that transform in the vector representation of an $SO(5)_R$ R-symmetry group. In the simply connected case, such reduction was considered in [26, 27]. Unless explicitly noted otherwise, we shall assume throughout the paper that the homology $H_*(M_4, \mathbb{Z})$ has no torsion. In principle, it is not hard to relax this assumption, as we shall illustrate in 2.1.2, where a class of examples with torsion is considered.

Under the partial topological twist[3] that corresponds to embedding the fivebrane in a G_2-manifold (1.3), three out of five scalar fields combine into components of a self-dual 2-forms on M_4, whose Kaluza–Klein modes contribute b_2^+ real scalars to the spectrum of $T[M_4]$. By McLean's theorem, these modes can be identified with the moduli of the coassociative 4-manifold M_4, i.e. with SUSY-preserving displacements of the M5-brane inside the G_2-manifold (1.3). Since $\mathcal{N} = (0,2)$ supersymmetry requires right-moving scalars to be complex-valued, these b_2^+ real scalar fields must be in pairs with other Kaluza–Klein modes. And, indeed, there are precisely b_2^+ right-moving compact scalars that come from

[2] This analysis was done jointly with S. Schafer-Nameki and J. Wong, whom we wish to thank for many enjoyable discussions on this topic.

[3] Described in detail e.g. in [21, sec. 5].

Kaluza–Klein modes of the self-dual 2-form gauge field B. Its field strength $H = dB$ satisfies the self-duality equation $H = *H$ and, therefore, after reduction on M_4, gives rise to b_2^+ right-moving compact scalars X_R^i, and b_2^- left-moving compact scalars X_L^i, as well as $b_1 = b_3$ vector fields A^k:

$$H = \sum_{j=1}^{b_2^-} \partial X_L^j \wedge \omega_j^- + \sum_{i=1}^{b_2^+} \bar{\partial} X_R^i \wedge \omega_i^+ + \sum_{k=1}^{b_1} dA^k \wedge \omega_k^1 + \sum_{k=1}^{b_1} n^k \wedge \omega_k^3, \tag{2.2}$$

where $\omega_i^\pm, \omega_k^1, \omega_k^3$ are generators of $H^{2,\pm}(M_4,\mathbb{R})$, $H^1(M_4,\mathbb{R})$ and $H^3(M_4,\mathbb{R})$, respectively, and $n^k \in \mathbb{Z}$ denote fluxes of 2d gauge fields A^k. Finally, the only remaining part of the bosonic Kaluza–Klein spectrum are the two real scalars of the 6d theory not affected by the topological twist. They parametrize transverse displacements of the fivebrane (1.3) inside $\mathbb{R}^4$ and can be naturally combined into a complex scalar of a standard 2d $\mathcal{N} = (0,2)$ chiral multiplet. In what follows, we refer to it as the 'centre of mass' multiplet Φ_0. The complete Kaluza–Klein spectrum of $T[M_4]$, including fermions, is summarized in Table 1.

To complete the description of $T[M_4]$, we also need to specify the operator product expansion between compact bosons. Note, this is not necessary for non-compact bosons since they can always be rescaled. Suppose that the fields $X_{L,R}$ are normalized such that, if $X_{L,R}$ is considered an element of $H^{2,\pm}(M_4) \equiv P_\pm H^2(M_4,\mathbb{R})$ with $P_\pm = (1 \pm *)/2$, the compactness is realized by periodicity with respect to elements of $H^2(M_4,\mathbb{Z}) \subset H^2(M_4,\mathbb{R})$. Much like the non-abelian version of the theory [1], our $T[M_4]$ here depends in a crucial way on the intersection form on M_4:

$$\begin{array}{rccc} Q: & H_2(M_4,\mathbb{Z}) \otimes H_2(M_4,\mathbb{Z}) & \longrightarrow & \mathbb{Z} \\ & \mu \otimes \nu & \longmapsto & \#(\mu \cap \nu). \end{array} \tag{2.3}$$

Its inverse is a bilinear form on the dual lattice:

$$\begin{array}{rccc} Q^{-1}: & H^2(M_4,\mathbb{Z}) \otimes H^2(M_4,\mathbb{Z}) & \longrightarrow & \mathbb{Z} \\ & [\mu] \otimes [\nu] & \longmapsto & \int_{M_4} \mu \wedge \nu, \end{array} \tag{2.4}$$

Table 1 The field content of 2d theory $T[M_4]$ for $G = U(1)$.

	Field	Chirality	Description
X_R^i	$i = 1,\ldots,b_2^+$	right-moving	compact real bosons
σ^i	$i = 1,\ldots,b_2^+$	non-chiral	non-compact real bosons
ψ_+^i	$i = 1,\ldots,b_2^+$	right-moving	complex Weyl fermions
$\phi_0, \overline{\phi}_0$		non-chiral	non-compact complex boson
χ_+		right-moving	complex Weyl fermion
X_L^j	$j = 1,\ldots,b_2^-$	left-moving	compact real bosons
A^k	$k = 1,\ldots,b_1$	non-chiral	vector fields
γ_-^k	$k = 1,\ldots,b_1$	left-moving	complex Weyl fermions

which can be extended to a real-valued bilinear form on $H^2(M_4,\mathbb{R}) \cong \mathcal{H}^2(M_4)$. On the other hand, the OPE of fields $X_{L,R}$ is determined by the metric on $H^2(M_4,\mathbb{R})$:

$$\begin{array}{rcl} G: \; H^2(M_4,\mathbb{R}) \otimes H^2(M_4,\mathbb{R}) & \longrightarrow & \mathbb{R} \\ \mu \otimes \nu & \longmapsto & \int_{M_4} \mu \wedge *\nu. \end{array} \tag{2.5}$$

The intersection form Q depends only on the topology of M_4, whereas the Hodge $*$ operator acts $H^2(M_4,\mathbb{R})$ and therefore G and $P_\pm$ all depend on the conformal structure of M_4.

Note that G and Q^{-1} coincide on the self-dual subspace but differ by a sign on anti-self-dual classes. What appears in the OPE is G^{-1}, and so we can write

$$\partial X_L(z) \otimes \partial X_L(0) \;\sim\; -\frac{(P_- \otimes P_-)(Q)}{z^2}, \tag{2.6}$$

$$\partial X_R(\bar{z}) \otimes \partial X_R(0) \;\sim\; \frac{(P_+ \otimes P_+)(Q)}{\bar{z}^2}, \tag{2.7}$$

where Q from (2.3) is understood as an element of $H^2(M_4,\mathbb{Z})^{\otimes 2} \subset H^2(M_4,\mathbb{R})^{\otimes 2}$, so that $(P_\pm \otimes P_\pm)(Q) \in H^{2,\pm}(M_2)^{\otimes 2}$.

Normally, non-abelian 2d $\mathcal{N} = (0,2)$ gauge theories exhibit a rich structure of phases [28, 29]: conformal, confining, dynamical SUSY breaking, etc. In part for this reason, the infrared theory $T[M_4]$ is usually very different from the result of a naive Kaluza–Klein reduction. However, in the abelian case, i.e. for a single fivebrane, the theory is free and, therefore, a careful Kaluza–Klein reduction is expected to correctly capture the physics of $T[M_4]$. To verify that this is indeed the case, we can write down the central charges. Combining the contributions of all left-moving fields in Table 1, we find

$$c_L \;=\; \underbrace{2 + b_2^+(M_4)}_{\substack{\text{non-compact}\\ \text{non-chiral bosons}}} + \underbrace{b_2^-(M_4)}_{\substack{\text{compact}\\ \text{chiral bosons}}} - \underbrace{3b_1(M_4)}_{\text{vectors}} + \underbrace{b_1(M_4)}_{\substack{\text{Weyl}\\ \text{fermions}}} \;=\; \chi(M_4), \tag{2.8}$$

and, similarly, in the right-moving sector,

$$c_R \;=\; \underbrace{2 + b_2^+(M_4)}_{\substack{\text{non-compact}\\ \text{non-chiral bosons}}} + \underbrace{\frac{1}{2}(2 + 2b_2^+(M_4))}_{\substack{\text{right-moving}\\ \text{real fermions}}} + \underbrace{b_2^+(M_4)}_{\substack{\text{compact}\\ \text{chiral bosons}}} - \underbrace{3b_1(M_4)}_{\text{vectors}} \;=\; \frac{3}{2}(\chi + \sigma), \tag{2.9}$$

where χ is the Euler characteristic of the 4-manifold M_4, and σ is its signature. These expressions agree with the infrared central charges of $T[M_4]$ computed from the anomaly polynomial of a single fivebrane [30, 31]:

$$A_{M5} \;=\; \frac{1}{48}\Big[p_2(N) - p_2(T) + \frac{1}{4}(p_1(T) - p_1(N))^2\Big], \tag{2.10}$$

where T and N denote, respectively, the tangent and normal bundle of the fivebrane world volume in the eleven-dimensional space-time. In our set-up (1.3), we have $T = T\Sigma \oplus TM_4$

and $N = R \oplus \Lambda^{2,+}(M_4)$, where R is the $U(1)_R$ R-symmetry bundle. Substituting these into (2.10) and integrating over M_4 as in [32], we obtain the standard form of the anomaly polynomial in a 2d $\mathcal{N} = (0,2)$ theory:

$$A_{2d} = \frac{c_R}{6} c_1(R)^2 + \frac{c_L - c_R}{24} p_1(T\Sigma), \tag{2.11}$$

with the values of (c_L, c_R) in (2.8) and (2.9). This gives a further justification to the result of the Kaluza–Klein reduction.

Still, there is something very peculiar about the fields listed in Table 1. While (ϕ_0, χ_+) combine into a standard 2d $\mathcal{N} = (0,2)$ chiral superfield $\Phi_0 = \phi_0 + \theta^+ \chi_+ - i\theta^+ \overline{\theta}^+ \overline{D}\phi_0$, the other fields form somewhat unusual representations of 2d $\mathcal{N} = (0,2)$ supersymmetry algebra. For example, the left-moving compact scalars X_L^j live in a trivial representation of the supersymmetry algebra. This is consistent because they satisfy $\overline{\partial} X_L^j = 0$, so one indeed can impose $Q_+ X_L^j = \overline{Q}_+ X_L^j = 0$.[4] These are not off-shell multiplets, however.

Similarly, the right-moving compact scalars X_R^i belong to peculiar multiplets $\Phi_i = (\phi^i, \psi_+^i)$ whose lowest components are complex scalar fields $\phi^i := \sigma^i + iX_R^i$. The latter has the unusual property that its imaginary part is a chiral (right-moving) scalar, while the real part is an ordinary scalar. We can still treat Φ_i as an off-shell $\mathcal{N} = (0,2)$ chiral multiplet if we regard the chirality condition of X_R^i as part of the equations of motion.

At this point, it is instructive to put our theories $T[M_4]$ in the context of general 2d supersymmetric sigma-models used, e.g. as world-sheet theories in string compactifications. Such sigma-models describe maps

$$\phi : \Sigma \to X \tag{2.12}$$

into a target space X, which in theories with $\mathcal{N} = (2,2)$ or $\mathcal{N} = (0,2)$ supersymmetry must be Kähler (if there is no B-field) [33]. In the case of 2d $\mathcal{N} = (0,2)$ sigma-models that are directly relevant to us here, the geometric data also involves a holomorphic bundle $\mathcal{E}$ over X, such that

$$\begin{aligned} c_1(\mathcal{E}) &= c_1(TX) \mod 2, \\ c_2(\mathcal{E}) &= c_2(TX). \end{aligned} \tag{2.13}$$

These anomaly cancellation conditions are trivially satisfied for our theories $T[M_4]$ which, for simply connected M_4, have trivial bundle $\mathcal{E}$ and

$$X = \mathbb{C} \times (\mathbb{C}^*)^{b_2^+}. \tag{2.14}$$

Actually, our theories are slightly more subtle than traditional sigma-models with this target due to asymmetry between left- and right-moving bosons (see Table 1). We will return to

[4] This can be contrasted with the Fermi multiplets that contain only left-moving on-shell degrees of freedom, yet are acted on non-trivially by right-moving supersymmetries.

this class of theories shortly; for the moment, though, let us continue with the general sigma-model based on a Kähler target space X.

Following the standard conventions, we introduce local complex coordinates ϕ^i, $i = 1,\ldots,\dim_{\mathbb{C}}(X)$ and their complex conjugate $\phi^{\bar{i}} := \overline{\phi_i}$, and denote by ψ_+^i and $\psi_+^{\bar{i}}$ projections of the right-moving fermions—which couple to the tangent bundle TX—into $K_\Sigma^{1/2} \otimes \phi^*(T^{1,0}X)$ and $K_\Sigma^{1/2} \otimes \phi^*(T^{0,1}X)$, respectively:

$$\psi_+^i \in \Gamma\Big(K_\Sigma^{1/2} \otimes \phi^*(T^{1,0}X)\Big), \qquad \psi_+^{\bar{i}} \in \Gamma\Big(K_\Sigma^{1/2} \otimes \phi^*(T^{0,1}X)\Big). \tag{2.15}$$

Here, K_Σ is the canonical line bundle of Σ, the bundle of 1-forms of type $(1,0)$.

The half-twisted model is obtained by modifying the Lorentz transformations of the fields by the $U(1)_R$ R-symmetry, under which spin-$\frac{1}{2}$ fields ψ_+^i have charge $+1$ and $\psi_+^{\bar{i}}$ have charge -1. As a result, in the half-twisted model, they transform as spin-0 and spin-1 fields:

$$\psi_+^i \in \Gamma\Big(\phi^*(T^{1,0}X)\Big), \qquad \psi_+^{\bar{i}} \in \Gamma\Big(K_\Sigma \otimes \phi^*(T^{0,1}X)\Big). \tag{2.16}$$

Their zero modes contribute to the 'ghost number anomaly' which, according to the Hirzebruch–Riemann–Roch theorem, on a genus g surface Σ is given by (see e.g. [34–36])

$$\begin{aligned} &\#(\psi_+^i \text{ zero-modes}) - \#(\psi_+^{\bar{i}} \text{ zero-modes}) \\ &\qquad = \dim_{\mathbb{C}}(X) \cdot (1-g) - \int_\Sigma \phi^*\left(c_1(X)\right). \end{aligned} \tag{2.17}$$

Note, to ensure that the twisted model has non-anomalous Lorentz symmetry, one should twist by a non-anomalous combination of global currents. For theories $T[M_4]$, this is not an issue since the $U(1)_R$ symmetry that acts on the fermions ψ_+ as described above is non-anomalous. Indeed, from (2.14) it is clear that $c_1(X) = 0$ and, moreover, the ghost number anomaly (2.17) is

$$\Delta R_{T[M_4]} = (1-g)(1+b_2^+) = (1-g)\frac{\chi+\sigma}{2}. \tag{2.18}$$

This calculation can be easily generalized to non-simply connected 4-manifolds.

Note, another consistency check of the R-charge assignments in the theory $T[M_4]$ is the relation to the right-moving central charge, which follows from 2d $\mathcal{N} = (0,2)$ superconformal algebra:

$$c_R = 3\mathrm{Tr}\gamma^3 R^2. \tag{2.19}$$

Since, among the fields in Table 1, only complex Weyl fermions carry non-trivial R-charge, namely $R = 1$, the right-hand side is equal to $3(1+b_2^+)$. This agrees with the value of central charge c_R in (2.9).

Before we turn the page, let us make a few brief remarks on the 2d $\mathcal{N}=(0,2)$ theory $T[M_4]$ with $G=U(N)$ or $SU(N)$. In particular, the chiral ring of such theory for $G=SU(2)$ or, equivalently, the ground ring of its half-twisted model is expected [1] to contain information about all Donaldson invariants of M_4. While it does not require any additional impurity operators that are crucial for Seiberg–Witten invariants, one serious drawback of the higher-rank version is that it can not be obtained by a simple Kaluza–Klein reduction. Nevertheless, as in the abelian case (2.10), one can derive the central charges (c_L, c_R) from the anomaly polynomial of N fivebranes [31, 37, 38]:

$$A_{N\,\text{fivebranes}} = NA_{M5} + \frac{N(N^2-1)}{24} p_2(N). \tag{2.20}$$

Integrating over M_4 and comparing with (2.11) gives (cf. [32])

$$\begin{aligned} c_L &= \chi N + (2\chi + 3\sigma)(N^3 - N), \\ c_R &= \frac{3}{2}(\chi + \sigma)N + (2\chi + 3\sigma)(N^3 - N). \end{aligned} \tag{2.21}$$

Now it is easy to verify that, for $N > 1$, these central charges of $T[M_4]$ do not agree with the naive Kaluza–Klein spectrum; in particular, one needs to use alternative routes to find $T[M_4]$ (some of such routes were offered in [1]). Moreover, unitarity of $T[M_4]$ requires $c_L \geq 0$ and $c_R \geq 0$. In the abelian case, these bounds automatically follow from positivity of the Betti numbers. In the non-abelian case, though, they appear to imply a non-trivial condition on $c(M_4) = 2\chi + 3\sigma$, which must be non-negative for the large-N limit to preserve superconformal symmetry.

It is interesting to compare the large-N limit of $T[M_4; U(N)]$ with a particular class of supergravity solutions constructed in [39]. The latter exists when M_4 admits a conformally half-flat Einstein metric, which is a relatively strong condition. In particular, the virtual dimension of the moduli space of half-flat structures on M_4 is [40]:

$$\text{expected dim}\,\mathcal{M}_{\text{half-flat}} = 26\chi_h(M_4) - 7c(M_4), \tag{2.22}$$

where, for balance, instead of χ and σ we use

$$\begin{aligned} \chi_h(M_4) &= \frac{\chi(M_4) + \sigma(M_4)}{4}, \\ c(M_4) &= 2\chi(M_4) + 3\sigma(M_4) \qquad (= K^2_{M_4} \quad \text{when } M_4 \text{ is a complex surface}). \end{aligned} \tag{2.23}$$

It would be interesting to explore the relation between the space $\mathcal{M}_{\text{half-flat}}$ and the conformal manifold of 2d $\mathcal{N}=(0,2)$ theory $T[M_4]$ in the large-N limit.

2.1.1. Example: $M_4 = S^2 \times S^2$

The discussion so far has been fairly general. To make it a bit more concrete, let us consider an example of a 4-manifold which, on the one hand, is basic enough to show up a number of times through our study and, on the other hand, plays an important role in classification of smooth structures:[5]

$$M_4 = S^2 \times S^2. \tag{2.24}$$

This is a simply connected 4-manifold with $b_2^+ = b_2^- = 1$. Its intersection form is represented by the following matrix:

$$Q^{-1} = Q = \begin{pmatrix} 0 & 1 \\ 1 & 0 \end{pmatrix}. \tag{2.25}$$

Let $\omega^\pm \in H^2(M_4, \mathbb{R}) \cong \mathbb{R}^2$ be eigenvectors of $*$:

$$*\omega^\pm = \pm\omega^\pm, \tag{2.26}$$

which are normalized such that

$$\int_{S^2 \times S^2} \omega^\pm \wedge \omega^\pm = \pm 1. \tag{2.27}$$

They are related to the basis elements e_I of $H^2(M_4, \mathbb{Z})$ as follows:

$$\omega^\pm = \frac{e_1}{R} \pm \frac{e_2 R}{2}, \tag{2.28}$$

where R depends on the conformal structure of the $S^2 \times S^2$ and describes the ratio of the sizes of the two S^2's. The linear combination $X_L^1 \omega^- + X_R^1 \omega^+ \in H^2(M_4, \mathbb{R})$ is periodic with respect to shifts by $e_{1,2}$. Therefore, $X_L^1 + X_R^1$ can be understood as a non-chiral compact boson valued in a circle with radius R. The T-duality $R \leftrightarrow 2/R$ corresponds to the exchange of two S^2's in $M_4 = S^2 \times S^2$.

It follows that the full theory $T[M_4, U(1)]$ can be described as a 2d $\mathcal{N} = (0,2)$ sigma-model with target space $X = \mathbb{C} \times \mathbb{C}^*$ parametrized by *non-chiral* bosons. The same result follows from the twisted compactification on S^2 of the 4d theory $T[S^2, U(1)]$, the sigma-model with the Nahm 1-monopole moduli space $S^1 \times \mathbb{R}^3$ [41]. The result of such compactification is a $\mathcal{N} = (0,4)$ sigma-model with the target space $S^1 \times \mathbb{R}^3$ and trivial bundle of left-moving fermions [42].

Our next class of examples is more delicate and involves theories akin to the ones which play a role in classification of topological phases of matter.

[5] In classification of smooth structures, taking a connected sum with $S^2 \times S^2$ plays the role of a 'nilpotent' operation which, if repeated a sufficient number of times, can relate any pair of smooth structures. In particular, according to Wall's stable h-cobordism theorem, *if M_4 and M_4' are h-cobordant simply connected smooth 4-manifolds, then there exists an integer $n \geq 0$ such that $M_4 \# n(S^2 \times S^2)$ is diffeomorphic to $M_4' \# n(S^2 \times S^2)$.*

2.1.2. Examples with torsion in homology

Consider first a 3-manifold $M_3(\Gamma)$ associated to a (genus zero) plumbing graph Γ. There are various ways to define M_3 for a given Γ. One of them is to say that M_3 is obtained by a Dehn surgery on the corresponding link $\mathcal{L}(\Gamma)$ of unknots, illustrated in Figure 2. For example, Seifert fibrations over S^2 correspond to star-shaped plumbing graphs. Denote by L the number of vertices of Γ. It is equal to the number of components of the link $\mathcal{L}(\Gamma)$. We will also need an $L \times L$ linking matrix of $\mathcal{L}(\Gamma)$ which, somewhat suggestively, will be called Q:

$$Q_{v_1,v_2} = \begin{cases} 1, & v_1, v_2 \text{ connected}, \\ a_v, & v_1 = v_2 = v, \\ 0, & \text{otherwise}. \end{cases} \qquad v_i \in \text{Vertices of } \Gamma \cong \{1,\ldots,L\}. \tag{2.29}$$

We will assume that Q is non-degenerate, unless explicitly noted otherwise. A plumbed 3-manifold M_3 can also be viewed as a link of a singularity. The plumbing graph then plays the role of the resolution graph. The resolution is a smooth 4-manifold M_4' bounded by M_3 which has matrix Q as the intersection form. In [1] it was argued that $T[M_3, U(1)]$ is (an appropriate $\mathcal{N} = 2$ version of) $U(1)^L$ Chern–Simons theory with level matrix Q. Note that the 2d theory $T[M_4', U(1)]$ described in the previous section indeed naturally couples to 3d theory $T[M_3, U(1)]$ via gauging $U(1)^L$ global symmetry of $T[M_4', U(1)]$ in the 3d bulk. The $U(1)^L$ global symmetry acts as translations in the target space of compact bosons. This defines a gauge-invariant boundary condition for $U(1)^L$ Chern–Simons theory. Note that a 3d Wilson line of charge $h \in \mathbb{Z}^L/Q\mathbb{Z}^L$ ending on a boundary must couple to a boundary field of appropriate charge $\lambda \in H_2(M_4', \mathbb{Z}) \cong \mathbb{Z}^L$, such that $\lambda = h \mod Q\mathbb{Z}^L$, e.g. a vertex operator $e^{2\pi i\lambda\cdot(X_L+X_R)}$.

The theory $T[M_4]$ for $M_4 = M_3 \times S^1$ then can be easily obtained by the dimensional reduction of 3d theory $T[M_3]$ to two dimensions. Note that Q is the intersection form for M_4' described earlier, but not for M_4. Moreover, $b_2(M_4) = 0$ when Q is non-degenerate. The non-trivial sector of the theory is given by compactification of abelian Chern–Simons theory on a circle. This is a $(U(1)^L/U(1)^L)_Q$ gauged WZW with $U(1)^L$ WZW fields $e^{i\varphi_i} = \exp i \int_{S^1} A_i$. Equivalently, the effective 2d theory, considered on surface Σ, can be described as a BF theory with level matrix Q:

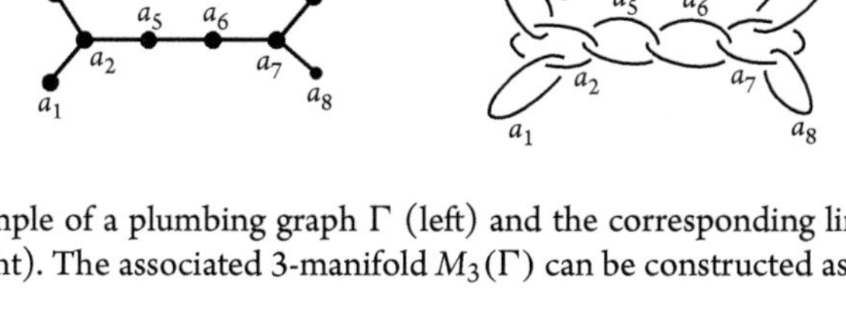

Figure 2 An example of a plumbing graph Γ (left) and the corresponding link $\mathcal{L}(\Gamma)$ of framed unknots in S^3 (right). The associated 3-manifold $M_3(\Gamma)$ can be constructed as a Dehn surgery on $\mathcal{L}(\Gamma)$.

$$\frac{1}{2}\sum_{ij} Q_{ij} \int_{\Sigma \times S^1} A_i dA_j = \sum_{ij} Q_{ij} \int_{\Sigma} \varphi_i dA_j, \tag{2.30}$$

which, in turn, can be described as a discrete gauge theory (of Dijkraaf–Witten type with trivial cocycle) with gauge group Coker Q. Note that one could also obtain such a desciption by first compactifying the M5-brane on S^1 and then compactifying the 5d $U(1)$ gauge theory describing dynamics of the D4-brane on M_3. From this point of view,

$$\varphi_i = \int_{e_i \subset M_3} A^{5\text{d}}, \tag{2.31}$$

where e_i is the corresponding generator of $\mathbb{Z}^L$ in $\mathbb{Z}^L/Q\mathbb{Z}^L \cong H_1(M_3)$, and $A^{5\text{d}}$ is the 5d $U(1)$ gauge field.

The fields A_j on the right-hand side of (2.30) then play the role of Lagrange multipliers imposing the constraint that $e_i \in H_1(M_3, \mathbb{Z})$ is a torsion element (contrary to a free generator, which would produce an ordinary free field via (2.31)). Integrating out A_j (together with summation over fluxes through Σ) on the right-hand side of (2.30) localizes on the configurations of the form

$$\varphi_i = \text{const along } \Sigma, \qquad \sum_i Q_{ij}\varphi_i = 0 \mod 2\pi. \tag{2.32}$$

The fact that $T[M_4]$ is topological for such class of 4-manifolds is consistent with the fact that $c_L = c_R = 0$, which follows from $\chi = \sigma = 0$.

2.2. *Holomorphic Differentials and Fluxes*

After describing general features of the theory $T[M_4]$, let us study its flux vacua, in particular those that preserve supersymmetry. We already mentioned in (2.1) that one of the conditions needed to preserve supersymmetry is $F^+ = 0$. Because we are mostly working[6] with 4-manifolds that have $b_2^+ > 1$, one can always pick a generic metric on M_4 for which the space of harmonic anti-self-dual forms $\mathcal{H}^{2,-}$ has no integral points, so there are no solutions to $F^+ = 0$ with $F/2\pi$ representing an integral cohomology class.

This suggests that generic $T[M_4]$ does not have any supersymmetric flux vacua, just like string compactifications with generic values of moduli. Adjusting moduli to special values that allow supersymmetric flux vacua corresponds to choosing non-generic metrics on M_4 that allow solutions to $F^+ = 0$. In order to learn how one can circumvent this condition, we need to understand better its two-dimensional interpretation in $T[M_4]$. Recall from [43] that, in twisted compactification of the 6d (2,0) theory on a Riemann surface Σ, four-dimensional gauge fields originate from the 6d self-dual 2-form gauge field, and they correspond to normalizable holomorphic differentials on Σ. If Σ is closed and has genus g,

[6] Part of the reason for this assumption is to avoid wall-crossing phenomena, whose 2d interpretation in the theory $T[M_4]$ we hope to address elsewhere.

the space of such differentials has complex dimension g, and we pick a basis h_I, $I = 1, \ldots g$. The equation that relates 6d and 4d gauge field strengths is

$$H = \sum_{I=1}^{g} F^{I-} \wedge h_I + F^{I+} \wedge \overline{h}_I, \tag{2.33}$$

and it is a close analogue of (2.2). From the 4d point of view, the supersymmetry equation $F^+ = 0$ simply follows from the SUSY variation of a gaugino. Now we can see the two-dimensional interpretation of fluxes by expanding $F^{I\pm}$ in the basis of (anti-)self-dual harmonic 2-forms $\omega_j^{\pm}$ and comparing (2.33) to (2.2).

From the 2d point of view, ∂X_L^j becomes a non-trivial linear combination of basic holomorphic differentials h_I, and $\overline{\partial} X_R^i$ becomes a linear combination of anti-holomorphic differentials $\overline{h}_I$. This means that scalar fields X_L^j and X_R^i are only locally defined, while globally they might have monodromies. Their derivatives ∂X_L^j and $\overline{\partial} X_R^i$ are, of course, globally defined holomorphic and anti-holomorphic differentials, respectively.

We can say more about two-dimensional flux vacua if we recall that there are $\mathcal{N} = (0,2)$ chiral multiplets $\Phi_i = (\phi^i, \psi_+^i)$ with $\phi^i = \sigma^i + iX_R^i$. Supersymmetry variation of ψ_+^i implies another BPS equation:

$$\overline{\partial}\left(\sigma^i + iX_R^i\right) = 0, \tag{2.34}$$

which means that $\sigma^i + iX_R^i$ is locally holomorphic on Σ. In the absence of topologically non-trivial fluxes, we would say that this is also true globally and use it (as well as chirality of X_R^i) to argue that σ^i and X_R^i have to be constant separately. In the presence of fluxes, however, we already know that $\overline{\partial} X_R^i$ becomes a non-trivial anti-holomorphic differential, so this equation implies the same for $\overline{\partial}\sigma^i$. In other words, in the presence of fluxes, σ^i is only locally defined too. Globally, it has monodromies around cycles of Σ.

While the fact that X_L^j and X_R^i are only locally defined is natural—after all, we can think of them as originating from the terms $\sum_{j=1}^{b_2^-} X_L^j \omega_j^- + \sum_{i=1}^{b_2^+} X_R^i \omega_i^+$ in the B-field, and B, being a 2-form gauge field, is certainly only locally defined—the fact that σ^i are also only locally defined on Σ comes as a surprise. The fields σ^i parametrize deformations of coassociative 4-cycles in a G_2-manifold that locally looks like $\Lambda^{2,+}(M_4)$. The point where all $\sigma^i = 0$ corresponds to the choice of $M_4 \subset \Lambda^{2,+}(M_4)$ as a zero section of $\Lambda^{2,+}(M_4)$. Non-zero σ^i describe deformations of this $M_4 \subset \Lambda^{2,+}(M_4)$ that preserve its calibrated property. Therefore, vacuum expectation values of σ^i describe the choice of M_4 in a G_2 manifold, and so describe the geometry of the fivebrane world volume.

The fact that, for supersymmetric flux vacua, σ^i has monodromies on Σ means that, as we go around cycles of Σ, σ^i get shifted by constants. This means that the M5-brane world volume is not really a compact manifold $M_4 \times \Sigma$, as was assumed from the very beginning, but rather its non-compact covering: every time we go around a cycle on Σ, we end up on a different copy of M_4 inside of $\Lambda^{2,+}(M_4)$. We interpret this contradiction with the initial assumptions as a statement that the theory $T[M_4]$ (which was *defined* by twisted compactification on M_4) does not have supersymmetric flux vacua.

This problem can be avoided by introducing our next ingredient, the impurity vertex operators in 2d theory $T[M_4]$, which correspond to extra matter fields in the 4d theory on M_4 and contribute to the right-hand side of the BPS equation $F^+ = 0$.

3. Seiberg–Witten Invariants and the Kondo Problem

The classical version of the Kondo problem has to do with the structure of a ground state in a system of free fermions coupled to a localized magnetic impurity. In particular, it explains a peculiar rise in resistivity (observed in 1930s) that some metals exhibit at low temperature. The model, proposed by Jun Kondo in 1964, is based on the so-called s-d Hamiltonian,

$$H = \psi_\alpha^\dagger \left(-\frac{\nabla^2}{2m} - \epsilon_F \right) \psi_\alpha + J\delta(\vec{x})\vec{\mathcal{S}} \cdot \frac{\vec{\tau}_{\alpha\beta}}{2} \psi_\alpha^\dagger \psi_\beta, \tag{3.1}$$

which describes the conduction band (s-band) electrons ψ_α of a non-magnetic metal interacting with a magnetic impurity (unfilled d-level) represented here by the spin-$\frac{1}{2}$ operator $\vec{\mathcal{S}}$. Here, $\alpha = \uparrow$ or $\downarrow$ is the spin index, ϵ_F is the Fermi energy, and J is the Kondo coupling. In the renormalization group approach, the Kondo problem is reduced to the effective theory of massless left-moving Dirac fermions in $1+1$ dimensions, a close cousin of $T[M_4]$, interacting with a localized impurity.[7]

When the Kondo interaction is ferromagnetic, $J < 0$, the effective coupling flows to zero and, at low energies, the impurity completely decouples from the conduction electrons. On the other hand, when the interaction is anti-ferromagnetic ($J > 0$), the effective Kondo coupling grows logarithmically at low temperatures, so that the impurity is completely screened by forming a singlet with the electrons around it. In the description where impurity is represented by auxiliary 'slave' fermions, the formation of a Kondo singlet manifests itself as a condensation of a $(0+1)$-dimensional charged scalar, made of an electron and a slave fermion.

The physics of this simple model and its generalizations is closely related to the 2d and 3d realization of Seiberg–Witten invariants discussed below, where $T[M_4]$ or $T[M_3]$ play the role of conduction electrons, and interesting physics comes from interaction with a certain 'impurity' operator $\mathcal{S}$. Similar generalizations of the Kondo model, where free electrons are replaced by an interacting CFT, describe e.g. impurities in Luttinger liquids. Another natural generalization involves a version of the model (3.1) where fermions carry an extra index $i = 1, \ldots, N_f$. This 'multi-channel' Kondo model exhibits interesting phases: in the underscreened regime ($N_f < 2S_{\text{imp}}$), the ground state has non-trivial degeneracy, and the physics is similar to that of a single-channel Kondo model whereas, in the overscreened

[7] The s-wave reduction and the doubling trick lead to the effective $1+1$ model with the Hamiltonian:

$$H_{\text{eff}} = \frac{i}{2\pi} \int_{-\infty}^{+\infty} dr\, \psi_L^\dagger \frac{d}{dr} \psi_L + \lambda \vec{\mathcal{S}} \cdot \psi_L^\dagger(0) \frac{\vec{\tau}}{2} \psi_L(0). \tag{3.2}$$

case ($N_f > 2S_{\text{imp}}$), the RG flow leads to a non-trivial fixed point and the ground state is not described by standard Fermi-liquid theory.

3.1. *Electric and Magnetic Impurities*

As our first example of a supersymmetric Kondo problem, let us consider a simple three-dimensional theory:

$$3\text{d}\,\mathcal{N} = 2 \text{ super-Chern–Simons theory } U(1)_p. \tag{3.3}$$

This theory by itself is rather boring; it has only one supersymmetric ground state on a 2-sphere S^2 and, correspondingly, the Q-cohomology $\mathcal{H}_{\text{BPS}}$ is one-dimensional.[8]

Now, let us see how this changes when we introduce an impurity localized in space [21]: a pair of chiral multiplets ϕ and $\widetilde{\phi}$ with $U(1)$ gauge charges $+1$ and -1, respectively. The Lagrangian of this combined 1d–3d system is similar to that of a 3d $\mathcal{N} = 2$ supersymmetric quantum electrodynamics (SQED), except the chiral multiplets ϕ and $\widetilde{\phi}$ are restricted to a $0+1$ dimensional world line of the impurity. Then, in the presence of the impurity, the space of BPS states is simply the space of states of a harmonic oscillator, which are excitations of the 'meson' $\phi\widetilde{\phi}$. We give it a name:

$$\mathcal{T}^+ \quad = \tag{3.5}$$

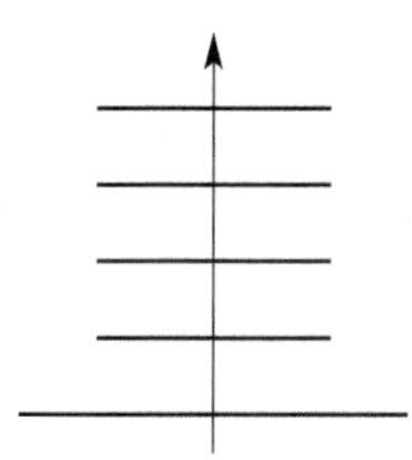

What we just produced is the Heegaard Floer homology $HF^+(M_3)$ of a particular simple 3-manifold M_3.

This simple example can be generalized in a number of ways, in particular, making it possible to describe homological invariants of an arbitrary 3-manifold (possibly with knots) as a space of supersymmetric ground states in a 3d $\mathcal{N} = 2$ theory with a certain half-BPS impurity localized in space (see e.g. [44]). The prototypical example of a 3d $\mathcal{N} = 2$ theory $T[M_3]$ is the three-dimensional SQED. In fact, for many 3-manifolds, the an theory $T[M_3]$

[8] And it has p isolated supersymmetric vacua on a 2-torus T^2. There are several ways to see this, e.g. from the supersymmetric index $\text{Tr}(-1)^F = |p|$ or by reducing on one circle at a time. Indeed, on a circle it gives a 2d $\mathcal{N} = (2,2)$ theory of a single twisted chiral multiplet with a twisted superpotential $\widetilde{W} = p\sigma^2/2$. The vacua of such theory are the critical points, i.e. solutions to

$$\exp\left(\frac{\partial \widetilde{W}}{\partial \sigma}\right) = 1, \tag{3.4}$$

and, in our case at hand, it is easy to see that there are p of them.

admits duality frames where its UV Lagrangian involves only abelian gauge group coupled to matter fields. In such duality frames, half-BPS impurities are basically electric or magnetic impurities of the 3d $\mathcal{N} = 2$ SQED studied, e.g. in [45, 46].

Let us consider an abelian 3d $\mathcal{N} = 2$ gauge theory with the gauge group $U(1)$, which is a basic building block of many 3d $\mathcal{N} = 2$ theories $T[M_3]$. A half-BPS electric impurity in this theory creates an equal source for the gauge field as well as for the real scalar field σ in the $U(1)$ vector multiplet (see Figure 3):

$$S_{\text{impurity}} = -\frac{i}{2\pi} \int d^3x \, (A_0 - \sigma) \rho(x). \tag{3.6}$$

Since A_0 and σ are both sourced by the electric charge density $\rho(x)$ and obey the same equations of motion, the BPS (resp. anti-BPS) solutions have $A_0 = \sigma$ (resp. $A_0 = -\sigma$). Indeed, the supersymmetry variation of the gaugino leads to the BPS equation

$$(\partial_i A_0 \gamma^0 - \partial_i \sigma)\epsilon = 0, \tag{3.7}$$

which is solved by $A_0 = \pm\sigma$ and the spinor ϵ that obeys $(1 \mp \gamma_0)\epsilon = 0$. In the presence of the source term (3.6), the equation of motion for the temporal component of the gauge field looks like

$$\frac{2\pi}{g^2} \partial_i^2 A_0 = \rho(x), \tag{3.8}$$

and σ obeys the same equation, since $A_0 = \sigma$. Later on, it will be convenient to replace the gauge coupling constant g^2 by the parameter $2\pi R = g^2$.

The simplest example of a half-BPS electric impurity is the familiar supersymmetric Wilson line. If we assume that a Wilson line (of charge q_0) is extended along the 'time'

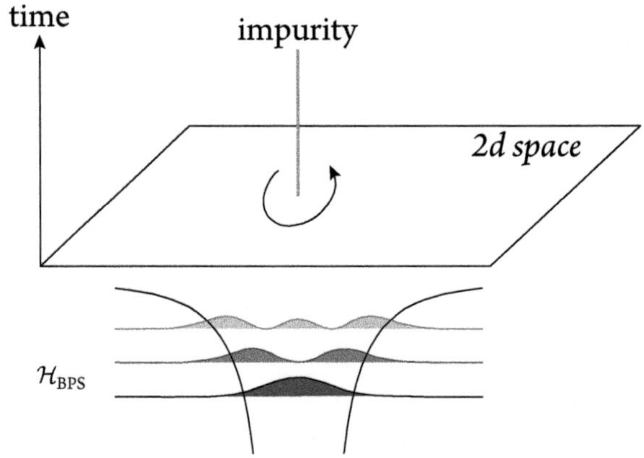

Figure 3 An illustration (from [44]) of a supersymmetric spectrum in 3d $\mathcal{N} = 2$ theory with an impurity.

direction and localized at the origin of the two-dimensional 'space' Σ, it can be represented by the charge density

$$\rho(z,\bar{z}) = 2\pi q_0 \delta^2(z,\bar{z}), \tag{3.9}$$

where z is a complex coordinate on Σ. (For the purposes of the present discussion, we can take Σ to be simply a complex plane, $\Sigma \cong \mathbb{C}$.) From the above equations, it is easy to see that such a delta-function source creates an equal profile for the fields A_0 and σ:

$$A_0 = \sigma = -\frac{1}{4\pi} g^2 q_0 \log |z|^2. \tag{3.10}$$

We can write it in a more convenient form, which will be suggestive of generalizations in 2d $\mathcal{N} = (0,2)$ theory $T[M_4]$, by introducing the dual photon X,

$$F_{\mu\nu} = \epsilon_{\mu\nu\rho} \partial^\rho X, \tag{3.11}$$

or, in holomorphic coordinates,

$$\partial_z X = i\partial_z A_0, \qquad \partial_{\bar{z}} X = -\partial_{\bar{z}} A_0. \tag{3.12}$$

Then, the charge quantization implies that $X \sim X + g^2$ is periodic with period g^2, and the solution (3.10) can be written as:

$$\sigma + iX = -\frac{g^2}{2\pi} q_0 \log(z). \tag{3.13}$$

In other words, in terms of $\sigma + iX$, the BPS condition is simply $\bar{\partial}(\sigma + iX) = 0$ away from the impurity. This is precisely the form of the BPS equations (2.34) in a 2d half-twisted (0,2) model on Σ that will be relevant to us in what follows; its solutions are the holomorphic maps (2.12).

Moreover, by considering $z \mapsto e^{i\theta} z$, it is easy to see that, going around an electric impurity of charge q_0, the dual photon has a monodromy $X \to X + q_0 g^2$. In other words, a half-BPS electric impurity creates a 'winding' of the dual photon and, therefore, in the language of a two-dimensional theory on Σ, it is a winding-state operator. In D-brane physics, a half-BPS solution (3.13) is known as the 'BIon' spike [47, 48], which describes charges carried by strings ending on D-branes and related configurations.

Magnetic impurities can be treated in a similar way and correspond to 'momentum' operators for the dual photon. For example, a half-BPS magnetic impurity (or, equivalently, an external vortex) in a 3d $\mathcal{N} = 2$ SQED is a source that equally couples to the magnetic field B and to the auxiliary field D. Thus, a delta-function source $B = 2\pi q_0 \delta^2(z,\bar{z})$ has $A = q_0 d\theta$ and defines a magnetic impurity of charge q_0. The 'mirror symmetry' of 3d $\mathcal{N} = 2$ gauge theories maps it to an electric impurity of the dual theory (see e.g. [45, 46] for more details).

Dimensional reduction of electric and magnetic impurities discussed here gives local impurity operators in 2d $\mathcal{N} = (2,2)$ theories. For example, dimensional reduction of BPS vortex operators gives the supersymmetric completions of the 'momentum' operators for the dual photon:

$$\exp\left(\pm 2\pi \frac{\sigma + iX}{g^2}\right), \tag{3.14}$$

which play an important role in mirror symmetry of 2d $\mathcal{N} = (2,2)$ theories [49].

In fact, our next goal will be to consider two-dimensional versions of such electric and magnetic impurities in theories with $\mathcal{N} = (0,2)$ supersymmetry, first in general and then focusing more closely on applications to $T[M_4]$.

3.2. Half-Twisted Model with Target Space $\mathbb{C}^* \simeq \mathbb{R} \times S^1$

Let us consider a simple 2d $\mathcal{N} = (0,2)$ theory which consists of a single chiral superfield Φ with period $2\pi iR$. (A reader less familiar with 2d $\mathcal{N} = (0,2)$ supersymmetry may find it convenient to think of a chiral superfield either in 3d $\mathcal{N} = 2$ theory or, via dimensional reduction, in 2d $\mathcal{N} = (2,2)$ theory; this will also help to make contact with the above discussion of impurities in 3d.) This simple theory is basically a model for $T[M_4]$ when $M_4 = S^2 \times S^2$ (cf. Section 11.2.1.1).

Let $\sigma + iX$ be a complex scalar in a 2d free chiral superfield Φ, such that X is periodic,

$$X \sim X + 2\pi R, \tag{3.15}$$

and, as usual, we can introduce its left-moving and right-moving components on-shell (see Figure 4):

$$X(z,\bar{z}) = X_L(z) + X_R(\bar{z}). \tag{3.16}$$

To construct winding and momentum operators in this theory, we can consider general vertex operators of the form

$$e^{i(k_L X_L + k_R X_R) + p\sigma}, \tag{3.17}$$

where, for a moment, we suppress fermions. Note, since X is periodic, the momenta k_L and k_R are quantized:

$$(k_L, k_R) = \left(\frac{n}{R} + \frac{wR}{2}, \frac{n}{R} - \frac{wR}{2}\right), \qquad n, w \in \mathbb{Z}. \tag{3.18}$$

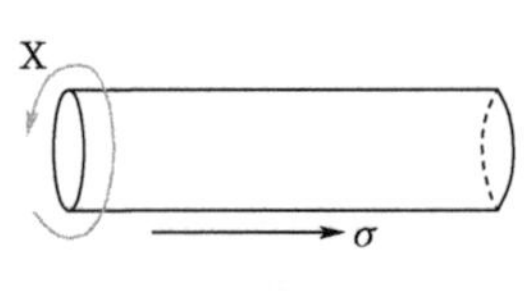

Figure 4 Target space $\mathbb{C}^* \simeq S^1 \times \mathbb{R}$ parametrized by $\sigma + iX$.

Moreover, as in our previous discussion, we are interested in BPS operators. In 2d $\mathcal{N}=(0,2)$ theory, these are the operators in $\overline{Q}_+$-cohomology of the right-moving $\mathcal{N}=2$ supersymmetry algebra. In particular, in our simple example, this imposes a condition on the right-moving components of the fields X and σ that they must appear in a holomorphic combination $\sigma + iX_R$ in order to be in $\overline{Q}_+$-cohomology. Such operators have $p = k_R$, and we conclude that (3.17) can be a bosonic representative of a BPS vertex operator when it is right-holomorphic,

$$e^{ik_L X_L + k_R(\sigma + iX_R)}. \tag{3.19}$$

Another way to arrive at the same conclusion is to note that general operators (3.17) have the scaling dimensions

$$h = \frac{1}{2}k_L^2 - \frac{1}{2}p^2, \qquad \overline{h} = \frac{1}{2}k_R^2 - \frac{1}{2}p^2, \tag{3.20}$$

and the BPS bound $\overline{h}=0$ also leads to $p = k_R$. (Vertex operators with $p = -k_R$ are anti-BPS.)

In what follows, we denote the supersymmetric completion of the right-holomorphic vertex operators (3.19) by V_λ and write

$$V_\lambda = e^{ik_L X_L + k_R \Phi}, \tag{3.21}$$

where $\lambda := (k_L, k_R)$, and Φ is a peculiar version of a 2d $\mathcal{N}=(0,2)$ chiral multiplet discussed in section 2; the lowest component of Φ is a complex scalar field $\sigma + iX_R$ whose real part is an ordinary, non-chiral scalar and the imaginary part is a chiral right-moving scalar. Note, this form of winding and momentum operators has an obvious generalization to a theory with multiple fields σ, X_L and X_R, as in Table 1. The only modification is that λ takes values in the winding/momentum lattice Γ, which in theories $T[M_4]$ can be identified with the cohomology lattice of the 4-manifold M_4:

$$\lambda := (k_L, k_R) \in \Gamma := H^2(M_4, \mathbb{Z}). \tag{3.22}$$

By construction, the supersymmetric operators V_λ have $\overline{h}=0$, and $h = \frac{1}{2}k_L^2 - \frac{1}{2}k_R^2$, which we sometimes write as

$$h = \lambda_-^2 - \lambda_+^2, \tag{3.23}$$

using the fact (2.2) that self-dual and anti-self-dual projections of λ correspond to projecting on the right-moving and left-moving modes, respectively. The supersymmetric completions of momentum and winding operators play an important role in mirror symmetry [49] and in various extensions of the sine-Liouville theory.[9]

[9] Note that, in Liouville theory, a correlation function of operators V_{λ_i} is finite if and only if these operators satisfy the Seiberg bound [50]:

$$\lambda_i < Q, \qquad \sum_i \lambda_i > 2Q. \tag{3.24}$$

One of the main goals of this paper is to relate Seiberg–Witten invariants of M_4 (and their generalizations) to correlation functions of the operators V_λ in the half-twisted theory $T[M_4]$. In the absence of a background charge, all such correlators are subject to a 'neutrality condition' which states that correlation functions vanish unless the total charge λ of the operators is equal to zero.[10] Sometimes, it will be convenient to consider the corresponding winding and momentum states:

$$|\lambda\rangle \equiv |k_L, k_R\rangle := V_\lambda|0\rangle \sim e^{ik_L X_L(0)+k_R\sigma(0)+ik_R X_R(0)}|0\rangle. \tag{3.26}$$

So far, our discussion of BPS winding-momentum operators V_λ has been fairly general and applies to a broad class of 2d $\mathcal{N}=(0,2)$ theories. Our next task is to gain a more detailed understanding of vertex operators in theories $T[M_4]$ by using the fivebrane set-up (1.3).

3.3. *Anomalies: M2-Branes Ending on M5-Branes and Embedded Surfaces*

In this section, we propose to identify vertex operators V_λ in 2d theory $T[M_4]$ with (non-dynamical) half-BPS self-dual strings in 6d (2,0) fivebrane theory—or, equivalently, with M2-branes ending on M5-branes, as illustrated in Figure 5—such that

$$\lambda = [D] \tag{3.27}$$

is the Poincaré dual cohomology class of the string world sheet, the embedded surface $D \subset M_4$. Then, we discuss important consequences of this identification, including e.g. application to knot cobordisms.

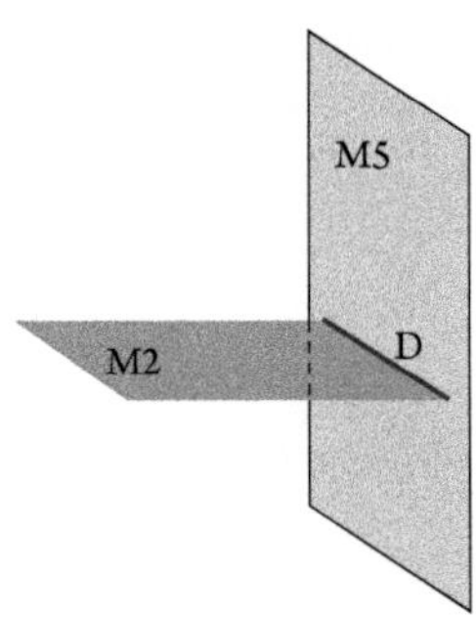

Figure 5 M2-branes ending on M5-branes.

In Liouville theory, normalizable states are called non-local, and non-normalizable states are called local; the latter correspond to curvature singularity on Σ.

10 Recall that, in a theory of a chiral boson X, vertex operators $V_\lambda = e^{i\lambda X(z)}$ have conformal weight $h = \frac{1}{2}\lambda^2$ and charge $Q(V_\lambda) = \lambda$ with respect to the $U(1)$ current $J = \partial X$:

$$Q = \frac{1}{2\pi}\oint dw J(w). \tag{3.25}$$

In this case, the 'neutrality condition' means that the total $U(1)$ charge must vanish.

First, since an M2-brane ending on a fivebrane acts as a source for the 3-form field H,

$$dH = \pi\delta^{(4)}(D \hookrightarrow M_4 \times \Sigma), \tag{3.28}$$

it follows from the Kaluza–Klein reduction (2.2) that scalar fields X_L^j and X_R^i are no longer single-valued, but rather have prescribed monodromies. And, V_λ are precisely the vertex operators that create such monodromies.

Second, both V_λ described above as well as M2-branes ending on $D \subset M_4$ preserve the same supersymmetry, namely they are both BPS with respect to the same supercharge $\overline{Q}_+$, regardless of the choice of D. We already demonstrated this for the vertex operators V_λ and it is also true for the non-dynamical self-dual strings in 6d $(2,0)$ fivebrane theory. The reason is that 6d theory is topologically twisted along the 4-manifold M_4, so that any choice of metric on M_4 and any shape of $D \subset M_4$ preserve the topological supercharge.

From the vantage point of the brane system (1.3), the topological twist along the 4-manifold M_4 is implemented by embedding this part of the fivebrane world volume in the G_2 manifold $\Lambda^{2,+}(M_4)$. In this set-up, non-dynamical self-dual strings supported on $D \subset M_4$ are realized by M2-branes supported on associative 3-manifolds in $\Lambda^{2,+}(M_4)$ that meet M_4 along D. In other words, given a surface $D \subset M_4$, we wish to associate to it an associative submanifold inside $\Lambda^{2,+}(M_4)$ that meets M_4 along D. The geometry of such associative submanifolds was briefly discussed in [51] and follows the 'conormal bundle' construction of [52] when $M_4 = \mathbb{R} \times M_3$.

Recall that, in general, for a submanifold $K \subset M$, the conormal bundle $N^*K \subset T^*M$ is a subbundle defined as

$$N^*K := \{(x,p) \in T^*M \mid x \in K, T_xK \subset \ker(p)\}. \tag{3.29}$$

Similarly, given a two-dimensional surface $D \subset M_4$, the span of $\mathrm{vol}_D + *_4\mathrm{vol}_D$ defines a line bundle $L_D \subset \Lambda^{2,+}(M_4)$, where vol_D is the induced volume form on D. The total space of this line bundle defines an associative submanifold, which meets the coassociative 4-manifold M_4 along D. Indeed, by examining[11] the associative 3-form Φ and the coassociative 4-form $*\Phi$, it is easy to see that associative and coassociative submanifolds can meet only over subsets of dimension 0 or 2.

Now, it is easy to see that M2-branes supported on this associative submanifold indeed preserve one supercharge $\overline{Q}_+$. In the set-up (1.3), the M-theory reduced on $\Lambda^{2,+}(M_4)$

[11] Locally,

$$\Phi = e^{567} - (e^{12} + e^{34}) \wedge e^5 - (e^{13} + e^{42}) \wedge e^6 - (e^{14} + e^{23}) \wedge e^7, \tag{3.30}$$

$$*\Phi = e^{1234} + (e^{12} + e^{34}) \wedge e^{67} + (e^{13} + e^{42}) \wedge e^{75} + (e^{14} + e^{23}) \wedge e^{56}. \tag{3.31}$$

Associative 3-manifolds are defined by the condition that Φ restricts to a volume form and coassociative 4-manifolds are defined by a similar condition with respect to $*\Phi$. Equivalently, an associative submanifold of X can be characterized by vanishing of the restriction of the TX-valued 3-form $\chi \in \Omega^3(X, TX)$ defined by the identity [53]

$$\langle \chi(u,v,w), z \rangle = *\Phi(u,v,w,z). \tag{3.32}$$

preserves; in the remaining $\mathbb{R}^4$, the 4d $\mathcal{N} = 1$ supersymmetry was parametrized by a spinor ϵ. The fivebrane supported on a coassociative cycle in a G_2 manifold $\Lambda^{2,+}(M_4)$ preserves half of it, namely

$$\gamma_{01}\epsilon = \epsilon, \tag{3.33}$$

where $x^{0,1}$ are the directions along the non-compact part of the M5-brane that we call Σ. This is precisely 2d $\mathcal{N} = (0,2)$ supersymmetry of the theory $T[M_4]$ on Σ. Introducing an extra M2-brane supported on an associative 3-manifold breaks supersymmetry further. The corresponding condition (where γ_5 is the 4d chirality gamma-matrix),

$$\gamma_5\epsilon = \epsilon, \tag{3.34}$$

leaves one Weyl spinor in 4d (on which the full 4d Lorentz algebra is still represented) and then (3.33) leaves one of its components which has the required spin.

Now, once we covered the 'charge' $\lambda \in \Gamma$ and supersymmetry of the M2-branes ending on fivebranes, our next task is to discuss anomalies of the 2d degrees of freedom localized on the membrane boundary $D \subset M_4$. For applications to Seiberg–Witten invariants, we are mostly interested in the set-up (1.3) with a single fivebrane (i.e. $N = 1$) and a single M2-brane ending on it, as illustrated in Figure 5. Let $(x^0, x^1, x^2, x^3, x^4, x^5)$ be the coordinates along the M5-brane world volume and (x^4, x^5, x^6) be the coordinates along the M2-brane world volume:

	T				D			N			
Brane	0	1	2	3	4	5	6	7	8	9	10
$M5$	x	x	x	x	x	x					
$M2$					x	x	x				
	Σ		M_4				$\Lambda^{2,+}$			R	

(3.35)

Before we implement partial topological twists along Σ and M_4 by embedding this brane system in a curved background, we can start with branes in flat space. Such a configuration has a symmetry group $SO(2)_D \times SO(4)_T \times SO(4)_N$ where, following [54], we denote

$$\begin{aligned} SO(2)_D &:= SO(2)_{45}, \\ SO(4)_T &:= SO(4)_{0123}, \\ SO(4)_N &:= SO(4)_{78910}. \end{aligned} \tag{3.36}$$

After the topological twist, N and T are replaced by non-trivial $SO(4)$ bundles. For example, in our system, $T = T\Sigma \oplus N_D$, where N_D is the normal bundle to D in M_4.

In general, the anomaly polynomial of the self-dual string with world-sheet $D \subset M_4$, i.e. M2-brane ending on a fivebrane, is given by [54, 55] (see also [56]):

$$A_{M2} = e(N) - e(T), \tag{3.37}$$

where $e(T)$ is the Euler class of the $SO(4)_T$ bundle T, and similarly for the 'normal bundle' N. For simplicity, let us assume that M_4 is a complex surface. Then, using the information about the topological twist along the M_4 in our system, we can write

$$A_{M2} = 2c_1(TM_4)c_1(R) - c_1(N_D)c_1(T\Sigma), \tag{3.38}$$

where R is the $U(1)_R$ symmetry bundle and $T\Sigma$ is the $U(1)_q$ line bundle. In identifying $c_1(R)$ with the first Chern class of the $SO(2)_{910}$ bundle, we introduced a factor of 2 in order to match the standard normalization of R-charges. Integrating over D, we conclude that local BPS operators on Σ constructed from M2-branes ending on a fivebrane carry the R-charge

$$R(D) = 2\int_D c_1(TM_4) \tag{3.39}$$

and 2d spin

$$\deg_q(D) = -\int_D c_1(N_D) = \chi(D) - \int_D c_1(TM_4), \tag{3.40}$$

where we used $c_1(TD) + c_1(N_D) = c_1(TM_4)$. Note, since $\deg_q = h - \overline{h}$ and BPS operators in 2d $\mathcal{N} = (0,2)$ theory have $\overline{h} = \frac{R}{2}$, it follows that M2-branes ending on fivebranes along surface D have $\overline{h} = \int_D c_1(TM_4)$ and

$$h = \chi(D). \tag{3.41}$$

Equivalently, this expression gives the value of $h - \overline{h}$ in the half-twisted model on Σ.

As a consistency check of this anomaly calculation, we now consider 'surface operators' (or 'foams') in application of this formalism to knot homology. In particular, using this approach, we make a concrete prediction for the q-degree of foams coloured by symmetric representations of $sl(2)$; this prediction, to the best of our knowledge, has not appeared in math or physics literature so far.

3.3.1. Foams and knot cobordisms

Although somewhat outside our main line of development, it is instructive to consider the configuration (1.3) with multiple fivebranes ($N > 1$). When considered on a 4-manifold of the form $M_4 = \mathbb{R} \times M_3$, it provides a physical realization of the Khovanov–Rozansky $sl(N)$ homology and its generalizations (see e.g. [44] and references therein). For the purposes of the following discussion we can simply take $M_4 = \mathbb{R}^4$.

In application to knot homologies, the type of the surface operator or M2-brane boundary determines the 'colour' of the foam D, which can be either a knot cobordism or a closed surface. In part to establish the dictionary, let us start with a relatively well-understood *uncoloured* version and then make predictions for *coloured* foams and knot cobordisms.[12] In

[12] The terminology is such that 'uncoloured' refers to knots and their cobordisms coloured by the fundamental representation, whereas invariants associated with more general representations are called 'coloured'.

the case of a fundamental N-dimensional representation of $sl(N)$, the 2d TQFT (Frobenius algebra) that lives on D is a B-model based on the Landau-Ginzburg potential $W = x^{N+1}$ [6]. In particular, a pair-of-pants cobordism $D =$ 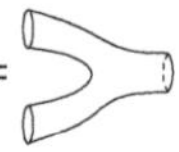 defines a product on the cohomology of the unknot,

$$\mathcal{H}(\bigcirc) = \mathbb{C}[x]/\langle x^N \rangle. \tag{3.42}$$

Note, this is the *classical* cohomology of $\mathbf{CP}^{N-1}$. The chiral ring of the 2d TQFT on the surface operator D is generated by the 'observable' x. In the convention $\deg_q(x) = 2$, a genus g surface D has $\deg_q(D) = 2(N-1)(g-1)$, so that the closed surfaces which have non-zero evaluation are either of genus zero with $N-1$ observables, or genus one with no extra insertions [15–20]:

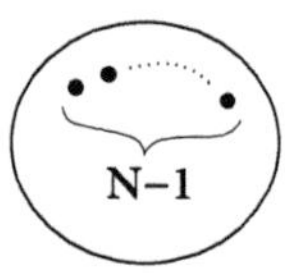

$= 1,$  $= N. \quad (3.43)$

For a generalization to foams coloured by antisymmetric representations of $sl(N)$ see e.g. [57, 58].

This structure is in perfect agreement with anomalies of M2-branes ending on fivebranes. Consider, for example, the case of $N = 2$, just as in the ordinary Khovanov homology. This means we have to study M2-brane boundaries/self-dual strings in A_1 fivebrane theory. Using the anomaly polynomial of such strings (with multiplicity r) [59],

$$A_{M2} = r^2 e(T) - re(N), \tag{3.44}$$

we can repeat the steps that led to (3.39) and (3.40) and conclude that a multiplicity-r closed surface in $SU(2)$ theory has zero R-charge (= homological grading) and

$$\deg_q(D) = r^2 \chi(D). \tag{3.45}$$

3.4. Anomalies: Intersecting M5-Branes and Basic Classes

Now we are ready to introduce the second key ingredient needed for realizing Seiberg–Witten invariants in 2d theory $T[M_4]$, namely a vertex operator localized at a point on Σ which, if the order of compactification was reversed, would produce a charged hypermultiplet in 4d $\mathcal{N} = 2$ theory on M_4.

Indeed, if we start with 6d (2,0) tensor multiplet on a single fivebrane and first reduce it on Σ, we obtain precisely the construction [43] of 4d $\mathcal{N} = 2$ low-energy effective physics with Seiberg–Witten curve Σ. From the perspective of this low-energy 4d $\mathcal{N} = 2$ gauge theory, a further topological twist along the 4-manifold M_4 (which corresponds to embedding it as a coassociative 4-cycle in G_2 geometry) can be identified with the standard topological twist which leads to Donaldson or Seiberg–Witten invariants (see e.g. [21, sec. 5]):

$$\begin{array}{ccc} & \text{6d (2,0) theory} & \\ & \text{on } \Sigma \times M_4 & \\ \swarrow & & \searrow \\ \text{2d half-twisted} & = & \text{4d } \mathcal{N}=2 \text{ theory} \\ \text{model on } \Sigma & & \text{twisted on } M_4. \end{array} \tag{3.46}$$

After reduction on Σ, the resulting 4d $\mathcal{N}=2$ theory has R-symmetry $U(1)_R \times SU(2)_R$. Let R and E denote the corresponding $U(1)_R$ and $SU(2)_R$ bundles. Then, much like its simpler version (2.19) that we encountered earlier, the anomaly polynomial of a general 4d $\mathcal{N}=2$ superconformal theory is a degree 6 polynomial in characteristic classes of R and E, whose coefficients are determined by the conformal central charges, a and c (see e.g. [60, 61]):

$$A_{4d} = (n_v - n_h)\left(\frac{c_1(R)}{12}p_1(TM_4) - \frac{c_1(R)^3}{3}\right) - n_v c_1(R)c_2(E) + \frac{k}{2}c_1(R)\mathrm{ch}_2(F). \tag{3.47}$$

Here, F is the curvature of a G-bundle over M_4, $n_v = 8a - 4c$ and $n_h = 20c - 16a$ denote the effective number of vector and hypermultiplets, respectively,[13] and the central charge k of the symmetry group G is defined via OPE of two G-symmetry currents:

$$J^a_\mu(x)J^b_\nu(0) = \frac{3k}{4\pi^4}\delta^{ab}\frac{x^2 g_{\mu\nu} - 2x_\mu x_\nu}{x^8} + \cdots. \tag{3.48}$$

Since the R-symmetry $U(1)_R$ of the 4d $\mathcal{N}=2$ theory on M_4 is precisely the R-symmetry (under the same name) of 2d $\mathcal{N}=(0,2)$ theory $T[M_4]$ on Σ, we can determine $U(1)_R$ anomaly of the latter by integrating (3.47) over M_4, with $G = U(1)$ and $E = \Lambda^{2,+}(M_4)$. In particular, if we choose Σ such that 4d $\mathcal{N}=2$ theory is a theory of a single vector multiplet, i.e. $n_v = 1$ and $n_h = 0$, we find precisely the $U(1)_R$ anomaly (3.49) of the theory $T[M_4]$, computed earlier from the 2d perspective:

$$\Delta R_{T[M_4]} = -\frac{\chi + \sigma}{2}, \tag{3.49}$$

where we used

$$\frac{1}{12}\int_{M_4} p_1(TM_4) = \frac{\sigma}{4}, \qquad \int_{M_4} c_2(E) = \frac{2\chi + 3\sigma}{4}, \tag{3.50}$$

when $E = \Lambda^{2,+}(M_4)$.

[13] In terms of the conformal anomaly coefficients, $c = \frac{1}{6}n_v + \frac{1}{12}n_h$ and $a = \frac{5}{24}n_v + \frac{1}{24}n_h$, the same expression reads

$$A_{4d} = (a-c)\left(2c_1(R)p_1(TM_4) - 8c_1(R)^3\right) - (8a-4c)c_1(R)c_2(E) + \frac{k}{2}c_1(R)\mathrm{ch}_2(F).$$

Similarly, an impurity on Σ that adds a charged hypermultiplet to 4d $\mathcal{N}=2$ theory on M_4 has $(n_v, n_h) = (0,1)$ and, according to (3.47), changes the $U(1)_R$ anomaly by

$$\Delta R_{\text{impurity}} = \frac{\lambda^2 - \sigma}{4}. \tag{3.51}$$

In what follows, we denote this kind of impurity by $\mathcal{S}_+$ to distinguish it from the impurity $\mathcal{S}$ defined by the intersection of two M5-branes. Combining this with (3.49), we learn that a Riemann surface Σ that, via reduction (3.46), engineers N_f topologically twisted hypermultiplets charged under the $U(1)$ vector multiplet on M_4 has $U(1)_R$ anomaly:

$$\Delta R_{N_f} = \frac{N_f(\lambda^2 - \sigma) - 2(\chi + \sigma)}{4}. \tag{3.52}$$

In particular, in the special case $N_f = 1$, we recover the well-known expression for the virtual dimension of the moduli space of solutions to the Seiberg–Witten equations [62, 63]:

$$\text{VirDim}\, \mathcal{M}_{\text{SW}}(\lambda) = \frac{\lambda^2 - 2\chi - 3\sigma}{4}, \tag{3.53}$$

whereas the general formula (3.52) gives a (less familiar) expression [64] for the dimension of the moduli space of generalized multi-monopole solutions that will be the subject of section 4. Note, in gauge theory on a 4-manifold M_4, the expression (3.53) is (minus) the index of the deformation complex for the Seiberg–Witten monopole equations:

$$\begin{aligned} 0 \longrightarrow \Omega^0(M_4) &\longrightarrow \Omega^1(M_4) \oplus \Gamma(M_4, S^+ \otimes L^{1/2}) \\ &\longrightarrow \Omega^{2,+}(M_4) \oplus \Gamma(M_4, S^- \otimes L^{1/2}) \longrightarrow 0. \end{aligned} \tag{3.54}$$

It splits into a complex for the Dirac operator, with the familiar index $\frac{\lambda^2-\sigma}{4}$, and a simple complex for the anti-self-duality equation (2.1), with index $-b_0 + b_1 - b_2^+ = -\frac{1}{2}(\chi + \sigma)$. Note, these two contributions are precisely the contributions to $U(1)_R$ anomaly from the impurity (3.51) and from 2d theory (3.49), respectively. Adding them together, we obtain (3.53).

As a by-product of deriving (3.52), we also obtain the anomaly polynomial of two intersecting fivebranes:

$$A_{\text{M5} \cap \text{M5}'} = \frac{1}{2} c_1(R) c_2(E) + \frac{1}{2} c_1(T\Sigma) c_2(E). \tag{3.55}$$

Indeed, the half-BPS impurity operator $\mathcal{S}$ is realized by a codimension 2 defect in 6d (2,0) theory or, equivalently, by intersecting M5-branes in the M-theory set-up (1.3) (cf. (3.35)):

	Σ		M_4				E			R	
Brane	0	1	2	3	4	5	6	7	8	9	10
M5	x	x	x	x	x	x					
M5′			x	x	x	x				x	x

(3.56)

By using (3.46) and analysing how the Kaluza–Klein spectrum of 4d $\mathcal{N}=2$ theory on M_4 depends on the topology of Σ, it is easy to see that a puncture on Σ effectively carries $n_v = -\frac{1}{2}$ and $n_h = -\frac{1}{2}$. Then, from (3.47) and the symmetry between two fivebranes in (3.56) we obtain the anomaly polynomial (3.55).

Note that, for $E = \Lambda^{2,+}(M_4)$, integrating $c_2(E)$ over M_4 in the first term of (3.55) gives precisely the 'constant' term $2\chi + 3\sigma$ in (3.53). Alternatively, from the second term in (3.55) we can find the 2d spin of the impurity operator $\mathcal{S}$ defined as the intersection of two M5-branes. Comparing it with the spin of V_λ, we obtain the condition $\lambda^2 = 2\chi + 3\sigma$ on basic classes in Seiberg–Witten theory.

In this section, we described the R-charges of the impurity operators $\mathcal{S}$ and $\mathcal{S}_+$ that will play an important role in the rest of this paper. While it would be desirable to give a more detailed 2d characterization of these vertex operators (and their various generalizations, e.g. to Argyres–Douglas theories), what we have so far suffices to describe structural properties of their correlation functions. By matching the latter with a known answer in one simple case, say for $G = U(1)$ and $N_f = 1$, then determines more general correlators of $\mathcal{S}$ and $\mathcal{S}_+$, which in turn lead to new concrete predictions for 4-manifold invariants. This will be our strategy in the next subsection.

3.5. *Gauge Theoretic Invariants of 4-Manifolds as 2d Correlators*

We are ready to make a proposal for the structure of 4-manifold invariants, old and new, by putting together all of the above ingredients: half-twisted theory $T[M_4]$ summarized in section 2, winding-momentum vertex operators (section 3.3) and BPS impurities (section 3.4).

As in [21], it will be useful to consider the following generating function of Seiberg–Witten invariants:

$$\widehat{\mathrm{SW}}(t) \;=\; \sum_{\lambda \in H^2(M_4,\mathbb{Z})} \mathrm{SW}(\lambda)\, t^\lambda, \tag{3.57}$$

where the expansion parameter t takes values in $\mathrm{Hom}(H^2(M_4,\mathbb{Z}),\mathbb{C}^*)$. One can use it to define the following local BPS operator in $T[M_4, U(1)]$:

$$\mathcal{S}(z) \;:=\; \widehat{\mathrm{SW}}(e^{iX(z)}) \equiv \sum_\lambda \mathrm{SW}(\lambda)\, V_\lambda(z), \tag{3.58}$$

where

$$X(z) \equiv (X_L(z), X_R(z) + i\sigma(z)) \in H^{2,-}(M_4) \oplus (H^{2,+}(M_4) \otimes \mathbb{C}) \subset H^2(M_4,\mathbb{C}) \overset{Q}{\cong} H_2(M_4,\mathbb{C}), \tag{3.59}$$

so that $e^{iX(z)} \in \exp iH_2(M_4,\mathbb{C}) \subset \mathrm{Hom}(H^2(M_4,\mathbb{Z}),\mathbb{C}^*)$. Since we are interested in half-twisted correlators and the chiral ring of $T[M_4]$, we always consider $\mathcal{S}(z)$ and other BPS vertex operators modulo $\overline{Q}_+$-exact terms. With this proviso, the insertion of the local operator $\mathcal{S}(z)$ at a point in the 2d space-time Σ of $T[M_4]$ corresponds to the insertion of the fivebrane that also wraps M_4 but has two directions orthogonal to Σ, as in the set-up (3.56):

$$\mathcal{S}(z) \quad \sim$$ 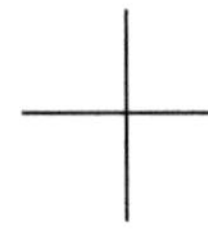 , (3.60)

so that (3.58) can be interpreted as a 'decomposition' of a codimension 2 defect in terms of codimension 4 defects in the fivebrane theory on M_4, again, up to $\overline{Q}_+$-exact terms. This can be justified as follows.

3.5.1. Seiberg–Witten invariants from half-twisted correlators

Consider the standard brane realization of the 4d $\mathcal{N}=2$ gauge theory with gauge group $U(1)$ and one charged hypermultiplet shown on Figure 6 (left panel). In flat space, the $SU(2)_R$ triplet of the FI parameters corresponds to the relative position of two NS5 branes in the directions 678. After the topological twist along a 4-manifold M_4, the directions 678 become the fibre directions of the $\Lambda^{2,+}(M_4)$ bundle. And the FI parameter η becomes an element of $H^{2,+}(M_4) \subset H^2(M_4,\mathbb{R}) \overset{Q}{\cong} H_2(M_4,\mathbb{R})$ (complexified). In the case of $M_4 = M_3 \times S^1$, this set-up was considered in [21, 65]. Setting $t = e^{i\eta}$, the partition function of the 4d Seiberg–Witten theory then reads

$$\sum_{\lambda \in H^2(M_4,\mathbb{Z})} \mathrm{SW}(\lambda)\, e^{i\eta\cdot\lambda} \equiv \widehat{\mathrm{SW}}(t), \tag{3.61}$$

where the sum represents the sum over all Spin^c structures in the path integral.[14]

Note, once we turn on the FI parameter, the relative position of the NS5 brane in the direction x^9 becomes irrelevant, and all supersymmetric configurations that contribute to the partition function (3.61) are represented by D2 branes stretched between NS5$'$ and D4 branes, as shown on the right side of Figure 6. In the Higgs phase of the 4d $\mathcal{N}=2$ gauge theory, these D2 branes correspond to supersymmetric vortex solutions [66] supported on embedded surfaces inside M_4. The mathematical counterpart of this statement is that Seiberg–Witten invariants of M_4 are equal to Gromov invariants that count (vortices localized on) embedded surfaces. These two phases are related by a configuration with vanishing FI parameters, in which supersymmetric contributions to the partition function (3.61) are localized at the intersection of NS5$'$ and D4 branes. In M-theory, it is lifted to the intersection of M5 and M5$'$ described in (3.56).

The same partition function is given by a correlator of local operators in $T[M_4]$. In the M-theory lift, the brane NS5$'$ is a fivebrane on $M_4 \times \Sigma$, where $\Sigma = \mathbb{C}$ spans directions 01. Take $X(z)$ to be the collection of fields (3.59) of $T[M_4]$ living on Σ. Then, the components of $X(z)$ valued in $H^{2,+}(M_4) \otimes \mathbb{C}$ describe the local (complexified) position of the brane

[14] For simplicity, we assume that $w_2 = 0$.

in the fibre directions of $\Lambda^{2,+}(M_4)$. Namely, 'position' refers to a choice of the harmonic section of $\Lambda^{2,+}(M_4)$. This is same space where the FI parameter takes values. Although we can have local fluctuations of the position of the NS5′ brane, the brane is infinite and its position at infinity should be fixed (since the movement of the whole brane would require infinite energy). As was discussed earlier, its position relative to other branes is given by the FI parameter η. Similarly, we need to fix values of other fields at infinity of $\Sigma = \mathbb{C}$, in particular $X_L(\infty) = 0$. So, the partition function should be given by the following 2d correlator:

$$\begin{aligned} &\langle \mathcal{S}(z_1)\, \delta(X(\infty) - \eta) \rangle \\ &\quad = \sum_{\lambda} \mathrm{SW}(\lambda) \sum_{\mu} \langle e^{i\lambda \cdot X(z_1)} e^{-i\mu\cdot(X(\infty)-\eta)} \rangle = \sum_{\lambda} \mathrm{SW}(\lambda)\, t^{\lambda}, \end{aligned} \tag{3.62}$$

where z_1 is the position of the brane intersection on Σ and we used

$$\langle e^{i\lambda\cdot X(z_1)}\, e^{-i\mu\cdot X(\infty)} \rangle = \delta(\mu - \lambda), \tag{3.63}$$

up to normalization. This indeed reproduces (3.61). The insertion of $e^{-i\mu\cdot X(\infty)}$ in the correlation function can be interpreted as the 'background charge' at infinity.

Note, when the FI parameter is turned off (as on the left panel of Figure 6), there is a continuous Coulomb branch modulus that corresponds to the position of the D4 brane stretched between two NS5 brane. When theory is put on a compact space, one has to integrate over it. Turning on the FI parameter corresponds to localization of this integral to the origin of the Coulomb branch, the point where the two D4 branes align.

3.5.2. Multi-monopole invariants

Consider now $U(1)$ theory with N_f hypermultiplets with non-zero masses z_i. The corresponding brane construction is shown in Figure 7. Turning on the FI parameter localizes the Coulomb branch (u-plane) integral on one of N_f possible brane configurations where the D4 brane stretched between the NS5 branes aligns with one of the semi-infinite D4 branes with mass z_i. In M-theory, the NS5′ brane is lifted again to an M5 brane on $M_4 \times \Sigma$, where $\Sigma = \mathbb{C}$ spans directions (x^0, x^1). The masses of hypermultiplets correspond to positions z_j of semi-infinite D4 branes on Σ. The partition function of the fivebrane then has the following realization in terms of $T[M_4]$ on Σ:

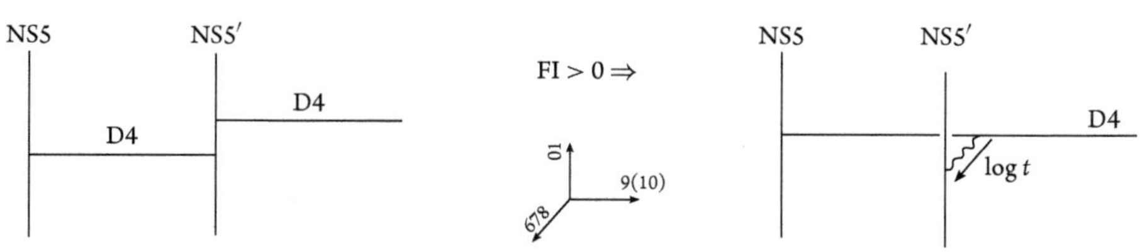

Figure 6 Brane construction à la [43] of an ordinary Seiberg–Witten theory (with $N_f = 1$) and its Higgs phase with non-zero FI parameter.

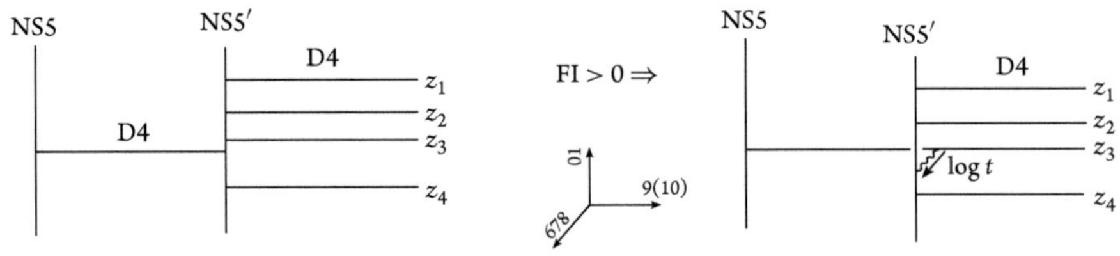

Figure 7 Brane construction of $U(1)$ theory with N_f hypermultiplets and its Higgs phase with non-zero FI parameter.

$$\langle \mathcal{S}(z_i) \prod_{j\neq i} \mathcal{S}_+(z_j)\, \delta(X(\infty)-\eta)\rangle = \sum_{\mu\in H^2(M_4,\mathbb{Z})} \langle \mathcal{S}(z_i) \prod_{j\neq i} \mathcal{S}_+(z_j)\, e^{-i\mu\cdot X(\infty)}\rangle\, t^{\mu}, \tag{3.64}$$

where $\mathcal{S}_+(z)$ is the operator that corresponds to adding a semi-infinite D4 brane located at z. Schematically,

$$\mathcal{S}_+(z) \quad \sim \tag{3.65}$$

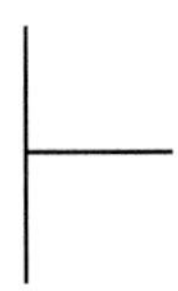

In M-theory lift, it corresponds to creating a semi-infinite neck (cusp) on Σ. Inserting such operators at points z_j, $j \neq i$ is equivalent to deforming the geometry of Σ from $\mathbb{C} \cong \{y = \text{const}\} \subset \mathbb{C} \times \mathbb{C}_*$ to

$$\left\{ y = \text{const} \prod_{j\neq i} (z - z_j) \right\} \subset \mathbb{C} \times \mathbb{C}^*, \tag{3.66}$$

where $z = x_0 + ix_1$ and $y = e^{x_9 + ix_{10}}$ are the coordinates parametrizing $\mathbb{C}$ and $\mathbb{C}^*$, respectively.

The total partition function of the 4d theory on M_4 should be given by the sum over N_f such configurations:

$$\sum_{i=1}^{N_f} \langle \mathcal{S}(z_i) \prod_{j\neq i} \mathcal{S}_+(z_j)\, \delta(X(\infty)-\eta)\rangle. \tag{3.67}$$

In particular, using (3.64), we obtain the structure of the multi-monopole invariant of M_4 with a given Spin^c structure λ:

$$\sum_{i=1}^{N_f} \langle \mathcal{S}(z_i) \prod_{j\neq i} \mathcal{S}_+(z_j)\, e^{-i\lambda\cdot X(\infty)}\rangle. \tag{3.68}$$

From the vantage point of the topological theory on M_4, the parameters z_i can be understood as the equivariant parameters for the maximal torus of $SU(N_f)$ symmetry. This will

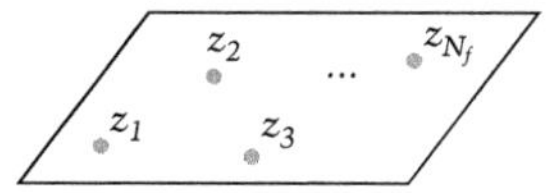

Figure 8 Impurity vertex operators at points z_i, $i = 1,\dots,N_f$.

be the subject of section 4, where we also present a verification of the prediction (3.68) by direct calculation in gauge theory.

Note, the 2d correlator (3.68) that computes the desired topological invariant of M_4 has another representation, which is completely symmetric in all N_f impurity operators and corresponds to the Coulomb phase/brane configuration in the left panel of Figure 7:

$$\langle \oint \frac{dz}{2\pi i} J(z)\, \mathcal{S}_+(z_1)\cdots \mathcal{S}_+(z_{N_f})\, e^{-i\lambda\cdot X(\infty)} \rangle. \tag{3.69}$$

Here, the z-integral is performed along a large contour that encloses all of the points z_i, $i = 1,\dots,N_f$, and

$$J(z)\mathcal{S}_+(w) \sim \frac{1}{z-w}\mathcal{S}(w) + \text{regular}. \tag{3.70}$$

In other words, $\mathcal{S}(w) = [\mathcal{Q}, \mathcal{S}_+(w)]$, where the charge $\mathcal{Q} = \frac{1}{2\pi i}\oint J(z)\, dz$. Deforming the integration contour into N_f smaller contours, each of which encircles only one of the vertex operators $\mathcal{S}_+(z_i)$, $i = 1,\dots,N_f$, we obtain (3.68)

$$\langle \oint \frac{dz}{2\pi i} J(z)\, \mathcal{S}_+(z_1)\cdots \mathcal{S}_+(z_{N_f})\, e^{-i\lambda\cdot X(\infty)} \rangle = \sum_{i=1}^{N_f} \langle \mathcal{S}(z_i) \prod_{j\neq i} \mathcal{S}_+(z_j)\, e^{-i\lambda\cdot X(\infty)} \rangle. \tag{3.71}$$

It would be interesting to give a more intrinsic 2d characterization of the impurity vertex operators $\mathcal{S}$ and $\mathcal{S}_+$, along with an explanation of the relation (3.70) between them. We hope to return to this problem in the future work.

3.5.3. Multiple $U(1)$ groups and multiple monopoles

Consider another example, a $U(1)^n$ quiver gauge theory, which is realized by a brane configuration with $n+1$ parallel NS5 branes shown in Figure 9. The matter consists of $(n-1)$ bifundamental hypermultiplets and one fundamental hyper. Denote the FI parameters for $U(1)$ gauge factors by η_j and their exponentiated versions by $t_j = e^{i\eta_j}$. They correspond to relative positions of adjacent NS5 branes in the directions 678.

As in our previous examples, turning on the FI parametes for all of the $U(1)$ gauge factors gives us $T[M_4]$ on n disconnected copies of $\mathbb{C}$, each with one insertion of the impurity operator $\mathcal{S}(z)$. Therefore, the partition function should be given by the product of n copies of the partition function (3.61), with FI parameters given by the positions of NS5 branes relative to the semi-infinite D4 brane:

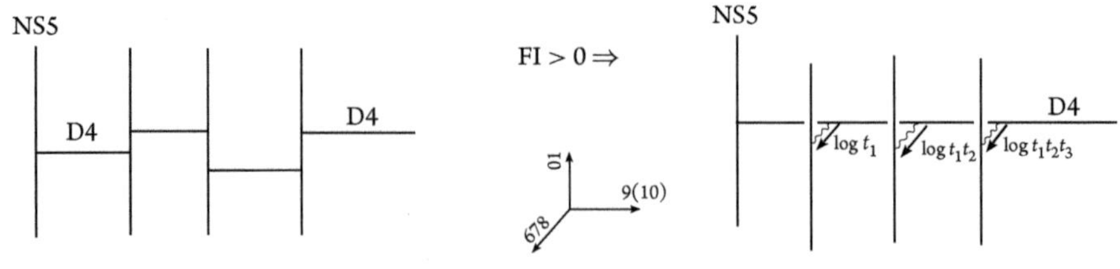

Figure 9 Brane construction of a $U(1)^n$ quiver gauge theory with $n = 3$, and its Higgs phase with non-zero FI parameters.

$$\widehat{\mathrm{SW}}(t_1)\,\widehat{\mathrm{SW}}(t_1t_2)\cdots\widehat{\mathrm{SW}}(t_1t_2\cdots t_n). \tag{3.72}$$

This prediction is consistent with the fact that the quiver theory described here is dual to the tensor product of n copies of $U(1)$ gauge theory with $N_f = 1$. The equivalence between theories and the corresponding identification of the FI parameters can be seen by integrating out $U(1)$ gauge multiplets one by one, starting from the left end of the quiver.

Hopefully, by now it is clear how to use the basic rules (3.60) and (3.65) to convert gauge theoretic invariants of 4-manifolds to correlators of impurity operators $\mathcal{S}(z)$ and $\mathcal{S}_+(z)$. We present one more illustration in section 5, with a proposal for the structural properties of non-abelian gauge theoretic invariants, and now focus on verification of the abelian ones.

4. Equivariant Multi-Monopole Invariants

In this section, we explore 4-manifold invariants which, on the one hand, share some tractability of the Seiberg–Witten invariants and, on the other hand, provide the simplest context in which one can concretely see the connection with 2d correlators. We then compute the new equivariant invariants for many 4-manifolds and confirm that they indeed have the expected structure (3.71).

4.1. *A Cure for Non-Compactness*

Very much like 1-monopole Seiberg–Witten equations, the multi-monopole equations on M_4 are formulated in terms of a $U(1)$ connection A and N_f commuting left-handed (i.e. positive chirality) spinors Ψ_i, $i = 1,\ldots,N_f$. As usual, one can combine the Spin-connection and ($\frac{1}{2}$ times) the $U(1)$ connection into a single Spin^c connection and thus see that only a Spin^c structure is required, which exists on an arbitrary 4-manifold M_4. Of course, in all local expressions we are free to separate Spin^c connection into a Spin and a $U(1)$ part, remembering that only their sum makes sense globally, unless M_4 is a Spin manifold.

The Spin^c structure on M_4 is given by a pair of $U(2)$ bundles $W^\pm$ with a Clifford multiplication map $c : \Omega^1(M_4) \otimes \Gamma(M_4, W^\pm) \to \Gamma(M_4, W^\mp)$. The determinant bundle of W^+ is denoted L, and A is a connection on L. Also, we denote $c_1(L) = \lambda \in H^2(M_4)$ and abbreviate $Q^{-1}(\lambda,\lambda) \equiv \lambda^2$. (This is the same λ as in the previous two sections.) With $\Psi_i,\ldots,\Psi_{N_f} \in \Gamma(M_4, W^+)$, the N_f-monopole Seiberg–Witten equations are written as

$$F_A^+ = i\sum_{i=1}^{N_f} (\Psi_i \overline{\Psi}_i)_0, \qquad (4.1)$$
$$\not{D}\Psi_i = 0, \qquad i = 1, \ldots, N_f,$$

where the notation $(\Psi\overline{\Psi})_0$ means the traceless part of $\Psi \otimes \overline{\Psi}$, i.e. $\Psi \otimes \overline{\Psi} - \frac{1}{2}(\overline{\Psi}\Psi)\mathrm{id}$, and $(\overline{\Psi}\Psi)$ is an inner product on spinors. With the spinor indices $\alpha, \beta = 1, 2$ made explicit, these objects are simply given by $(\overline{\Psi}\Psi) = \varepsilon^{\alpha\beta}\overline{\Psi}_\alpha \Psi_\beta$ and $[(\Psi\overline{\Psi})_0]_{\alpha\beta} = \Psi_{(\alpha}\overline{\Psi}_{\beta)}$.

We denote by $\mathcal{M}_{N_f}(M_4; \lambda)$, or simply by $\mathcal{M}_{N_f}$, the moduli space of solutions to these equations modulo gauge transformations. By index theorem, this moduli space has virtual dimension equal to (3.52), which we have already identified as the ghost number anomaly of $T[M_4]$ with N_f copies of the impurity operator. We will also study the perturbed version of these equations, in which the first equation is replaced by

$$F_A^+ + \eta = i\sum_{i=1}^{N_f} (\Psi_i \overline{\Psi}_i)_0, \qquad (4.2)$$

where $\eta \in \mathcal{H}^{2,+}(M_4)$ is a generic self-dual harmonic perturbation. In such situations, we will call the moduli space $\mathcal{M}_{N_f}(\lambda, \eta)$, or again simply $\mathcal{M}_{N_f}$ whenever it creates no confusion.

Although the system of equations (4.1) is well motivated in physics (see e.g. [62, 63, 67] where closely related questions were studied), its mathematical study presents a largely uncharted territory, with only a few brave ventures in this direction [64, 68, 69]. Part of the reason is that $\mathcal{M}_{N_f}$ can be non-compact when $N_f > 1$, and naively following the same steps as in the $N_f = 1$ case can lead to an ill-defined integral:

$$\int_{\mathcal{M}_{N_f}} c_1(\mathcal{L})^{d/2}, \qquad (4.3)$$

where $\mathcal{L} \to \mathcal{M}_{N_f}$ is a universal line bundle over the moduli space (to be defined shortly).

4.1.1. Moduli space

In this subsection, we briefly review the key ingredients in the construction of moduli space. The N_f-monopole case parallels the 1-monopole theory in most aspects, except that now the moduli space can become non-compact. Non-compactness surely makes the definition of invariants more subtle. Also, in this non-compact space, there are sequences of irreducible solutions that would converge to reducible solutions had we included the latter in the moduli space, which is another potential subtlety.

Denote the space of $U(1)$ connections on L by $\mathcal{A}$. The configuration space of the multi-monopole problem is defined as

$$\mathcal{C} = \mathcal{A} \times \Gamma(M_4, W^+)^{N_f}. \qquad (4.4)$$

The group of gauge transformations $\mathcal{G} = \mathrm{Hom}(M_4, U(1))$ acts on $\mathcal{C}$, but this action is not free. So, the most general moduli space of connections and spinors, the one given by

$$\mathcal{M} = \mathcal{C}/\mathcal{G}, \tag{4.5}$$

is singular, with singularities at reducible solutions. However, since we work under the assumption that $b_2^+ > 1$, generic metrics do not admit abelian instantons (i.e. solutions to $F_A^+ = 0$), except for trivial solutions. (Trivial solutions include $A = 0$, $\Psi_i = 0$ and, if $H_1(M_4) \neq 0$, flat connections.) Therefore, equations (4.1) only have solutions with some Ψ_i non-zero, and an additional locus of trivial solutions. Hence, it makes sense to consider the subspace

$$\mathcal{C}^* = \{(A, \Psi_1, \ldots, \Psi_{N_f}) \in \mathcal{C} \,|\, \exists i : \Psi_i \neq 0\}, \tag{4.6}$$

on which $\mathcal{G}$ acts freely. It gives rise to a smooth moduli space of spinors and connections:

$$\mathcal{M}^* = \mathcal{C}^*/\mathcal{G}. \tag{4.7}$$

The multi-monopole moduli space, with or without perturbation, is a subspace $\mathcal{M}_{N_f} \subset \mathcal{M}^*$ of this moduli space determined by the equations (4.1) (or their perturbed version). For the perturbed problem, this $\mathcal{M}_{N_f}$ is really all we have because, in the case of generic perturbation, there are no reducible solutions. For the unperturbed problem (4.1), in addition to $\mathcal{M}_{N_f}$, there is a locus of reducible solutions that have all $\Psi_i = 0$ and A flat.

Under the standard technical assumption (that various infinite-dimensional spaces we are working with can be completed in appropriate Sobolev norms), which are known to hold for the problem at hand [64], we can study the moduli spaces locally. The linearization of the multi-monopole equations, with or without perturbation, is then described by the familiar deformation complex

$$0 \to \Omega^0(M_4) \xrightarrow{C} \Omega^1(M_4) \oplus \Gamma(M_4, W^+)^{N_f} \xrightarrow{ds} \Omega^{2,+}(M_4) \oplus \Gamma(M_4, W^-)^{N_f} \to 0, \tag{4.8}$$

where the first map $C(\chi) = (-d\chi, i\chi\Psi_i)$ describes gauge transformations, and the second one, $ds(\psi,\mu) = ((d\psi)^+_{\alpha\beta} - i(\overline{\Psi}^i_{(\alpha}\mu_{\beta)i} + \overline{\mu}^i_{(\alpha}\Psi_{\beta)i}), D_{\alpha\dot\alpha}\mu_i^\alpha + i\psi_{\alpha\dot\alpha}\Psi_i^\alpha)$ is the linearization of the multi-monopole equations. An equivalent complex is

$$0 \to \Omega^1(M_4) \oplus \Gamma(M_4, W^+)^{N_f} \xrightarrow{C^\dagger \oplus ds} \Omega^0(M_4) \oplus \Omega^{2,+}(M_4) \oplus \Gamma(M_4, W^-)^{N_f} \to 0, \tag{4.9}$$

where $C^\dagger(\psi,\mu) = d^*\psi + \frac{i}{2}(\overline{\Psi}^i\mu_i - \overline{\mu}^i\Psi_i)$. The index of this complex gives the virtual dimension of $\mathcal{M}_{N_f}$:

$$\mathrm{VirDim}\, \mathcal{M}_{N_f}(M_4; \lambda) = \frac{N_f(\lambda^2 - \sigma) - 2(\chi + \sigma)}{4}. \tag{4.10}$$

We will work under the usual assumption that, for generic metric (or, possibly, generic perturbation, when needed), the actual dimension of $\mathcal{M}_{N_f}$ will coincide with its virtual dimension. Any possible topological invariants of M_4 should be formulated in terms of such generic metrics, and it should be possible to connect two generic metrics by a family of generic metrics, which works well for $b_2^+ > 1$.

There is one new effect specific to the $N_f > 1$ case: $\mathcal{M}_{N_f}$ can become non-compact. Such non-compactness is controlled by non-zero solutions to the following system of equations:

$$\sum_{i=1}^{N_f} (\Psi_i^{nc}\overline{\Psi}_i^{nc})_0 = 0,$$
$$\not{D}\Psi_i^{nc} = 0, \qquad i = 1, \ldots, N_f. \tag{4.11}$$

This point has been studied in the 3d case in [69, 70].[15] The way it works is as follows. Suppose we have a sequence of solutions (A, Ψ_i). Non-compactness means there exists such a sequence with no convergent subsequence, so the norms $||\Psi_i||_{L^2} \to \infty$ (at least for some subsequence). After renormalizing $\Psi_i \to \widetilde{\Psi}_i = \frac{\Psi_i}{\sqrt{\sum_j ||\Psi_j||^2_{L^2}}}$, we get a sequence with bounded (unit) norm of $\widetilde{\Psi}_i$'s. Its subsequence will converge to some Ψ_i^{nc}, which has to satisfy equations (4.11). The reason is that $\widetilde{\Psi}_i$ satisfy Seiberg–Witten equations, with the first one replaced by $\frac{1}{\sum_j ||\Psi_j||^2_{L_2}}(F^+ + \eta) = i\sum_{i=1}^{N_f}\left(\widetilde{\Psi}_i\overline{\widetilde{\Psi}}_i\right)_0$. In the limit, since norms of Ψ_i diverge, the left-hand side tends to zero, and we get the first equation of (4.11).

More precise treatment has to take into account that convergence to Ψ^{nc} takes place in a complement of some codimension 2 locus $\mathcal{Z} \subset \mathcal{M}_{N_f}$ of the moduli space. We are not going to investigate this question in any details, postponing the 4d case to the other studies, as well as referring to the existing literature on the 3d case [69, 70]. For now, we will only describe a qualitative (and conjectural) picture of how $\mathcal{M}_{N_f}$ becomes non-compact.

Suppose that equations (4.11) have some non-trivial solutions Ψ_i, $i = 1, \ldots, N_f$ with flat connections, in particular $F_A^+ = 0$. Such solutions will also satisfy Seiberg–Witten equations without perturbation (4.1). Moreover, if Ψ is a solution, then $t\Psi$ is also a solution. By taking $t \to 0$, we get a sequence of solutions which converges to the reducible solution, with $\Psi = 0$ and A a flat connection. Such solutions, when they exist, form a cone inside $\mathcal{M}_{N_f}(\lambda = 0, \eta = 0)$.

When we consider a non-zero flux sector $\lambda \neq 0$ or turn on a generic perturbation η, the multi-monopole Seiberg–Witten equations do not have reducible solutions anymore. However, the asymptotic structure of the moduli space is still controlled by solutions to (4.11), albeit they cannot satisfy Seiberg–Witten equations now. In particular, if such solutions exist, the moduli space is still non-compact, and we expect that it is due to asymptotic cones, as in Figure 10*b*. Since a reducible point is not a solution any more, we expect that the origin of this cone is smoothed out, so that the moduli space remains regular.

For M_4 admitting a Spin structure, in the $\lambda = 0$ sector with the vanishing perturbation $\eta = 0$, we can build solutions forming a cone, as in Figure 10*a*, explicitly. (We thank A. Haydys for the following example). Put $A = 0$. Suppose $\Xi_1, \ldots, \Xi_\ell$ are harmonic spinors

[15] We thank A. Haydys for illuminating explanations.

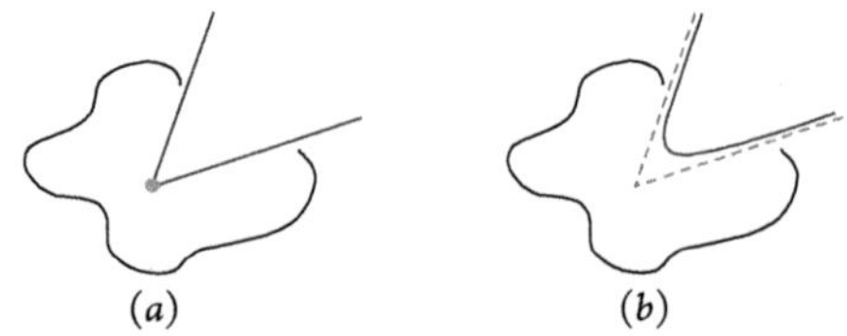

Figure 10 (*a*) The cone is a non-compact subspace in $\mathcal{M}_{N_f}(\lambda = 0, \eta = 0)$; its origin (represented by the dot) is a singular point corresponding to the reducible solution and therefore not included in a smoothly defined $\mathcal{M}_{N_f}(\lambda = 0, \eta = 0)$. (*b*) For non-trivial flux $\lambda \neq 0$ or in the presence of perturbation $\eta \neq 0$, the cone looks the same at infinity, but smoothes out near its origin because reducibles do not satisfy Seiberg–Witten equations anymore.

on M_4, i.e. they satisfy

$$\not{D}\, \Xi_j = 0. \tag{4.12}$$

There is a quaternionic structure J acting on spinors by an anti-linear endomorphism[16] which commutes with $\not{D}$. Thanks to that, any Ξ_j with $J\Xi_j$ form an orthogonal basis for positive-chirality spinors at any point of M_4 where Ξ_j is non-zero. Using this, we can construct $N_f = 2$ solutions by taking $\Psi_1 = \Xi_j$ and $\Psi_2 = J\Xi_j$. They will satisfy $(\Psi_1\overline{\Psi}_1)_0 + (\Psi_2\overline{\Psi}_2)_0 = 0$ and, indeed, solve the multi-monopole equations with zero gauge field $A = 0$. It is trivial to build such solutions for any $N_f > 1$, e.g. for $N_f = 3$ we could simply complete the $N_f = 2$ solution by $\Psi_3 = 0$, and for $N_f = 4$ we could do the same or, alternatively, pick $\Psi_3 = \Xi_k$, $\Psi_4 = J\Xi_k$ for some k. All these solutions can be rescaled $\Psi_i \to t\Psi_i$, with $t \in \mathbb{C}^*$. Therefore, their space forms a cone, illistrated in Figure 10*a*.

Such an example depends on the existence of harmonic spinors, which, for generic metric, is determined by the index $\mathrm{Ind}(\not{D})$. For $\lambda = 0$, $\mathrm{Ind}_{\mathbb{C}}(\not{D}) = -\frac{1}{8}\sigma$. If it is positive, harmonic spinors exist for any metric. If it is non-positive, harmonic spinors appear in the codimension $(1 - \mathrm{Ind}_{\mathbb{C}}(\not{D}))$ subspaces of the space of metrics. In particular, for vanishing index, they appear in codimension 1, and the space of metrics is divided into chambers of compactness by the walls at which $\mathcal{M}_{N_f}$ might become non-compact. This kind of 'wall crossing' might take place in 3d case [69, 70] (although its consequences are not clear yet). Its existence and possible implications in 4d have to be investigated elsewhere.

Fortunately, this non-compactness does not affect the equivariant quantities that we are going to define later. The reason is that they will be related to fixed points of the maximal torus $U(1)^{N_f-1}$ of the flavour symmetry $SU(N_f)$ acting on $\mathcal{M}_{N_f}$. As we will see, fixed points are given by solutions that have Ψ_i non-zero only for one value of i. Such solutions cannot satisfy (4.11): indeed, $(\Psi_i\overline{\Psi}_i)_0 = 0$ (without summation over i) implies $\Psi_i = 0$, because

16 In physics language, it is given by charge conjugation, i.e. complex conjugation composed with the multiplication by the matrix $\begin{pmatrix} 0 & 1 \\ -1 & 0 \end{pmatrix}$.

$\left((\Psi_i\overline{\Psi}_i)_0\right)^2 \propto \left(|\Psi_i|^2\right)^2$. Hence, fixed points do not satisfy the 'non-compactness' equation (4.11), and so they cannot 'run away' to infinity along such asymptotic cones.

4.1.2. A vanishing theorem

Arguably, the most popular vanishing theorem in the ordinary Seiberg–Witten theory is the statement that, on a 4-manifold with $b_2^+ > 0$ that admits a metric of positive scalar curvature, all Seiberg–Witten invariants must vanish [62]. The same vanishing theorem holds for multi-monopole invariants ($N_f > 1$) and also follows from the Weitzenböck formula:

$$\int_{M_4} \frac{1}{4}|F_A^+ - i\sum_{i=1}^{N_f}(\Psi_i\overline{\Psi}_i)_0|^2 + \sum_{i=1}^{N_f}|\not{D}\Psi_i|^2$$

$$= \int_{M_4} \frac{1}{4}|F_A^+|^2 + \sum_{i=1}^{N_f}\left(|\nabla_A\Psi_i|^2 + \frac{s}{4}|\Psi_i|^2\right) + \frac{1}{4}\left|\sum_{i=1}^{N_f}(\Psi_i\overline{\Psi}_i)_0\right|^2, \tag{4.13}$$

which, for simplicity, we write without perturbation (see [64] for a version with perturbation). When the scalar curvature s is strictly positive, all terms on the right-hand side are non-negative and, as a result, there are no non-zero solutions to (4.1). An example of a 4-manifold that admits a positive curvature metric and, therefore, vanishing multi-monopole invariants is

$$M_4 = \ell\mathbb{C}\mathbf{P}^2 \# m\overline{\mathbb{C}\mathbf{P}}^2, \tag{4.14}$$

which is homeomorphic but not diffeomorphic to a degree $d > 4$ hypersurface in $\mathbb{C}\mathbf{P}^3$, with

$$\ell = \frac{d^3 - 6d^2 + 11d - 3}{3}, \qquad m = \frac{2d^3 - 6d^2 + 7d - 3}{3} \tag{4.15}$$

and non-trivial Seiberg–Witten invariants.

Note, since $\sum_{i=1}^{N_f}(\Psi_i\overline{\Psi}_i)_0 \neq \sum_{i=1}^{N_f}|\Psi_i|^2$ when $N_f > 1$, we can not immediately establish separate bounds on λ_+^2 and λ_-^2 which, in the case of the ordinary Seiberg–Witten theory, also follow from (4.13) and the virtual dimension formula (3.52). However, we can still use a version of this argument when M_4 admits a metric of zero scalar curvature. Then, from (4.13), we learn that $\lambda_+ = 0$, which, as discussed in section 2, also implies that $\lambda = 0$ when $b_2^+ > 1$ and when the metric on M_4 is generic. Zero scalar curvature metrics are not generic, however. A good example is a $K3$ surface with its Ricci-flat metric, which has the geometric genus $p_g = h^{2,0} = 1$ and

$$\Gamma = U \oplus U \oplus U \oplus (-E_8) \oplus (-E_8), \tag{4.16}$$

where E_8 is the unique even unimodular positive definite lattice of rank 8, and the *hyperbolic plane* U is a rank 2 lattice with a bilinear form:

$$U = \begin{pmatrix} 0 & 1 \\ 1 & 0 \end{pmatrix}. \tag{4.17}$$

By the Lefschetz theorem on $(1,1)$-classes, the Néron–Severi lattice can be realized as $\mathrm{NS}(M_4) = H^2(M_4,\mathbb{Z}) \cap H^{1,1}(M_4)$, and its rank $\rho(M_4) = \operatorname{rank} \mathrm{NS}(M_4)$ is called the Picard number of M_4. All values of $0 \leq \rho \leq 20$ can be realized by complex K3 surfaces. Moreover, when K3 is algebraic, the Néron–Severi lattice has signature $(1, \rho - 1)$. Note, when combined with (3.52), the fact that the K3 surface admits a metric of zero scalar curvature implies $\lambda_+ = 0$, and $\lambda_-^2 \leq 16\big(1 - \frac{1}{N_f}\big)$.

4.1.3. The universal bundle and its Chern class

The space $\mathcal{M}^* \times M_4$ is canonically equipped with a line bundle $\mathcal{L}$ called the universal line bundle. It is constructed in the following way. Since we have a line bundle L on M_4, we also have the line bundle

$$\begin{array}{c} \mathcal{C}^* \times L \\ \downarrow \\ \mathcal{C}^* \times M_4. \end{array} \tag{4.18}$$

The group of gauge transformations $\mathcal{G}$ acts on $\mathcal{C}^*$ and (fibrewise) on L, so we can take a quotient:

$$\begin{array}{c} \mathcal{L} \cong \mathcal{C}^* \times_{\mathcal{G}} L \\ \downarrow \\ \mathcal{M}^* \times M_4. \end{array} \tag{4.19}$$

This $\mathcal{L}$—the universal line bundle[17]—allows us, in the theory of Seiberg–Witten invariants, just as in the Donaldson theory, to define a map $\mu : H_i(M_4) \to H^{2-i}(\mathcal{M}^*)$ by integrating the first Chern class $c_1(\mathcal{L})$ over an i-cycle of M_4. The result is a form on $\mathcal{M}_{N_f}$ of degree $(2 - i)$ that can be integrated over $\mathcal{M}_{N_f}$ to yield topological invariants (at least in the 1-monopole case, where $\mathcal{M}_1$ is compact).

Proceeding in this direction, we need to construct a differential form on $\mathcal{M}^* \times M_4$ that represents $c_1(\mathcal{L})$. This can be done using the Chern–Weil theory, because the canonical line bundle $\mathcal{L}$ can be naturally equipped with the connection. Since $\mathcal{L}$ is defined as a quotient of the bundle (4.18), the connection on $\mathcal{L}$ can be constructed as a quotient connection. It requires two pieces of data: one is the connection on (4.18), and the other one is the connection on the $\mathcal{G}$-bundle:

$$\begin{array}{c} \mathcal{C}^* \times M_4 \\ \downarrow \pi \\ \mathcal{M}^* \times M_4. \end{array} \tag{4.20}$$

17 In the literature, $\mathcal{L}$ is sometimes defined in an equivalent way: for fixed $x \in M_4$, define the total space of $\mathcal{L}$ as the framed moduli space trivialized over x. (That is, one factors over gauge transformations that are trivial at x.)

The first connection is canonically defined: at a point $(A, \Psi_1, \ldots, \Psi_{N_f}, x) \in \mathcal{C}^* \times M_4$, the connection 1-form is simply given by A. One can write a covariant derivative $\widehat{\nabla}$ acting on the section $\widehat{s} : \mathcal{C}^* \times M_4 \to \mathcal{C}^* \times L$ by the formula

$$\widehat{\nabla}\widehat{s} = \int_{M_4} d^4x \left[\delta A_\mu(x) \frac{\delta \widehat{s}}{\delta A_\mu(x)} + \delta \Psi_i(x) \frac{\delta \widehat{s}}{\delta \Psi_i(x)} \right] + \nabla^A \widehat{s}, \tag{4.21}$$

where δ represents the de Rham differential on the infinite-dimensional functional space $\mathcal{C}^*$, and ∇^A is the connection on L.

To define the second ingredient—the connection on (4.20), which is a principal $\mathcal{G}$-bundle—we have to pick a horizontal subspace in every tangent space to $\mathcal{C}^* \times M_4$, i.e. a direction transversal to the fibre (i.e. transversal to the gauge orbit). Such connection will be represented by a 1-form $\theta \in \Omega^1(\mathcal{C}^* \times M_4, \mathfrak{g})$, where $\mathfrak{g} = Lie(\mathcal{G}) = iC^\infty(M_4)$. The kernel of θ at each point of $\mathcal{C}^* \times M_4$ is the horizontal subspace and, for the vertical vector field ξ_X corresponding to the Lie algebra element $X \in \mathfrak{g}$, one has $\theta(\xi_X) = X$.

In physics, the choice of such a connection θ goes under the name of gauge fixing. A convenient choice in the multi-monopole case is as follows. For infinitesimal variations $\delta A, \delta\Psi_i$ representing a tangent vector to $\mathcal{C}^*$, we require

$$d^*\delta A + \frac{i}{2}\left(\overline{\Psi}_i \delta \Psi_i - \delta \overline{\Psi}_i \Psi_i\right) = 0. \tag{4.22}$$

Since the group $\mathcal{G}$ acts on fields according to $A \to A - d\chi(x)$, $\Psi_j \to e^{i\chi(x)}\Psi_j$, where $\chi(x) \in C^\infty(M_4)$, for the vertical direction in tangent space we have $\delta A = -d\chi$ and $\delta\Psi_j = i\chi\Psi_j$. The left-hand side of the above equation becomes simply $-d^*d\chi - \sum_i \overline{\Psi}_i \Psi_i \chi$. To build θ that satisfy condition $\theta(\xi_X) = X$, we need to invert the operator $d^*d + \sum_i \overline{\Psi}_i\Psi_i$ (which is positive and hence invertible, whenever at least one $\Psi_i \neq 0$), i.e. introduce the Green's function:

$$\left(d_x^* d_x + \sum_i \overline{\Psi}_i(x)\Psi_i(x) \right) G(x,y) = \delta^{(4)}(x-y). \tag{4.23}$$

In terms of it, the connection is defined as

$$\theta = -i \int_{M_4} d^4y\, G(x,y) \left(d^*\delta A + \frac{i}{2}\left(\overline{\Psi}_i \delta\Psi_i - \delta\overline{\Psi}_i \Psi_i\right)\right). \tag{4.24}$$

It is a 1-form on $\mathcal{C}^* \times M_4$, so the right-hand side depends on $x \in M_4$ as well as $(A, \Psi_1, \ldots, \Psi_{N_f}) \in \mathcal{C}^*$ (even though it actually is A-independent). The connection θ is a $\mathcal{G}$-invariant connection on the principal bundle $\mathcal{C}^* \times M_4 \to \mathcal{M}^* \times M_4$, so it allows us to lift vector fields $X \in \Gamma(T(\mathcal{M}^* \times M_4))$ to horizontal vector fields on $\mathcal{C}^* \times M_4$, which we denote X^h.

Now we can define the quotient connection [71] on the universal line bundle $\mathcal{L}$. A section $s : \mathcal{M}^* \times M_4 \to \mathcal{L}$ can be lifted to a $\mathcal{G}$-equivariant section $\widehat{s} : \mathcal{C}^* \times M_4 \to \mathcal{C}^* \times L$, and a vector field $X \in \Gamma(T(\mathcal{M}^* \times M_4))$—to the horizontal vector field $X^h \in \Gamma(T(\mathcal{C}^* \times M_4))$. Since $\widehat{\nabla}$ is a $\mathcal{G}$-invariant connection, $\widehat{\nabla}_{X^h}\widehat{s}$ is also a $\mathcal{G}$-equivariant section of $\mathcal{C}^* \times L$. It then corresponds to some section of $\mathcal{L}$, which we define to be $\nabla_X s$. So,

$$\widehat{\nabla_X s} = \widehat{\nabla}_{X^h}\widehat{s}. \tag{4.25}$$

A simple calculation (see Appendix A) shows that the curvature is

$$F^{\nabla} = dA + \delta A - 2\Omega, \tag{4.26}$$

where $dA \in \Omega^2(M_4)$ is the usual field strength, $\delta A = \delta A_\mu(x) \wedge dx^\mu$ and Ω is a 2-form on $\mathcal{M}^*$ depending on the point $x \in M_4$ and defined in the following way. For any two vectors $U, V \in T(\mathcal{M}^*)$, we take their horizontal lifts U^h, V^h and define $\Omega(U,V) = \Theta(U^h, V^h)$, where Θ is a 2-form on $\mathcal{C}^*$ given by

$$\Theta = \int_{M_4} d^4y\, G(x,y) \sum_i \left(\delta\overline{\Psi}_i(y) \wedge \delta\Psi_i(y)\right), \tag{4.27}$$

where the parentheses on the right denote the inner product on spinors, as before. The form Ω is a closed 2-form on $\mathcal{M}^*$, even though it is not manifest from this expression. (See Appendix A for details.)

4.1.4. Equivariant form and the integral

In the 1-monopole problem, which has been studied a lot in the literature, one simply takes this Ω, raises it to the power $\frac{1}{2}\dim\mathcal{M}_1$ and integrates over $\mathcal{M}_1 \subset \mathcal{M}^*$. The result is usually taken as a definition of the Seiberg–Witten invariant. Because for $N_f > 1$ the space $\mathcal{M}_{N_f}$ is non-compact, we cannot really follow this route any more.

A way out is related to the group $SU(N_f)$ that acts naturally (through its defining representation) on $\mathcal{M}_{N_f}$. What we can do is extend all the cohomology classes we are dealing with to $SU(N_f)$-equivariant cohomology, in particular construct an equivariant version of Ω that we denote $\Omega(\xi)$, which represents the equivariant Chern class of $\mathcal{L}$. Then, we can perform an equivariant integration over $\mathcal{M}_{N_f}$, which is possible even if $\mathcal{M}_{N_f}$ is non-compact. All we need is that the relevant fixed point sets of the maximal torus action of $SU(N_f)$ on $\mathcal{M}_{N_f}$ are compact. Then the equivariant integration produces a Laurent polynomial in the equivariant parameters as the answer. This will be our definition of the multi-monopole invariants.

The group $SU(N_f)$ starts its life through the action on $\mathcal{C}^*$ by the vector field

$$v_a = \sum_{i,j} (T_a)^{ij} \int_{M_4} d^4x \left(\Psi_{j\alpha} \frac{\delta}{\delta\Psi_{i\alpha}(x)} - \overline{\Psi}_i^{\alpha}(x) \frac{\delta}{\delta\overline{\Psi}_j^{\alpha}(x)} \right), \tag{4.28}$$

where T_a is a generator of $SU(N_f)$ in the fundamental representation. Using the projection $\pi : \mathcal{C}^* \to \mathcal{M}^*$, we get a vector field $d\pi(v_a)$ on $\mathcal{M}^*$. Denoting the equivariant parameters of $SU(N_f)$ by ξ^a, we introduce the equivariant differential of the Cartan model on $\mathcal{M}^*$:

$$D = d + \xi^a \iota_{d\pi(v_a)}. \tag{4.29}$$

If Ω is $SU(N_f)$-invariant, one can construct the D-closed extension of Ω as

$$\Omega(\xi) = \Omega + \xi^a H_a, \tag{4.30}$$

where the Hamiltonian

$$H_a = \int_{M_4} d^4y\, G(x,y) \sum_{i,j} \left(\overline{\Psi}_i(y) T_a^{ij} \Psi_j(y) \right) \tag{4.31}$$

is a function on $\mathcal{M}^* \times M_4$. The proof of this expression for H_a, as well as the $SU(N_f)$-invariance of Ω, can be found in Appendix A.

Our goal is to define equivariant multi-monopole invariants as

$$\text{equivariant} \int_{\mathcal{M}_{N_f}} f[\Omega(\xi)], \tag{4.32}$$

where f is some function of $\Omega(\xi)$, e.g. $(\Omega(\xi))^{d/2}$ or $\exp[\Omega(\xi)]$. In general, the result is an $SU(N_f)$-invariant rational function of the equivariant parameters, i.e. the element of $\mathbb{C}(\mathfrak{g}^*)^{SU(N_f)} = \mathbb{C}(\mathfrak{t}^*)^W \subset \mathbb{C}(\mathfrak{t}^*)$, where $\mathfrak{g}$ is the Lie algebra of $SU(N_f)$, $\mathfrak{t}$ is the Cartan subalgebra of $\mathfrak{g}$, and W is the Weil group. Moreover, following [23, 24], we can identify the coordinates on the Cartan subalgebra of $\mathfrak{g}$ with the hypermultiplet masses in the topologically twisted 4d $\mathcal{N} = 2$ gauge theory on M_4. It is enough to think of the integral as equivariant with respect to the maximal torus $U(1)^{N_f-1}$, the answer will automatically lie in the W-invariant subspace $\mathbb{C}(\mathfrak{t}^*)^W \subset \mathbb{C}(\mathfrak{t}^*)$.

The maximal torus acts on Ψ_j as $\Psi_j \mapsto e^{i\varphi_j}\Psi_j$, where $\sum_i \varphi_i = 0$. At the fixed point set of the $U(1)^{N_f-1}$ action on the moduli space, this should be equivalent to the gauge transformation $\Psi_j \mapsto e^{i\varphi}\Psi_j, j = 1,\ldots,N_f$. This is possible only if Ψ_j vanish for all j except $j = i$ with some $1 \le i \le N_f$. In other words, only one of the N_f monopoles Ψ_i is non-zero. We call the corresponding component of the fixed point set F_i. There are N_f such disjoint components, and each one is isomorphic to the 1-monopole moduli space, $F_i \cong \mathcal{M}_1$. Denote the inclusion of the i-th component as

$$s_i : \mathcal{M}_1 \hookrightarrow \mathcal{M}_{N_f}. \tag{4.33}$$

Recall that the equivariant integral for non-compact spaces is *defined* by the Atiyah–Bott localization formula. So it is given by

$$\sum_{i=1}^{N_f} \int_{\mathcal{M}_1} \frac{(s_i)^* f[\Omega(\xi)]}{Eul(N_i)(\xi)}, \tag{4.34}$$

where $Eul(N_i)(\xi)$ is the equivariant Euler class of the normal bundle to $s_i(\mathcal{M}_1) \subset \mathcal{M}_{N_f}$.

4.2. Computation for M_4 of Simple Type

To move further, recall that there exist notions of Kronheimer–Mrowka and Seiberg–Witten simple type. We need the latter notion, the Seiberg–Witten simple type, which requires the manifold to have vanishing Seiberg–Witten invariants whenever the 1-monopole moduli space $\mathcal{M}_1(M_4;\lambda)$ has non-zero expected dimension. More precisely, we need a closely related and slightly stronger version—which one might call[18] 'strong simple type'—which requires that the manifold has only zero-dimensional 1-monopole moduli spaces $\mathcal{M}_1(M_4;\lambda)$ for all λ. The Simple Type Conjecture states that every closed simply connected oriented Riemannian 4-manifold with $b_2^+ > 1$ is of simple type—both Kronheimer–Mrowka and Seiberg–Witten.

No matter whether the Simple Type Conjecture holds or not, if we have a 4-manifold of strong simple type, we can compute the above equivariant integral explicitly. In this case, the space $\mathcal{M}_1$ consists of isolated points with signs. The normal bundle N_i to each of these points is trivial. Denote equivariant parameters for $U(1)^{N_f}$ acting naturally on $\mathcal{M}_{N_f}$ by $z_1, z_2, \ldots, z_{N_f}$, and require that[19]

$$\sum_{i=1}^{N_f} z_i = 0, \tag{4.35}$$

so these are really equivariant parameters for the maximal torus $U(1)^{N_f-1} \subset SU(N_f)$ acting on $\mathcal{M}_{N_f}$. Then the equivariant parameters for the maximal torus $U(1)^{N_f-1}$ acting on N_i are given by $z_j - z_i$ for $j \neq i$.

Since M_4 is of strong simple type (and we use generic metric), $\dim \mathcal{M}_1 = 0 = \frac{1}{4}(\lambda^2 - \sigma) - \frac{1}{2}(\chi + \sigma)$. In this case,

$$\dim_{\mathbb{C}} \mathcal{M}_{N_f} = \frac{N_f - 1}{8}(\lambda^2 - \sigma) = (N_f - 1)\,\mathrm{Ind}_{\mathbb{C}}(\not{D}). \tag{4.36}$$

So, every N_i is a direct sum of $N_f - 1$ copies of a trivial complex bundle whose rank[20] is $\mathrm{Ind}_{\mathbb{C}}(\not{D})$. Each $U(1)$ factor in the maximal torus $U(1)^{N_f-1}$ acts on the corresponding copy of this trivial bundle, with the equivariant parameter $z_j - z_i$. Thus, the equivariant Euler class is

$$Eul(N_i)(\xi) = \prod_{j\neq i}(z_j - z_i)^{\mathrm{Ind}_{\mathbb{C}}(\not{D})} = \prod_{j\neq i}(z_j - z_i)^{\frac{1}{8}(\lambda^2-\sigma)}. \tag{4.37}$$

[18] We thank C. Manolescu for suggesting this notion.

[19] In physics literature, mass parameters are traditionally denoted m_i. Nevertheless, we remain faithful to the two-dimensional perspective, where a standard notation for these parameters would be z_i, as they denote positions of impurity vertex operators on Σ. A physicist more familiar with gauge theory may find it comforting to read 'z_i' as 'm_i'.

[20] We use a subscript $\mathbb{C}$ to emphasize that it is the complex dimension that is relevant here; when we omit $\mathbb{C}$, we always mean real dimension.

As for the pullback of $\Omega(\xi)$, since $\dim F_i = 0$, we have $(s_i)^*(\Omega + \xi^a H_a) = (s_i)^*(\xi^a H_a)$. Because at F_i only $\Psi_i \neq 0$, we have

$$(s_i)^* H_a = T_a^{ii} \int_{M_4} d^4y\, G(x,y) \overline{\Psi}_i(y) \Psi_i(y), \qquad \text{no sum over } i, \tag{4.38}$$

and, from

$$\int_{M_4} d^4y\, G(x,y) \overline{\Psi}_i(y) \Psi_i(y) = \int_{M_4} d^4y\, G(x,y) \left(d_y^* d_y + \overline{\Psi}_i(y) \Psi_i(y) \right) 1 = 1, \tag{4.39}$$

we conclude

$$(s_i)^* H_a = T_a^{ii}, \qquad \text{no sum over } i, \tag{4.40}$$

and this gives simply

$$\sum_a \xi^a s_i^* H_a = \sum_a \xi^a T_a^{ii} = \sum_{j=1}^{N_f} (z_j - z_i) \frac{-1}{N_f} = z_i. \tag{4.41}$$

So the pullback is simply $(s_i)^* \Omega(\xi) = z_i$. We are almost done; all we have to compute is the zero-dimensional integral:

$$\sum_{i=1}^{N_f} \frac{f[z_i]}{\prod_{j \neq i} (z_j - z_i)^{\frac{1}{8}(\lambda^2 - \sigma)}} \int_{\mathcal{M}_1} 1. \tag{4.42}$$

The only non-trivial step left here is to recall that isolated points in $\mathcal{M}_1$ come with signs. These signs are induced from the orientation of the moduli space $\mathcal{M}_{N_f}$, which in turn is induced by the orientation of $\mathcal{M}^*$ in which $\mathcal{M}_{N_f}$ is embedded. The space $\mathcal{M}_{N_f}$ might have several connected components, each with its own orientation. This is why points in the fixed set $F_i \cong \mathcal{M}_1$ might contribute with different signs.

Of course, points in F_i and F_j that are identified by the isomorphism $F_i \cong \mathcal{M}_1 \cong F_j$ belong to the same connected component and thus come with the same orientation. This is because they correspond to the same solution $\Psi^{(1)}$ of the 1-monopole problem and can be connected by the path

$$\begin{aligned} &\Psi_i = \Psi^{(1)} \cos\alpha, \quad \Psi_j = \Psi^{(1)} \sin\alpha, \quad \alpha \in [0, \pi/2], \\ &\Psi_k = 0, \quad k \neq i \text{ and } k \neq j. \end{aligned} \tag{4.43}$$

However, different points inside F_i might belong to different connected components of $\mathcal{M}_{N_f}$ and come with different signs. The way these signs should be determined is precisely as in the 1-monopole problem [62]. In fact, thanks to the equivariant localization, we actually have completely eliminated the extra $(N_f - 1)$ spinors and have reduced the N_f-monopole equations to the 1-monopole problem. Therefore, the numbers

$$\mathrm{SW}(\lambda) = \int_{\mathcal{M}_1} 1 \tag{4.44}$$

are nothing else but the 1-monopole Seiberg–Witten invariants. Even if hard-boiled skeptics may find this claim not completely convincing, in Appendix B we present another derivation of it that does not assume M_4 to be of simple type. There, it will be very clear that, after applying equivariant localization, we are left precisely with the 1-monopole version of the problem.

Note also that, for M_4 of the Seiberg–Witten simple type, there is no real value in keeping function f in the equivariant integral arbitrary. We can just choose $f = 1$ and define

$$\mathrm{ESW}_{M_4}(\lambda, z_i) \ := \ \text{equivariant} \int_{\mathcal{M}_{N_f}} 1. \tag{4.45}$$

The above computation shows that the answer is

$$\mathrm{ESW}_{M_4}(\lambda, z_i) \ = \ \mathrm{SW}(\lambda) \sum_{i=1}^{N_f} \frac{1}{\prod_{j\neq i}(z_j - z_i)^{\frac{1}{8}(\lambda^2-\sigma)}}. \tag{4.46}$$

This result has precisely the expected structure (3.68), where two vertex operators $\mathcal{S}_+(z)$ and $\mathcal{S}_+(w)$ have non-singular OPE and

$$\mathcal{S}_+(z)\mathcal{S}(w) \ \sim \ \frac{1}{(z-w)^{\frac{1}{8}(\lambda^2-\sigma)}}\mathcal{S}(w), \tag{4.47}$$

with the background charge λ.

4.3. *Simply connected Kähler Surfaces*

In this subsection, we review the structure of multi-monopole moduli spaces for M_4 that is Kähler and simply connected, following [64], as well as discuss the perturbed problem and some further properties of $\mathcal{M}_{N_f}$.

Any Kähler manifold has the canonical Spin^c structure

$$\begin{aligned} W_0^+ &= \Lambda^0 \oplus \Lambda^{0,2}, \\ W_0^- &= \Lambda^{0,1}, \end{aligned} \tag{4.48}$$

with the Levi–Civita connection ∇_0 playing the role of the Spin^c connection, and the Dirac operator $\partial\!\!\!/_0 = \sqrt{2}(\overline{\partial} + \overline{\partial}^*)$. Every other Spin^c structure with connection is determined by a unique line bundle with connection (E, A):

$$\begin{aligned} W_E^+ &= W_0^+ \otimes E, \\ W_E^- &= W_0^- \otimes E, \\ \nabla_E &= \nabla_0 \otimes A. \end{aligned} \tag{4.49}$$

The multi-monopole equations are written in terms of pairs $(\alpha_i, \beta_i), i = 1 \dots N_f$, where $\alpha_i \in \Omega^0(E)$ and $\beta_i \in \Omega^{0,2}(E)$:

$$\begin{aligned} \overline{\partial}_A \alpha_i + \overline{\partial}_A^* \beta_i &= 0, \\ 2F_A^{0,2} + \overline{\eta} &= \sum_{i=1}^{N_f} \alpha_i^* \beta_i, \\ -2i\Lambda_\omega F_A &= \frac{1}{2}\sum_{j=1}^{N} (|\alpha_j|^2 - |\beta_j|^2) - i\Lambda_\omega F_{\nabla_0} - r, \end{aligned} \tag{4.50}$$

where Λ_ω is a dual of the Hodge operator $\alpha \mapsto \omega \wedge \alpha$, and ω is a Kahler form. We slightly abuse notations for the perturbation here, and η is (twice) a $(2,0)$-part of what was a general self-dual harmonic perturbation η before, while the $(1,1)$-part of that perturbation is now chosen to be $r\omega$ with $r \in \mathbb{R}$.

Following [64], the moduli space at $\eta = 0$ has the following description. Define $V_1 = \oplus_{N_f} H^0(E) \cong \mathbb{C}^{N_f h^0(\lambda)}$ and $V_2 = \oplus_{N_f} H^0(K \otimes E^*) \cong \mathbb{C}^{N_f h^2(\lambda)}$, where we denote

$$h^0(\lambda) \;=\; \dim_{\mathbb{C}} H^0(E), \qquad\qquad h^2(\lambda) \;=\; \dim_{\mathbb{C}} H^0(K \otimes E^*). \tag{4.51}$$

Define a map $S : V_1 \oplus V_2 \to H^0(K)$ by the equation

$$S(\alpha_1, \dots, \alpha_{N_f}, \beta_1^*, \dots, \beta_{N_f}^*) = \sum_{i=1}^{N} \alpha_i \beta_i^*, \tag{4.52}$$

and define $Z \subset V_1 \oplus V_2$ as a zero set of S. Next, for $2\pi c_1(L) \cdot [\omega] + r[\omega] \cdot [\omega] > 0$, define $Z^s \subset Z$ as a subset with $(\alpha_1, \dots, \alpha_N) \neq (0, \dots, 0)$, i.e. $Z^s \cong Z \cap [(V_1 \setminus \{0\}) \oplus V_2]$, while, for $2\pi c_1(L) \cdot [\omega] + r[\omega] \cdot [\omega] < 0$, the role of α_i and β_i^* is interchanged: $Z^s \cong Z \cap [V_1 \oplus (V_2 \setminus \{0\})]$. Finally, there is a $\mathbb{C}^*$ action on V_1 and V_2 via multiplication by λ and λ^{-1}, respectively. Then, the moduli space at $\eta = 0$ is given by

$$\mathcal{M}_{N_f}(\lambda, \eta = 0) = Z^s / \mathbb{C}^*. \tag{4.53}$$

It is straightforward to generalize this to non-zero $\eta \in H^0(K)$. At $\eta = 0$, as was shown in [64], the monopole equations imply $F_A^{0,2} = 0$. In other words, for Kähler M_4, only bundles with $c_1(L) = \lambda$ of type $(1,1)$ contribute to the moduli space $\mathcal{M}_{N_f}$. This property must hold after turning on the perturbation as well, even if it is not entirely obvious from the equations: for λ not of type $(1,1)$, the moduli space must be empty. Therefore, we might restrict to λ of type $(1,1)$ from the very beginning. Then,

$$\int_{M_4} \eta \wedge F^{0,2} = \int_{M_4} \overline{\eta} \wedge F^{2,0} = 0. \tag{4.54}$$

Using this and repeating manipulations from [64], one arrives at the following description:

$$\mathcal{M}_{N_f}(\lambda, \eta \neq 0) = S^{-1}(\eta)/\mathbb{C}^*. \tag{4.55}$$

Now we wish to discuss various properties of $\mathcal{M}_{N_f}(\lambda, \eta = 0)$ and $\mathcal{M}_{N_f}(\lambda, \eta \neq 0)$.

4.3.1. Dimensionality

As we change η from a generic non-zero value (which is possible only if $b_2^+ > 1$) to zero, the dimensionality of the space $\mathcal{M}_{N_f}$ might change. Since Kähler metrics are non-generic, we expect that the actual dimension of $\mathcal{M}_{N_f}$ might be higher than the virtual dimension, and the dimension at $\eta = 0$ might be higher than that at generic η.

The space $V_1 \oplus V_2$ has dimension $N_f(h^0(\lambda) + h^2(\lambda))$, and we denote the dimension of $H^0(K)$ by $h^{2,0}$. The condition $S(\alpha_1, \ldots \alpha_{N_f}, \beta_1^*, \ldots, \beta_{N_f}^*) = \eta$ consists of $h^{2,0}$ equations. So, whenever η is a regular value of the map S, the complex dimension of $\mathcal{M}_{N_f}$ is given by what we call regular dimension:

$$\text{RegDim}_{\mathbb{C}}\, \mathcal{M}_{N_f} = N_f(h^0(\lambda) + h^2(\lambda)) - h^{2,0} - 1, \tag{4.56}$$

where '-1' comes from taking a quotient by $\mathbb{C}^*$. For example, for a $K3$ surface with a Kähler metric and $\lambda = 0$, we have $h^0(\lambda) = h^2(\lambda) = 1$, as well as $h^{2,0} = 1$. Therefore, the (complex) regular dimension is $4(N_f - 1)$ (cf. (3.52)).

On the other hand, consider $\eta = 0$ and $2\pi c_1(L) \cdot [\omega] + r[\omega] \cdot [\omega] > 0$. In this case, $\mathcal{M}_{N_f}$ was constructed as a subvariety in $[(V_1 \setminus \{0\}) \oplus V_2]/\mathbb{C}^*$. Let us look at the geometry of $\mathcal{M}_{N_f}$ near the fixed point locus of the maximal torus $U(1)^{N_f - 1} \subset SU(N_f)$. One component of such a fixed point locus is given, as we know, by solutions with only α_i and β_i^* possibly non-zero, while for $j \neq i$, $\alpha_j = \beta_j^* = 0$. The equation $S = 0$ becomes simply $\alpha_i \beta_i^* = 0$. Since $\alpha_i \neq 0$ for $2\pi c_1(L) \cdot [\omega] + r[\omega] \cdot [\omega] > 0$, we conclude $\beta_i = 0$ and $\alpha_i \in \mathbb{C}^{h^0(\lambda)}$ has an arbitrary non-zero value. Factoring by $\mathbb{C}^*$, the fixed point locus is isomorphic to $\mathbb{C}P^{h^0(\lambda)-1}$.

Let us determine the normal bundle to this fixed point set inside $\mathcal{M}_{N_f}(\lambda, \eta = 0)$, just for fun and to have a better understanding of the geometry of $\mathcal{M}_{N_f}$. Points of this $\mathbb{C}P^{h^0(\lambda)-1}$ are given by homogeneous coordinates $(0 : \cdots : 0 : \alpha_i : 0 : \cdots : 0)$. Moving away from this point by $\delta\alpha_j$, $\delta\beta_j^*$, the equation $S = 0$ only implies that $\alpha_i \delta\beta_i^* = 0$, hence $\delta\beta_i^* = 0$ and $\delta\alpha_j$, $\delta\beta_j^*$ with $j \neq i$ are not constrained. We also put $\delta\alpha_i = 0$ since we do not want to move in the tangent direction. Recalling that $\mathbb{C}^*$ acts on $\delta\alpha_j$ in the same way as on α_i, and in the opposite way on $\delta\beta_j^*$, we can find that the normal bundle looks as follows:

$$\begin{array}{c} \mathcal{O}(1)^{\oplus(N_f-1)h^0(\lambda)} \oplus \mathcal{O}(-1)^{\oplus(N_f-1)h^2(\lambda)} \\ \downarrow \\ \mathbb{C}P^{h^0(\lambda)-1}. \end{array} \tag{4.57}$$

This describes the geometry of $\mathcal{M}_{N_f}$ near the fixed point set, and one can read off $\dim_{\mathbb{C}} \mathcal{M}_{N_f}(\lambda, \eta = 0)$ from this, at least near the fixed point set:

$$\begin{aligned}\dim_{\mathbb{C}} \mathcal{M}_{N_f}(\lambda, \eta = 0) &= (N_f - 1)\left(h^0(\lambda) + h^2(\lambda)\right) + h^0(\lambda) - 1 \\ &= N_f\left(h^0(\lambda) + h^2(\lambda)\right) - h^2(\lambda) - 1.\end{aligned} \tag{4.58}$$

This differs from (4.56) by $h^{2,0} - h^2(\lambda)$. It is not hard to determine the sign of this difference. Since $h^0(\lambda) > 0$, there exists a section $\alpha \in H^0(E)$. Multiplication by α defines a map of sheaves $E^* \to \mathcal{O}$ (where we identify the bundle E with its sheaf of holomorphic sections), which can be further completed into the short exact sequence:

$$0 \longrightarrow E^* \longrightarrow \mathcal{O} \longrightarrow \mathcal{O}/E^* \longrightarrow 0. \tag{4.59}$$

Taking a tensor product with K (which is locally free) and taking a long exact sequence in the cohomology implies an injective map $H^0(K \otimes E^*) \to H^0(K)$. Therefore, it must be that $h^{2,0} = \dim_{\mathbb{C}} H^0(K) \geq \dim_{\mathbb{C}} H^0(K \otimes E^*) = h^2(\lambda)$. So, we conclude $h^{2,0} - h^2(\lambda) \geq 0$, i.e. the actual dimension $\dim_{\mathbb{C}} \mathcal{M}_{N_f}(\lambda, \eta = 0)$ can only jump upward compared to the expected dimension (4.56), which was derived for η, a regular value of S.

Let us also compare this to the virtual dimension. From the Riemann–Roch theorem,

$$\begin{aligned}h^0(\lambda) + h^2(\lambda) - h^1(\lambda) &= 1 + h^{2,0} + \frac{1}{2}\left[c_1(E)^2 - c_1(E) \cdot c_1(K)\right] \\ &= 1 + h^{2,0} + \frac{1}{8}\left[c_1(L)^2 - c_1(K)^2\right] \\ &= 1 + h^{2,0} + \mathrm{VirDim}_{\mathbb{C}} \mathcal{M}_1,\end{aligned} \tag{4.60}$$

where $h^1(\lambda) = \dim_{\mathbb{C}} H^1(E)$. From this and $\frac{1}{4}(\chi + \sigma) = 1 + h^{2,0}$, one can write the complex virtual dimension of $\mathcal{M}_{N_f}$ as follows:

$$\begin{aligned}\mathrm{VirDim}_{\mathbb{C}} \mathcal{M}_{N_f} &= N_f \mathrm{VirDim}_{\mathbb{C}} \mathcal{M}_1 + (N_f - 1)\frac{\chi + \sigma}{4} \\ &= N_f(h^0(\lambda) + h^2(\lambda) - h^1(\lambda)) - h^{2,0} - 1.\end{aligned} \tag{4.61}$$

Because $h^1(\lambda) \geq 0$, this cannot be larger than $\mathrm{RegDim}_{\mathbb{C}} \mathcal{M}_{N_f}$ from (4.56). So far, we have found the following inequalities:

$$\mathrm{VirDim}_{\mathbb{C}} \mathcal{M}_{N_f} \leq \mathrm{RegDim}_{\mathbb{C}} \mathcal{M}_{N_f} \leq \dim_{\mathbb{C}} \mathcal{M}_{N_f}, \tag{4.62}$$

where the first one follows from the definition, and the second one holds because RegDim counts dimension in the assumption that $(h^{2,0} + 1)$ equations in S are independent. If they happen to be dependent, the actual dimension of $\mathcal{M}_{N_f}$ can only be larger than RegDim. To make one more estimate, we recall the following result:

Proposition 1 [62] Pick a non-zero $\eta \in H^0(K)$ and consider the following equation on $\alpha \in H^0(E)$, $\beta^* \in H^0(K \otimes E^*)$:

$$\alpha\beta^* = \eta. \tag{4.63}$$

Its space of solution modulo the $\mathbb{C}^*$ action, $(\alpha, \beta^*) \to (t\alpha, t^{-1}\beta^*)$ is either empty or consists of isolated points.

Using this, we can make one more estimate on the dimension of the space of solutions to $\sum_{i=1}^{N_f} \alpha_i\beta_i^* = \eta$. Arbitrarily pick generic $\alpha_1, \ldots, \alpha_{N_f-1}$ and $\beta_1^*, \ldots, \beta_{N_f-1}^*$, which are just $(N_f - 1)\left(h^0(\lambda) + h^1(\lambda)\right)$ arbitrary complex numbers. Then, the equation

$$\alpha_N\beta_N^* = \eta - \sum_{i=1}^{N_f-1} \alpha_i\beta_i^*, \tag{4.64}$$

as an equation on (α_N, β_N^*), assuming that the right-hand side is non-zero, either has (after modding out $\mathbb{C}^*$) isolated solutions, in which case we say that $\dim_{\mathbb{C}} \mathcal{M}_{N_f} = (N_f - 1)\left(h^0(\lambda) + h^1(\lambda)\right)$, or has no solutions at all. In the latter case, it is still possible that it has solutions for $\alpha_1, \ldots, \alpha_{N_f-1}$ and $\beta_1^*, \ldots, \beta_{N_f-1}^*$ not arbitrary but chosen from a certain subspace. In such a situation, it must be that $\dim_{\mathbb{C}} \mathcal{M}_{N_f} < (N_f - 1)\left(h^0(\lambda) + h^1(\lambda)\right)$. Combining this with the previous inequalities, we get the following proposition:

Proposition 2 If $\mathcal{M}_{N_f}$ is non-empty, it is true that

$$\mathrm{VirDim}_{\mathbb{C}} \mathcal{M}_{N_f} \leq \mathrm{RegDim}_{\mathbb{C}} \mathcal{M}_{N_f} \leq \dim_{\mathbb{C}} \mathcal{M}_{N_f} \leq (N_f - 1)\left(h^0(\lambda) + h^2(\lambda)\right). \tag{4.65}$$

This has an important corollary:

Corollary If $\mathcal{M}_{N_f}$ is non-empty and $\mathrm{VirDim}_{\mathbb{C}} \mathcal{M}_1 \geq 0$, then, in fact, $\mathrm{VirDim}_{\mathbb{C}} \mathcal{M}_1 = h^1(\lambda) = 0$, and all inequalities in Proposition 2 become equalities.

Indeed, the Riemann–Roch estimate combined with $\mathrm{VirDim}_{\mathbb{C}} \mathcal{M}_1 \geq 0$ implies $h^0(\lambda) + h^2(\lambda) \geq 1 + h^{2,0}$, with the equality only for $\mathrm{VirDim}_{\mathbb{C}} \mathcal{M}_1 = h^1(\lambda) = 0$. But then $\mathrm{RegDim}_{\mathbb{C}} \mathcal{M}_{N_f} = N_f(h^0(\lambda) + h^2(\lambda)) - 1 - h^{2,0} \geq (N_f - 1)(h^0(\lambda) + h^2(\lambda))$; due to Proposition 2, this is only possible if $h^{2,0} + 1 = h^0(\lambda) + h^2(\lambda)$, so $\mathrm{VirDim}_{\mathbb{C}} \mathcal{M}_1 = h^1(\lambda) = 0$, and all inequalities turn into equalities.

Recall from the previous subsection (see also Appendix B) that our equivariant multi-monopole invariants were determined by solutions to the 1-monopole Seiberg–Witten equations. For Kähler manifolds, those exist only when $\mathcal{M}_1$ is zero-dimensional. Here, we have shown a bit more, but only for simply connected M_4: moduli space $\mathcal{M}_{N_f}$ is non-empty only if $\mathrm{VirDim}_{\mathbb{C}} \mathcal{M}_1 \leq 0$ and $h^1(\lambda) = 0$.

4.3.2. Connectedness

According to Proposition 1, for Kähler M_4, the 1-monopole moduli space consists of isolated points. In particular, it can be disconnected. The multi-monopole moduli space $\mathcal{M}_{N_f}$ with $N_f > 1$ and simply connected M_4, however, is connected or, more precisely, its locus, which contains the fixed points of the $U(1)^{N_f-1} \subset SU(N_f)$ action, is definitely connected. Pick $\alpha^{(1)}, \beta^{(1)*}$ and $\alpha^{(2)}, \beta^{(2)*}$, two (possibly equal) solutions to $\alpha\beta^* = \eta$. Consider the following family of solutions:

$$\begin{aligned} \alpha_1 &= \alpha^{(1)} \cos\theta, \quad \beta_1^* = \beta^{(1)*} \cos\theta, \\ \alpha_2 &= \alpha^{(2)} \sin\theta, \quad \beta_2^* = \beta^{(2)*} \sin\theta, \\ \alpha_i &= \beta_i^* = 0, \text{ for } i > 2. \end{aligned} \tag{4.66}$$

This obviously solves $\sum_{i=1}^{N_f} \alpha_i \beta_i^* = \eta$. As we change θ from 0 to $\pi/2$, it connects solution $(\alpha^{(1)}, \beta^{(1)*}, 0, \ldots, 0)$ to $(0, 0, \alpha^{(2)}, \beta^{(2)*}, 0, \ldots, 0)$. It is clear that, using similar paths within $\mathcal{M}_{N_f}$, we can connect all fixed points of the $U(1)^{N_f-1}$ action. In particular, they all lie in the same connected component.

As we discussed before, this means that all these fixed points contribute with the same sign to the N_f-monopole invariant, as well as to the 1-monopole Seiberg–Witten invariant (to which the equivariant N_f-monopole problem reduces). This might look surprising, but actually it already follows from the description in [62] that, for simply connected Kähler M_4, all solutions to $\alpha\beta^* = \eta$ contribute with the same sign.[21]

Connectedness of $\mathcal{M}_{N_f}$ might fail for a non-simply connected M_4.

4.4. *Multi-Monopole Homology of 3-Manifolds*

Even though one of our main goals, starting with (1.2), was to bring the categorification of 3-manifold and knot invariants closer to its roots, namely to the corresponding invariants of 4-manifolds, we conclude this section by going back and considering 4-manifolds of the form

$$M_4 = \mathbb{R} \times M_3 =$$

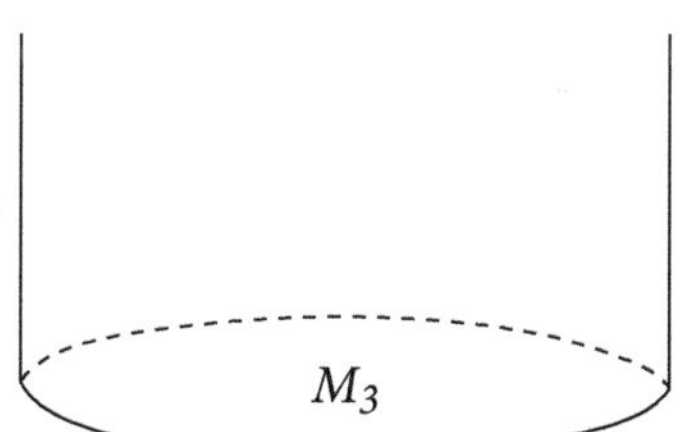

. (4.67)

This leads to a version of Floer homology $\mathcal{H}_{N_f}(M_3)$ based on multi-monopole equations (4.1), equivariant with respect to $SU(N_f)$ action.

Note, the $N_f = 1$ version is the familiar theory, the so-called monopole Floer homology $HM(M_3) \cong HF(M_3) \cong ECH(M_3)$, based on the ordinary Seiberg–Witten equations. As a module over $\mathbb{Z}[U]$, this homology is naturally a part of the 4d TQFT that associates

[21] In [62], for $c_1(L) \cdot [\omega] > 0$, the sign of each solution was argued to be $(-1)^{\dim_{\mathbb{C}} H^0(M_4, R)}$, where R is a sheaf which fits into the short exact sequence $0 \to \mathcal{O} \to E \to R \to 0$, and the map $\mathcal{O} \to E$ is a multiplication by α. Taking a long exact sequence in the cohomology and using $H^1(M_4, \mathcal{O}) = 0$, we see that $\dim_{\mathbb{C}} H^0(R) = h^0(\lambda) - 1$, and hence all solutions contribute with the same sign. For $c_1(L) \cdot [\omega] < 0$, one replaces α with β^*.

graded vector spaces to 3-manifolds (equipped with a choice of Spin^c structure) and, more importantly for us here, it admits an equivariant interpretation, with respect to a circle action, such that $H^*_{S^1}(\text{pt}) \cong \mathbb{Z}[U]$. Our equivariant multi-monopole homology $\mathcal{H}_{N_f}(M_3)$ is a natural generalization of that, where the role of U is played by N_f equivariant parameters z_i, $i = 1, \ldots, N_f$.

In general, there are various ways to compute $\mathcal{H}_{N_f}(M_3)$ via different compactifications of 6d fivebrane theory on $\mathbb{R} \times \Sigma \times M_3$ and, possibly, using additional dualities. For example, first compactifying on M_3, one can compute $\mathcal{H}_{N_f}(M_3)$ as a Q-cohomology of 3d $\mathcal{N} = 2$ theory $T[M_3]$ on $\mathbb{R} \times \Sigma$ or, reversing the order of the compactification, as a Q-cohomology of a 4d TQFT on $M_4 = \mathbb{R} \times M_3$ (cf. (3.46)). Furthermore, when M_3 is a Seifert manifold, both of these routes lead to a computation of a certain partition function of 3d $\mathcal{N} = 2$ theory, which we illustrate below for a small sample of simple 3-manifolds.

A simple example of a Seifert manifold is the total space of a circle fibration (with no singular fibres) over a genus g Riemann surface C_g. Further simplifications can be achieved by setting either the degree p or the genus g to zero. The corresponding 3d theory $T[S^1 \xrightarrow{p} C_g]$, that describes the low-energy physics of N fivebranes on M_3 is $\mathcal{N} = 2$ level-p super-Chern–Simons, coupled to $2g + 1$ adjoint chiral multiplets [44]. In the case of the single fivebrane ($N = 1$) that we are interested in here, all adjoint chiral multiplets are neutral:

	$U(1)_{\text{gauge}}$	R-charge	$U(1)_\beta$
chiral	0	2	1
g chirals	0	0	-1
g chirals	0	0	0

(4.68)

The BPS spectrum of this theory in the presence of N_f impurity operators $\mathcal{S}_+(z_i)$ on Σ gives the desired multi-monopole homology $\mathcal{H}_{N_f}(M_3)$. Due to a phenomenon of 'homological-flavour locking' that holds for Seifert 3-manifolds, its Poincaré polynomial is equal to the 'refined index' (= graded Euler characteristic) of $\mathcal{H}_{N_f}(M_3)$, graded by the extra flavour symmetry $U(1)_\beta$. In [21], this computation was performed for $N_f = 1$ and many simple 3-manifolds, and we leave it to an interested reader to generalize it to $N_f > 1$.

Instead, we compute the refined index and, therefore, the Poincaré polynomial of $\mathcal{H}_{N_f}(M_3)$, by interpreting it as a fivebrane partition function on $S^1 \times \Sigma \times M_3$ and reducing on $S^1 \times \Sigma$ first. This gives a 3d theory $T[S^1 \times \Sigma]$ topologically twisted on M_3. In our case, $T[S^1 \times \Sigma]$ is a 3d $\mathcal{N} = 4$ vector multiplet coupled to N_f charged hypermultiplets or, in $\mathcal{N} = 2$ language,

	$U(1)_{\text{gauge}}$	R-charge
chiral	0	0
N_f chirals	1	1
N_f chirals	-1	1

(4.69)

Its topological partition function on $S^1 \xrightarrow{p} C_g$ can be computed as in [21] or [72, 73]. For simplicity, let us set $p = 0$ and start with $g = 0$, i.e. $M_3 = S^1 \times S^2$. Then, the topologically twisted index of 3d theory $T[S^1 \times \Sigma]$ on a 2-sphere is

$$\left(\frac{y^{-1}}{1-y^{-2}}\right)\sum_{h\in\mathbb{Z}}(-q)^{-h}\int\frac{dx}{2\pi ix}\prod_{i=1}^{N_f}\left(\frac{x^{1/2}z_i^{1/2}y^{1/2}}{1-xz_iy}\right)^{h}\left(\frac{x^{-1/2}z_i^{-1/2}y^{1/2}}{1-x^{-1}z_i^{-1}y}\right)^{-h}, \tag{4.70}$$

where we follow the conventions of [21, sec. 3] and, in particular, $h \in \mathbb{Z}$ labels the choice of Spin^c structure on $M_3 = S^1 \times S^2$. Using the Jeffrey–Kirwan residue prescription and picking up the residues of negatively charged fields, we obtain (with $h < 0$)

$$\frac{1}{2\pi i}\sum_{j=1}^{N_f}\mathrm{Res}_{x=y/z_j}\left(\frac{1}{x}\prod_{i=1}^{N_f}\left(\frac{xz_i-y}{1-xz_iy}\right)^{h}\right) = (-1)^{hN_f}\left(y^{-hN_f}-y^{hN_f}\right). \tag{4.71}$$

Incorporating the factor $\frac{1}{y-y^{-1}}$ from (4.70), we come to a prediction that a multi-monopole analogue of the homology $HF^+(M_3)$ has dimension hN_f for $M_3 = S^1 \times S^2$ with Spin^c structure h.

Similarly, the genus g topologically twisted index of 3d theory (4.69) can be expressed as a sum:

$$Z_{S^1\times C_g} = \sum_{x=x(\alpha)} Z_{\text{cl,1-loop}}|_{h=0}\left(i\frac{\partial B}{\partial \log x}\right)^{g-1}, \tag{4.72}$$

taken over solutions to the Bethe ansatz equation $1 = e^{iB} := \exp\left(\frac{\partial \log Z_{\text{cl,1-loop}}}{\partial h}\right)$,

$$1 = q\cdot\prod_{i=1}^{N_f}\left(\frac{xz_i-y}{1-xz_iy}\right), \tag{4.73}$$

where

$$Z_{\text{cl,1-loop}} = q^h\prod_{i=1}^{N_f}\left(\frac{x^{1/2}z_i^{1/2}y^{1/2}}{1-xz_iy}\right)^{h}\left(\frac{x^{-1/2}z_i^{-1/2}y^{1/2}}{1-x^{-1}z_i^{-1}y}\right)^{-h}. \tag{4.74}$$

For example, when $N_f = 2$, the Bethe ansatz equation is quadratic in x and, for $g = 2$, gives the following generating series for the equivariant multi-monopole invariants:

$$\sum_h q^h P_{N_f=2}(z_1,z_2) = \frac{(y^2+1)\left(4z_1z_2(q-y^2)(qy^2-1)-(q-1)^2y^2(z_1+z_2)^2\right)}{q(y^2-1)(z_1-y^2z_2)(z_2-y^2z_1)}. \tag{4.75}$$

On the other hand, in the genus zero case, we recover the result (4.71) obtained earlier by a different method.

Note, if we restore the overall factor $\frac{1}{y-y^{-1}}$ which was present in (4.70) but for brevity omitted in (4.71) and (4.75), both of these expressions have a well-defined unrefined

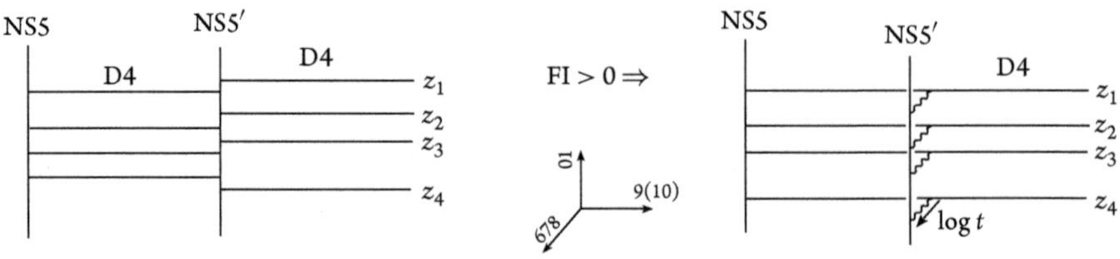

Figure 11 Brane construction of $U(N)$ theory with $N_f = N$ fundamental hypermultiplets and its Higgs phase with non-zero FI parameters.

limit ($y \to 1$). In this limit, the dependence on the equivariant parameters z_i disappears, as expected from the fact that, for $M_4 = M_3 \times S^1$, the 2d theory $T[M_4]$ has enhanced $\mathcal{N} = (2,2)$ supersymmetry, and the half-twisted correlator on Σ is, in fact, topological.

5. Non-Abelian Generalizations

One of the main goals in this paper was to explain how $\text{VOA}_G[M_4]$ or, equivalently, the $\overline{Q}_+$-cohomology of 2d $\mathcal{N} = (0,2)$ theory $T[M_4, G]$ knows about traditional 4-manifold invariants, such as Seiberg–Witten invariants. In the process, we had to establish a dictionary between gauge theory on M_4 and half-twisted correlators in $T[M_4, G]$. This dictionary, then, can be used as a very effective tool to study the structure of 4-manifold invariants, old and new, when G is abelian or non-abelian.

Examples of new invariants which are relatively simple but nonetheless non-trivial are the equivariant multi-monopole invariants introduced in the previous section. Based on $G = U(1)$, they can be viewed as a stepping stone towards a more powerful invariant $\text{VOA}_G[M_4]$ for non-abelian G, which we expect to be at least as strong as Donaldson invariants and, hopefully, even stronger.

Relegating a more thorough study of such non-abelian invariants to future work, here we illustrate how the rules of section 3.5 can predict the structure of these invariants, thanks to the dictionary with 2d correlators. Consider topological gauge theory on M_4 with gauge group $G = U(N)$ and $N_f = N$ fundamental hypermultiplets of mass z_i, $i = 1, \ldots, N$. The corresponding brane set-up is shown in Figure 11. Turning on the FI parameter forces all N D4 branes stretched between NS5 branes to align with semi-infinite D4 branes leading to the 'colour–flavour locking.'

In order to give a 2d dual formulation of this problem, as in section 3, we lift it to a fivebrane configuration in M-theory and then reduce to two dimensions (cf. (3.46)). In particular, the NS5′ brane is lifted to a fivebrane on $M_4 \times \Sigma$, with $\Sigma = \mathbb{C}$. From the viewpoint of $T[M_4]$ on Σ, we have insertions of impurity operators $\mathcal{S}(z_i)$, the operators that correspond to fivebrane intersections at points z_i (see Figure 8):[22]

[22] The NS5 brane is also lifted to a fivebrane on $M_4 \times \widetilde{\Sigma}$, with $\widetilde{\Sigma} = \mathbb{C}$ as well. Since there are D4 branes ending on NS5, the theory $T[M_4]$ on $\widetilde{\Sigma}$ could contribute another factor to the answer that would look like $\langle \mathcal{S}_+(z_1) \cdots \mathcal{S}_+(z_N) \rangle$. Whether such a contribution is really present will be studied in future work.

$$\sum_\lambda \langle \mathcal{S}(z_1) \cdots \mathcal{S}(z_N) \, e^{-i\lambda \cdot X(\infty)} \rangle \, t^\lambda. \tag{5.1}$$

It would be interesting to test this prediction by calculating the $SU(N_f)$-equivariant invariants directly in 4d gauge theory, here with $N_f = N$. Note, in the derivation of (5.1), it was important that the gauge group be $U(N)$ and not $SU(N)$.

Similarly, one can consider the case with $N_f > N$ fundamental hypermultiplets. Turning on the FI parameter then localizes the Coulomb branch ('u-plane') integral on $\binom{N_f}{N}$ configurations where N finite D4 branes are distributed among N_f semi-infinite D4 branes. This leads to a Higgs phase of 4d gauge theory with the following 'colour–flavour locking' pattern:

$$U(N)_{\text{gauge}} \times SU(N_f) \quad \to \quad S[U(N)_{\text{diag}} \times U(N_f - N)]. \tag{5.2}$$

We leave it as an exercise for an interested reader to write down the corresponding 2d correlators.

Returning to the case $N_f = N$, let us make the proposed structure (5.1) a little bit more explicit for minimal surfaces of general type (with $b_2^+ > 1$). Such M_4 have only one basic class (up to a sign),

$$\lambda = \pm K, \tag{5.3}$$

where K is the canonical line bundle. The corresponding Seiberg–Witten invariants are

$$\text{SW}(\lambda) \;=\; 1 \text{ and } (-1)^{\chi_h}, \tag{5.4}$$

where $\chi_h = \frac{1}{4}(\chi + \sigma)$ was introduced in (2.23). Note, for a simply connected M_4, we have $\frac{\chi+\sigma}{4} = \frac{1+b_2^+}{2}$, which is an integer, since the definition of Seiberg–Witten invariants requires $b_2^+ - b_1$ to be odd. Therefore, for minimal surfaces of general type, the impurity operator (3.58) is in the same $\overline{Q}_+$-cohomology class as the sum of two winding-momentum vertex operators of the form (3.21):

$$V_\lambda(z) \;\sim\; e^{ik_L X_L(z) + k_R \sigma(z)}, \tag{5.5}$$

where '$\sim$' means that we focus on the left-moving sector and ignore the right-moving sector. These operators have the following chiral correlators:

$$\langle V_{\lambda_1}(z_1) \cdots V_{\lambda_n}(z_n) \rangle_\lambda \;=\; \prod_{i<j} (z_i - z_j)^{k_L^i k_L^j - k_R^i k_R^j} \;=\; \prod_{i<j} (z_i - z_j)^{\lambda_i \lambda_j}, \tag{5.6}$$

which must satisfy the 'neutrality' condition $\lambda_1 + \cdots + \lambda_n = -\lambda$. Here, λ is the 'background charge' at infinity,

$$\mathcal{O}_{\text{bkgr}} \;=\; \lim_{z \to \infty} z^{\lambda^2} e^{-i\lambda X(z)}. \tag{5.7}$$

It corresponds to the choice of Spin^c structure in the generating function (5.1), which we can now explicitly evaluate for each given rank.
$\underline{N=2}$:

$$\langle \mathcal{S}(z_1)\mathcal{S}(z_2)\rangle_\lambda = \langle (V_K(z_1)+(-1)^{\chi_h}V_{-K}(z_1))(V_K(z_2)+(-1)^{\chi_h}V_{-K}(z_2))\rangle_\lambda, \tag{5.8}$$

$$\langle \mathcal{S}(z_1)\mathcal{S}(z_2)\rangle_\lambda = \begin{cases} (z_1-z_2)^c, & \text{if } \lambda = 2K, \\ 2(-1)^{\chi_h}(z_1-z_2)^{-c}, & \text{if } \lambda = 0, \\ (z_1-z_2)^c, & \text{if } \lambda = -2K. \end{cases} \tag{5.9}$$

$\underline{N=3}$:

$$\langle (V_K(z_1)+(-1)^{\chi_h}V_{-K}(z_1))(V_K(z_2)+(-1)^{\chi_h}V_{-K}(z_2))(V_K(z_3)+(-1)^{\chi_h}V_{-K}(z_3))\rangle_\lambda,$$

$$= \begin{cases} (-1)^{\chi_h}(z_1-z_2)^c(z_2-z_3)^c(z_1-z_3)^c, & \text{if } \lambda = 3K, \\ (z_1-z_2)^c(z_2-z_3)^{-c}(z_1-z_3)^{-c} + \text{permutations}, & \text{if } \lambda = K, \\ (-1)^{\chi_h}(z_1-z_2)^c(z_2-z_3)^{-c}(z_1-z_3)^{-c} + \text{permutations}, & \text{if } \lambda = -K, \\ (z_1-z_2)^c(z_2-z_3)^c(z_1-z_3)^c, & \text{if } \lambda = -3K. \end{cases}$$

Following (2.23), here we use $c = K^2$.

5.1. *Coulomb-Branch Index and New 4-Manifold Invariants*

As noted in [10], the most interesting homological invariants point in the direction of 4d gauge theories with non-anomalous R-symmetry $U(1)_R$. Indeed, the Khovanov–Rozansky knot homology and its recent generalization to 3-manifolds are equipped with a $\mathbb{Z}$-valued homological grading,[23] which in the physical realization [8, 21] corresponds to the R-charge. Therefore, formulating such homological invariants as Floer homologies in 4d TQFTs (in the sense of Figure 1 and discussion around it) suggests that most powerful 4-manifold invariants come from theories with $\mathbb{Z}$-valued $U(1)_R$ charge.

While fivebranes compactified on $\Sigma = \mathbb{R}^2_q$ (='cigar') give rise to a rather exotic 4d $\mathcal{N}=2$ theory with non-anomalous $U(1)_R$ symmetry, its simpler analogues can be obtained by considering 4d $\mathcal{N}=2$ superconformal theories (possibly non-Lagrangian). Their topological twists give rise to TQFTs with $\mathbb{Z}$-valued homological grading. And it is natural to ask whether any of these simpler variants produce 4-manifold invariants which are stronger than Donaldson or Seiberg–Witten invariants.

As a practical way to answer this question, below we offer a quick 'detector' based on a simple and highly computable quantity. In any 4d $\mathcal{N}=2$ theory with non-anomalous R-symmetry $U(1)_R$, one can define a topological $S^3 \times S^1$ index:

$$\mathcal{I} = \mathrm{Tr}_{\mathcal{H}(S^3)}(-1)^F e^{-\beta H} t^R. \tag{5.10}$$

[23] As opposed, say, to $\mathbb{Z}_8$ grading in the original Donaldson–Floer theory.

It is basically a partition function of the topologically twisted 4d $\mathcal{N}=2$ theory on $M_4 = S^3 \times S^1$, refined by the $U(1)_R$ charge.[24] It can be viewed as a four-dimensional analogue of the 3d topologically twisted $S^2 \times S^1$ index [74, 75] or, alternatively, as a topological cousin of various $S^3 \times S^1$ superconformal indices in four dimensions [76]. In fact, it is closely related to the Coulomb branch index, as we shall see next.

In a 4d $\mathcal{N}=2$ theory with a Lagrangian description, it is easy to determine contributions of vector and hypermultiplets to the index (5.10) or a closely related quantity: the corresponding Floer homology $\mathcal{H}(S^3)$. Among fields in a vector multiplet (summarized in Appendix B) only the scalar ϕ and the gluino ρ have zero modes. After the topological twist, all fields in a hypermultiplet transform as spinors and have no zero modes. Therefore, only vector multiplets contribute to the topological $S^3 \times S^1$ index. Furthermore, the contribution of vector multiplets can be evaluated using localization, as in the case of 3d topological index [74, 75] or as in 4d superconformal index [76]; schematically, the result looks like

$$\mathcal{I} = \int_{\mathcal{M}_{\text{BPS}}} \mathcal{D}\varphi_0 \, Z_{\text{1-loop}} \, e^{-S[\varphi_0]}, \tag{5.11}$$

where

$$Z_{\text{1-loop}} = \frac{\det \mathcal{O}_{\text{Fermi}}}{\det \mathcal{O}_{\text{Bose}}} \tag{5.12}$$

is a ratio of 1-loop determinants. Concretely, for $G = SU(N)$, the contribution of a vector multiplet is given by the expression

$$\mathcal{I}_{\text{vector}} = \frac{1}{N!(1-\mathfrak{t})^{N-1}} \oint_{\mathbb{T}^{N-1}} \prod_{i=1}^{N-1} \frac{da_i}{2\pi i a_i} \prod_{i \neq j} \frac{1 - a_i/a_j}{1 - \mathfrak{t} a_i/a_j}, \tag{5.13}$$

in which the reader familiar with 4d superconformal index can recognize the Coulomb-branch index of a 4d $\mathcal{N}=2$ vector multiplet [76]. (See also [77, 78] for a connection between the Coulomb-branch index and the equivariant Verlinde formula that counts graded dimensions of the Hilbert space in complex Chern–Simons theory.)

What can the index $\mathcal{I}$ tell us about 4-manifold invariants? It can give us some useful clues about the behaviour of the invariants computed by the topologically twisted 4d $\mathcal{N}=2$ theory (not necessarily Lagrangian) under the operation of connected sum:

$$M_4 = M_4^+ \# M_4^- . \tag{5.14}$$

The standard, by now neck-stretching argument [79] shows that both Donaldson and Seiberg–Witten [62] invariants of M_4 vanish when $b_2^+(M_4^\pm) \geq 1$, simply because there are no non-trivial solutions in the neck $\cong \mathbb{R} \times S^3$. In fact, we have already touched upon this

24 It is the same 'refinement' as in the study of refined BPS invariants of refined topological strings. This point is explained in many of the above mentioned papers and plays a role in many concrete calculations.

a few times: first when we computed the Floer homology of a 3-sphere (3.5) from physics, and then in the equivariant setting of section 4, when we interpreted $\mathcal{T}^+ \cong H^*_{S^1}(\text{pt})$ as the contribution of the (trivial) reducible solution, treated equivariantly.

Now, we can compare this to the index of $SU(2)$ vector multiplet (whose topological twist gives Donaldson–Witten TQFTs) to see that $\mathcal{I} = \frac{1}{1-t^2} = 1 + t^2 + \cdots$ is basically the graded dimension of $\mathcal{T}^+$. Therefore, factors like this will signal that a 4d $\mathcal{N} = 2$ theory in question has trivial (reducible) solutions in the neck $\mathbb{R} \times S^3$, and the more non-trivial the structure of $\mathcal{I}$ is, the more likely we are to find new 4-manifold invariants. In particular, the topological twist of a 4d $\mathcal{N} = 2$ theory that has non-trivial solutions on $\mathbb{R} \times S^3$ may lead to 4-manifold invariants that do not enjoy the above vanishing theorem and, therefore, may be stronger than Donaldson or Seiberg–Witten invariants.[25]

To summarize, in search of new 4-manifold invariants, we should focus special attention on theories with non-trivial structure of $\mathcal{I}$. Let us consider a few simple examples. Among 4d $\mathcal{N} = 2$ superconformal theories, one of the simplest yet interesting examples is the original Argyres–Douglas theory, often denoted (A_1, A_2), whose Coulomb branch index is

$$\mathcal{I}_{(A_1,A_2)} = \frac{1}{1 - t^{6/5}}. \tag{5.15}$$

The somewhat non-trivial structure of this expression suggests that a topological twist of the Argyres–Douglas theory may be a good candidate for constructing new invariants. Note, this theory has no global flavour symmetry. Similarly, the Coulomb branch index of E_6 SCFT can be obtained via Argyres–Seiberg duality [80], as in [81]:

$$\mathcal{I}_{E_6} = \frac{1}{1 - t^3}. \tag{5.16}$$

Our next example is more advanced.

As we mentioned earlier, for applications to refined BPS states and homological invariants of knots and 3-manifolds, one of the most interesting 4d $\mathcal{N} = 2$ theories is a 6d (2,0) fivebrane theory 'compactified' on the cigar $\Sigma = \mathbb{R}^2_q$. Even though this exotic theory has all the features of the 4d $\mathcal{N} = 2$ superconformal theory, strictly speaking, it is not four-dimensional (since $\mathbb{R}^2_q$ is non-compact). Nevertheless, its states can be graded by $U(1)_q$ symmetry that acts as a rotation of $\mathbb{R}^2_q$—which explains the use of the subscript—and each graded component behaves much like a standard 4d $\mathcal{N} = 2$ SCFT.[26] In particular, for a theory $T[\text{cigar}]$ of N fivebranes on $\mathbb{R}^2_q$, the following equivariant version of the topological index (5.10) is well defined:

$$\mathcal{I}_{T[\text{cigar}]} = \text{Tr}_{\mathcal{H}(S^3)}(-1)^F e^{-\beta H} t^R q^n. \tag{5.17}$$

25 Of course, the index $\mathcal{I}$ is a rather simple characteristic, which has its advantages—e.g. it is relatively easy to compute, even in non-Lagrangian theories—and disadvantages. For example, even an interesting theory that leads to new 4-manifold invariants may have an index as simple as $\frac{1}{1-t^2}$.

26 Note, here we use the same cure for non-compactness as in section 4. This time, equivariant technique regulates non-compactness of space-time, bringing us even closer to the origins of such methods [23, 24].

As far as we know, the superconformal index of this theory (or its limits) has not been studied in the literature. Here, we propose the answer for the Coulomb-branch index, based on the same trick we used earlier, in sections 3 and 4.4. Namely, exchanging the order of the compactification, we can compute it as a supersymmetric partition function of the 3d $\mathcal{N}=2$ theory $T[S^3]$ on $S^1 \times \mathbb{R}^2_q$:

$$\begin{aligned}\mathcal{I}_{T[\text{cigar}]} &= Z_{S^1\times\mathbb{R}^2_q}(T[S^3]) \\ &= \prod_{i=1}^{N} \frac{1}{(\mathfrak{t}^i q^i; q)_\infty} = 1 + \mathfrak{t}q + (2\mathfrak{t}^2+\mathfrak{t})q^2 + (3\mathfrak{t}^3+2\mathfrak{t}^2+\mathfrak{t})q^3 + \cdots . \end{aligned} \tag{5.18}$$

To be more precise, this is the 'unreduced' version of the partition function [44]; the 'reduced' version $\prod_{i=1}^{N} \frac{(\mathfrak{t}q;q)_\infty}{(\mathfrak{t}^i q^i;q)_\infty}$ does not include the contribution from the Cartan of the adjoint chiral multiplet. It would be interesting to find other independent checks of (5.18).

Clearly, (5.18) is a good example of a non-trivial index, indicating that the corresponding 4-manifold invariants may not vanish under the connected sum operation (5.14) when $b_2^+(M_4^\pm) \geq 1$. Note, the choice $\Sigma = \mathbb{R}^2_q$ (= cigar) is precisely the one that gives 4d TQFT with surface operators whose spaces of states are identified with $sl(N)$ Khovanov–Rozansky homology.

Acknowledgments

We would like to thank J. Bryan, A. Dabholkar, A. Gadde, A. Haydys, C. Manolescu, M. Mariño, V. Mikhaylov, J.W. Morgan, H. Nakajima, H. Ooguri, J. Rasmussen, S. Schafer-Nameki, A. Soldatenkov, M. Stosic, E. Verlinde, H. Verlinde, J. Wong and K. Ye for useful discussions and comments. The work of M.D. and S.G. is supported in part by the U.S. Department of Energy, Office of Science, Office of High Energy Physics, under Award Number DE-SC0011632. In addition, the work of S.G. is supported in part by the ERC Starting Grant no. 335739 'Quantum fields and knot homologies' funded by the European Research Council under the European Union Seventh Framework Programme. P.P. gratefully acknowledges the support from Marvin L. Goldberger Fellowship and the DOE Grant DE-SC0009988. Opinions and conclusions expressed here are those of the authors and do not necessarily reflect the views of funding agencies.

A. Curvature of the Canonical Connection on the Universal Bundle

In this appendix, we are going to compute the curvature of the quotient connection ∇ on the universal bundle $\mathcal{L} \to \mathcal{M}^* \times M_4$. The quotient connection was defined in (4.25), using equivariant lift of sections and horizontal lift of vector fields. So, we perform the following calculation:

$$\begin{aligned}\left(F^{\nabla}(X,Y)s\right)^{\wedge} &= ((\nabla_X\nabla_Y - \nabla_Y\nabla_X - \nabla_{[X,Y]})s)^{\wedge} = (\widehat{\nabla}_{X^h}\widehat{\nabla}_{Y^h} - \widehat{\nabla}_{Y^h}\widehat{\nabla}_{X^h} - \widehat{\nabla}_{[X,Y]^h})\widehat{s} \\ &= \left(\widehat{\nabla}_{X^h}\widehat{\nabla}_{Y^h} - \widehat{\nabla}_{Y^h}\widehat{\nabla}_{X^h} - \widehat{\nabla}_{[X^h,Y^h]} + \widehat{\nabla}_{[X^h,Y^h]} - \widehat{\nabla}_{[X,Y]^h}\right)\widehat{s} \\ &= F^{\widehat{\nabla}}(X^h,Y^h)\widehat{s} - \widehat{\nabla}_{[X,Y]^h-[X^h,Y^h]}\widehat{s}.\end{aligned} \tag{A.1}$$

In this expression, $F^{\widehat{\nabla}} = \widehat{\nabla}^2 = d_x A + \delta A = \frac{1}{2}F_{\mu\nu}dx^\mu \wedge dx^\nu + \delta A_\mu(x) \wedge dx^\mu$, and also we notice that

$$[X,Y]^h - [X^h,Y^h] = hor([X^h,Y^h]) - [X^h,Y^h] = 2\Theta(X^h,Y^h), \tag{A.2}$$

where $hor(.)$ denotes the horizontal projection of the vector field, $\Theta = (d_x + \delta)\theta$ is the curvature 2-form on the principal $\mathcal{G}$-bundle $\mathcal{C}^* \times M_4$ and we have identified the Lie algebra of $\mathcal{G}$ (where Θ takes values) with the vertical vector fields (where $hor([X^h,Y^h]) - [X^h,Y^h]$ takes values). So, taking the derivative along the vertical vector field $\widehat{\nabla}_{[X,Y]^h-[X^h,Y^h]}$ is the same as acting with the Lie algebra element $2\Theta(X^h,Y^h)$, which acts simply by multiplication. So, we can write

$$\left(F^{\nabla}(X,Y)s\right)^{\wedge} = F^{\widehat{\nabla}}(X^h,Y^h)\widehat{s} - 2\Theta(X^h,Y^h)\widehat{s}. \tag{A.3}$$

This gives the following expression for F^{∇}:

$$F^{\nabla} = d_x A + \delta A - 2\Omega, \tag{A.4}$$

where Ω is understood to be a 2-form on $\mathcal{M}^* \times M_4$ that corresponds to the form $\Theta = (d_x + \delta)\theta$ on $\mathcal{C}^* \times M_4$; in a sense $\Omega(X,Y) = \Theta(X^h,Y^h)$.

To calculate $\delta\theta$ more explicitly, we first note that (as can be found by varying the defining equation for the Green's function)

$$\delta G(x,y) = -\int_{M_4} d^4u\, G(x,u)\left(\overline{\Psi}(u)\delta\Psi(u) + \delta\overline{\Psi}(u)\Psi(u)\right)G(u,y), \tag{A.5}$$

so we find

$$\begin{aligned}\delta\theta = &\int_{M_4} d^4y\, G(x,y)\delta\overline{\Psi}(y) \wedge \delta\Psi(y) \\ &+ i\int_{M_4\times M_4} d^4y\, d^4u\, G(x,u)G(u,y)\left[\overline{\Psi}(u)\delta M(u) + \delta\overline{\Psi}(u)\Psi(u)\right] \\ &\wedge \left[d^*\delta A(y) + \frac{i}{2}\left(\overline{\Psi}(y)\delta\Psi(y) - \delta\overline{\Psi}(y)\Psi(y)\right)\right].\end{aligned} \tag{A.6}$$

When we substitute a pair of horizontal vector fields into $(d_x + \delta)\theta$, because $d_y^*\delta A(y) + \frac{i}{2}\left(\overline{\Psi}(y)\delta\Psi(y) - \delta\overline{\Psi}(y)\Psi(y)\right)$ vanishes on horizontal vectors by definition, the only part which survives is $\int_{M_4} d^4y\, G(x,y)\delta\overline{\Psi}(y) \wedge \delta\Psi(y)$. In particular, if we pick a point $x \in M_4$, consider the inclusion

$$\mathcal{M}^* \times \{x\} \subset \mathcal{M}^* \times M_4, \tag{A.7}$$

and pull back $\mathcal{L}$ to $\mathcal{M}^* \times x$, we get a line bundle on $\mathcal{M}^*$, which we also call $\mathcal{L}$. It has a pullback connection, and its curvature Ω evaluated at point $m \in \mathcal{M}^*$ on vectors $U, V \in T_m\mathcal{M}^*$ is given by $\Theta(U^h, V^h)$ evaluated at some point $c \in \mathcal{C}^*$ lying above m, where

$$\Theta = \int_{M_4} d^4y\, G(x,y)\delta\overline{\Psi}(y) \wedge \delta\Psi(y) + (\text{terms that vanish on horizontal vectors}), \tag{A.8}$$

and $U^h, V^h \in T_c\mathcal{C}^*$ are horizontal lifts of U, V. This is the expression for Θ that was advertised in (4.27).

By construction, Ω has to be closed, but it is slightly tricky to see it explicitly from the above equation. A way to check it is to use the following formula:

$$\begin{aligned}(d\Omega)(U,V,W) &= U\cdot\Omega(V,W) + c.p. - \Omega([U,V],W) - c.p.\\ &= U^h\cdot\Theta(V^h,W^h) + c.p. - \Theta([U,V]^h,W^h) - c.p.\\ &= U^h\cdot\Theta(V^h,W^h) + c.p. - \Big(\Theta([U^h,V^h],W^h) + c.p.\Big)\\ &\quad - \Big(\Theta([U,V]^h - [U^h,V^h],W^h) + c.p.\Big)\\ &= (\delta\Theta)(U,V,W) - 2\Big(\Theta(\Theta(U^h,V^h),W^h) + c.p.\Big),\end{aligned} \tag{A.9}$$

where *c.p.* stands for cyclic permutations. Since Θ is exact on $\mathcal{C}^*$, the first term is zero. Then we look more closely at

$$\Theta(\Theta(U^h,V^h),W^h), \tag{A.10}$$

where $\Theta(U^h,V^h)$, as an element of the Lie algebra of $\mathcal{G}$, is identified with the vertical vector fields. Using expression (A.6) for Θ, one can explicitly check (through a two-step computation) that $\Theta(\Theta(U^h,V^h),W^h) = 0$.

A.1. $SU(N_f)$ Invariance and the Moment Map

To check that Ω is invariant with respect to the $SU(N_f)$ action given by the vector field v_a from (4.28), we need to show that $\mathcal{L}_{d\pi(v_a)}\Omega = 0$, where $\pi : \mathcal{C}^* \times M_4 \to \mathcal{M}^* \times M_4$ is our principal $\mathcal{G}$-bundle. Since $d\Omega = 0$, we have to prove that $d\iota_{d\pi(v_a)}\Omega = 0$. According to our construction of Ω, to find $\iota_{d\pi(v_a)}\Omega$, we have to compute $\iota_{V_a^h}\Theta$, where V_a^h is a horizontal lift of $d\pi(v_a)$. By definition, V_a^h is a vector field on $\mathcal{C}^*$, and it differs from v_a by a vertical vector field in such a way that V_a^h is horizontal. Write V_a^h as v_a plus a general vertical vector field parametrized by $\chi(x) \in C^\infty(M_4)$:

$$\begin{aligned}V_a^h = \int_{M_4}\mathrm{Vol}_{M_4}\Bigg[& i(T^a)_i^{\ j}\left(\Psi_{\alpha j}\frac{\delta}{\delta\Psi_{\alpha i}} - \overline{\Psi}^{\alpha i}\frac{\delta}{\delta\overline{\Psi}^{\alpha j}}\right)\\ &+ i\chi(x)\left(\Psi_{\alpha i}\frac{\delta}{\delta\Psi_{\alpha i}} - \overline{\Psi}^{\alpha i}\frac{\delta}{\delta\overline{\Psi}^{\alpha i}}\right) - \partial_\mu\chi(x)\frac{\delta}{\delta A_\mu(x)}\Bigg].\end{aligned} \tag{A.11}$$

The horizontality condition is that this vector field is in the kernel of 1-form $d_x^*\delta A + \frac{i}{2}(\overline{\Psi}\delta\Psi - \delta\overline{\Psi}\Psi)$. This gives the following equation on $\chi(x)$:

$$(d_x^* d_x + \overline{\Psi}(x)\Psi(x))\chi(x) = -\overline{\Psi}(x)T^a\Psi(x), \tag{A.12}$$

which is solved by

$$\chi(x) = -\int_{M_4} d^4y G(x,y)\overline{\Psi}(y)T^a\Psi(y). \tag{A.13}$$

With this $\chi(x)$, equation (A.11) gives a horizontal lift of v_a. A computation shows

$$\begin{aligned} \iota_{V_a^h}\Theta &= -i\int_{M_4} d^4y \Big[G(x,y)\delta(\overline{\Psi}(y)T^a\Psi(y)) \\ &\quad - \int_{M_4} d^4u\, G(x,u)\delta(\overline{\Psi}(u)\Psi(u))G(u,y)\overline{\Psi}(y)T^a\Psi(y)\Big] \\ &= -\delta\int_{M_4} d^4y\, G(x,y)\overline{\Psi}(y)iT^a\Psi(y). \end{aligned} \tag{A.14}$$

In particular, this implies that $\delta\iota_{V_a^h}\Theta = 0$. To prove that $d\iota_{d\pi(v_a)}\Omega = 0$, we do the following standard computation:

$$\begin{aligned} (d\iota_{d\pi(v_a)}\Omega)(U,V) &= U\cdot(\iota_{d\pi(v_a)}\Omega)(V) - V\cdot(\iota_{d\pi(v_a)}\Omega)(U) - (\iota_{d\pi(v_a)}\Omega)([U,V]) \\ &= U^h\cdot(\iota_{V_a^h}\Theta)(V^h) - V^h\cdot(\iota_{V_a^h}\Theta)(U^h) - (\iota_{V_a^h}\Theta)([U,V]^h) \\ &= (\delta\iota_{V_a^h}\Theta)(U^h,V^h) - (\iota_{V_a^h}\Theta)(hor[U^h,V^h] - [U^h,V^h]) \\ &= -2(\iota_{V_a^h}\Theta)(\Theta(U^h,V^h)) = 0, \end{aligned} \tag{A.15}$$

where the Lie algebra element $\Theta(U^h,V^h)$ is identified with the vertical vector field. Therefore, substituting the vertical vector field $\Theta(U^h,V^h)$ into the 1-form $\iota_{V_a^h}\Theta$ gave zero in the last step. This proves that, indeed, $d\iota_{d\pi(v_a)}\Omega = 0$. Notice that, actually, we have shown more:

$$\iota_{d\pi(v_a)}\Omega = -idH_a, \tag{A.16}$$

where

$$H_a = \int_{M_4} d^4y\, G(x,y)\overline{\Psi}(y)T_a\Psi(y). \tag{A.17}$$

B. Topological Twist of SQED

It is possible to write a Lagrangian 4d TQFT that gives multi-monopole equations (4.1) as its localization equations. It was done in [82] using the Mathai–Quillen formalism, and the result was a topologically twisted 4d $\mathcal{N}=2$ gauge theory with gauge group $U(1)$ and one charged hypermultiplet. It is trivial to generalize it to N_f hypers, and here we summarize such a theory.

A topological $\mathcal{N}=2$ vector multiplet contains a local connection 1-form A_μ, a complex bosonic scalar ϕ, an auxiliary bosonic self-dual 2-form $D_{\mu\nu}$, a fermionic scalar ρ, a fermionic 1-form ψ_μ and a fermionic self-dual 2-form $\chi_{\mu\nu}$. A topological $\mathcal{N}=2$ hypermultiplet contains a complex bosonic spinor field of left (or positive) chirality Ψ_α, an auxiliary complex bosonic spinor field of right (or negative) chirality $h_{\dot\alpha}$, a pair of left-handed fermion spinors $\mu_\alpha, \overline{\mu}^\alpha$ and a pair of right-handed fermion spinors $\nu_{\dot\alpha}$, $\overline{\nu}^{\dot\alpha}$. (The latter four spinors are independent because we work in Euclidean space; in Minkowski space, they would be related by complex conjugation, which would flip chirality.) In the N_f-monopole case, all hypermultiplets are equipped with the additional index $i = 1, \ldots, N_f$. The Q-transformations of these fields are[27]

$$
\begin{aligned}
&[Q, A_\mu] = \psi_\mu, \quad \{Q, \psi_\mu\} = \partial_\mu \phi, \quad \{Q, \phi\} = 0, \\
&[Q, \overline{\phi}] = \rho, \quad \{Q, \rho\} = 0, \\
&\{Q, \chi_{\mu\nu}\} = D_{\mu\nu}, \quad [Q, D_{\mu\nu}] = 0, \\
&[Q, \Psi_{\alpha i}] = \mu_{\alpha i}, \quad \{Q, \mu_{\alpha i}\} = -i(\phi + z_i)\Psi_{\alpha i}, \\
&[Q, \overline{\Psi}^{\alpha i}] = \overline{\mu}^{\alpha i}, \quad \{Q, \overline{\mu}^{\alpha i}\} = -i(\overline{\phi} + \overline{z}_i)\overline{\Psi}^{\alpha i}, \\
&\{Q, \nu_{\dot\alpha i}\} = h_{\dot\alpha i}, \quad [Q, h_{\dot\alpha i}] = -i(\phi + z_i)\nu_{\dot\alpha i}, \\
&\{Q, \overline{\nu}^{\dot\alpha i}\} = \overline{h}^{\dot\alpha i}, \quad [Q, \overline{h}^{\dot\alpha i}] = -i(\overline{\phi} + \overline{z}_i)\overline{\nu}^{\dot\alpha i}.
\end{aligned} \tag{B.1}
$$

The action is defined as

$$
S = \{Q, \Omega + \Omega_\eta\} + S_{\text{top}}, \tag{B.2}
$$

where Ω is given by

$$
\begin{aligned}
\Omega = \int_{M_4} \mathrm{Vol}_{M_4} \Big[&- \chi^{\alpha\beta} \left(\frac{i}{\sqrt{2}} (F^+_{\alpha\beta} - i\overline{\Psi}^i_{(\alpha} \Psi_{\beta) i}) - \frac{1}{4} D_{\alpha\beta} \right) - \frac{i}{2} (\overline{\nu}^{i\dot\alpha} D_{\alpha\dot\alpha} \Psi^\alpha_i + \overline{\Psi}^{\alpha i} D_{\alpha\dot\alpha} \nu^{\dot\alpha}_i) \\
&+ \frac{1}{8} (\overline{\nu}^{\dot\alpha i} h_{\dot\alpha i} - \overline{h}_{\dot\alpha i} \nu^{\dot\alpha i}) + \overline{\phi} d^* \psi - \frac{i}{2} \overline{\mu}^{\alpha i} (\overline{\phi} + \overline{z}_i) \Psi_{\alpha i} + \frac{i}{2} \overline{\Psi}^{\alpha i} (\overline{\phi} + \overline{z}_i) \mu_{\alpha i} \Big]
\end{aligned} \tag{B.3}
$$

and Ω_η gives a topological FI term determined in terms of a fixed harmonic $\eta \in \mathcal{H}^{2,+}(M_4)$:

$$
\Omega_\eta = -\frac{i}{\sqrt{2}} \int_{M_4} \chi \wedge \eta, \tag{B.4}
$$

and S_{top} is a topological term that is Q-closed but not necessarily Q-exact. It might include a θ term (which we choose not to add) and, for our purposes, has the following form:

$$
S_{\text{top}} = -\frac{1}{2\pi} \int_{M_4} F \wedge \log t, \tag{B.5}
$$

where $\log t \in \mathcal{H}^{2,+}(M_4)$. From the topological theory point of view, this t is simply a fugacity for the first Chern class of our Spin^c structure. However, if we think of our theory as obtained by twisting of the physical model (which we do, since we rely on brane constructions), then t is not a free parameter but rather $t = \exp i\eta$.

Of course, this η is the same perturbation in multi-monopole equations that we introduced before. For reference, we also provide the full action of the model with auxiliary fields integrated out:

[27] With apologies to QFT practitioners, we remind thre reader that z_i denote mass parameters of the hypermultiplets, $i = 1, \ldots, N_f$.

$$S = \int_{M_4} \mathrm{Vol}_{M_4} \Big[\frac{1}{2} \left(F^+_{\alpha\beta} - i\overline{\Psi}^i_{(\alpha} \Psi_{\beta)i} + \eta_{\alpha\beta} \right)^2 + D^{\dot{\alpha}}_{\alpha} \overline{\Psi}^{\alpha i} D_{\beta\dot{\alpha}} \Psi^\beta_i + \overline{\phi} d^* d\phi + \rho d^* \psi$$
$$+ \frac{i}{\sqrt{2}} \chi^{\alpha\beta} \left((d\psi)^+_{\alpha\beta} - i\overline{\Psi}^i_{(\alpha} \mu_{\beta)i} - i\overline{\mu}^i_{(\alpha} \Psi_{\beta)i} \right) + \frac{i}{2} (\overline{\nu}^{i\dot{\alpha}} D_{\alpha\dot{\alpha}} \mu^\alpha_i - \overline{\mu}^{\alpha i} D_{\alpha\dot{\alpha}} \nu^{\dot{\alpha}}_i)$$
$$- \frac{1}{2} (\overline{\nu}^{i\dot{\alpha}} \psi_{\alpha\dot{\alpha}} \Psi^\alpha_i - \overline{\Psi}^{\alpha i} \psi_{\alpha\dot{\alpha}} \nu^{\dot{\alpha}}_i) - \frac{i}{2} \rho \overline{\mu}^{\alpha i} \Psi_{\alpha i} + \frac{i}{2} \rho \overline{\Psi}^{\alpha i} \mu_{\alpha i}$$
$$+ \frac{i}{4} \overline{\nu}^{\dot{\alpha} i} (\phi + z_i) \nu_{\dot{\alpha} i} + \overline{\Psi}^{\alpha i} (\overline{\phi} + \overline{z}_i)(\phi + z_i) \Psi_{\alpha i} + i\overline{\mu}^{\alpha i} (\overline{\phi} + \overline{z}_i) \mu_{\alpha i} \Big] + S_{\mathrm{top}}. \quad \text{(B.6)}$$

One can see that computations in this model go in parallel with our previous analysis based on the moduli space geometry and its universal bundle. Since the action is Q-exact, one multiplies it by κ and takes the limit $\kappa \to \infty$ without affecting the answer for path integral. In this way, computation localizes to configurations which, in the case of zero masses $z_i = 0$, are given by the perturbed multi-monopole equations and an additional condition:

$$\phi \Psi_i = 0. \quad \text{(B.7)}$$

This equation has two branches of solutions. One is the Coulomb branch with all $\Psi_i = 0$ and ϕ arbitrary. In the case of generic metric and/or perturbation (and $b_2^+ > 1$), this branch is empty because one of the equations reduces to $F^+ + \eta = 0$, which generically has no solutions. Another one is the Higgs branch, where $\phi = 0$ and Ψ_i are determined from the multi-monopole equations. This branch contributes non-trivially, and the answer is expressed as the integral over the multi-monopole moduli space $\mathcal{M}_{N_f}$. The integrand consists of determinants for fluctuations of various fields around the localization locus, as well as possible insertions of observables.

One can argue in a multitude of ways that, in order to get a sensible answer, we actually have to insert an observable in the path integral, the simplest option being

$$\phi(x)^{\frac{1}{2} \mathrm{VirDim}\, \mathcal{M}_{N_f}}. \quad \text{(B.8)}$$

Of course, the most obvious way to argue this is by using the R-symmetry anomaly (which was already used in section 3). Alternatively (which is equivalent to anomaly considerations), one could, in the limit $\kappa \to \infty$, expand the fields as field $=$ field$_0 + \frac{1}{\sqrt{\kappa}}$field$_1$, where field$_0$ is the localization locus value, and field$_1$ is the fluctuation. By accurately counting powers of κ, one gets $\kappa^{\frac{1}{2}\dim \mathcal{M}_{N_f}}$ from the path integral measure, $\kappa^{-\frac{1}{4}(\dim \mathcal{M}_{N_f} + \dim \mathrm{coker}\, ds)}$ from fermion zero modes (here, ds is a map in the linearization complex of multi-monopole equations) and $\kappa^{-\frac{1}{4}\mathrm{VirDim}\, \mathcal{M}_{N_f}}$ from the above insertion. These three contributions cancel.

In order to get a non-zero answer, we need to bring down $\frac{1}{2}\mathrm{VirDim}\, \mathcal{M}_{N_f}$ powers of $\overline{\phi}$ from the action to contract it with (B.8). For contractions, we use the Green's function $\langle \phi(x) \overline{\phi}(y) \rangle_{free}$, which is given by

$$G(x,y) = \langle x | \frac{1}{d^* d + \sum_i \overline{\Psi}_i \Psi_i} | y \rangle. \quad \text{(B.9)}$$

This is the same Green's function as we had in (4.23). Each power of $\overline{\phi}$ comes from the interaction term $i\overline{\mu}^{\alpha i} \overline{\phi} \mu_{\alpha i}$ in the action and, after contracting ϕ's and leaving only zero modes for μ's, we end up with the following insertion in the action:

$$\left(\int_{M_4} d^4y\, G(x,y) i\overline{\mu}^{\alpha i}(y) \mu_{\alpha i}(y) \right)^{\frac{1}{2}\mathrm{VirDim}\,\mathcal{M}_{N_f}}. \tag{B.10}$$

We recognize the expression in the parentheses as the curvature form Θ representing the first Chern class of the universal bundle, as computed in (4.27), if we identify $\delta\Psi_i$ with the zero mode of μ_i. So, indeed, as expected, the answer is given by an integral $\int_{\mathcal{M}_{N_f}} (c_1(\mathcal{L}))^{d/2}$, where $d = \mathrm{VirDim}\,\mathcal{M}_{N_f}$. As we said, this integral is, in general, divergent due to the non-compactness of $\mathcal{M}_{N_f}$.

This problem is resolved by switching on masses, which are nothing else but equivariant parameters for the maximal torus of the flavour symmetry group $SU(N_f)$ [23, 24]. With masses, equation (B.7) replaced by

$$(\phi + z_i)\Psi_i = 0. \tag{B.11}$$

Again, the Coulomb branch would have all $\Psi_i = 0$ and arbitrary ϕ, but it is generically empty. The Higgs branch in the presence of masses gets partially lifted, and we have only fixed points of the maximal torus $U(1)^{N_f - 1} \subset SU(N_f)$ left on it. They occur at

$$\begin{aligned} \phi &= -z_i, \\ \Psi_j &= 0, \quad j \neq i, \end{aligned} \tag{B.12}$$

where $i = 1, \ldots, N_f$, and, for each i, Ψ_i and the connection satisfy the 1-monopole Seiberg–Witten equations. We call the ith component of the localization locus F_i, and this is precisely the equivariant localization discussed in the main text. One advantage of such a TQFT approach is that reduction to the 1-monopole problem might be somewhat more transparent for some readers.

The contribution of the ith fixed point set F_i is simply given by the path integral of the 1-monopole problem (localized to the 1-monopole Seiberg–Witten moduli space), coupled to the extra $(N_f - 1)$ hypermultiplets, but this coupling vanishes in the $t \to \infty$ limit, as one can see by studying the topological action more carefully. Thus, the final result factorizes into the 1-monopole answer and the determinant for $(N_f - 1)$ topological hypers.

The 1-monopole answer is simply $\mathrm{SW}(\lambda)$, the standard Seiberg-–Witten invariant. To compute the determinant of the $(N_f - 1)$ hypers, we write the quadratic part of the action which describes their fluctuations:

$$\begin{aligned} \int_{M_4} \mathrm{Vol}_{M_4} \Bigg[& \sum_{j\neq i} D^{\dot\alpha}{}_{\alpha} \overline{\Psi}^{\alpha j} D_{\beta\dot\alpha} \Psi_j^{\beta} + \frac{i}{2} \sum_{j\neq i} (\overline{\nu}^{j\dot\alpha} D_{\alpha\dot\alpha} \mu_j^{\alpha} - \overline{\mu}^{\alpha j} D_{\alpha\dot\alpha} \nu_j^{\dot\alpha}) \\ & + \sum_{j\neq i} |z_j - z_i|^2 \overline{\Psi}^{\alpha j} \Psi_{\alpha j} + \sum_{j\neq i} \frac{i}{4} \overline{\nu}^{\dot\alpha j} (z_j - z_i) \nu_{\dot\alpha j} + \sum_{j\neq i} i\overline{\mu}^{\alpha j} (\overline{z}_j - \overline{z}_i) \mu_{\alpha j} \Bigg]. \end{aligned} \tag{B.13}$$

The Dirac operator matches its own non-zero modes of positive and negative chirality, so their eigenvalues coincide and cancel between bosonic and fermionic contributions. This is not true for the zero modes of $\not{D}$, and their contribution brings down factors of masses in the following way:

$$\frac{\prod_{j\neq i} (z_j - z_i)^{\#R_0} (\overline{z}_j - \overline{z}_i)^{\#L_0}}{\prod_{j\neq i} |z_j - z_i|^{2\#L_0}} = \frac{1}{\prod_{j\neq i} (z_j - z_i)^{\#L_0 - \#R_0}}, \tag{B.14}$$

where $\#L_0$ and $\#R_0$ are numbers of left and right zero modes of $\not{D}$. Of course, $\#L_0 - \#R_0 = \mathrm{Ind}_{\mathbb{C}}(\not{D}) = \frac{1}{8}(\lambda^2 - \sigma)$, and the above expression gives the determinant due to the $(N_f - 1)$ hypers. Combining this with the 1-monopole answer $\mathrm{SW}(\lambda)$ and summing over $i = 1, \ldots, N_f$, which labels localization loci in this case, we finally obtain

$$\mathrm{SW}(\lambda) \sum_{i=1}^{N_f} \frac{1}{\prod_{j \neq i} (z_j - z_i)^{\frac{1}{8}(\lambda^2 - \sigma)}}, \tag{B.15}$$

which is the same answer as we got for $\mathrm{ESW}_{M_4}(\lambda, z_i)$ in section 4.2 by performing equivariant integration over the moduli space. This is very satisfying, and also shows that the answer does not really depend on the (strong) simple type assumption that we had in section 4.2.

The partition function of the topological theory is given by the sum over fluxes λ and includes the fugacity $t = \exp i\eta$:

$$Z = \sum_{\lambda} t^{\lambda} \, \mathrm{ESW}_{M_4}(\lambda, z_i). \tag{B.16}$$

REFERENCES

[1] A. Gadde, S. Gukov and P Putrov, *Fivebranes and 4-manifolds*, arXiv:1306.4320 (2015).

[2] H. Nakajima, *Instantons on ale spaces, quiver varieties, and Kac–Moody algebras*, *Duke Math. J.* **76** (1994) 365–416.

[3] H. Nakajima, *Instantons and affine Lie algebras*, *Nucl. Phys. Proc. Suppl.* **46** (1996) 154–61.

[4] L. Crane and I. B. Frenkel, *Four-dimensional topological quantum field theory, Hopf categories, and the canonical bases*, *J. Math. Phys.* **35** (1994) 5136–54.

[5] M. Khovanov, *A categorification of the Jones polynomial*, *Duke Math. J.* **101** (2000) 359–426.

[6] M. Khovanov and L. Rozansky *Matrix factorizations and link homology*, *Fund. Math.* **199** (2008) 1–91.

[7] M. Khovanov and L. Rozansky *Matrix factorizations and link homology. II*, *Geom. Topol.* **12** (2008) 1387–425.

[8] S. Gukov, A. S. Schwarz and C. Vafa, *Khovanov–Rozansky homology and topological strings*, *Lett. Math. Phys.* **74** (2005) 53–74.

[9] M. Aganagic, T. Ekholm, L. Ng and C. Vafa, *Topological strings, D-model, and knot contact homology*, *Adv. Theor. Math. Phys.* **18** (2014) 827–956.

[10] S. Gukov, *Gauge theory and knot homologies*, *Fortsch. Phys.* **55** (2007) 473–490 [0706.2369].

[11] E. Witten, *Fivebranes and Knots*, arXiv:1101.3216 (2011).

[12] N. Carqueville and D. Murfet, *Computing Khovanov–Rozansky homology and defect fusion*, *Algebr. Geom. Topol.* **14** (2014) 489–537.

[13] S. Chun, S. Gukov and D. Roggenkamp, *Junctions of surface operators and categorification of quantum groups*, arXiv:1507.06318 (2016).

[14] S. Nawata and A. Oblomkov, *Lectures on knot homology* arXiv:1510.01795 (2015).

[15] M. Jacobsson, *An invariant of link cobordisms from Khovanov homology*, *Algebr. Geom. Topol.* **4** (2004) 1211–51.

[16] D. Bar-Natan, *Khovanov's homology for tangles and cobordisms, Geom. Topol.* **9** (2005) 1443–99.
[17] J. S. Carter, M. Saito and S. Satoh, *Ribbon-moves for 2-knots with 1-handles attached and Khovanov–Jacobsson numbers, Proc. Amer. Math. Soc.* **134** (2006) 2779–83.
[18] M. Khovanov *An invariant of tangle cobordisms, Trans. Amer. Math. Soc.* **358** (2006) 315–27.
[19] J. Rasmussen, *Khovanov's invariant for closed surfaces,* arXiv:math/0502527 (2005).
[20] P. B. Kronheimer and T. S. Mrowka, *Gauge theory and Rasmussen's invariant, J. Topol.* **6** (2013) 659–74.
[21] S. Gukov, P Putrov and C. Vafa, *Fivebranes and 3-manifold homology,* arXiv:1602.05302 (2016).
[22] S. Gukov, M. Marino and P Putrov, *Resurgence in complex Chern–Simons theory,* arXiv:1605.07615 (2016).
[23] G. W. Moore, N. Nekrasov and S. Shatashvili, *Integrating over Higgs branches, Commun. Math. Phys.* **209** (2000) 97–121.
[24] N. A. Nekrasov, *Seiberg–Witten prepotential from instanton counting, Adv. Theor. Math. Phys.* **7** (2004) 831–64.
[25] A. Losev, G. W. Moore and S. L. Shatashvili, *M & m's, Nucl. Phys. B* **522** (1998) 105–24.
[26] E. P. Verlinde, *Global aspects of electric–magnetic duality, Nucl. Phys. B* **455** (1995) 211–28.
[27] O J. Ganor, *Compactification of tensionless string theories,* arXiv:hep-th/9607092 (1996).
[28] A. Gadde, S. Gukov and P Putrov, *Exact solutions of 2d supersymmetric gauge theories,* arXiv:1404.5314 (2014).
[29] A. Gadde and P Putrov, *Exact solutions of (0,2) Landau-Ginzburg models,* arXiv:1608.07753 (2016).
[30] E Witten, *Five-brane effective action in M theory, J. Geom. Phys.* **22** (1997) 103–33.
[31] J. A. Harvey R. Minasian and G. W. Moore, *Non-Abelian tensor multiplet anomalies, JHEP* **09** (1998) 004.
[32] L. F. Alday F. Benini and Y. Tachikawa, *Liouville/Toda central charges from M5-branes, Phys. Rev. Lett.* **105** (2010) 141601.
[33] C. M. Hull and E. Witten, *Supersymmetric sigma models and the heterotic string, Phys. Lett.* **B160** (1985) 398–402.
[34] S. H. Katz and E. Sharpe, *Notes on certain (0,2) correlation functions, Commun. Math., Phys.* **262** (2006) 611–44.
[35] E. Witten, *Two-dimensional models with (0,2) supersymmetry: Perturbative aspects, Adv. Theor. Math. Phys.* **11** (2007), 1–63.
[36] I. Melnikov, S. Sethi and E. Sharpe, *Recent developments in (0,2) mirror symmetry, SIGMA* **8** (2012) 068.
[37] K. A. Intriligator, *Anomaly matching and a Hopf–Wess–Zumino term in 6d, N=(2,0) field theories, Nucl. Phys. B* **581** (2000) 257–73.
[38] P. Yi, *Anomaly of (2,0) theories, Phys. Rev. D* **64** (2001) 106006.
[39] J. P Gauntlett, N. Kim and D. Waldram, *M-Fivebranes wrapped on supersymmetric cycles, Phys. Rev. D* **63** (2001) 126001.
[40] M. Itoh, *Moduli of half conformally flat structures, Math. Ann.* **296** (1993) 687–708.
[41] B. Assel, S. Schafer-Nameki and J.-M. Wong, *M5-branes on S^2 M_4: Nahms equations and 4d topological sigma-models, JHEP* **09** (2016) 120.
[42] P. Putrov, J. Song and W. Yan, *(0,4) dualities, JHEP* **03** (2016) 185.
[43] E. Witten, *Solutions of four-dimensional field theories via M theory, Nucl. Phys. B* **500** (1997) 3–42.
[44] S. Gukov, D. Pei, P Putrov and C. Vafa, *BPS spectra and 3-manifold invariants,* arXiv:1701.06567 (2017).
[45] A. Hook, S. Kachru and G. Torroba, *Supersymmetric defect models and mirror symmetry, JHEP* **11** (2013) 004 [1308.4416].

[46] D. Tong and K. Wong, *Vortices and impurities, JHEP* **01** (2014) 090.
[47] C. G. Callan and J. M. Maldacena, *Brane death and dynamics from the Born–Infeld action, Nucl. Phys. B* **513** (1998) 198–212.
[48] G. W. Gibbons, *Born–Infeld particles and Dirichlet p-branes, Nucl. Phys. B* **514** (1998) 603–39.
[49] K. Hori and C. Vafa, *Mirror symmetry*, arXiv:hep-th/0002222 (2000).
[50] N. Seiberg, *Notes on quantum Liouville theory and quantum gravity, Prog. Theor. Phys., Suppl.* **102** (1990) 319–49.
[51] E. Gorsky S. Gukov and M. Stosic, *Quadruply-graded colored homology of knots*, arXiv:1304.3481 (2013).
[52] H. Ooguri and C. Vafa, *Knot invariants and topological strings, Nucl. Phys. B* **577** (2000) 419–38.
[53] R. Harvey and H. B. Lawson, Jr., *Calibrated geometries, Acta Math.* **148** (1982) 47–157.
[54] D. S. Berman and J. A. Harvey *The self-dual string and anomalies in the M5-brane, JHEP* **11** (2004) 015.
[55] P. Brax and J. Mourad, *Open supermembranes coupled to M theory five-branes, Phys. Lett. B* **416** (1998) 295–302.
[56] B. Haghighat, A. Iqbal, C. Kozaz, G. Lockhart and C. Vafa, *M-Strings, Commun. Math., Phys.* **334** (2015) 779–842.
[57] M. Mackaay M. Stosic and P Vaz, *sl(N)-link homology* $(N \geq 4)$ *using foams and the Kapustin–Li formula, Geom. Topol.* **13** (2009) 1075–28.
[58] H. Queffelec and D. E. V. Rose, *The* sl_n *foam 2-category: A combinatorial formulation of Khovanov–Rozansky homology via categorical skew Howe duality, Adv. Math.* **302** (2016) 1251–339.
[59] H. Shimizu and Y. Tachikawa, *Anomaly of strings of 6d* $N = (1, 0)$ *theories, JHEP* **11** (2016) 165.
[60] S. M. Kuzenko and S. Theisen, *Correlation functions of conserved currents in N=2 superconformal theory, Class. Quant. Grav.* **17** (2000) 665–96.
[61] A. D. Shapere and Y. Tachikawa, *Central charges of N=2 superconformal field theories in four dimensions, JHEP* **09** (2008) 109.
[62] E. Witten, *Monopoles and four manifolds, Math. Res. Lett.* **1** (1994) 769–96.
[63] J. Labastida and M. Marino, *Topological quantum field theory and four manifolds*, Mathematical Physics Studies, 25. Springer, Dordrecht (2005).
[64] J. A. Bryan and R. Wentworth, *The multi-monopole equations for Käahler surfaces, Turkish J. Math.* **20** (1996) 119–28.
[65] V. Mikhaylov, *Analytic torsion, 3d mirror symmetry and supergroup Chern–Simons theories*, arXiv:1505.03130 (2015).
[66] A. Hanany and D. Tong, *Vortices, instantons and branes, JHEP* **07** (2003) 037.
[67] A. Losev, N. Nekrasov and S. L. Shatashvili, *Issues in topological gauge theory, Nucl. Phys. B* **534** (1998) 549–611, [hep-th/9711108].
[68] A. Haydys, *Generalized Seiberg–Witten equations and hyperkähler geometry, PhD dissertation, Universitat Gottingen, Gottingen* (2006).
[69] A. Haydys and T. Walpuski, *A compactness theorem for the Seiberg–Witten equation with multiple spinors in dimension three, Geom. Funct. Anal.* **25** (2015) 1799–821.
[70] A. Haydys, *Topology of the blow-up set for the Seiberg–Witten equation with multiple spinors*, arXiv:1607.01763 (2018).
[71] S. Donaldson and P Kronheimer, *The geometry of four-manifolds* Oxford Mathematical Monographs. Oxford University Press, New York (2004).
[72] F. Benini and A. Zaffaroni, *Supersymmetric partition functions on Riemann surfaces*, arXiv:1605.06120 (2017).
[73] C. Closset and H. Kim, *Comments on twisted indices in 3d supersymmetric gauge theories, JHEP* **08** (2016) 059.

[74] F. Benini and A. Zaffaroni, *A topologically twisted index for three-dimensional supersymmetric theories*, arXiv:1504.03698 (2015).
[75] C. Closset, S. Cremonesi and D. S. Park, *The equivariant A-twist and gauged linear sigma models on the two-sphere, JHEP* **06** (2015) 076.
[76] A. Gadde, L. Rastelli, S. S. Razamat and W. Yan, *Gauge theories and Macdonald polynomials, Commun. Math. Phys.* **319** (2013) 147–93.
[77] S. Gukov, D. Pei, W. Yan and K. Ye, *Equivariant Verlinde algebra from superconformal index and Argyres–Seiberg duality*,l arXiv:1605.06528 (2018).
[78] L. Fredrickson, D. Pei, W. Yan and K. Ye, *Argyres–Douglas theories, chiral algebras and wild Hitchin characters*, arXiv:1701.08782 (2018).
[79] S. K. Donaldson, *Polynomial invariants for smooth four-manifolds, Topology* **29** (1990) 257–315
[80] P. C. Argyres and N. Seiberg, *S-duality in N=2 supersymmetric gauge theories, JHEP* **12** (2007) 088.
[81] A. Gadde, L. Rastelli, S. S. Razamat and W. Yan, *The superconformal index of the E_6 SCFT, JHEP* **08** (2010) 107.
[82] J. M. F. Labastida and M. Marino, *A topological Lagrangian for monopoles on four manifolds, Phys. Lett. B* **351** (1995) 146–52.

12

Hyperfunctions, the Duistermaat–Heckman Theorem and Loop Groups

LISA C. JEFFREY
JAMES A. MRACEK

Department of Mathematics, University of Toronto, Bahen Centre, 40 St. George Street, Room 6290, Toronto, ON, Canada M5S2E4
Email: jmracek@math.toronto.edu, jeffrey@math.toronto.edu

ABSTRACT. In this article, we investigate the Duistermaat–Heckman theorem using the theory of hyperfunctions. In applications involving Hamiltonian torus actions on infinite-dimensional manifolds, this more general theory seems to be necessary in order to accommodate the existence of the infinite-order differential operators which arise from the isotropy representations on the tangent spaces to fixed points. We will quickly review the theory of hyperfunctions and their Fourier transforms. We will then apply this theory to construct a hyperfunction analogue of the Duistermaat–Heckman distribution. Our main goal will be to study the Duistermaat–Heckman hyperfunction of $\Omega SU(2)$ but, in getting to this goal, we will also characterize the singular locus of the moment map for the Hamiltonian action of $T \times S^1$ on ΩG. The main goal of this paper is to present a Duistermaat–Heckman hyperfunction arising from a Hamiltonian action on an infinite-dimensional manifold.

1. Introduction

For finite-dimensional compact symplectic manifolds equipped with a Hamiltonian torus action with moment map μ, the Duistermaat–Heckman theorem gives an explicit formula for an oscillatory integral over the manifold in terms of information about the fixed point set of the torus action, and the action of the torus on the normal bundle to the fixed point set. Furthermore, the Fourier transform of this integral controls the structure of the cohomology

Date: 1 May 2017.

Jeffrey, L. C., Mracek, J. A., *Hyperfunctions, the Duistermaat–Heckman Theorem and Loop Groups.* In: *Geometry and Physics: A Festschrift in Honour of Nigel Hitchin,* Jørgen Ellegaard Andersen, Andrew Dancer, Oscar García-Prada (Eds): Oxford University Press (2018). © Oxford University Press. DOI: 10.1093/oso/9780198802006.003.0012

rings of the various symplectic reductions. For Hamiltonian actions on infinite-dimensional symplectic manifolds, little is known is known about the behaviour of their corresponding Duistermaat–Heckman distributions. In this paper, we define the same oscillatory integral for the natural Hamiltonian torus action on the based loop group, as introduced by Atiyah and Pressley, in order to give an expression for a Duistermaat–Heckman *hyperfunction*. The essential reason for introducing hyperfunction theory is that the local contribution to the Duistermaat–Heckman polynomial near the image of a fixed point is a Green's function for an infinite-order differential equation. Since infinite-order differential operators do not act on Schwarz distributions, we are forced to use this more general theory. After this work had been completed, we learned of the related work of Roger Picken [Pic89].

The layout of this paper is as follows. In Section 2, we review the theory of hyperfunctions, following [KS99, Kan89]. In Section 3, we study hyperfunctions that arise naturally from Hamiltonian group actions via localization. Section 4 reviews the based loop group and its Hamiltonian action (introduced by Atiyah and Pressley [AP83]). Section 5 describes the fixed point set of any 1-parameter subgroup of this torus. In Section 6, we demonstrate the theorems of Section 5 for the based loop group of $SU(2)$. In Section 7, we compute the isotropy representations of the torus that acts on the based loop group on the tangent spaces to each of the fixed points. Finally, Section 8 applies the hyperfunction localization theorem to $\Omega SU(2)$.

2. Introduction to Hyperfunctions

In this section, we will quickly review the elements of hyperfunction theory which are needed in order to make sense of the fixed point localization formula for a Hamiltonian action on an infinite-dimensional manifold. We will assume that the reader is familiar with Hamiltonian group actions, but not necessarily with hyperfunctions. Our exposition will follow a number of sources. The bulk of the background material follows [KS99, Kan89], while the material on the Fourier transform of hyperfunctions is covered in [Kan89] as well as the original paper of Kawai [Kaw70]. The original papers of Sato also give great insight into the motivation for introducing this theory [Sat59]. The lecture notes of Kashiwara, Kawai and Sato also give useful insight into why hyperfunction and microfunction theory are needed to solve problems in linear partial differential equations [KKS].

We will let $\mathcal{O}$ be the sheaf of holomorphic functions on $\mathbb{C}^n$. Points in $\mathbb{C}^n$ will be denoted $z=(z_1,\dots,z_n)=(x_1+iy_1,\dots,x_n+iy_n)$, and we will write $\mathrm{Re}(z)=(x_1,\dots,x_n)\in\mathbb{R}^n$ and $\mathrm{Im}(z)=(y_1,\dots,y_n)\in i\mathbb{R}^n$. An open convex cone $\gamma\subseteq\mathbb{R}^n$ is a convex open set such that, for every $c\in\mathbb{R}_{>0}$, if $x\in\gamma$ then $cx\in\gamma$. We allow $\mathbb{R}^n$ itself to be an open convex cone. If γ is an open convex cone, we will denote its polar dual cone by γ°. Let Γ denote the set of all open convex cones in $\mathbb{R}^n$. If $\gamma\in\Gamma$ and $\Omega\subseteq\mathbb{R}^n$ is an open set, then we denote $\Omega\times i\gamma=\left\{(x_j+iy_j)\in\mathbb{C}^n\,|\,\mathrm{Re}(x)\in\Omega,\mathrm{Im}(z)\in\gamma\right\}$. An *infinitesimal wedge*, denoted $\Omega\times i\gamma 0$, is a choice of an open subset $U\subseteq\Omega\times i\gamma$ which is asymptotic to the cone opening (we will not need the precise definition, so we omit it). We will denote the collection of germs of holomorphic functions on the wedge $\Omega\times i\gamma$ by $\mathcal{O}(\Omega\times i\gamma 0)$; that is, we take an inverse limit of the holomorphic functions varying over the collection of all infinitesimal wedges $\Omega\times i\gamma 0$:

$$\mathcal{O}(\Omega \times i\gamma 0) = \varprojlim_{U} \mathcal{O}(U)$$

We will use the notation $F(z+i\gamma 0)$ to denote an element of $\mathcal{O}(\Omega \times i\gamma 0)$.

Definition *A hyperfunction on $\Omega \subseteq \mathbb{R}^n$ is an element*

$$\sum_{i=1}^{n} F(z+i\gamma_i 0) \in \bigoplus_{\gamma \in \Gamma} \mathcal{O}(\Omega \times i\gamma 0) / \sim$$

where the equivalence relation is given as follows. If $\gamma_1, \gamma_2, \gamma_3 \in \Gamma$ are such that $\gamma_3 \subseteq \gamma_1 \cap \gamma_2$ and $F_i \in \mathcal{O}(\Omega \times i\gamma_i)$, then $F_1(z) + F_2(z) \sim F_3(z)$ if and only if $(F_1(z) + F_2(z))|_{\gamma_3} = F_3(z)$. If $\Omega \subseteq \mathbb{R}^n$, we will denote the collection of hyperfunctions on Ω by $\mathcal{B}(\Omega)$.

When we wish to keep track of the cones, we will use the notation $f(x) = \sum_j F(z+i\gamma_j 0)$; we call such a sum a boundary value representation of $f(x)$. Alternatively, we will sometimes also use the notation $F(z+i\gamma 0) = b_\gamma(F(z))$ when the expression for $F(z)$ makes it notationally burdensome to include the text $+i\gamma 0$. The association $\Omega \mapsto \mathcal{B}(\Omega)$ forms a flabby sheaf on $\mathbb{R}^n$, although we will not make use of the sheaf theoretical nature of hyperfunctions in this article. Actually, what is more is that this is a sheaf of D-modules on $\mathbb{C}^n$; the sheaf of differential operators acts termwise on each element of a sum $\sum_\gamma F(z+i\gamma 0)$.

The relation defining the sheaf of hyperfunctions allows us to assume that the cones appearing in the sum are disjoint. Indeed, if we have a hyperfunction $f(x) = F_1(z+i\gamma_1 0) + F_2(z+i\gamma_2 0)$ such that $\gamma_1 \cap \gamma_2 \neq \emptyset$, then we simply observe that we have an equality of equivalence classes:

$$F_1(z+i\gamma_1 0) + F_2(z+i\gamma_2 0) = (F_1+F_2)(z+i\gamma_1 \cap \gamma_2 0)$$

Similarly, if $\gamma_1 \subseteq \gamma_2$ and $F(z)$ is an analytic function on the wedge $\Omega \times i\gamma_1 0$ that admits an analytic extension to $\Omega \times i\gamma_2 0$, then $F(z+i\gamma_1 0) = F(z+i\gamma_2 0)$ as hyperfunctions. A particular example of this says that two hyperfunctions $f(x) = F_+(z+i0) + F_-(z-i0)$, $g(x) = G_+(z+i0) + G_-(z-i0) \in \mathcal{B}(\mathbb{R})$ are equal when the function

$$F(z) = \begin{cases} F_+(z) - G_+(z) & \mathrm{Im}(z) > 0 \\ F_-(z) - G_-(z) & \mathrm{Im}(z) < 0 \end{cases}$$

admits an analytic extension across the real axis.

The following definition is necessary to define the product of hyperfunctions. We say that a hyperfunction $f(x)$ is *microanalytic at* $(x,\xi) \in T^*\mathbb{R}^n$ if and only if there exists a boundary value representation

$$f(x) = \sum_{j=1}^{n} F(z+i\gamma_j 0)$$

such that $\gamma_j \cap \{y \in \mathbb{R}^n \mid \xi(y) < 0\} \neq \emptyset$ for every $j \in 1, \ldots, n$. The *singular support* of a hyperfunction $f(x)$, denoted $SS(f) \subseteq T^*\mathbb{R}^n$, is defined to be the set of points $(x,\xi) \in$

$T^*\mathbb{R}^n$ such that $f(x)$ is not microanalytic at (x,ξ). If $S \subseteq T^*\mathbb{R}^n$, then we denote $S^\circ = \{(x,\xi) \in T^*\mathbb{R}^n : (x,-\xi) \in S\}$.

Definition *Suppose that $f,g \in \mathcal{B}(\Omega)$ are two hyperfunctions such that $SS(f) \cap SS(g)^\circ = \emptyset$; then, the product $f(x) \cdot g(x)$ is the hyperfunction defined by*

$$f(x) \cdot g(x) = \sum_{j,k} (F_j \cdot G_k)(x + i(\gamma_j \cap \Delta_k)0)$$

where we have chosen appropriate boundary value representations

$$f(x) = \sum_j F_j(z + i\gamma_j 0)$$

$$g(x) = \sum_k G_k(z + i\Delta_k 0)$$

such that $\gamma_j \cap \Delta_k \neq \emptyset$ for all j,k.

In the above definition, the condition on singular support is simply ensuring the existence of boundary value representations of f and g such that, for all pairs j,k, the intersection $\gamma_j \cap \Delta_k \neq \emptyset$ [Kan89, Theorem 3.2.5].

Under restricted circumstances, we may define an infinite product of hyperfunctions. This result will be necessary to define the equivariant Euler class of the normal bundle to a fixed point in ΩG as a hyperfunction.

Lemma 1 *If $\{F_k(z + i\gamma_k 0)\}_{k=1}^\infty$ is a sequence of hyperfunctions on Ω such that*

(1) For all pairs $j \neq k$, $SS(F_k(z + i\gamma_k 0)) \cap SS(F_j(z + i\gamma_k 0))^\circ = \emptyset$

(2) $\gamma = \bigcap_{k=1}^{\infty} \gamma_k$ is open

(3) The infinite product $\prod_{k=1}^{\infty} F_k(z)$ is uniformly convergent on compact subsets of $\Omega \times i\gamma$

then there exists a hyperfunction $F(z + i\gamma 0)$ such that

$$F(z + i\gamma 0) = \prod_{k=1}^{\infty} F(z + i\gamma_k 0)$$

Proof The condition on singular support is necessary to define any product of the F_k. Since the intersection of the cones is open, the wedge $\Omega \times i\gamma$ is a well-defined open set in $\mathbb{C}^n$, and the convergence condition on the infinite product ensures that the following limit is a holomorphic function on $\Omega \times i\gamma$:

$$F(z) = \lim_{N\to\infty} \prod_{k=1}^{N} F_k(z)$$

This result has shown that the infinite product of the hyperfunctions $F_k(z + i\gamma_k 0)$ is well defined and equal to $F(z + i\gamma 0)$. □

We now describe how to define the Fourier transform of a hyperfunction. The following two definitions are central to the theory of hyperfunction Fourier transforms. We will restrict our attention to the class of *Fourier hyperfunctions*, also known as *slowly increasing hyperfunctions*.

Definition *A holomorphic function $F \in \mathcal{O}(\mathbb{R}^n \times i\gamma 0)$ is called slowly increasing if and only if, for every compact subset $K \subseteq i\gamma_j 0$ and for every $\epsilon > 0$, there exist constants $M, C > 0$ such that, for all $z \in \mathbb{R}^n \times iK$, if $|Re(z)| > M$, then $|F_j(z)| \leq C\exp(\epsilon Re(z))$.*

A holomorphic function $F \in \mathcal{O}(\mathbb{R}^n \times i\gamma 0)$ is called exponentially decreasing on the (not necessarily convex) cone $\Delta \subseteq \mathbb{R}^n$ if and only if there exists $\delta > 0$, such that, for every compact $K \subseteq i\gamma_j 0$ and for every $\epsilon > 0$, there exist constants $M, C > 0$ such that, for every $z \in \Delta \times iK$, if $|Re(z)| > M$, then $|F_j(z)| \leq C\exp(-(\delta - \epsilon)Re(z))$.

Remarks on the definition:

(1) A hyperfunction will be called slowly increasing (resp. exponentially decreasing) on Δ if and only if it admits a boundary value representation

$$f(x) = \sum_{j=1}^{n} F_j(z + i\gamma 0)$$

such that each of the $F_j(z)$ is slowly increasing (resp. exponentially decreasing) on Δ.

(2) If $F(z)$ is slowly increasing and $G(z)$ is exponentially decreasing on Δ, then $F(z) \cdot G(z)$ is exponentially decreasing on Δ.

(3) The class of exponentially decreasing functions is closed under the classical Fourier transform (see [Kaw70]). The Fourier transform of slowly increasing hyperfunctions will be defined to be dual to this operation via a pairing between slowly increasing hyperfunctions and exponentially decreasing holomorphic functions.

Intuitively, a hyperfunction is slowly increasing when, after fixing the imaginary part of z inside of $i\gamma_j$, its asymptotic growth along the real line is slower than every exponential function. A hyperfunction is exponentially decreasing on the cone γ when the holomorphic functions in a boundary value representation decay exponentially in the real directions which are inside of the cone γ.

As previously mentioned, there exists a pairing between slowly increasing hyperfunctions and exponentially decreasing holomorphic functions. Let $f(x) = F(z + i\gamma 0)$ be a slowly increasing hyperfunction, $G(z)$ an exponentially decreasing analytic function, and S a contour of integration chosen so that $\text{Im}(z) \in i\gamma 0$ for all $z \in S$. The pairing is given by

$$\langle f, G \rangle = \int_S F(x + iy)G(x + iy)\, dx$$

Convergence of the integral is guaranteed by the condition that $F(z)G(z)$ is exponentially decreasing. That the pairing does not depend on the choice of contour follows from the Cauchy integral formula. The pairing allows us to identify the slowly increasing hyperfunctions as the topological dual space to the space of exponentially decreasing holomorphic functions. The Fourier transform of a slowly increasing hyperfunction is then defined by a duality with respect to this pairing:

$$\langle \mathscr{F}(f), G\rangle := \langle f, \mathscr{F}(G)\rangle$$

In practice, the Fourier transform of a hyperfunction is not computed directly from the definition. Let us now introduce the practical method by which one normally computes the Fourier transform of a slowly increasing hyperfunction. Suppose that $F(z)$ is a holomorphic function which is exponentially decreasing outside of a closed convex cone Δ. Let $z = x + iy$ and $\zeta = \sigma + i\tau$, and suppose that $x \in \Delta$. We have the following estimate:

$$|\exp(-i\zeta \cdot z)| = \exp(y \cdot \sigma + x \cdot \tau)$$

The above estimate shows that $\exp(-i\zeta z)$ will be exponentially decreasing on Δ, so long as we fix $\tau \in -\Delta^\circ$. It then follows that the product $e^{-i\zeta z}F(z)$ is exponentially decreasing on $\mathbb{R}^n$. If $f(x) = F(z + i\gamma 0)$, then its Fourier transform is the hyperfunction given by

$$\mathscr{F}(f) = G(\zeta - i\Delta^\circ) = b_{-\Delta^\circ}\left(\int_S e^{-i\zeta z}F(z)\,dz\right)$$

This can be extended to an arbitrary boundary value expression $f(x) = \sum_j F_j(z + i\gamma_j 0)$ by linearity, assuming that each of the $F_j(z)$ decreases exponentially outside of some cone.

We must now deal with the case that $f(x) = F(z + i\gamma 0)$ is a slowly increasing hyperfunction, but that it does not decrease exponentially on any cone.

Definition *Let Σ be a finite collection of closed convex cones. A holomorphic partition of unity is a collection of holomorphic functions $\{\chi_\sigma(z)\}_{\sigma\in\Sigma}$ such that*

(1) $\displaystyle\sum_{\sigma\in\Sigma} \chi_\sigma(z) = 1$

(2) $\chi_\sigma(z)$ *is exponentially decreasing outside of any open cone* $\sigma' \supset \sigma$

(3) $\displaystyle\bigcup_{\sigma\in\Sigma} \sigma = \mathbb{R}^n$

Example of a holomorphic partition of unity:

Let Σ denote the collection of orthants in $\mathbb{R}^n$. If $\sigma = (\sigma_1, \ldots, \sigma_n)$ is a multi-index whose entries are ± 1 (clearly, such objects are in bijection with the orthants), we will denote the corresponding orthant by γ_σ. Consider the following two functions:

$$\chi_+(t) = \frac{1}{1 + e^{-t}}$$
$$\chi_-(t) = \frac{1}{1 + e^{t}}$$

where $t \in \mathbb{C}$ is a complex variable. We notice that $\chi_+(t)$ is exponentially decreasing on $\mathrm{Re}(t) < 0$, and $\chi_-(t)$ is exponentially decreasing on $\mathrm{Re}(t) > 0$. For a fixed orthant $\sigma \in \Sigma$, define the holomorphic function $\chi_\sigma(z)$ by

$$\chi_\sigma(z) = \prod_{i=1}^{n} \frac{1}{1+e^{\sigma_i z_i}}$$

This function exponentially decreases on the complement of γ_σ. The collection $\{\chi_\sigma(z)\}_{\sigma\in\Sigma}$ is a holomorphic partition of unity.

We have introduced holomorphic partitions of unity as an abstract concept, but we will only ever use this example in our computations. The reason we have done this, as we will see later, is that the computations can be made easier or harder by a clever choice of holomorphic partition of unity (although the actual result of the computation is, of course, independent of any such choices). Our main result on the Duistermaat–Heckman hyperfunction of $\Omega SU(2)$ will remain in an integral form, but it is possible that the computation of the Fourier transform could be completed by redoing the computation with a judicious choice of holomorphic partition of unity.

We are now ready to explain how to compute the Fourier transform of a general slowly increasing hyperfunction. Again, by linearity of the Fourier transform, we may assume our hyperfunction takes the form $f(x) = F(z+i\gamma 0)$, and that $F(z)$ is a slowly increasing holomorphic function. Choose a holomorphic partition of unity $\{\chi_\sigma(z)\}_{\sigma\in\Sigma}$; we then observe that

$$F(z) = \sum_{\sigma\in\Sigma} F(z)\chi_\sigma(z)$$

where now $F(z)\chi_\sigma(z)$ is exponentially decreasing outside of σ. By our previous observations,

$$\mathscr{F}(f) = \sum_{\sigma\in\Sigma} b_{-\sigma^\circ}\left(\int_S e^{-i\zeta z}F(z)\chi_\sigma(z)\,dz\right) \tag{1}$$

Equation 1 exactly tells us how to compute the Fourier transform of a general slowly increasing hyperfunction.

3. Hyperfunctions Arising from Localization of Hamiltonian Group Actions

Let (M,ω) be a finite-dimensional compact symplectic manifold with a Hamiltonian action of a d-dimensional compact torus T; call the moment map $\mu : M \to \mathfrak{t}^*$. The symplectic form ω gives us the Liouville measure $\omega^n/n!$ on M, which we we may push forward to $\mathfrak{t}^*$ using the moment map μ. We let $\mathcal{F}$ denote the connected components of the fixed point set for the T action on M; furthermore, if $q\in\mathcal{F}$, we denote by e_q^T the equivariant Euler class of the normal bundle to the fixed point set. We can identify $e_q^T \in H^*(BT) \simeq \mathrm{Sym}(\mathfrak{t}^*)$ with the product of the weights appearing in the isotropy representation of T on T_qM.

Theorem 1 [DH82] *The measure $\mu_*(\omega^n/n!)$ has a piecewise polynomial density function. Furthermore, the inverse Fourier transform of $\mu_*(\omega^n/n!)$ has an exact expression:*

$$\int_M e^{i\mu(p)(X)}\omega^n/n! = \frac{1}{(2\pi i)^d}\sum_{q\in\mathcal{F}}\frac{e^{i\mu(q)(X)}}{e_q^T(X)} \tag{2}$$

where $X \in \mathfrak{t}$ is such that $e_q^T(X) \neq 0$ for all $q \in \mathcal{F}$.

The Duistermaat–Heckman theorem applies to the case where M is finite-dimensional and compact. We are interested in finding some version of a Duistermaat–Heckman distribution in the setting where M is an infinite-dimensional manifold with a Hamiltonian group action. There are some immediate technical obstructions to producing such a distribution. Most notably, the inability to take a top exterior power of ω prevents us from defining a suitable Liouville measure. There are signification analytic challenges in properly defining the left-hand side of Equation 2; a related problem is defining a rigorous measure of integration for the kinds of path integrals which appear in quantum field theory. We will not attempt to answer this question in this article. Nevertheless, it is possible to make sense of the right-hand side of Equation 2.

The main goal for this section is explain how Hamiltonian actions of compact tori yield, in a natural way, hyperfunctions on $\mathfrak{t}$. The hyperfunction one gets in this way should be a substitute for the reciprocal of the equivariant Euler class which appears in the localization formula. We then reinterpret the sum over the fixed points in the localization formula as a hyperfunction on $\mathfrak{t}$, and define the *Duistermaat–Heckman hyperfunction* to be its Fourier transform as a hyperfunction.

We will start by considering the local picture. Suppose that T has a Hamiltonian action on a (finite-dimensional, for now) complex vector space with weights λ_i. Let the weights of the action be given by $W = \{\lambda_i\}_{i\in I}$. The weights of the action are linear functionals $\mathfrak{t}_\mathbb{C} \to \mathbb{C}$. For every weight $\lambda \in W$ we get a corresponding half space $H_\lambda = \{y \in \mathfrak{t} \,|\, \lambda(y) < 0\}$, as well as a hyperfunction:

$$f_\lambda(x) = \frac{1}{\lambda(z) + iH_\lambda 0}$$

The singular support of $f_\lambda(x)$ is given by

$$SS(f_\lambda) = \{(x,\xi) \in T^*(\mathfrak{t}) \,|\, \lambda(x) = 0, \exists c > 0, \xi = c\,d\lambda(x)\}$$

Proposition 1 *If $\mu : V \to \mathfrak{t}^*$ is proper, then, for all pairs of weights λ, λ', $SS(f_\lambda) \cap SS(f_{\lambda'})^\circ = \emptyset$.*

Proof If the moment map is proper, then all of the weights are contained in a half space [GLS88]. There exists $X \in \mathfrak{t}$ such that, for any pair of weights λ, λ', we have both $\lambda(X) > 0$ and $\lambda'(X) > 0$. Suppose that $(x,\xi) \in SS(f_\lambda) \cap SS(f'_\lambda)^\circ$. This means that

(1) $\lambda(x) = \lambda'(x) = 0$

(2) $\exists c, c' > 0$ such that $\xi = c\,d\lambda = -c'd\lambda'$

Rearranging the second condition implies that the function $L = \lambda' + \frac{c}{c'}\lambda$ is constant.
However, we have obtained a contradiction as $L(x) = 0$, while $L(X) > 0$. □

The following is immediate from the proposition.

Corollary 1 *Let* $\gamma = \bigcap_{\lambda \in W} H_\lambda$. *If* $\mu : V \to \mathfrak{t}^*$ *is proper, then the following product of hyperfunctions is well defined:*

$$\frac{1}{e^T(x)} = \prod_{\lambda \in W} f_\lambda(x) = b_\gamma \left(\prod_{\lambda \in W} \frac{1}{\lambda(z)} \right)$$

When $V = T_pM$ is the isotropy representation for a Hamiltonian action of T on M, we call

$$\frac{1}{e_p^T(x)} = b_{\gamma_p} \left(\prod_{\lambda \in W} \frac{1}{\lambda(z)} \right)$$

the reciprocal of the equivariant Euler class to the normal bundle of p.

We can use the reciprocals of the equivariant Euler classes in an expression which imitates the sum over the fixed points in the Duistermaat–Heckman formula.

Definition *Suppose that* (M, ω) *has a Hamiltonian action of a compact, dimension* d *torus* T *such that all the fixed points are isolated; let* M^T *denote the fixed point set, and* $\mu : M \to \mathfrak{t}^*$ *the moment map. We then call the following expression the Picken hyperfunction:*

$$L(x) = \frac{1}{(2\pi i)^d} \sum_{p \in M^T} \frac{e^{i\mu(p)(x)}}{e_p^T(x)}$$

Example 3.1 S^2 *with a circle action by rotation*

We will first use a simple example to demonstrate that the formalism of hyperfunctions reproduces the results one would expect from the Duistermaat–Heckman function. For the usual Hamiltonian circle action on S^2 by counterclockwise rotation about the z-axis, there are fixed points at the north and south poles, N and S, respectively. The torus acts on T_NS^2 with weight $+1$, while it acts on T_SS^2 with weight -1. Let's compute the reciprocal of the equivariant Euler class to the normal bundle of N (as a hyperfunction). There is only one weight at this fixed point, and the corresponding half space is $H_+ = \{x \in i\mathfrak{t} \mid x < 0\}$. The north pole contributes the following hyperfunction as a summand of the Picken hyperfunction, which we denote pictorially in Figure 1:

$$J_+(x) = b_{H_+}\left(\frac{e^{iz}}{z}\right)$$

The contribution to the Picken hyperfunction coming from the south pole is computed similarly. The end result is that the Picken hyperfunction of this Hamiltonian group action is

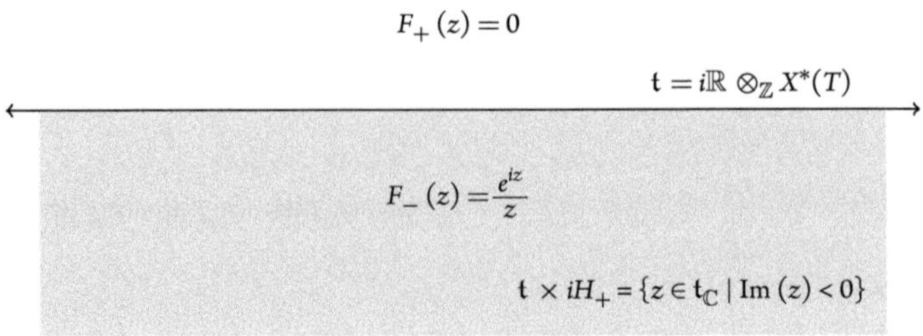

Figure 1 A depiction of the hyperfunction $J_+(x) \in \mathcal{B}(\mathfrak{t})$

$$2\pi i L(x) = b_{H_+}\left(\frac{e^{iz}}{z}\right) + b_{H_-}\left(\frac{e^{-iz}}{-z}\right)$$

The Duistermaat–Heckman hyperfunction is the Fourier transform of the Picken hyperfunction. We will now compute it according to the formula in Equation 1. We choose the holomorphic partition of unity given by the functions

$$\chi_+(z) = \frac{1}{1+e^{-z}}$$
$$\chi_-(z) = \frac{1}{1+e^{z}}$$

which gives a decomposition of the Picken hyperfunction into four parts:

$$2\pi i L(x) = b_{H_+}\left(\frac{e^{iz}}{z}\chi_+(z)\right) + b_{H_-}\left(\frac{e^{-iz}}{-z}\chi_+(z)\right) + b_{H_+}\left(\frac{e^{iz}}{z}\chi_-(z)\right) + b_{H_-}\left(\frac{e^{-iz}}{-z}\chi_-(z)\right)$$

The Fourier transform can now be computed termwise, noticing that the first two terms in the above expression are exponentially decreasing on $\mathrm{Re}(z) < 0$, while the third and fourth terms are exponentially decreasing on the cone $\mathrm{Re}(z) > 0$. Let $1 \gg \delta > 0$; then, we may write the Fourier transform $\mathscr{F}(L(x)) = G_+(\zeta + i0) + G_-(\zeta - i0)$, where:

$$G_+(\zeta) = \int_{-\infty - i\delta}^{\infty - i\delta} \frac{e^{-i(\zeta-1)z}}{z(1+e^{z})}\,dz + \int_{-\infty + i\delta}^{\infty + i\delta} -\frac{e^{-i(\zeta+1)z}}{z(1+e^{z})}\,dz$$

$$G_-(\zeta) = \int_{-\infty - i\delta}^{\infty - i\delta} \frac{e^{-i(\zeta-1)z}}{z(1+e^{-z})}\,dz + \int_{-\infty + i\delta}^{\infty + i\delta} -\frac{e^{-i(\zeta+1)z}}{z(1+e^{-z})}\,dz$$

Each of these integrals can be computed by completing to a semicircular contour in the lower half plane and applying the residue theorem (noting that, as the contour is oriented clockwise, we must include an extra minus sign). The contour we use for the first integral appearing in $G_+(\zeta)$ is depicted in Figure 2, along with the locations of the poles.

We show how to compute the first integral in the expression for $G_+(\zeta)$; the rest are similar, but one must be careful that the pole at zero will be included on the interior of the contour for the integrals where the contour runs above the real axis.

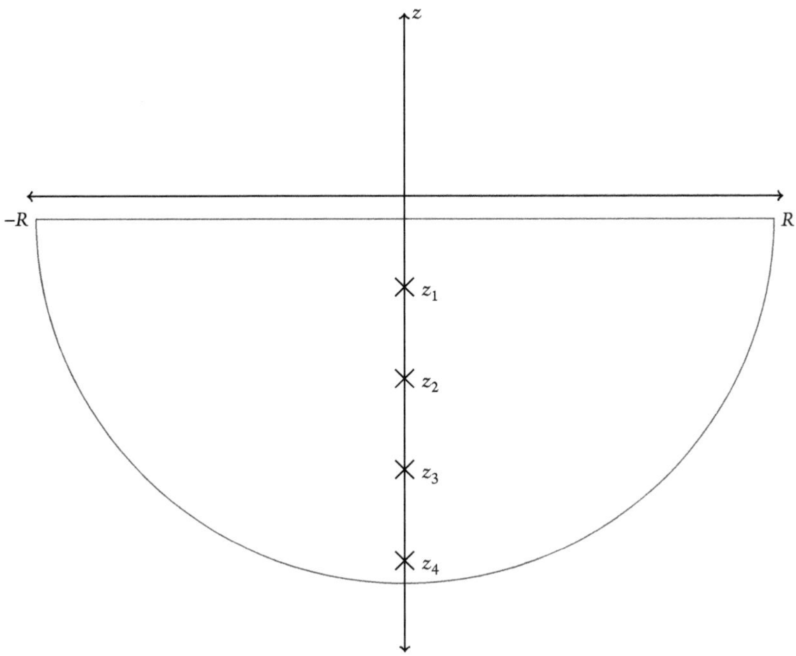

Figure 2 Integration contour for the first integral in $G_+(\zeta)$

The integrand of the first integral in $G_+(\zeta)$ has poles at $z_0 = 0$ and $z_k = -(2k+1)\pi i$ for $k \in \mathbb{Z}$; however, the only poles inside our contour (in the limit as the radius of the semicircle tends to infinity) are the poles at z_k for $k \geq 0$. Also, in the limit as the radius of the semicircle gets large, we see that the contribution to the integral coming from the semicircular part of the contour vanishes because the integrand is exponentially decreasing in $\mathrm{Re}(z)$, and decreasing exponentially in $\mathrm{Im}(z)$ when $\mathrm{Im}(z) < 0$. By the residue theorem,

$$\begin{aligned}
\int_{-\infty-i\delta}^{\infty-i\delta} \frac{e^{-i(\zeta-1)z}}{z(1+e^z)}\,dz &= -2\pi i \sum_{k=0}^{\infty} \mathrm{Res}\left(\frac{e^{-i(\zeta-1)z}}{z(1+e^z)}, z = z_k\right) \\
&= -2\pi i \sum_{k=0}^{\infty} \frac{e^{-(\zeta-1)(2k+1)\pi}}{-(2k+1)\pi i(-1)} \\
&= 2\pi \sum_{k=0}^{\infty} \int_c^{\zeta} e^{-(\zeta'-1)(2k+1)\pi}\,d\zeta' \\
&= 2\pi \int_c^{\zeta} \sum_{k=0}^{\infty} e^{-(\zeta'-1)(2k+1)\pi}\,d\zeta'
\end{aligned}$$

$$
\begin{aligned}
&= 2\pi \int_c^{\zeta} \frac{d\zeta'}{e^{\pi(\zeta'-1)} - e^{-\pi(\zeta'-1)}} \\
&= \pi \int_c^{\zeta} \frac{d\zeta'}{\sinh(\pi(\zeta'-1))} d\zeta' \\
&= \operatorname{Log}\left(\tanh\left(\frac{\pi(\zeta-1)}{2}\right)\right) \\
&= \pi i + \operatorname{Log}\left(\tanh\left(\frac{\pi(1-\zeta)}{2}\right)\right) \qquad \text{valid for } \operatorname{Im}(\zeta) > 0
\end{aligned}
$$

From the second to the third line, we found a primitive function for the summand. From the third to the fourth line, we applied the monotone convergence theorem to interchange the order of summation and integration. A similar computation yields the result

$$
\int_{-\infty+i\delta}^{\infty+i\delta} -\frac{e^{-i(\zeta+1)z}}{z(1+e^z)} dz = -\pi i - \operatorname{Log}\left(\tanh\left(\frac{\pi(1+\zeta)}{2}\right)\right)
$$

Summarizing, to this point we have computed

$$
\begin{aligned}
G_+(\zeta) &= -\operatorname{Log}\left(\tanh\left(\frac{\pi(1+\zeta)}{2}\right)\right) + \operatorname{Log}\left(\tanh\left(\frac{\pi(1-\zeta)}{2}\right)\right) \\
G_-(\zeta) &= \operatorname{Log}\left(\tanh\left(\frac{\pi(1+\zeta)}{2}\right)\right) - \operatorname{Log}\left(\tanh\left(\frac{\pi(1-\zeta)}{2}\right)\right)
\end{aligned}
$$

The above expressions can be simplified. We notice that the holomorphic function $\operatorname{Log}(\tanh(\zeta)) - \operatorname{Log}(\zeta)$ admits an analytic extension across a neighbourhood of the real axis, and is therefore zero as a hyperfunction. This means that all of the tanh factors may be ignored for the purposes of computing the hyperfunction Fourier transform. Therefore, the final result of our computation is

$$
\mathscr{F}(L(x)) = b_+\left(-\frac{1}{2\pi i}\operatorname{Log}\left(\frac{1+\zeta}{1-\zeta}\right)\right) - b_-\left(-\frac{1}{2\pi i}\operatorname{Log}\left(\frac{1+\zeta}{1-\zeta}\right)\right)
$$

which we recognize as the standard defining hyperfunction of $\chi_{[-1,1]}(x)$ (see [Kan89] Example 1.3.11, p. 29). This has shown that the Fourier transform of the Picken hyperfunction gives the standard defining hyperfunction of the Duistermaat–Heckman distribution.

Jeffrey and Kirwan, building on work of Witten [Wit92], formalized the notion of a residue in symplectic geometry [JK95b]. They fruitfully applied this construction to compute relations in the cohomology ring of the moduli space of stable holomorphic bundles on a Riemann surface [JK95a]. The properties that uniquely characterize the residue (c.f. Proposition 8.11, [JK95b]) can be recovered from the usual notion of a residue [GH14] of a multivariable complex meromorphic function, using our construction of the Picken hyperfunction.

4. ΩG and its Hamiltonian Group Action

Let G be a compact connected real Lie group, and call its Lie algebra $\mathfrak{g}$. If H is any compact Lie group, we will denote its complexification by $H_{\mathbb{C}}$. In this section, we will consider the space of smooth loops $LG = C^\infty(S^1, G)$. LG is itself an infinite-dimensional Lie group, with the group operation taken to be multiplication in G pointwise along a loop. The Lie algebra of LG is easily seen to consist of the space of smooth loops into the Lie algebra, which we denote $L\mathfrak{g}$.

We will also consider its quotient $\Omega G = LG/G$, where the quotient is taken with respect to the subgroup of constant loops. One may alternatively identify ΩG as the collection of loops such that the identity in S^1 maps to the identity in G:

$$\Omega G = \{\gamma \in LG : \gamma(1) = e\}$$

Its Lie algebra can be identified with the subset $\Omega\mathfrak{g} = \left\{X : S^1 \to \mathfrak{g} \,|\, X(0) = 0\right\}$.

ΩG has a lot of extra structure, which essentially comes from its realization as a coadjoint orbit of a central extension of LG [KW08]. We can give ΩG a symplectic structure as follows. Since G is a compact Lie group, there exists a non-degenerate symmetric bilinear form $\langle\cdot,\cdot\rangle : \mathfrak{g} \times \mathfrak{g} \to \mathbb{R}$. This form induces an antisymmetric form:

$$\begin{aligned} \omega_e &: L\mathfrak{g} \times L\mathfrak{g} \to \mathbb{R} \\ (X, Y) &\mapsto \frac{1}{2\pi} \int_0^{2\pi} \langle X(\theta), Y'(\theta)\rangle \, d\theta \end{aligned}$$

This bilinear form is an antisymmetric, non-degenerate form when restricted to $\Omega\mathfrak{g}$, and extends to a symplectic form on ΩG using a left trivialization of the tangent bundle of ΩG, That is, for every $\gamma \in \Omega G$, we fix the isomorphism

$$\begin{aligned} T_\gamma \Omega G &\simeq \Omega\mathfrak{g} \\ X &\mapsto \left(\theta \mapsto \gamma^{-1}(\theta)X(\theta)\right) \end{aligned}$$

This choice allows us to define a form on ΩG as

$$\begin{aligned} \omega_\gamma &: T_\gamma \Omega G \times T_\gamma \Omega G \to \mathbb{R} \\ (X, Y) &\mapsto \omega_e(\gamma^{-1}X, \gamma^{-1}Y) \end{aligned}$$

The form so defined is symplectic; a proof can be found in [PS86].

Consider the following group action on ΩG. Fix $T \subseteq G$ a maximal compact torus, and let $\mathfrak{t}$ be its Lie algebra. Pointwise conjugation by elements of T defines a T action on ΩG:

$$\begin{aligned} T \times \Omega G &\to \Omega G \\ t \cdot \gamma &= \left(\theta \mapsto t\gamma(\theta)t^{-1}\right) \end{aligned}$$

There is also an auxiliary action of S^1 on ΩG, which comes about by descending the loop rotation action on LG to the quotient LG/G. Explicitly,

$$S^1 \times \Omega G \to \Omega G$$
$$\exp(i\psi) \cdot \gamma = \left(\theta \mapsto \gamma(\theta + \psi)\gamma(\psi)^{-1}\right)$$

These actions commute with one another, so define an action of $T \times S^1$ on ΩG. We will let $\mathrm{pr}_{\mathfrak{t}} : \mathfrak{g} \to \mathfrak{t}$ denote the orthogonal projection coming from the Cartan–Killing form. We now define two functions on ΩG:

$$p(\gamma) = \frac{1}{2\pi}\mathrm{pr}_{\mathfrak{t}}\left(\int_0^{2\pi} \gamma^{-1}(\theta)\gamma'(\theta)\, d\theta\right)$$
$$E(\gamma) = \frac{1}{2\pi}\int_0^{2\pi} ||\gamma'(\theta)||^2\, d\theta$$

Proposition 2 [AP83] *The $T \times S^1$ action on ΩG is Hamiltonian. The moment map is given by*

$$\mu : \Omega G \to Lie(T \times S^1)$$
$$\gamma \mapsto \begin{pmatrix} p(\gamma) \\ E(\gamma) \end{pmatrix}$$

Furthermore, the Hamiltonian vector fields associated to the group action are given by

$$(X_E)_\gamma = \gamma'(\theta) - \gamma(\theta)\gamma'(0)$$
$$(X_{p_\tau})_\gamma = \tau\gamma(\theta) - \gamma(\theta)\tau$$

where $\tau \in \mathfrak{t}$.

If $\beta \in \mathrm{Lie}(T \times S^1)$, then we let $(X_\beta)_\gamma$ denote the Hamiltonian vector field evaluated at the loop γ.

5. Fixed Points Sets of Rank 1 Subtori

We will now proceed to identify the fixed point sets of dimension 1 subtori of $T \times S^1$ acting on ΩG. The moment map image of the fixed point submanifolds should correspond to the locus where the Duistermaat–Heckman density function is not differentiable. Using the exponential map, we identify $X_*(T \times S^1) \simeq P \times \mathbb{Z}$, where P is the coweight lattice of $\mathrm{Lie}(T)$. Fix an element $\beta = (\lambda, m) \in X_*(T \times S^1)$ and call the cocharacter it generates by T_β. Let $\Lambda \in X_*(T)$ be the cocharacter generated by λ,

$$\Lambda(\theta) = \exp(i\lambda\theta)$$

We will say the fixed point set of T_β is trivial when $\Omega G^{T_\beta} = \mathrm{Hom}_{\mathrm{Grp}}(S^1, T)$. In this section, we say that $L \subseteq G$ is a *Levi subgroup* if and only if there exists a parabolic subgroup $Q \subseteq G_{\mathbb{C}}$ such that $L_{\mathbb{C}}$ is a Levi factor of Q. Every Levi subgroup of G is the centralizer of a subtorus $S \subseteq T$.

If we have two groups K and N, together with a map $\varphi : K \to \operatorname{Aut} N$, then we can construct the semidirect product group $N \rtimes K$ whose point set is the Cartesian product $N \times K$, but the group operation is $(n,k) \cdot (n',k') = ((\phi(k') \cdot n)n', kk')$. In our specific context, if we fix any Levi subgroup $L \subseteq G$, we can construct a group homomorphism:

$$\varphi_\beta : S^1 \to \operatorname{Aut} L$$

$$\varphi_\beta(\psi) \cdot x = \Lambda\left(\frac{\psi}{m}\right)^{-1} x \Lambda\left(\frac{\psi}{m}\right)$$

Remarks:

(1) Since $\varphi_\beta(1) = \mathrm{id}_L$ and S^1 is connected then we may consider $\varphi_\beta : S^1 \to \operatorname{Inn}(L)$. We identify $\operatorname{Inn}(L) \simeq L_{\mathrm{ad}}$, which may further be identified with $[L,L]/Z(L) \cap [L,L]$. Under these identifications, $\varphi_\beta \in X_*(T_{\mathrm{ad}})$ is a cocharacter of the maximal torus in L_{ad}.

(2) This homomorphism is well defined if and only if $\Lambda(\frac{2\pi}{m}) \in Z(L)$. In particular, λ/m must be an element of the coweight lattice for the Levi subgroup L, mod $\mathfrak{z}(L)$.

(3) $\varphi_\beta = \varphi_{\beta'}$ if and only if $\lambda/m - \lambda'/m' \in \mathfrak{z}(L)$

We will denote the resulting semidirect product group as $L \rtimes_\beta S^1$.

It can be easily seen that for any Levi subgroup L, $T \times S^1$ is a maximal torus of $L \rtimes_\beta S^1$. Any 1-parameter subgroup of $L \rtimes_\beta S^1$ is abelian, and is therefore contained in a maximal torus conjugate to $T \times S^1$. We can obtain all 1-parameter subgroups by considering one of the form $(\eta(\theta),\theta)$ for $\eta \in \operatorname{Hom}(S^1,T)$, then conjugating by an element of $L \rtimes_\beta S^1$:

$$\gamma(\theta) = \Lambda\left(\frac{\psi-\theta}{m}\right) g \Lambda\left(\frac{\theta}{m}\right) \eta(\theta) g^{-1} \Lambda\left(\frac{\psi}{m}\right)^{-1} \tag{3}$$

Proposition 3 *For any $\beta \in P \times \mathbb{Z}$, there exists a Levi subgroup $T \subseteq L_\beta \subseteq G$ such that $\gamma \in \Omega G^{T_\beta}$ if and only if $(\gamma(\theta),\theta)$ is a 1-parameter subgroup of $L_\beta \rtimes_\beta S^1$.*

Proof Fix $\beta \in P \times \mathbb{Z}$ and set $L_\beta = Z_G(\Lambda(2\pi/m))$; that $T \subseteq L_\beta$ follows, since $\Lambda(2\pi/m) \in T$, and T is abelian.

Suppose we have a loop γ fixed by T_β. Recall how T_β acts on a loop $\gamma \in \Omega G$. For every $(\Lambda(\psi), \exp(im\psi)) \in T_\beta$, the action is

$$(\Lambda(\psi), \exp(im\psi)) \cdot \gamma(\theta) = \Lambda(\psi)\gamma(\theta + m\psi)\Lambda^{-1}(\psi)\gamma(m\psi)^{-1} \quad \forall \psi,\theta \in [0,2\pi)$$

Let's rescale the ψ variable; then, by periodicity, we may write the condition to be fixed under T_β as

$$\gamma(\theta+\psi) = \Lambda\left(\frac{\psi}{m}\right)^{-1} \gamma(\theta) \Lambda\left(\frac{\psi}{m}\right) \gamma(\psi) \qquad \forall \theta,\psi \in [0,2\pi)$$

When $\psi = 2\pi$ in the above equation, we get the condition $\gamma(\theta) \in L_\beta$ for all θ. That $(\gamma(\theta), \theta)$ is a 1-parameter subgroup of $L \rtimes_\beta S^1$ follows immediately from the multiplication rule for the semidirect product.

Now suppose conversely that $(\gamma(\theta), \theta)$ is a 1-parameter subgroup of $L \rtimes_\beta S^1$. There exists $\eta \in X_*(T)$, $g \in L$ and $\psi \in S^1$ such that γ can be written as in Equation 3. To show that γ is fixed by T_β it suffices to prove that the Hamiltonian vector field corresponding to β vanishes at γ. This is a straightforward (but tedious) verification. □

A consequence of the previous proposition is that, for any such β, there exists a Levi subgroup L_β such that $\Omega G^{T_\beta} = \Omega L_\beta^{T_\beta}$. This follows, since the semidirect product formula forces any loop fixed under T_β to have its image be contained in L_β.

Proposition 4 *Every connected component of the fixed point set of T_β is a translate of an adjoint orbit in Lie$(L_\beta) \subseteq \mathfrak{g}$.*

Proof Fix a loop γ in some connected component of the fixed point set of T_β. Using the exponential map on $L \rtimes_\beta S^1$, it can be seen that $(\gamma(\theta), \theta)$ is a 1-parameter subgroup of $L_\beta \rtimes_\beta S^1$ if and only if γ is a solution to the differential equation

$$\frac{d\gamma}{d\theta} = \left[\gamma(\theta), \frac{\lambda}{m}\right] + \gamma(\theta)\gamma'(0)$$

Compactness of G (and therefore, of L_β, since it is a closed subgroup) and the Picard–Lindelöf theorem allow us to identify the loops in the fixed point set of T_β with their initial conditions $\gamma'(0) \in \mathfrak{g}$. We can use Equation 3 to compute $\gamma'(0)$:

$$\gamma'(0) = \mathrm{Ad}_{\Lambda(\frac{\psi}{m})g}\left[\frac{\lambda}{m} + \eta'(0)\right] - \frac{\lambda}{m}$$

Any other loop in the same connected component of the fixed point set of T_β can be obtained by varying $g \in L_\beta$ and $\psi \in [0, 2\pi)$. □

Notice that, by fixing $\lambda = 0$ in the preceding discussion, we recover the result that the fixed point set of the loop rotation action consists of the group homomorphisms $S^1 \to G$ [PS86].

The last result of this section characterizes exactly when two rank 1 subtori have the same fixed point sets.

Proposition 5 *Let $\beta = (\lambda, m)$ and $\beta' = (\lambda', m')$ be generators of rank 1 subgroups T_β, $T_{\beta'}$ of $T \times S^1$, and let $L_\beta, L_{\beta'}$ be the Levi subgroups provided by Proposition 3. Then $\Omega G^{T_\beta} = \Omega G^{T_{\beta'}}$ if and only if $\lambda/m - \lambda'/m' \in \mathfrak{z}(L_\beta)$*

Remarks *If $\lambda/m - \lambda'/m' \in \mathfrak{z}(L_\beta)$, then $L_\beta = L_{\beta'}$. This is due to the fact that L_β was defined to be the G-centralizer of* $\exp(2\pi i\lambda/m)$ *(and similarly for $L_{\beta'}$).*

Proof Suppose that the fixed point sets of T_β and $T_{\beta'}$ are equal. Then for any γ, we have $(X_\beta)_\gamma = 0$ if and only if $(X_{\beta'})_\gamma = 0$. These conditions yield two differential equations:

$$0 = m\frac{d\gamma}{d\theta} - \gamma(\theta)\gamma'(0) + \lambda\gamma(\theta) - \gamma(\theta)\lambda$$

$$0 = m'\frac{d\gamma}{d\theta} - \gamma(\theta)\gamma'(0) + \lambda'\gamma(\theta) - \gamma(\theta)\lambda'$$

We may subtract these, and left translate back to $\mathfrak{g}$ to get the condition:

$$\forall \gamma \in \Omega G^{T_\beta}, \theta \in [0,2\pi), \quad \mathrm{Ad}_{\gamma(\theta)}\left(\frac{\lambda}{m} - \frac{\lambda'}{m'}\right) = \frac{\lambda}{m} - \frac{\lambda'}{m'}$$

The derivative of this condition at the identity is

$$\left[\gamma'(0), \frac{\lambda}{m} - \frac{\lambda'}{m'}\right] = 0$$

so the statement is proved if, for every element $Y \in \mathrm{Lie}([L_\beta, L_\beta])$, there exists $\gamma \in \Omega G^{T_\beta}$ and $c \in \mathbb{R}$ such that $Y = c\gamma'(0)$. By Proposition 4, we can identify the set of all such $\gamma'(0)$ with a translated adjoint orbit. This can be achieved by choosing a cocharacter $\eta(\theta)$ such that $\eta'(0) + \frac{\lambda}{m}$ is regular for the Ad_{L_β}-action and $\eta'(0)$ is sufficiently large so that the translated adjoint orbit intersects every ray through the origin.

Conversely, if $\lambda/m - \lambda'/m' \in \mathfrak{z}(L_\beta)$, then, by the above remark, $L_\beta = L'_\beta$ and, furthermore, β and β' yield identical automorphisms $\varphi_\beta = \varphi_{\beta'} : S^1 \to \mathrm{Aut}(L_\beta)$. Then, by Proposition 3, we have $\Omega G^{T_\beta} = \Omega G^{T_{\beta'}}$. □

6. An Explicit Example: The Loop Space of $SU(2)$

When $G = SU(2)$, the general theory of the previous section can be understood in a very explicit way.

The way to do this is to translate the condition of being fixed under the group action into a solution of a system of differential equations for the matrix parameters. Let's work through this derivation. We can describe an element $\gamma(t) \in \Omega SU(2)$ by

$$\gamma(t) = \begin{pmatrix} \alpha(t) & -\beta(t)^* \\ \beta(t) & \alpha(t)^* \end{pmatrix}$$

subject to the constraints $|\alpha(t)|^2 + |\beta(t)|^2 = 1$ for all $t \in [0, 2\pi]$, $\alpha(0) = 1$, and $\beta(0) = 0$; 1-parameter subgroups correspond bijectively with elements of the Lie algebra of $T \times S^1$. In that spirit, fix some element $(\theta, \psi) \in \mathrm{Lie}(T \times S^1)$, exponentiate to the group and act on our loop $\gamma(t)$

$$\left(\begin{pmatrix} e^{i\theta} & 0 \\ 0 & e^{-i\theta} \end{pmatrix}, e^{i\psi}\right) \cdot \gamma(t) = e^{i\psi} \cdot \left(\begin{pmatrix} e^{i\theta} & 0 \\ 0 & e^{-i\theta} \end{pmatrix} \begin{pmatrix} \alpha(t) & -\beta(t)^* \\ \beta(t) & \alpha(t)^* \end{pmatrix} \begin{pmatrix} e^{-i\theta} & 0 \\ 0 & e^{i\theta} \end{pmatrix}\right)$$

$$= e^{i\psi} \cdot \begin{pmatrix} \alpha(t) & -e^{i2\theta}\beta(t)^* \\ e^{-i2\theta}\beta(t) & \alpha(t)^* \end{pmatrix}$$

$$
\begin{aligned}
&= \begin{pmatrix} \alpha(t+\psi) & -e^{i2\theta}\beta(t+\psi)^* \\ e^{-i2\theta}\beta(t+\psi) & \alpha(t+\psi)^* \end{pmatrix} \begin{pmatrix} \alpha(\psi)^* & e^{i2\theta}\beta(\psi)^* \\ -e^{-i2\theta}\beta(\psi) & \alpha(\psi) \end{pmatrix} \\
&= \begin{pmatrix} \alpha(t) & -\beta(t)^* \\ \beta(t) & \alpha(t)^* \end{pmatrix} \quad \text{when } \gamma(t) \text{ is a fixed loop}
\end{aligned}
$$

so, by rearranging slightly,

$$
\begin{aligned}
\begin{pmatrix} \alpha(t+\psi) & -e^{i2\theta}\beta(t+\psi)^* \\ e^{-i2\theta}\beta(t+\psi) & \alpha(t+\psi)^* \end{pmatrix} &= \begin{pmatrix} \alpha(t) & -\beta(t)^* \\ \beta(t) & \alpha(t)^* \end{pmatrix} \begin{pmatrix} \alpha(\psi) & -e^{i2\theta}\beta(\psi)^* \\ e^{-i2\theta}\beta(\psi) & \alpha(\psi)^* \end{pmatrix} \\
&= \begin{pmatrix} \alpha(t)\alpha(\psi) - e^{-i2\theta}\beta(t)^*\beta(\psi) & -\alpha(\psi)^*\beta(t)^* - e^{i2\theta}\alpha(t)\beta(\psi)^* \\ \alpha(\psi)\beta(t) + e^{-i2\theta}\alpha(t)^*\beta(\psi) & \alpha(t)^*\alpha(\psi)^* - e^{i2\theta}\beta(t)\beta(\psi)^* \end{pmatrix}
\end{aligned}
$$

this yields the finite difference relations

$$
\begin{aligned}
\alpha(t+\psi) &= \alpha(t)\alpha(\psi) - e^{-i2\theta}\beta(t)^*\beta(\psi) \\
\beta(t+\psi) &= e^{i2\theta}\alpha(\psi)\beta(t) + \alpha(t)^*\beta(\psi)
\end{aligned}
$$

We use the infinitesimal form of these relations to get the necessary system of differential equations. Set $\theta = ns$ and $\psi = ms$ so that we can vary the group element along a fixed 1-parameter subgroup:

$$
\begin{aligned}
m\frac{d\alpha}{dt} &= \lim_{s\to 0} \frac{\alpha(t+ms) - \alpha(t)}{s} \\
&= \lim_{s\to 0} \frac{\alpha(t)\alpha(ms) - e^{-i2ns}\beta(t)^*\beta(ms) - \alpha(t)}{s} \\
&= \alpha(t) \lim_{s\to 0} \frac{\alpha(ms) - 1}{s} - \beta(t)^* \lim_{s\to 0} \frac{e^{-i2ns}\beta(ms)}{s} \\
&= m\alpha(t)\alpha'(0) - m\beta(t)^*\beta'(0)
\end{aligned}
$$

And similarly, for $\beta(t)$,

$$
\begin{aligned}
m\frac{d\beta}{dt} &= \lim_{s\to 0} \frac{\beta(t+ms) - \beta(t)}{s} \\
&= \beta(t) \lim_{s\to 0} \frac{e^{i2ns}\alpha(ms) - 1}{s} + m\alpha(t)^*\beta'(0) \\
&= \beta(t)\left[i2ne^{i2ns}\alpha(ms) + me^{i2ns}\alpha'(ms)\right]\Big|_{s=0} + m\beta'(0)\alpha(t)^* \\
&= (i2n + m\alpha'(0))\beta(t) + m\beta'(0)\alpha(t)^*
\end{aligned}
$$

so the system of differential equations we must solve (for $m \neq 0$, when $m = 0$ the problem is trivial) is given by

$$\frac{d\alpha}{dt} = \alpha(t)\alpha'(0) - \beta(t)^*\beta'(0)$$
$$\frac{d\beta}{dt} = \beta'(0)\alpha(t)^* + \left(i2\frac{n}{m} + \alpha'(0)\right)\beta(t)$$

These differential equations are exactly the ones we could have gotten by searching for zeroes of the Hamiltonian vector field corresponding to $(n,m) \in \mathrm{Lie}(T \times S^1)$ (c.f. the differential equation given in Proposition 4). The system we have described depends on four parameters: n, m, $\alpha'(0)$ and $\beta'(0)$. Once we fix these parameters, the solutions $\alpha(t)$ and $\beta(t)$ are uniquely determined. The parameters n and m are fixed from the start, so we are only free to vary $\alpha'(0)$ and $\beta'(0)$. The choices that will turn out to yield periodic solutions will be exactly those loops whose derivatives at the identity are elements of the translated adjoint orbits of Proposition 4.

An explicit analytic solution to the system of differential equations can be found by expanding $\alpha(t)$ and $\beta(t)$ in Fourier series.

$$\alpha(t) = \sum_{k=-\infty}^{\infty} \alpha_k e^{-ikt}$$
$$\beta(t) = \sum_{k=-\infty}^{\infty} \beta_k e^{-ikt}$$

Plugging these expressions into the system of differential equations yields a system of algebraic relations for each k:

$$0 = (\alpha'(0) + ik)\alpha_k - \beta'(0)\beta^*_{-k} \tag{4}$$
$$0 = \beta'(0)\alpha^*_{-k} + \left(i(2\frac{n}{m} + k) + \alpha'(0)\right)\beta_k \tag{5}$$

We can solve by taking $ik - i2\frac{n}{m} + \alpha'(0)^*$ times the first equation above and substituting into the conjugate of the second equation (replacing k by $-k$). For each k, this yields the expression

$$\left(|\alpha'(0)|^2 + |\beta'(0)|^2 - k^2 + \frac{2n}{m}\alpha'(0) + \frac{2n}{m}k\right)\alpha_k = 0$$

which implies that either $\alpha_k = 0$ or (after completing the square and setting $\alpha'(0) = iA$, which is necessary for $\gamma \in \Omega SU(2)$)

$$\left(k - \frac{n}{m}\right)^2 = \left(A + \frac{n}{m}\right)^2 + |\beta'(0)|^2 \tag{6}$$

The purpose of Equation 6 is to characterize the set of initial conditions for the differential equations above which yield periodic solutions; in other words, Equation 6 exactly identifies to the fixed point set of the subtorus generated by (n,m) with a disjoint union of translated adjoint orbits of $SU(2)$, as in Proposition 4. It is evident from Equation 6 that, for any loop fixed under the subgroup (n,m), at most two Fourier modes can be non-zero. These two modes correspond to precisely the values of k that satisfy $k - \frac{n}{m} = \pm C$ for some constant C, for which we require integer solutions of k. We can get two distinct solutions only if $n + Cm = ml$, and $n - Cm = ml'$, which implies that $C = (l - l')/2$ is a half-integer, and $n/m = (l + l')/2$ is a half-integer.

We should contextualize this result in the language of Proposition 3. For $SU(2)$, only two Levi subgroups are possible: the maximal torus T or $SU(2)$ itself. The former case arises when $n/m \notin \frac{1}{2}\mathbb{Z}$, and the latter case arises when $n/m \in \frac{1}{2}\mathbb{Z}$. Stated slightly differently, when $n/m \in P \subseteq \mathfrak{t}$ is in the coweight lattice of $SU(2)$, then $\exp(2\pi i n/m) \in Z(SU(2))$ and the Levi subgroup corresponding to (n,m) is $G = SU(2)$ (and is the maximal torus otherwise).

7. Isotropy Representation of $T \times S^1$

Whenever a group G acts on a manifold M and $x \in M$ is a fixed point of the action, one obtains a representation of G on T_xM by taking the derivative of the action map at x. In this section, we compute this representation on the tangent space at any fixed point of the $T \times S^1$ action on ΩG. As we are considering the action of torus on a vector space, we present a splitting of the representation in terms of its weight vectors.

Proposition 6 *Let γ be fixed by $T \times S^1$ and suppose that $(t,\psi) \in T \times S^1$; then, after identifying $T_\gamma \Omega G \simeq \Omega\mathfrak{g}$, the isotropy representation of $T \times S^1$ on $T_\gamma \Omega G$ is given by*

$$\begin{aligned} (t, e^{i\psi})_* &: \Omega\mathfrak{g} \to \Omega\mathfrak{g} \\ X(\theta) &\mapsto \mathrm{Ad}_{t\gamma(\psi)} X(\theta + \psi) \end{aligned}$$

Proof By embedding G in $U(n)$, we may assume that G is a matrix group. Pick any variation $\delta\gamma \in T_\gamma \Omega G$ and write $\delta\gamma(\theta) = \gamma(\theta)X(\theta)$ for some $X \in \Omega\mathfrak{g}$. We compute the push-forward as follows:

$$\begin{aligned} (t, e^{i\psi})_*(\delta\gamma) &= \frac{d}{d\epsilon}\bigg|_{\epsilon=0} [(t,\psi)\cdot(\gamma(\theta) + \epsilon\gamma(\theta)X(\theta))] \\ &= \frac{d}{d\epsilon}\bigg|_{\epsilon=0} \left[t(\gamma(\theta+\psi) + \epsilon\gamma(\theta+\psi)X(\theta+\psi))(1+\epsilon X(\psi))^{-1}\gamma(\psi)^{-1}t^{-1}\right] \\ &= \frac{d}{d\epsilon}\bigg|_{\epsilon=0} \left[\sum_{j=0}^{\infty}(-1)^j\epsilon^j t(\gamma(\theta+\psi) + \epsilon\gamma(\theta+\psi)X(\theta+\psi))X(\psi)^j\gamma(\psi)^{-1}t^{-1}\right] \\ &= t\gamma(\theta+\psi)\,[X(\theta+\psi) - X(\psi)]\,\gamma(\psi)^{-1}t^{-1} \end{aligned}$$

But now, since γ is fixed under $T \times S^1$, we have $\gamma(\theta) = (t,\psi)\cdot\gamma(\theta) = t\gamma(\theta+\psi)\gamma(\psi)^{-1}t^{-1}$, which implies $\gamma(\theta)t\gamma(\psi) = t\gamma(\theta+\psi)$. Plugging this into the last line of the above yields the desired formula for the isotropy representation; notice that the constant term is equivalent to zero in the quotient $\Omega\mathfrak{g} \simeq L\mathfrak{g}/\mathfrak{g}$. □

The proposition above allows us to compute a weight basis for the isotropy representation, along with the corresponding weights.

Theorem 2 *If $\gamma(\theta) = \exp(\eta\theta) \in \Omega G$ is fixed by $T \times S^1$ (i.e. $\eta \in Q^\vee$), the $T \times S^1$ action on $T_\gamma \Omega G$ decomposes into non-trivial irreducible subrepresentations:*

$$T_\gamma \Omega G \simeq \Omega\mathfrak{g} \simeq \bigoplus_{k\in\mathbb{Z}} \left(\bigoplus_{\alpha\in R} V_{\alpha,k} \oplus \bigoplus_{i=1}^{n} V_{i,k} \right)$$

The weight of $T \times S^1$ on $V_{\alpha,k}$ is

$$\lambda_\alpha^k : Lie(T \times S^1)_\mathbb{C} \to \mathbb{C}$$
$$\lambda_\alpha^k(x_1, x_2) = \alpha(x_1 + \eta x_2) + kx_2$$

A basis of weight vectors for $V_{\alpha,k}$ is

$$X_{\alpha,k}^{(1)} = i\sigma_y^\alpha \cos(k\theta) \pm i\sigma_x^\alpha \sin(k\theta)$$
$$X_{\alpha,k}^{(2)} = i\sigma_x^\alpha \cos(k\theta) \mp i\sigma_y^\alpha \sin(k\theta)$$

where the plus or minus sign is taken depending on whether α is a positive or negative root, respectively. The weight of $T \times S^1$ on $V_{i,k}$ is

$$\lambda_i^k : Lie(T \times S^1)_\mathbb{C} \to \mathbb{C}$$
$$\lambda_i^k(x_1, x_2) = kx_2$$

A basis of weight vectors for $V_{i,k}$ is given by

$$X_{i,k}^{(1)} = i\sigma_z^\alpha \cos(k\theta)$$
$$X_{i,k}^{(1)} = i\sigma_z^\alpha \sin(k\theta)$$

Proof We will check that the pair $(X_{\alpha,k}^{(1)}, X_{\alpha,k}^{(2)})$ is a weight basis for $V_{\alpha,k}$ with the appropriate weight; the other cases are similar. Let $t = e^{x_1}$ for $x_1 \in \mathfrak{t}$, let $x_2 \in \mathrm{Lie}(S^1)$ and let $\Theta = \alpha(x_1 + \eta x_2)$. By Proposition 6,

$$\begin{aligned}
(t, e^{ix_2})_* X_{\alpha,k}^{(1)} &= \mathrm{Ad}_{t\gamma(x_2)} \left(i\sigma_y^\alpha \cos(k\theta) + i\sigma_x^\alpha \sin(k\theta) \right) \\
&= i\left(\sigma_y^\alpha \cos\Theta + \sigma_x^\alpha \sin\Theta\right) \cos(k(\theta + x_2)) \\
&\quad + i\left(\sigma_x^\alpha \cos\Theta - \sigma_y^\alpha \sin\Theta\right) \sin(k(\theta + x_2)) \\
&= i\left(\sigma_y^\alpha \cos\Theta + \sigma_x^\alpha \sin\Theta\right) (\cos kx_2 \cos k\theta - \sin kx_2 \sin k\theta) \\
&\quad + i\left(\sigma_x^\alpha \cos\Theta - \sigma_y^\alpha \sin\Theta\right) (\sin kx_2 \cos k\theta + \cos kx_2 \sin k\theta)
\end{aligned}$$

$$
\begin{aligned}
&= i\cos(\Theta + kx_2)\,\sigma_y^\alpha \cos k\theta + i\sin(\Theta + kx_2)\,\sigma_x^\alpha \cos k\theta \\
&\quad - i\sin(\Theta + kx_2)\,\sigma_y^\alpha \sin k\theta + i\cos(\Theta + kx_2)\,\sigma_x^\alpha \sin k\theta \\
&= \cos(\Theta + kx_2) X_{\alpha,k}^{(1)} + \sin(\Theta + kx_2) X_{\alpha,k}^{(2)}
\end{aligned}
$$

The computation for $X_{\alpha,k}^{(2)}$ is identical. This completes the proof. □

8. An Application of the Hyperfunction Fixed Point Localization Formula to $\Omega SU(2)$

In this section, we will present our approach to computing a regularized Duistermaat–Heckman distribution on $\mathrm{Lie}(T \times S^1)^*$ coming from the Hamiltonian action of $T \times S^1$ on ΩG. We will specialize to the case that $G = SU(2)$. This problem (and the work herein) was originally motivated by Atiyah's approach to a similar problem [Ati85]. In that paper, Atiyah showed that the Atiyah–Singer index theorem is a consequence of applying the Duistermaat–Heckman localization formula to the loop space of a Riemannian manifold. In [Ati85], Atiyah does also mention that similar methods can be applied to study ΩG; however, no further details or specific theorems are provided. Our original aim was to provide these details, as well as to study Duistermaat–Heckman distributions which come from Hamiltonian actions of compact tori on infinite-dimensional manifolds.

It was discovered after completing this project that some of these issues had already been considered [Pic89]. In this paper, Picken shows that the propagator for a quantum mechanical free particle moving on G (with the invariant Riemannian metric coming from the Killing form) can be exactly expressed by applying the fixed point localization formula for ΩG. In this case, the ill-defined left-hand side of the localization formula for ΩG is expressed as a path integral on G, while the right-hand side of the localization formula tells us exactly how to express the result of this path integral in terms of solutions to the classical equations of motion. We should highlight where our approach differs from his:

(1) Throughout, Picken uses a variable φ as a coordinate on $\mathfrak{t}$. We will be calling this coordinate x_1 in our work.

(2) Picken is implicitly setting $x_2 = 1$ throughout (i.e. he considers the slice $\mathfrak{t} \times \{1\} \subseteq \mathrm{Lie}(T \times S^1)$. This is evident in his choice of action functional, where the kinetic energy term

$$I_k[g] = \int \langle g^{-1}\dot{g}, g^{-1}\dot{g}\rangle \, d\theta$$

appears without a mass coefficient.

(3) We will directly apply a fixed point localization formula to ΩG with its $T \times S^1$ action, and interpret the result as a *hyperfunction* on $\mathrm{Lie}(T \times S^1)$. The advantage to this approach is that we will be able to Fourier transform this hyperfunction to

obtain a closed form of a density function for what one should expect is the push-forward of the 'Liouville measure' from ΩG to $\mathrm{Lie}(T \times S^1)^*$, using the moment map. Picken's formula is limited in this regard, since he does not use the localization formula to obtain a distribution on $\mathrm{Lie}(T \times S^1)$—he only obtains its restriction to a slice through $E = 1$. He also makes no use of hyperfunctions in his paper.

Definition *Let $\gamma \in \Omega G^{T\times S^1}$. The regularized equivariant Euler class of the normal bundle to γ is defined to be the holomorphic function on $\mathrm{Lie}(T \times S^1)_{\mathbb{C}}$ given by*

$$e_\gamma^{T\times S^1}(z_1,z_2) = \prod_{k=1}^{\infty}\left(\prod_{\alpha\in\Delta}\frac{\lambda_\alpha^k(z_1,z_2)}{kz_2}\right)$$

The difference between the 'usual' and the regularized equivariant Euler class of the normal bundle to γ is that we divide out by kz_2 on each weight. The regularization can be justified in a number of ways. We will see shortly that, when we include the regularizing terms, the resulting infinite product will converge to a useful functional expression for $e_\gamma^{T\times S^1}$. Without the regularization, the infinite product does not converge. Picken's work provides another justification for the regularization, since the resulting regularized localization formula provides an exact determination of the quantum mechanical propagator for a free particle moving on G.

For simplicity, let's examine the example $G = SU(2)$. We always use coordinates on $\mathrm{Lie}(T \times S^1)$ consisting of the coroot basis for $\mathfrak{t}$, and normalize the E-component of the moment map so that

$$E\begin{pmatrix} e^{i\theta} & 0 \\ 0 & e^{-i\theta}\end{pmatrix} = 1/2$$

Let $z = (z_1,z_2) \in \mathrm{Lie}(T \times S^1)$ and let $\gamma(\theta) = \exp(i\eta\theta) \in \Omega SU(2)$ be a fixed point of the $T \times S^1$ action. When we work with $G = SU(2)$, a choice of η is really just a choice of integer so, for $\alpha \in \Delta$ the non-zero positive root, we set $\alpha(\eta) = 2n$. For every k, we get four weights for the isotropy representation, corresponding to the two root vectors in $\mathfrak{sl}_2$ and the two weights coming from a non-zero element of the Cartan subalgebra:

$$\lambda_{h,i}^{(k)}(z_1,z_2) = kz_2 \quad i = 1,2$$
$$\lambda_e^{(k)}(z_1,z_2) = kz_2 + 2(z_2 n + z_1)$$
$$\lambda_f^{(k)}(z_1,z_2) = kz_2 - 2(z_2 n + z_1)$$

Proposition 7 *Let $G = SU(2)$. If $\gamma_n \in \Omega G^{T\times S^1}$, then the regularized equivariant Euler class of the normal bundle to γ_n is given by*

$$e_{\gamma_n}^T(z_1,z_2) = \frac{\sin(2\pi(n + z_1/z_2))}{2\pi(n + z_1/z_2)} \tag{7}$$

Proof Since the fixed points of the $T \times S^1$ action are isolated, we have that the normal bundle to the fixed point set is simply $T_{\gamma_n}\Omega G$. We can compute the regularized equivariant Euler class of $T_{\gamma_n}\Omega G$ by taking the product over the weights appearing in the isotropy representation of $T \times S^1$ on $T_{\gamma_n}\Omega G$, according to Theorem 2:

$$\begin{aligned} e^T_{\gamma_n}(z_1, z_2) &= \prod_{k=1}^{\infty} \prod_{\alpha \in \Delta} \lambda^{(k)}_\alpha / k z_2 \\ &= \prod_{k=1}^{\infty} \left[1 + \frac{2(z_2 n + z_1)}{z_2 k}\right]\left[1 - \frac{2(z_2 \eta + z_1)}{z_2 k}\right] \\ &= \prod_{k=1}^{\infty} \left[1 - \left(\frac{2(z_2 n + z_1)}{z_2 k}\right)^2\right] \\ &= \frac{\sin(2\pi(n + z_1/z_2))}{2\pi(n + z_1/z_2)} \end{aligned}$$

where the last line follows from the infinite product formula for $\sin(z)$. □

Remarks *In the more general case of $G = SU(n)$, each choice of positive root will give a difference of squares, which then translates to an extra* $\sin(z)/z$ *term in the final result. We would then take a product over all the positive roots.*

In what follows, we will write $e^{T\times S^1}_{\gamma_n}(z_1, z_2) = e_n(z_1, z_2)$ for notational simplicity. A formal application of the fixed point localization formula to ΩG would then yield the following expression, valid for $(x_1, x_2) \in \mathrm{Lie}(T \times S^1)$ such that $e_n(x_1, x_2) \neq 0$:

$$\int_{\Omega G} e^{\omega + i\langle \mu(\gamma), x\rangle} = \sum_{n \in \mathbb{Z}} e^{i(n x_1 + \frac{n^2}{2} x_2)} \frac{2\pi(n + x_1/x_2)}{\sin(2\pi(n + x_1/x_2))} \tag{8}$$

We have not addressed the types of objects that Equation 8 asserts an equality of. In the setting of a compact symplectic manifold with a Hamiltonian action of a compact torus, one is free to understand this to be an equality of distributions, and even an equality of density functions on some open set. But, for the purposes of ΩG, this perspective is insufficient. For instance, the Duistermaat–Heckman 'distribution' is supposed to be obtained by taking the Fourier transform of the right-hand side of Equation 8; however, we can see that the expression obtained from the localization formula is not even integrable, since it has poles and, even if we ignore the poles coming from the denominator, the numerator grows linearly in the ξ_1 variable. The terms appearing in the localization formula for ΩG also have unpleasant limiting behaviour as $x_2 \to 0$. The right-hand side of the localization formula should not be interpreted as a distribution (and, consequently, neither should the left-hand side).

There are further hints in [GLS88] which suggest that the localization formula for ΩG should be an expression positing an equality of two hyperfunctions. Suppose for a moment that we are considering a Hamiltonian action of a torus T on a finite-dimensional vector space with weights $\alpha_1,\ldots,\alpha_n$. To each weight we can associate a constant coefficient differential operator D_{α_i} on $\mathfrak{t}^*$. The Duistermaat–Heckman distribution is a solution to the differential equation:

$$D_{\alpha_1}\ldots D_{\alpha_n}(DH(x)) = \delta(x)$$

When V is infinite-dimensional and we have infinitely many weights (such as is the case for the isotropy representation of $T\times S^1$ on the tangent space to a fixed loop in ΩG), then we are forced to consider differential operators of infinite order. Infinite-order differential operators do not even act on distributions. For example, any infinite-order differential operator on $\mathbb{R}$ cannot act on the Dirac delta distribution because of the classical theorem which states that any distribution supported at the origin must be a finite sum of the Dirac delta distribution and its derivatives. Hyperfunctions (and the related concept of a microfunction) are a sheaf on which infinite-order differential operators do have a well-defined action. Furthermore, the entire classical theory of distributions is subsumed by the theory of hyperfunctions, so it makes more sense to study the Duistermaat–Heckman distribution as a hyperfunction, rather than as a distribution.

We now begin our construction of the Picken hyperfunction of $\Omega SU(2)$.

Lemma 8.1 *If $H^{(n)}_{k,\alpha} = \{(y_1,y_2)\in Lie(T\times S^1)\,|\,2y_1+(2n+k)y_2<0\}$ are the half-planes corresponding to the weights $\lambda_{\alpha,k}$ at the nth fixed point, then*

$$\gamma_n = \bigcap_{k=1}^{\infty}\left(H^{(n)}_{k,e}\cap H^{(n)}_{k,e}\right) = \left\{(y_1,y_2)\in Lie(T\times S^1)\,\Big|\,\left(\frac{2n-1}{2}\right)|y_2| < y_1 < \left(\frac{2n+1}{2}\right)|y_2|, y_2<0\right\}$$

Proof The lemma follows by noticing that the intersection $H^{(n)}_{1,e}\cap H^{(n)}_{1,f}$ is properly contained in each of the half-planes $H^{(n)}_{k,\alpha}$. □

We will interpret the right-hand side of Equation 8 as a Picken hyperfunction on Lie$(T\times S^1)$. We first need to show that each function appearing as a summand is holomorphic on an appropriate domain, and slowly increasing so that we may take its Fourier transform.

Proposition 8 *Let $\gamma_n\subseteq Lie(T\times S^1)$ be the open convex cone of Lemma 8.1. For every n, $1/e_n(z_1,z_2)$ is a slowly increasing holomorphic function on $\mathbb{R}^2\times i\gamma_n\subseteq Lie(T\times S^1)_{\mathbb{C}}$.*

Proof The polar structure of the function $1/e_n(z_1,z_2)$ is determined by the zeros of $\sin(2\pi z_1/z_2)$, which occur at $z_1/z_2=k/2$. The singularity at $z_1/z_2=-n$ is removable. We need only ensure that, under the condition $(y_1,y_2)\in\gamma_n$, we have that $z_1/z_2\neq k/2$ for any $k\neq -2n$. For fixed $(y_1,y_2)\in\gamma_n$, the image of the curves $x_1=mx_2$ $(m\in\mathbb{R})$ under the mapping $(z_1,z_2)\mapsto z_1/z_2$ are the parametric curves given by

$$\mathbb{R} \to \mathbb{C}$$
$$s \mapsto \frac{ms^2 + y_1 y_2}{s^2 + y_2^2} + i\frac{sy_1 - msy_2}{s^2 + y_2^2}$$

These are easily seen to be ellipsoidal arcs which cross the real axis at $\mathrm{Re}(z_1/z_2) = y_1/y_2$ when $s = 0$, and asymptotically approach the real axis from above (below) at $\mathrm{Re}(z_1/z_2) = m$ as $s \to \infty$ when $m > 0$ (and from below the axis if $m < 0$). This shows that the image of z_1/z_2 always avoids the poles of $\csc(2\pi z)$, except possibly when $x_1 = x_2 = 0$; however, since $(y_1, y_2) \in \gamma_n$, we have $(2n-1)/2 < y_1/|y_2| < (2n+1)/2$. This argument (and Figure 3) show that for any fixed $(y_1, y_2) \in \gamma_n$, the image of z_1/z_2 avoids the poles of $(n + z_1/z_2)\csc(2\pi z_1/z_2)$, so $1/e_n(z_1, z_2)$ is a holomorphic function on $\mathbb{R}^2 \times i\gamma_n$.

We now show that $1/e_n(z_1, z_2)$ is slowly increasing. Fix a compact set $K \subseteq \gamma_n$ and any $\epsilon > 0$. Since $(y_1, y_2) \mapsto y_1/y_2$ is continuous on K, it will achieve its maximum and minimum, so there is a $\delta > 0$ such that the estimate $(2n-1)/2 + \delta \le y_1/|y_2| \le (2n+1)/2 - \delta$ holds uniformly over K. We must show that

$$\left| \frac{e^{-\epsilon |\mathrm{Re}(z)|}}{\sin(2\pi z_1/z_2)} \right| \to 0$$

uniformly in K as $\mathrm{Re}(z) \to \infty$.

First, we notice that, if we fix y_1/y_2 as above, then, for every R sufficiently large, we have, for every (x_1, x_2) on the box $\max\{|x_1|, |x_2|\} = R$,

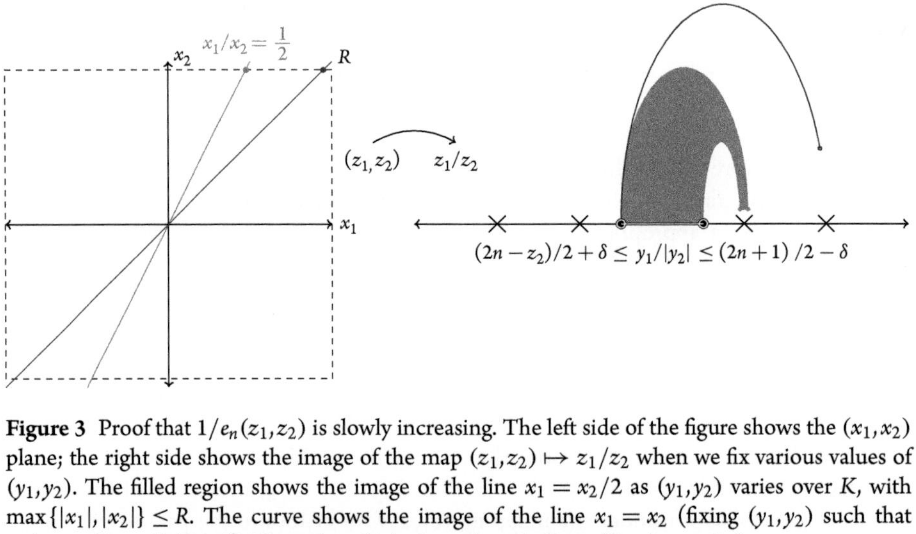

Figure 3 Proof that $1/e_n(z_1, z_2)$ is slowly increasing. The left side of the figure shows the (x_1, x_2) plane; the right side shows the image of the map $(z_1, z_2) \mapsto z_1/z_2$ when we fix various values of (y_1, y_2). The filled region shows the image of the line $x_1 = x_2/2$ as (y_1, y_2) varies over K, with $\max\{|x_1|, |x_2|\} \le R$. The curve shows the image of the line $x_1 = x_2$ (fixing (y_1, y_2) such that $y_1/y_2 = (2n-1)/2 + \delta$). The poles of $1/e_n(z_1, z_2)$ are indicated by the symbol $\times$.

$$\Rightarrow |\csc(2\pi z_1/z_2)| \leq \left|\csc\left(2\pi\frac{(2n+1)R^2/2+y_1y_2}{R^2+y_2^2}+i\frac{Ry_1-(2n+1)Ry_2/2}{R^2+y_2^2}\right)\right|$$

This estimate follows from the observation that the maximum of $\csc(2\pi z_1/z_2)$ on the box occurs at the point (x_1,x_2) such that the distance from z_1/z_2 to a pole of $\csc(2\pi z)$ is minimized; this condition is satisfied on the line $x_1=(2n+1)x_2/2$. A uniform bound over K can be found because of our previous estimate on y_1/y_2. Figure 3 demonstrates these estimates. The proof is completed by noticing that $\csc(2\pi z_1/z_2)$ has linear growth (which is dominated by any exponential) as $x_2\to\infty$, because all of its poles are simple. □

We use the cones γ_n of Lemma 8.1 in the construction of the Picken hyperfunction of $\Omega SU(2)$. For every n, we apply Lemma 1 to the set of hyperfunctions $\left\{f^{(n)}_{\lambda_{e,k}}(x),f^{(n)}_{\lambda_{f,k}}(x)\right\}_{k=1}^{\infty}$ (c.f. notation of Corollary 1), making sure to use the regularized weights to guarantee uniform convergence of the infinite product. The resulting hyperfunction is the regularized equivariant Euler class to the normal bundle of the nth fixed point:

$$\frac{1}{e_n(x_1,x_2)}=b_{\gamma_n}\left(\frac{2\pi(n+z_1/z_2)}{\sin(2\pi z_1/z_2)}\right)$$

Putting all of these results together, we obtain the Picken hyperfunction for the Hamiltonian $T\times S^1$ action on $\Omega SU(2)$:

$$L_{\Omega SU(2)}(x_1,x_2)=\frac{1}{(2\pi i)^2}\sum_{n\in\mathbb{Z}}b_{\gamma_n}\left(e^{iz_1n+iz_2n^2/2}\frac{2\pi(n+z_1/z_2)}{\sin(2\pi z_1/z_2)}\right)$$

By Proposition 8, $L_{\Omega SU(2)}(x_1,x_2)$ is a slowly increasing hyperfunction, so we may take its Fourier transform. Let S_n be a contour in $\mathrm{Lie}(T\times S^1)_{\mathbb{C}}$, chosen such that $(y_1,y_2)\in\gamma_n$. After choosing a holomorphic partition of unity $\chi_\sigma(z)$, we may write the following convergent expression for the Duistermaat–Heckman hyperfunction:

$$DH(\xi_1,\xi_2)$$
$$=\frac{1}{(2\pi i)^2}\sum_{\sigma\in\Sigma}\sum_{n\in\mathbb{Z}}b_{-\sigma^\circ}\left(\int_{S_n}e^{-i(\zeta_1-n)z_1-i(\zeta_2-n^2/2)z_2}\frac{2\pi(n+z_1/z_2)}{\sin(2\pi z_1/z_2)}\chi_\sigma(z_1,z_2)\,dz_1\,dz_2\right)$$

One might try and proceed with the computation of this integral, as in the example of Section 3; however, the computation does not seem to reduce to an elegant expression. If one uses the following partition of unity,

$$1=\frac{1}{1+e^{z_1}}\frac{1}{1+e^{\pi z_2}}+\frac{1}{1+e^{-z_1}}\frac{1}{1+e^{\pi z_2}}+\frac{1}{1+e^{z_1}}\frac{1}{1+e^{-\pi z_2}}+\frac{1}{1+e^{-z_1}}\frac{1}{1+e^{-\pi z_2}}$$

then the polar locus of the integrand defining the Fourier transform consists of isolated singularities which are locally cut out by a pair of equations. The residues near such singularities are readily computed, but do not appear to re-sum in any obvious way. We leave a further examination of the form of the Duistermaat–Heckman hyperfunction of $\Omega SU(2)$ as an open problem.

REFERENCES

[AP83] M.F. Atiyah and A.N. Pressley, *Convexity and loop groups*, in Arithmetic and Geometry (M. Artin and J. Tate, eds), Springer, 1983, pp. 33–63.

[Ati85] M.F. Atiyah, *Circular symmetry and stationary-phase approximation*, Astérisque **1985** (1985), no. 131, 43–59.

[DH82] J.J. Duistermaat and G.J. Heckman, *On the variation in the cohomology of the symplectic form of the reduced phase space*, Inventiones Mathematicae **69** (1982), no. 2, 259–68.

[GH14] P. Griffiths and J. Harris, *Principles of Algebraic Geometry*, John Wiley & Sons, 2014.

[GLS88] V. Guillemin, E. Lerman and S. Sternberg, *On the Kostant multiplicity formula*, Journal of Geometry and Physics **5** (1988), no. 4, 721–50.

[JK95a] L. Jeffrey and F. Kirwan, *Intersection pairings in moduli spaces of holomorphic bundles on a Riemann surface*, Electronic Research Announcements of the American Mathematical Society **1** (1995), no. 2, 57–71.

[JK95b] L. Jeffrey and F. Kirwan, *Localization for non-abelian group actions*, Topology **34** (1995), no. 2, 291–327.

[Kan89] A. Kaneko, *Introduction to the Theory of Hyperfunctions*, vol. 3, Springer Science & Business Media, 1989.

[Kaw70] T. Kawai, *On the theory of Fourier hyperfunctions and its applications to partial differential equations with constant coefficients.* Journal of the Faculty of Science, the University of Tokyo. Section 1A, Mathematics **17** (1970), no. 3, 467–517.

[KKS] M. Kashiwara, T. Kawai and M. Sato, *Microfunctions and pseudo-differential equations*, in Hyperfunctions and pseudo-differential equations: proceedings of a conference at Katata (H. Komatsu, ed.), Springer, 1973, pp. 265–529.

[KS99] G. Kato and D.C. Struppa, *Fundamentals of algebraic microlocal analysis*, vol. 217, CRC Press, 1999.

[KW08] B. Khesin and R. Wendt, *The Geometry of Infinite-Dimensional Groups*, vol. 51, Springer Science & Business Media, 2008.

[Pic89] R.F. Picken, *The propagator for quantum mechanics on a group manifold from an infinite-dimensional analogue of the Duistermaat–Heckman integration formula*, Journal of Physics A: Mathematical and General **22** (1989), no. 13, 2285.

[PS86] A.N. Pressley and G.B. Segal, *Loop Groups*, Clarendon Press, 1986.

[Sat59] M. Sato, *Theory of hyperfunctions, I.*, Journal of the Faculty of Science, the University of Tokyo. Section 1A, Mathematics **8** (1959), no. 1, 139–93.

[Wit92] E. Witten, *Two dimensional gauge theories revisited*, Journal of Geometry and Physics **9** (1992), no. 4, 303–68.

13

Quantization of the Quantum Hitchin System and the Real Geometric Langlands Correspondence

JÖRG TESCHNER

Department of Mathematics, University of Hamburg, Bundesstrasse 55, 20146 Hamburg, Germany, and: DESY theory, Notkestrasse 85, 20607 Hamburg, Germany

1. Introduction

1.1. Motivations

– *The Nekrasov-Shatashvili programme and its relation to gauge theory*

An important motivation comes from the programme initiated by Nekrasov and Shatashvili to investigate relations between supersymmetric field theories and quantum integrable models [1]. An interesting family of examples to which this programme can be applied is provided by a class of four-dimensional $\mathcal{N} = 2$-supersymmetric field theories associated to the choice of a pair $(C, \mathfrak{g})$ consisting of a (possibly punctured) Riemann surface C and a Lie-algebra $\mathfrak{g}$ of ADE-type [2, 13]. The integrable models relevant for this class of theories are known [13] to be the Hitchin systems [4]. Regularizing the supersymmetric field theories by means of the so-called omega deformation leads to the quantization of the corresponding integrable models [1, 5, 6].

1.1.1. Quantum integrable systems

Many quantum integrable models can be solved by the Bethe ansatz method. Whenever the Bethe ansatz is applicable, it is often useful to formulate the Bethe ansatz equations representing the quantization conditions in terms of a single, model-dependent function $\mathcal{Y}(\mathbf{a}, \mathbf{t})$ called Yang's function [7]. This function depends on two types of variables: $\mathbf{a} = (a_1, \ldots, a_d)$ and $\mathbf{t} = (\tau_1, \ldots, \tau_{d'})$. The parameters $\mathbf{t}$ are parameters of the commuting Hamiltonians,

Teschner, J., *Quantization of the Quantum Hitchin System and the Real Geometric Langlands Correspondence*. In: *Geometry and Physics: A Festschrift in Honour of Nigel Hitchin*, Jørgen Ellegaard Andersen, Andrew Dancer, Oscar García-Prada (Eds): Oxford University Press (2018). © Oxford University Press.
DOI: 10.1093/oso/9780198802006.003.0013

in the context of spin chains often called inhomogeneity parameters. The variables $\mathbf{a}$ are auxiliary, allowing us to represent the Bethe ansatz equations in the form

$$\frac{\partial}{\partial a_k}\mathcal{Y}(\mathbf{a},\mathbf{t}) = 2\pi i n_k, \qquad k = 1,\ldots,d. \tag{1.1}$$

In a non-degenerate situation, equation (1.1) has a unique solution $\mathbf{a} = \mathbf{a}_{\rm cr}(\mathbf{n})$ for given integers $\mathbf{n} = (n_1,\ldots,n_d)$. The eigenvalues E_r of a subset of the commuting conserved quantities H_r, $r = 1,\ldots,d'$, can the be obtained from $\mathcal{Y}(\mathbf{a},\mathbf{t})$ by taking the derivatives

$$E_r = \frac{\partial}{\partial \tau_r}\mathcal{Y}(\mathbf{a},\mathbf{t})\Big|_{\mathbf{a}=\mathbf{a}_{\rm cr}(\mathbf{n})}. \tag{1.2}$$

Beyond the class of quantum integrable systems soluble by Bethe ansatz techniques, there exists a large class of models where such techniques fail. One important outcome of the Nekrasov–Shatashvili programme is strong evidence for the proposal made in [1] that the quantization conditions in large classes of integrable models which cannot be solved by the Bethe ansatz method can nevertheless be described in terms of suitable Yang's functions.

However, for the models studied in this paper it will turn out that another type of condition formulated in terms of a single function $\mathcal{Y}(\mathbf{a},\mathbf{t})$ is appropriate. In general, it is not a priori obvious which type of condition is appropriate for a given model. The Nekrasov–Shatashvili scheme will be efficient for the solution of quantum integrable systems only if one knows exactly how a given quantization condition is represented in terms of the Yang's function $\mathcal{Y}(\mathbf{a},\mathbf{t})$. Answering this question for interesting integrable models may lead into fairly profound mathematical problems, as will be illustrated by the examples studied in this paper.

1.1.2. Geometric Langlands programme

The geometric Langlands correspondence is often loosely formulated as a correspondence which assigns $\mathcal{D}$-modules on Bun_G to ${}^LG$-local systems on a Riemann surface C (see [8] for a review of the aspects relevant here); ${}^LG$ is the Langlands dual group of a simple complex Lie group G. Most interesting for us is the special case considered in the original work of Beilinson and Drinfeld where the ${}^LG$-local systems are *opers*, pairs $(\mathcal{E},\nabla')$ in which ∇' is gauge equivalent to a certain standard form. The space of opers forms a Lagrangian subspace in the moduli space of all local systems. The corresponding $\mathcal{D}$-modules on Bun_G can be described more concretely as systems of partial differential equations taking the form of eigenvalue equations $\mathsf{H}_r f = E_r f$ for a family of differential operators H_r on Bun_G, quantizing the Hamiltonians of Hitchin's integrable system. The oper corresponding to such a $\mathcal{D}$-module in the geometric Langlands correspondence is the geometric object encoding the eigenvalues E_r.

1.1.3. Relation to conformal field theory

This paper is part of a larger programme outlined in [9, 10] on the relations between the quantization of the Hitchin system, supersymmetric field theories, conformal field theory and the geometric Langlands programme. Some of these relations will be briefly described at the end of this paper.

1.2. Main Results

We are going to propose a natural quantization condition for the Hitchin system, and explain how it can be reformulated in terms of a function $\mathcal{Y}(\mathbf{a},\mathbf{t})$. The function $\mathcal{Y}(\mathbf{a},\mathbf{t})$ relevant for this task is found to be the generating function for the variety of opers within the space of all local systems as predicted in [6, 9]. However, the condition on $\mathcal{Y}$ expressing the quantization condition turns out to be different from the types of conditions considered in [1]. Our derivation is essentially complete for Hitchin systems associated to the Lie algebra $\mathfrak{sl}_2$ in genus zero and one, which may be called the Gaudin and elliptic Calogero–Moser models associated to the group $\mathrm{SL}(2,\mathbb{C})$. It reduces to a conjecture of E. Frenkel [11] for $g > 1$, as will be discussed below.

Reformulating the quantization conditions in terms of $\mathcal{Y}$ can be done using the Separation of Variables (SOV) method pioneered by Sklyanin [12]. This method may be seen as a more concrete procedure for constructing the geometric Langlands correspondence relating opers to $\mathcal{D}$-modules (eigenvalue equations), as was pointed out in [11]. In our case, it will be found that the SOV method relates single-valued solutions of the eigenvalue equations to opers having real holonomy. The classification of opers or equivalently projective structures on C with real holonomy has been studied in [13]. Using complex Fenchel–Nielsen coordinates, we will reformulate this description in terms of the generating function for the variety of opers.

From the point of view of the geometric Langlands correspondence, we obtain a correspondence between opers with real holonomy and $\mathcal{D}$-modules admitting single-valued solutions. We expect that a generalization to more general local systems with real holonomy will exist. We propose to call such correspondences the real geometric Langlands correspondence.

2. Separation of Variables for the Classical Hitchin Integrable System

2.1. Integrability and Special Geometry

A complex symplectic manifold $\mathcal{M}$ with holomorphic symplectic form Ω is called an algebraic integrable system if it can be described as a Lagrangian torus fibration $\pi : \mathcal{M} \to \mathcal{B}$ with fibres being principally polarized abelian varieties. Algebraic integrability is equivalent to the fact that the base $\mathcal{B}$ is a special Kähler manifold satisfying certain integrality conditions [14].

These connections may be reformulated conveniently in terms of a covering of $\mathcal{M}$ with local charts carrying action-angle coordinates consisting of a tuple $\mathbf{a} = (a^1, \dots, a^d)$ of coordinates for the base $\mathcal{B}$, and complex coordinates $\mathbf{z} = (z_1, \dots, z_d)$ for the torus fibres $\Theta_b = \mathbb{C}^d/(\mathbb{Z}^d + \tau_b \cdot \mathbb{Z}^d)$, $b \in \mathcal{B}$, such that

$$\Omega = \sum_{r=1}^{d} da^r \wedge dz_r. \tag{2.3}$$

The transformation $\mathbf{z}^{\mathrm{D}} := \tau_b^{-1} \cdot \mathbf{z}$ gives an equivalent representation of the torus fibres Θ_b. It can be extended to a canonical transformation $(\mathbf{a},\mathbf{z}) \to (\mathbf{a}^{\mathrm{D}},\mathbf{z}^{\mathrm{D}})$ by introducing coordinates a_s^{D} satisfying $\frac{\partial}{\partial a^r} a_s^{\mathrm{D}} = \tau_{rs}$. As $\tau_{rs} = \tau_{sr}$, there exists a potential $\mathcal{F}(\mathbf{a})$ allowing us to represent a_r^{D} in the form $a_r^{\mathrm{D}} = \frac{\partial}{\partial a^r} \mathcal{F}(\mathbf{a})$. It follows that

$$\Omega = \sum_{r=1}^{d} da_r^{\mathrm{D}} \wedge dz_{\mathrm{D}}^r. \tag{2.4}$$

One may equivalently represent Θ_b as a *real* torus $\mathbb{R}^{2d}/\mathbb{Z}^{2d}$ using the coordinates $(\mathbf{w},\mathbf{w}_{\mathrm{D}})$, $\mathbf{w} = (w_1,\dots,w_d)$, $\mathbf{w}_{\mathrm{D}} = (w_{\mathrm{D}}^1,\dots,w_{\mathrm{D}}^d)$ such that $\mathbf{z} = \mathbf{w} + \tau \cdot \mathbf{w}_{\mathrm{D}}$. There exists a corresponding set of real action variables $(\mathbf{b},\mathbf{b}_{\mathrm{D}})$ such that

$$\mathrm{Re}(\Omega) = \sum_{r=1}^{d} (db^r \wedge dw_r + db_r^{\mathrm{D}} \wedge dw_{\mathrm{D}}^r). \tag{2.5}$$

The real action variables $(\mathbf{b},\mathbf{b}_{\mathrm{D}})$ are simply the real parts of $(\mathbf{a},\mathbf{a}_{\mathrm{D}})$. The coordinates above are only locally defined, in general. Different sets of coordinates are related by $\mathrm{Sp}(2d,\mathbb{Z})$-transformations acting in the standard fashion on the vectors $(\mathbf{w},\mathbf{w}_{\mathrm{D}})$.

2.2. *Integrability of the Hitchin System*

The phase space $\mathcal{M}_{\mathrm{H}}(C)$ of the Hitchin system [4] for $G = GL(2)$ on a Riemann surface C with genus $g > 1$ is the moduli space of stable pairs $(\mathcal{E},\varphi)$, where $\mathcal{E}$ is a holomorphic rank 2 vector bundle, and $\varphi \in H^0(C,\mathrm{End}(\mathcal{E}) \otimes K_C)$ is called the Higgs field, modulo gauge transformations. There is a natural stability condition for the pairs $(\mathcal{E},\varphi)$ allowing certain unstable bundles $\mathcal{E}$. The open dense subset of $\mathcal{M}_{\mathrm{H}}(C)$ consisting of pairs $(\mathcal{E},\varphi)$ with stable bundles $\mathcal{E}$ is isomorphic to the cotangent bundle $T^*\mathrm{Bun}_G(C)$. The moduli space $\mathcal{M}_{\mathrm{H}}(C)$ carries a natural holomorphic symplectic structure restricting to the canonical symplectic structure on the dense open subset $T^*\mathrm{Bun}_G(C)$. Considering bundles $\mathcal{E}$ with fixed determinant and Higgs fields φ with vanishing trace allows one to describe the Hitchin system for $G = SL(2)$ in a similar way.

The complete integrability of the Hitchin system is demonstrated using the so-called Hitchin map, in our case mapping a pair $(\mathcal{E},\varphi)$ to the coefficients $(\vartheta_1,\vartheta_2)$ of the characteristic polynomial $\det(v\,\mathrm{id} - \varphi(u)) = v^2 - \vartheta_1 v + \vartheta_2$. The coefficients $(\vartheta_1,\vartheta_2)$ can be identified with elements of the vector space $\mathcal{B} = H^0(C,K) \oplus H^0(C,K^2)$. Fixing bases $\{\rho_1,\dots,\rho_g\}$ and $\{q_1,\dots,q_{3g-3}\}$ for $H^0(C,K)$ and $H^0(C,K^2)$, respectively, allows us to define the Hamiltonians of the Hitchin system to be the coefficients in the expansions $\mathrm{tr}(\varphi(u)) = \sum_{i=1}^{g} \rho_i h_i$ and $\mathrm{tr}(\varphi^2(u)) = \sum_{r=1}^{3g-3} q_r H_r$. They form a maximal set of Poisson-commuting globally defined functions on $\mathcal{M}_{\mathrm{H}}(C)$. The Hitchin fibres Θ_b are the subvarieties of $\mathcal{M}_{\mathrm{H}}(C)$ associated to a point $b \in \mathcal{B}$.

In order to see that generic fibres Θ_b can be represented as abelian varieties (complex tori), one may first define the spectral curve Σ as

$$\Sigma = \left\{ (u,v) \in T^*C;\ \det(v\mathrm{id} - \varphi) = 0 \right\}. \tag{2.6}$$

To each pair $(\mathcal{E},\varphi)$ let us then associate a line bundle L on Σ, the bundle with fibres being the eigenlines of φ for a given eigenvalue v, defining a map from $(\mathcal{E},\varphi)$ to the pair (Σ,L). Conversely, given a pair (Σ,L), where $\Sigma \subset T^*C$ is a double cover of C, and L a holomorphic line bundle on Σ, one can recover $(\mathcal{E},\varphi)$ via

$$(\mathcal{E},\varphi) := \big(\pi_*(L),\pi_*(v)\big), \tag{2.7}$$

where π is the covering map $\Sigma \to C$, and π_* is the direct image. In this way, we may identify the Hitchin fibres Θ_b with the Jacobian of Σ parametrizing the choices of the line bundles L. This is how the space $\mathcal{M}_{\mathrm{H}}(C)$ gets described as a torus fibration, with the fibre over a point $b \in \mathcal{B}$ being the Jacobian.

For the case of $G = SL(2)$, one needs to impose the condition that the bundle $\mathcal{E}$ has a trivial determinant. The Jacobian is then replaced by the so-called Prym variety, parametrizing line bundles L such that $\det(\pi_*(L)) \simeq \mathcal{O}$.

It can furthermore be shown that the dynamics of the Hitchin system generated by the Hamiltonians with respect to the natural symplectic structure gets linear on the torus fibres [4], completing the proof of the complete integrability of the Hitchin system.

2.3. *Algebraic Integrability of Jacobian Fibrations*

Algebraic integrability is realized in a canonical fashion in terms of Jacobian or Prym fibrations of spectral curves. Indeed, given a spectral curve Σ, let us pick a canonical basis for the first homology of Σ, represented by mutually non-intersecting sets of cycles $\alpha_1,\dots,\alpha_h$ and $\beta_1,\dots,\beta_h$ satisfying $\alpha_r \cdot \beta_s = \delta_{r,s}$, where $h = 4g-3$ is the genus of Σ. A basic role is played by the periods

$$a^r = \int_{\alpha_r} \lambda, \qquad a_r^{\mathrm{D}} = \int_{\beta_r} \lambda \tag{2.8}$$

of the canonical differential $\lambda = vdu$ on Σ. The derivatives $\omega_r = \partial_{a^r}\lambda$ give a basis for the space of abelian differentials normalized as $\delta_{r,s} = \int_{\alpha_r} \omega_s$, The torus fibres may then be represented as $\Theta_{\mathbf{E}} = \mathbb{C}^h/(\mathbb{Z}^h + \tau \cdot \mathbb{Z}^h)$, with the period matrix τ having matrix elements $\tau_{rs} = \int_{\beta_s} \omega_r$. The Riemann bilinear relations give $\tau_{rs} = \tau_{sr}$. It follows that there exists a function $\mathcal{F}(\mathbf{a})$ giving the dual periods a_r^{D} as $a_r^{\mathrm{D}} = \partial_{a^r}\mathcal{F}(\mathbf{a})$.

When the integrable structure is represented in terms of a torus fibration over families of spectral curves which are branched coverings of an underlying curve C, one may alternatively represent the integrable structure in terms of a symmetric product $(T^*C)^{[h]}$ of the cotangent bundle of C. This relation is essentially canonical and most easily described when the torus fibres are the Jacobians of Σ. The Abel map from divisors $\hat{\mathbb{D}}_{\hat{\mathbf{u}}} = \sum_{r=1}^h \hat{u}_r$ on Σ to the Jacobian,

$$z_s(\mathbf{a},\mathbf{u}) = \sum_{r=1}^{h} \int^{\hat{u}_r} \omega_s, \tag{2.9}$$

can be inverted (Jacobi inversion problem), defining a divisor $\mathbb{D}_{\mathbf{u}} = \sum_{r=1}^{h} u_r$ on C by projection. The locally defined function

$$\mathcal{X}(\mathbf{a},\mathbf{u}) = \sum_{r=1}^{h} \int^{\hat{u}_r} \lambda \tag{2.10}$$

is a generating function for the change of variables from $(\mathbf{a},\mathbf{z})$ to $(\mathbf{v},\mathbf{u})$,

$$\frac{\partial}{\partial a_r}\mathcal{X}(\mathbf{a},\mathbf{u}) = z_r, \qquad \frac{\partial}{\partial u_r}\mathcal{X}(\mathbf{a},\mathbf{u}) = v_r. \tag{2.11}$$

It follows from the existence of the generating function $\mathcal{X}(\mathbf{a},\mathbf{u})$ that the coordinates $(\mathbf{v},\mathbf{u})$ are Darboux coordinates. Note that the points $(u_k, v_k) \in T^*C$ with $v_k = v_k(\mathbf{a},\mathbf{u})$ defined in (2.11) automatically satisfy

$$v_k^2 - \mathrm{tr}(\varphi(u_k)) + \mathrm{tr}(\varphi^2(u_k)) = 0 \quad \Leftrightarrow \quad (u_k, v_k) \in \Sigma, \tag{2.12}$$

for $k \in 1,\ldots,h$. A detailed explanation of the modifications of the Abel map that are necessary in the cases where the torus fibres are Prym varieties can be found in [15]. Only the subspace of $H_1(\Sigma)$ which is odd under the exchange of sheets is relevant in this case, reducing the number of relevant variables from h to $d = 3g - 3$.

The representation in terms of the symmetric product $(T^*C)^{[h]}$ will be called an SOV representation. We conclude that an SOV representation exists for the classical theory whenever there is a description in terms of pairs (Σ, L) as introduced above.

2.4. Separation of Variables

It may be necessary to describe the passage from the original description in terms of pairs $(\mathcal{E},\varphi)$ to either one of the two descriptions making the integrable structure manifest more explicitly. This requires constructing sections χ of the line bundle L as families of eigenvectors of the Higgs-field φ. The divisor $D_{\mathbf{u}}$ will be identified with the divisor of zeros of χ [16, 17].

To begin with, we need to represent the pairs $(\mathcal{E},\varphi)$ more concretely. This can be done by representing the bundles $\mathcal{E}$ as extensions:

$$0 \longrightarrow \mathcal{L}' \longrightarrow \mathcal{E} \longrightarrow \mathcal{L}'' \longrightarrow 0. \tag{2.13}$$

Describing such extensions by means of a covering $\mathcal{U}_\imath$ of C and transition functions $\mathcal{E}_{\imath\jmath}$ between patches $\mathcal{U}_\imath$ and $\mathcal{U}_\jmath$, one may assume that all $\mathcal{E}_{\imath\jmath}$ are upper triangular:

$$\mathcal{E}_{\imath\jmath} = \begin{pmatrix} \mathcal{L}'_{\imath\jmath} & 0 \\ 0 & \mathcal{L}''_{\imath\jmath} \end{pmatrix} \begin{pmatrix} 1 & \mathcal{E}'_{\imath\jmath} \\ 0 & 1 \end{pmatrix}. \tag{2.14}$$

This implies that the lower left matrix element $\varphi_-(y)$ of φ is a section of the line bundle $\mathcal{L}\otimes K_C$, with K_C being the canonical line bundle, and $\mathcal{L}=(\mathcal{L}')^{-1}\otimes\mathcal{L}''$. Without loss of generality, one may assume $\mathcal{L}'=\mathcal{O}$, $\mathcal{L}''=\mathcal{L}$, as can always by reached by tensoring $\mathcal{E}$ with a line bundle. Any holomorphic bundle can be represented as an extension (2.13). At least part of the moduli of the bundle $\mathcal{E}$ can be represented in terms of extension classes in $\mathbb{P}H^1(\mathcal{L}^{-1})$. Since $\dim H^1(\mathcal{L}^{-1})=g-1+\deg(\mathcal{L})$, this suffices to represent all moduli of $\mathrm{Bun}_{SL(2)}$ if $\deg(\mathcal{L})>2g-2$. To simplify the discussion, we shall assume $\deg(\mathcal{L})=2g-1$ in the following.

The matrix elements φ_- of φ represent elements of the vector space $H^0(C,\mathcal{L}\otimes K_C)$ dual to $H^1(\mathcal{L}^{-1})$ by Serre duality. The eigenvectors of $\varphi=\left(\begin{smallmatrix}\varphi_0' & \varphi_+\\ \varphi_- & \varphi_0''\end{smallmatrix}\right)$,

$$\chi=\begin{pmatrix} v-\varphi_0'' \\ \varphi_- \end{pmatrix}, \tag{2.15}$$

vanish at the zeros of $v-\varphi_0''$ which project to the $4g-3$ zeros $\mathbf{u}=(u_1,\ldots,u_h)$ of φ_- on C. The degree $4g-3$ line bundle $L=\mathcal{O}(\hat{\mathbb{D}}_{\hat{\mathbf{u}}})$ associated to the divisor $\hat{\mathbb{D}}_{\hat{\mathbf{u}}}=\sum_{r=1}^{h}\hat{u}_r$ represents the point in the Jacobian of Σ associated to $(\mathcal{E},\varphi)$. We thereby obtain the relation between pairs $(\mathcal{E},\varphi)$, where $\mathcal{E}$ is represented as extension of the form (2.13), and the tuples of points $(\mathbf{u},\mathbf{v})$ in $(T^*C)^{[h]}$ introduced above: $\mathbf{u}=(u_1,\ldots,u_h)$ is the collection of zeros of φ_-, while $\mathbf{v}=(v_1,\ldots,v_h)$ is defined by setting $v_k=\varphi_0''(u_k)$, $k=1,\ldots,h$.

In order to treat the case of the $G=SL(2)$ Hitchin system, one may consider the line bundle $\mathcal{L}\simeq\det(\mathcal{E})$ as fixed, which imposes g constraints on the positions of the $u_1,\ldots,u_h$. We furthermore have $\varphi_0'=-\varphi_0''\equiv\varphi_0$. Let σ be the sheet involution. The degree zero line bundle $L=\mathcal{O}(\hat{\mathbb{D}})$ associated to the divisor $\hat{\mathbb{D}}=\sum_{r=1}^{h}(\hat{u}_r-\sigma(\hat{u}_r))$ representing the point in the Prym variety of Σ associated to $(\mathcal{E},\varphi)$ has lines generated by

$$\chi=\frac{1}{v-\varphi_0}\begin{pmatrix} v+\varphi_0 \\ \varphi_- \end{pmatrix}. \tag{2.16}$$

Variants of this type of representation can be used to parametrize the pairs $(\mathcal{E},\varphi)$, and to describe the change of variables defining the tuples $(\mathbf{u},\mathbf{v})$, much more explicitly [18].

2.5. *Punctures*

It is possible to generalize the set-up by allowing n marked points on C. In the presence of marked points, one may also consider surfaces of genus zero or one. The resulting versions of the Hitchin integrable systems turn out to be related to the integrable models known as the Gaudin model ($g=0$), or the elliptic Calogero–Moser model ($g=1$). We will use the example of the Gaudin model as guidance for the quantization of the picture outlined above. The necessary ingredients will have clear analogues in this case, suggesting a path for the treatment of the general case. To this end, let us explain how the SOV is realized in this case.

The description of $\mathcal{E}$ as an extension amounts to a description in terms of a cover of $\mathbb{P}^1$ of the form $\{\mathbb{P}^1\setminus\{z_1,\ldots,z_n\},D_1,\ldots,D_n\}$, where $D_1,\ldots,D_n$ are small mutually non-intersecting discs around $z_1,\ldots,z_n$, with transition functions on $A_r = D_r \setminus \{z_r\}$ being of the form $\mathcal{E}_r = \left(\begin{smallmatrix} 1 & x_r \\ 0 & 1 \end{smallmatrix}\right)$. Assuming that φ has a regular singularity of the form $\frac{1}{y-z_r}\left(\begin{smallmatrix} l_r & 0 \\ p_r & -l_r \end{smallmatrix}\right)$ at z_r it follows that

$$\varphi(y) = \sum_{r=1}^{n} \frac{\varphi_r}{y - z_r}, \quad \varphi_r = \mathcal{E}_r \cdot \begin{pmatrix} l_r & 0 \\ p_r & -l_r \end{pmatrix} \cdot \mathcal{E}_r^{-1} = \begin{pmatrix} x_r p_r + l_r & x_r^2 p_r + 2 l_r x_r \\ p_r & -l_r - x_r p_r \end{pmatrix}. \tag{2.17}$$

Regularity of φ at infinity imposes three constraints:

$$\sum_{r=1}^{n} \left(x_r^{k+1} p_r + l_r (k+1) x_r^k \right) = 0, \qquad k = -1, 0, 1. \tag{2.18}$$

Identifying x_r with a coordinate on $\mathbb{P}^1$, and p_r with a coordinate on the cotangent fibre of $\mathbb{P}^1$, allows us to describe $\mathcal{M}_{\mathrm{H}}(C_{0,n})$ as a symplectic reduction of $(T^*\mathbb{P}^1)^n$ by the constraints (2.18). To this end, one needs to identify points of $(T^*\mathbb{P}^1)^n$ related by the Hamiltonian flows generated by the constraints. These flows generate the group $G = \mathrm{SL}(2)$ acting on the variables x_r as Möbius transformations $x_r \to \frac{a x_r + b}{c x_r + d}$. The quotient $(T^*\mathbb{P}^1)^n /\!/ G$ may be represented by fixing a slice $x_n = \infty$, $x_{n-1} = 1$ and $x_{n-2} = 0$ and using (2.18) to express p_n, p_{n-1} and p_{n-2} in terms of the remaining variables. This forces us to send $p_n \to 0$ such that $x_n p_n + 2 l_n = 0$.

The Hamiltonians of this integrable model are defined as the free parameters specifying the quadratic differential $\mathrm{tr}(\varphi^2)$, which can now be represented explicitly as

$$\mathrm{tr}(\varphi^2(y)) = \sum_{r=1}^{n} \left(\frac{l_r^2}{(y - z_r)^2} + \frac{H_r}{y - z_r} \right). \tag{2.19}$$

The change of variables $(\mathbf{x}, \mathbf{p}) \to (\mathbf{u}, \mathbf{v}, u_0)$ defined by

$$\varphi_-(y) = \sum_{r=1}^{n-1} \frac{p_r}{y - z_r} = u_0 \frac{\prod_{k=1}^{n-3} (y - u_k)}{\prod_{r=1}^{n-1} (y - z_r)}, \qquad v_r = \varphi_0(u_r), \tag{2.20}$$

gives the isomorphism $\mathcal{M}_H(C_{0,n}) \simeq (T^* C_{0,n})^{[n-3]}$ defined by the SOV method.

3. Quantization of Hitchin's Integrable System

We will now present an overview of known results on the quantization of the Hitchin system. Starting with the genus zero case, we will introduce a variant of the Gaudin model related to the non-compact group $\mathrm{SL}(2,\mathbb{C})$. Known results on the quantization of Hitchin's Hamiltonians in $g > 1$ and their relation to the geometric Langlands correspondence are reinterpreted from the point of view of this paper in the following subsection.

3.1. *Genus zero—The* **SL**$(2,\mathbb{C})$ *Gaudin Model*

The quantization of the Gaudin model is fairly simple on a purely algebraic level. It starts by turning the algebra of functions on $(T^*\mathbb{P}^1)^n$ with generators p_r, x_r, into a non-commutative algebra with generators p_r, x_r, $r = 1,\dots,n$, satisfying the relations $[\mathsf{p}_r,\mathsf{x}_s] = \epsilon_1\delta_{rs}$, $[\mathsf{p}_r,\mathsf{p}_s] = 0$, $[\mathsf{x}_r,\mathsf{x}_s] = 0$. The matrix elements φ_r^a, $a = -,0,+$ of the residues φ_r of φ get replaced by the generators of the Lie algebra $\mathfrak{sl}_2$ for all $r = 1,\dots,n$. The quantized algebra of functions $\mathfrak{A}_n$ on $(T^*\mathbb{P}^1)^n$ thereby gets identified with the direct sum of n copies of the Lie algebra $\mathfrak{sl}_2$.

When we are discussing the quantization of a phase space with complex coordinates, it is also natural to consider the conjugate algebra $\bar{\mathcal{A}}_n$ obtained by quantization of the complex conjugate coordinates $\bar{p}_r$, $\bar{x}_r$. The generators of $\bar{\mathcal{A}}_n$ will be denoted as $\bar{\mathsf{p}}_r$, $\bar{\mathsf{x}}_r$, $r = 1,\dots,n$.

Recall that we had represented $\mathcal{M}_{\mathrm{H}}(C_{0,n})$ as symplectic quotient of $(T^*\mathbb{P}^1)^n$ by the three constraints (2.18). The constraints become quantized to the 'diagonal' $\mathfrak{sl}_2$ embedded into the direct sum of n copies of $\mathfrak{sl}_2$ in the usual way. It is natural to define the quantized algebra $\mathcal{A}$ of global functions on $\mathcal{M}_{\mathrm{H}}(C_{0,n})$ to be the subalgebra of $\mathfrak{A}_n$ generated by the functions commuting with the diagonal $\mathfrak{sl}_2$. The algebra $\mathcal{A}$ contains the quantized Hamiltonians H_r,

$$\mathsf{H}_r \equiv \sum_{s\neq r} \frac{\mathsf{J}_{rs}}{z_r - z_s}, \tag{3.21}$$

where the differential operator $\mathcal{J}_{rs}$ is defined as

$$\mathsf{J}_{rs} := \eta_{aa'}\mathsf{J}_r^a\mathsf{J}_s^{a'} := \mathsf{J}_r^0\mathsf{J}_s^0 + \frac{1}{2}(\mathsf{J}_r^+\mathsf{J}_s^- + \mathsf{J}_r^-\mathsf{J}_s^+). \tag{3.22}$$

The generators H_r commute, $[\mathsf{H}_r,\mathsf{H}_s] = 0$ for all r,s. Similar statements hold for the conjugate algebra $\bar{\mathcal{A}}$, which commutes with $\mathcal{A}$ and contains the conjugate Hamiltonians $\bar{\mathsf{H}}_r$, $r = 1,\dots,n$.

A step towards the definition of suitable representations $\mathcal{R}_n$ of $\mathfrak{A}_n$ is to choose a polarization, a commutative subalgebra of $\mathfrak{A}_n$ that will be represented by multiplication operators on $\mathcal{R}_n$. In the present case, there are are two natural polarizations, defined by choosing either the sub-algbra generated by x_r, $r = 1,\dots,n$, or the one generated by p_r, $r = 1,\dots,n$. In both cases, one gets an n-fold tensor product $\mathcal{R}_n = \bigotimes_{r=1}^n \mathcal{P}_n$ of representations $\mathcal{P}_n$ of the Lie algebra $\mathfrak{sl}_2$. In the first case, one finds a representation realized by the differential operators $\mathcal{J}_r^{\pm}$, $\mathcal{J}_r^0$:

$$\mathcal{J}_r^- = \partial_{x_r}, \quad \mathcal{J}_r^0 = x_r\partial_{x_r} - j_r, \quad \mathcal{J}_r^+ = -x_r^2\partial_{x_r} + 2j_r x_r. \tag{3.23}$$

The parameters j_r appearing in (3.23) are related to the parameters l_r of the classical Gaudin model by $l_r = -\epsilon_1 j_r$. In the polarization generated by p_r, $r = 1,\dots,n$, we may choose the operators

$$\tilde{\mathcal{J}}_r^- = p_r, \quad \tilde{\mathcal{J}}_r^0 = -p_r\partial_{p_r}, \quad \tilde{\mathcal{J}}_r^+ = -p_r\partial_{p_r}^2 + \frac{j_r(j_r+1)}{p_r} \tag{3.24}$$

as generators for the representation on $\mathcal{P}_n$. The Casimir operator is, in both cases, represented as multiplication by $j_r(j_r+1)$.

In order to fully define the relevant representations of the Lie algebra $\mathfrak{sl}_2$, one needs to specify the spaces of functions that the differential operators defined in (3.23) and (3.24) should act on. In the Gaudin model, one usually considers finite-dimensional representations, restricting the choice of j_r to $j_r = 0, 1/2, 1, \dots$. The finite-dimensional representations can be realized via (3.23) on polynomial functions of the variables x_r. We will mostly be interested in infinite-dimensional representations realized by means of the differential operators (3.23) on suitable spaces of non-polynomial functions. One may, for example, consider representations defined by the differential operators $\mathcal{J}_r^a$ together with the conjugate operators $\bar{\mathcal{J}}_r^a$ obtained by $x_r \to \bar{x}_r$, $\partial_{x_r} \to \bar{\partial}_{\bar{x}_r}$ on certain (sub-)spaces of the space of smooth functions on $\mathbb{C}$. The class of such representations contains the Lie algebra representations associated to principal series representations $\mathcal{P}_n \equiv \mathcal{P}_{j_n}$ of SL(2, $\mathbb{C}$). The representations $\mathcal{P}_{j_n}$ are unitary if $j_n \in -\frac{1}{2} + i\mathbb{R}$.

The symplectic quotient of $(T^*\mathbb{P}^1)^n$ by the three constraints (2.18) is naturally described by considering the action of $\mathcal{A}$ and $\bar{\mathcal{A}}$ on the subspaces $\mathcal{R}_n^{\text{inv}} \subset \mathcal{R}_n$ of invariants under the diagonal $\mathfrak{sl}_2$-action. Representing the tensor product of representations $\mathcal{R}_n = \bigotimes_{r=1}^N \mathcal{P}_{j_n}$ in terms of functions $\Psi(\mathbf{x}, \bar{\mathbf{x}})$ with $\mathbf{x} = (x_1, \dots, x_n)$, one may represent the elements of $\mathcal{R}_{\text{inv}}$ as functions $\Psi(\mathbf{x}, \bar{\mathbf{x}})$ which are invariant under the diagonal action of SL(2, $\mathbb{C}$). We will find it more convenient to represent the elements of $\mathcal{R}_{\text{inv}}$ as functions $\Psi(\mathbf{x}, \bar{\mathbf{x}})$ of $(n-1)$ variables $\mathbf{x} = (x_1, \dots, x_{n-1})$ which are invariant under translations $x_r \to x_r + b$ and behave under dilatations $x_r \to a^2 x_r$ as

$$\Psi(a^2\mathbf{x}, a^2\bar{\mathbf{x}}) = a^{4J}\Psi(\mathbf{x}, \bar{\mathbf{x}}), \qquad J = -j_n + \sum_{r=1}^{n-1} j_r. \tag{3.25}$$

The two representations (3.23) and (3.24) are intertwined by the following slightly modified form of the Fourier transformation:

$$\Psi(\mathbf{x}, \bar{\mathbf{x}}) = \int d^2p_1 \cdots d^2p_{n-1}\; \Phi(\mathbf{p}, \bar{\mathbf{p}}) \prod_{r=1}^{n-1} e^{p_r x_r - \bar{p}_r \bar{x}_r} |p_r|^{-2j_r - 2}. \tag{3.26}$$

This map establishes an equivalence of the representation defined via (3.23) with a representation of the form (3.24) in which a nilpotent generator is represented as multiplication operator. We will refer to the representations defined on the functions $\Phi(\mathbf{p}, \bar{\mathbf{p}})$ via (3.24) as the Whittaker models for the representations $\bigotimes_{r=1}^{n-1} \mathcal{P}_{j_r}$. One may note that the conjugate operators $\bar{\mathcal{J}}_r^{\pm}$, $\bar{\mathcal{J}}_r^0$ get mapped to the complex conjugates of $\tilde{\mathcal{J}}_r^{\pm}$, $\tilde{\mathcal{J}}_r^0$.

3.2. *Quantization of Hitchin's Hamiltonians and the Geometric Langlands Correspondence*

Hitchin's Hamiltonians have been quantized in the work [19] of Beilinson and Drinfeld on the geometric Langlands correspondence. This means the following: there exist global differential operators H_i on the line bundle $K^{1/2}$ on Bun_G such that the following holds:

- the differential operators H_i generate the commutative algebra $\mathfrak{D}$ of global differential operators acting on $K^{1/2}$, and
- the symbols of the differential operators H_i coincide with generators of the algebra of functions on the Htichin base $\mathcal{B}$ defined via Hitchin's map.

The construction in [19] uses elements of conformal field theory and the representation-theoretic results of [20]. Our discussion follows the review [8].

Beilinson and Drinfeld put the quantization of the Hitchin in relation to the geometric Langlands correspondence, schematically represented as

$$\boxed{{}^{\mathrm{L}}\mathfrak{g}\text{-opers}} \quad \longleftrightarrow \quad \boxed{\mathcal{D}-\text{modules on } \mathrm{Bun}_G}, \tag{3.27}$$

as we shall now briefly explain. The relation between the geometric Langlands correspondence and the Gaudin model was described in [21].

3.2.1. Opers

Opers are a special class of holomorphic connections $(\epsilon_1 \partial_y + A(y))dy$ on C with $A(y)$ being gauge equivalent to the form $\left(\begin{smallmatrix} 0 & t \\ 1 & 0 \end{smallmatrix}\right)$. The equation defining horizontal sections s, $(\epsilon_1 \partial_y + A(y))s = 0$, reduces to the ordinary differential equation $(\epsilon_1^2 \partial_u^2 + t(u))s_2 = 0$ if $s = \left(\begin{smallmatrix} s_1 \\ s_2 \end{smallmatrix}\right)$. Covariance under changes of local coordinates requires that $t = t(u)$ transforms as

$$t(u) = (y'(u))^2 \tilde{t}_f(y(u)) + \frac{\epsilon_1^2}{2}\{y, u\}, \quad \{y, u\} = \left(\frac{y''}{y'}\right)' - \frac{1}{2}\left(\frac{y''}{y'}\right)^2, \tag{3.28}$$

identifying it as a *projective connection*. The underlying holomorphic bundle $\mathcal{E}_{\mathrm{op}}$ must be an extension of the form $0 \to K^{\frac{1}{2}} \to \mathcal{E}_{\mathrm{op}} \to K^{-\frac{1}{2}} \to 0$. As $\mathcal{E}_{\mathrm{op}}$ is uniquely defined thereby, an oper is completely specified by the choice of the projective connection t.

3.2.2. Geometric Langlands correspondence

One of the main results of Beilinson and Drinfeld is the existence of a canonical isomorphism of algebras

$$\mathrm{Fun}\,\mathrm{Op}_{{}^{\mathrm{L}}\mathfrak{g}}(C) \simeq \mathfrak{D}. \tag{3.29}$$

This result implies a special case of the geometric Langlands correspondence. Fixing an oper χ defines a homomorphism $\mathrm{Fun}\,\mathrm{Op}_{{}^{\mathrm{L}}\mathfrak{g}}(C) \to \mathbb{C}$. Using (3.29), one gets a homomorphism $\tilde{\chi} : \mathfrak{D} \to \mathbb{C}$. To each oper χ, one may assign a $\mathcal{D}$-module $\mathfrak{D}_\chi$ on Bun_G defined as

$$\mathfrak{D}_\chi = \mathfrak{D}/\ker\tilde{\chi} \cdot \mathfrak{D}. \tag{3.30}$$

The correspondence between ${}^{\mathrm{L}}\mathfrak{g}$-opers χ and $\mathcal{D}$-modules $\mathfrak{D}_\chi$ on Bun_G constructed in this way is an important part of what is called geometric Langlands correspondence.

This may be reformulated from the point of view of quantization of the Hitchin system as follows: to an oper χ, we may associate the following system of differential equations on Bun_G,

$$\mathsf{H}_i f = E_i f, \qquad E_i = \tilde{\chi}(\mathsf{H}_i). \tag{3.31}$$

This system of differential equations is regular on the open dense subset of Bun_G containing very stable bundles, bundles that do not admit a nilpotent Higgs field. On this locus, it defines a vector bundle with flat connection. Conjecturally, the vector bundle has regular singularities along the singular locus. Horizontal sections of the flat connection defined by the equations (3.31) will generically have non-trivial monodromy around the singular loci.

Observing that the differential equations (3.31) are the eigenvalue equations for Hitchin's Hamiltonians, it seems natural to interpret the results above as the statement that $\mathrm{Op}_{^L\mathfrak{g}}(C)$ represents the natural geometric 'home' for the eigenvalues of the quantized Hitchin Hamiltonians. The space of opers $\mathrm{Op}_{^L\mathfrak{g}}(C)$ on C represents the quantum analogue $\mathcal{B}_{\epsilon_1}$ of the base $\mathcal{B}$ of the Hitchin fibration.

4. Quantum Separation of Variables

We had noted in Section 3.2 that the geometric Langlands correspondence is related to the eigenvalue problem of the quantized Hitchin Hamiltonians. It characterizes the set of eigenvalues for which multi-valued analytic solutions can exist in terms of the opers associated to the Lie algebra $^L\mathfrak{g}$. In all the cases where the SOV approach has been developed, it gives a concrete realization of a correspondence between opers and eigenfunctions of the quantized Hitchin Hamiltonians. This has been fully realized when the surface C has genus $g = 0$ [11] or $g = 1$ [22–24] with any number of punctures. The SOV approach therefore offers an alternative approach to the geometric Langlands correspondence which is similar to the first construction of such a correspondence due to Drinfeld [25], as has been pointed out in [11]. It is natural to expect that the SOV approach can be extended to the cases with $g > 1$, furnishing a more concrete realization of the geometric Langlands correspondence in all cases.

In this section, we will briefly describe how the SOV approach works in the case of genus zero, and then formulate a conjecture about the generalization of the emerging picture to higher genus.

4.1. *Genus Zero*

The goal is to solve the eigenvalue problem

$$\mathsf{H}_r \Psi_{\mathbf{E}}(\mathbf{x}, \bar{\mathbf{x}}) = E_r \Psi_{\mathbf{E}}(\mathbf{x}, \bar{\mathbf{x}}), \qquad \bar{\mathsf{H}}_r \Psi_{\mathbf{E}}(\mathbf{x}, \bar{\mathbf{x}}) = \bar{E}_r \Psi_{\mathbf{E}}(\mathbf{x}, \bar{\mathbf{x}}), \tag{4.32}$$

where $\Psi_{\mathbf{E}}(\mathbf{x}, \bar{\mathbf{x}})$ is a function of the $(n-1)$ variables $\mathbf{x} = (x_1, \ldots, x_{n-1})$ and their complex conjugates which are invariant under translations $x_r \to x_r + b$ and behave under dilatations $x_r \to a^2 x_r$ as in (3.25).

The first step is to pass to the Whittaker model by means of the inverse of the Fourier transformation (3.26), expressing solutions $\Psi_{\mathbf{E}}(\mathbf{x},\bar{\mathbf{x}})$ in terms of the eigenfunctions $\Phi_{\mathbf{E}}(\mathbf{p},\bar{\mathbf{p}})$ in the Whittaker model. Let us then, following Sklyanin [12], perform the change of variables $\mathbf{p} \to (u_0, \mathbf{u})$ defined by the family of equations

$$\varphi_-(y) = \sum_{r=1}^{n-1} \frac{p_r}{y - z_r} = u_0 \frac{\prod_{k=1}^{n-3}(y - u_k)}{\prod_{r=1}(y - z_r)} \quad \Rightarrow \quad p_r(u) = u_0 \frac{\prod_{k=1}^{n-3}(z_r - u_k)}{\prod_{s\neq r}^{n-1}(z_r - z_s)}. \tag{4.33}$$

Abusing notation, we will denote $\Phi_{\mathbf{E}}(\mathbf{p}(u_0,\mathbf{u}),\bar{\mathbf{p}}(u_0,\mathbf{u}))$ by $\Phi_{\mathbf{E}}(\mathbf{u},\bar{\mathbf{u}})$. Using identities like

$$\partial_{u_k} = \sum_{r=1}^{n-1} \frac{\partial p_r}{\partial u_k} \partial_{p_r} = \sum_{r=1}^{n-1} \frac{1}{u_k - z_r} p_r \partial_{p_r}, \tag{4.34}$$

it becomes straightforward to show that the eigenvalue equation become equivalent to the set of ordinary differential equations

$$(\epsilon_1^2 \partial_{u_k}^2 + t(u_k))\Phi_{\mathbf{E}}(\mathbf{u},\bar{\mathbf{u}}) = 0, \qquad (\epsilon_1^2 \bar{\partial}_{\bar{u}_k}^2 + \bar{t}(\bar{u}_k))\Phi_{\mathbf{E}}(\mathbf{u},\bar{\mathbf{u}}) = 0, \tag{4.35}$$

which can be solved in factorized form from $\Phi_{\mathbf{E}}(\mathbf{u},\bar{\mathbf{u}}) = \prod_{k=1}^{n-3} \phi_k(u_k,\bar{u}_k)$. Further details can be found in [11, 12].

The transformation from eigenfunctions $\Psi_{\mathbf{E}}(\mathbf{x},\bar{\mathbf{x}})$ to the functions $\Phi_{\mathbf{E}}(\mathbf{u},\bar{\mathbf{u}})$ can be inverted explicitly [26]. The inverse may be represented as an integral transformation of the form

$$\Psi_{\mathbf{E}}(\mathbf{x},\bar{\mathbf{x}}) = N_J \int d^2u_1 \cdots d^2u_{n-3}\, \mathcal{K}^{\mathrm{SOV}}(x,u)\, \Phi_{\mathbf{E}}(\mathbf{u},\bar{\mathbf{u}}), \tag{4.36}$$

where the kernel $\mathcal{K}^{\mathrm{SOV}}(x,u)$ can be represented explicitly as

$$\mathcal{K}^{\mathrm{SOV}}(x,u) = \left| \sum_{r=1}^{n-1} x_r \frac{\prod_{k=1}^{n-3}(z_r - u_k)}{\prod_{s\neq r}^{n-1}(z_r - z_s)} \right|^{2J} \prod_{r=1}^{n-1} \left| \frac{\prod_{s\neq r}^{n-1}(z_r - z_s)}{\prod_{k=1}^{n-3}(z_r - u_k)} \right|^{2(j_r+1)} \prod_{k<l}^{n-3} |u_k - u_l|^2. \tag{4.37}$$

The integral transformation (4.36) with kernel (4.37) is manifestly well defined for generic $(\mathbf{x},\bar{\mathbf{x}})$ when the real parts of the parameters j_r are small enough. It may be defined for more general values of these parameters by analytic continuation. Integrable singularities of a specific type occur at certain loci in the space parametrize by the variables $\mathbf{x}$.

4.2. *Higher Genus*

The SOV approach appears to be less completely understood in the higher genus cases, but there is evidence that the qualitative picture remains essentially unchanged [11]. The first construction of the geometric Langlands correspondence due to Drinfeld [25] starts from a symmetric product $\mathcal{D}$-module represented by an oper. This $\mathcal{D}$-module can be seen as the result of the canonical quantization of the coordinates $(\mathbf{u},\mathbf{v})$ introduced in Section 2. Indeed, choosing a polarization where the coordinates u_k get represented as multiplication operators, and the coordinates v_k as derivatives $\epsilon_1\partial_{u_k}$, one may identify the differential equations $(\epsilon_1^2\partial_{u_k}^2+t(u_k))\psi(u_k)=0$ as a quantum counterpart of the equation $v_k^2+\mathrm{tr}(\varphi^2(u_k))=0$ defining the spectral curve Σ.

The description of Drinfeld's construction presented in [11] reveals the striking similarity of this construction with a quantum version of the SOV approach. It seems natural to conjecture that the resulting $\mathcal{D}$-modules on $\mathrm{Bun}_G(C)$ are isomorphic to the ones furnished by Beilinson and Drinfeld's constructions. The isomorphism of the $\mathcal{D}$-modules provided by Drinfeld's first, and Beilinson and Drinfeld's second construction of the geometric Langlands correspondence would imply the existence of a quantum version of the SOV for the Hitchin system [11].[1]

The resulting picture may be described a bit more concretely as follows. Using extensions to represent the bundles $\mathcal{E}$ allows us to introduce $3g-2$ coordinates $\mathbf{x}=(x_1,\ldots,x_{3g-2})$ for the vector space $H^1(\mathcal{L}^{-1})$ of extension classes. Eigenfunctions of the quantized Hitchin Hamiltonians may then be represented in terms of functions $\Psi_{\mathbf{E}}(\mathbf{x},\bar{\mathbf{x}})$ of the coordinates $\mathbf{x}$ and their complex conjugates which behave homogeneously of weight $-3g+2$ under dilatations $x_r\to a^2x_r$. A Fourier transformation similar to (3.26) will describe the passage to the Whittaker model in which the quantum counterpart $\mathsf{f}(y)$ of $\varphi_-(y)$ is realized as a multiplication operator. The eigenvalue $f(y)$ of $\mathsf{f}(y)$ is a section of the line bundle $\mathcal{L}\otimes K_C$. It is then natural to introduce the zeros u_k of $f(y)$ as new variables, and to rewrite the eigenvalue equations of the quantized Hitchin Hamiltonians in these variables. The discussion above motivates the following conjecture:

The quantized Hitchin system admits a Whittaker model representing eigenfunctions $\Phi_{\mathbf{E}}(\mathbf{u},\bar{\mathbf{u}})$ of the Hitchin Hamiltonians as sections of $(K\otimes\bar{K})^{\frac{3g-3}{2}}$ which satisfy

$$(\epsilon_1^2\partial_{u_k}^2+t(u_k))\Phi_{\mathbf{E}}(\mathbf{u},\bar{\mathbf{u}})=0,\qquad(\epsilon_1^2\bar{\partial}_{\bar{u}_k}^2+\bar{t}(\bar{u}_k))\Phi_{\mathbf{E}}(\mathbf{u},\bar{\mathbf{u}})=0,\tag{4.38}$$

with $t(u)$ representing an oper on C.

This would represent a more concrete realization of the geometric Langlands correspondence proven by Beilinson and Drinfeld. Note that no quantization condition has been imposed yet. Inverting the steps outlined above should enable us to construct an integral transformation similar to (4.36), allowing us to construct (generically multi-valued) eigenfunctions of Hitchin's Hamiltonians from opers on C.

5. Quantization Conditions

The main questions to be addressed in this paper concern exactly what is bypassed in the geometric Langlands correspondence by the use of the $\mathcal{D}$-module theory: what are

[1] The existence of such an isomorphism would follow from the uniqueness of the irreducible Hecke eigensheaf associated to an oper via the constructions in [25] and [19], which has not been established in the literature, as far as we know. According to E. Frenkel, the isomorphism can also be proved directly from the Hecke eigenvalue property. The author thanks E. Frenkel for pointing this out to him.

interesting spaces of functions or sections of suitable line bundles in which one can search for the solutions of the eigenvalue equations for Hitchin's Hamiltonians and, given a particular choice, what are the possible eigenvalues?

This issue will be referred to as the choice of quantization conditions. After identifying a choice of quantization condition that appears to be particularly natural for the Hitchin system, we will explain how this choice can be reformulated as a condition on the monodromy of the differential equations $(\epsilon_1^2\partial_u^2 + t(u))\psi(u) = 0$ representing the spectral problem in the SOV representation.

5.1. *Natural Choices of Quantization Conditions*

It may be instructive to compare the situation with the case of spectral problems for Schrödinger type operators $H = -\partial_y^2 + V(y)$, for simplicity restricting attention to functions on the real line. For sufficiently regular potentials, there will exist two linearly independent solutions of the eigenvalue equations $H\psi_E(y) = E\psi_E(y)$ for all real or even complex values of E. Physics usually motivates us to impose additional requirements on the solutions $\psi_E(y)$, such as square integrability, which often can only be satisfied for a discrete set of values of E. In other cases one may be interested in functions $\psi_E(y)$ which are periodic in y and which may again restrict the possible choices of E to a discrete set. The supplementary conditions used to define the spectral problem of interest precisely will henceforth be referred to as quantization conditions. Their mathematical content is to specify the exact class of functions which can be a solution to the spectral problem.

The Gaudin model is usually defined by considering functions of the variables x_r appearing in the definition of the Gaudin Hamiltonians which are polynomial in x_r with degrees being given in terms of the parameters j_r as $2j_r$. The representations defined on such functions via (3.23) correspond to the finite-dimensional representations of SU(2). This is one possible type of quantization condition defining what is called the SU(2) Gaudin model.

In this paper, we are interested in another type of quantization condition. Following the discussion above, we will consider the pair of eigenvalue equations

$$\mathsf{H}_r\Psi(\mathbf{x},\bar{\mathbf{x}}) = E_r\Psi(\mathbf{x},\bar{\mathbf{x}}), \qquad \bar{\mathsf{H}}_r\Psi(\mathbf{x},\bar{\mathbf{x}}) = \bar{E}_r\Psi(\mathbf{x},\bar{\mathbf{x}}), \tag{5.39}$$

where $\overline{\mathsf{H}}_r$ is the conjugate of H_r obtained by replacing $x_r \to \bar{x}_r$, $\partial_{x_r} \to \bar{\partial}_{\bar{x}_r}$. We are interested in solutions $\Psi(\mathbf{x},\bar{\mathbf{x}})$ which are real-analytic away from possible singularities of the differential operators H_r and which are, futhermore, *single-valued*. We had seen above that such quantization conditions are natural if one replaces the representations of SU(2) in the Gaudin model by principal series representations of $SL(2,\mathbb{C})$.

A similar type of quantization condition can be considered for the quantum Hitchin system on higher genus surfaces C. The quantum Hitchin Hamiltonians are conjectured to have regular singularities away from the locus within $\mathrm{Bun}_G(C)$ consisting of the very stable bundles. For generic choice of an oper χ, the solutions f of (3.31) will have non-trivial monodromies around the singular loci. For certain opers χ, there may exist an Hermitian form on the space of solutions which is invariant under the monodromy action, allowing us to construct single-valued solutions in the form of linear combinations of products of elements of a basis for the space of solutions to the eigenvalue equations multiplied by elements of the complex-conjugate basis.

Solutions to spectral problems establish generalized duality relations, in the simplest cases relating certain spaces of functions $\mathfrak{S}$ to the spaces of functions on the sets of eigenvalues of commuting differential operators acting on functions in $\mathfrak{S}$. Imposing additional conditions on one side will be reflected by additional restrictions occurring on the other side. From this point of view, we may view the geometric Langlands correspondence as the solution to a natural prequantization problem. It characterizes the space dual to the multi-valued solutions of the Hitchin eigenvalue equations as the space of opers. This sets the stage for the description of single-valued solutions to the Hitchin eigenvalue equation to be proposed below.

From a physicist's perspective, one might be tempted to look for a natural scalar product, and to look for normalizable solutions within the class of single-valued ones. This would be a natural next step. At the moment, we have little to say about it.

The reader may notice that the idea of combining holomorphic with anti-holomorphic functions into single-valued objects is familiar from conformal field theory. What we are proposing here is related to a particular limit of the so-called H_3^+-WZNW model [9, 27].

When this paper was undergoing final revisions, E. Frenkel pointed out to us that ideas similar to the ones presented above have been discussed in the talk he gave at MSRI in September 2014 [28].

5.2. *Quantization versus Classification of Real Projective Structures*

The SOV transformation maps single-valued common eigenfunctions of the Hitchin Hamiltonians to single-value functions having the factorized form

$$\Phi_{\mathbf{E}}(\mathbf{u},\bar{\mathbf{u}}) = \prod_{r=1}^{h} \phi_{\mathbf{E}}(u_r,\bar{u}_r). \tag{5.40}$$

Our goal is therefore to analyse the single-valuedness of the expression (5.40).

As preparation, let us now note that each oper defines a projective structure on C, an atlas of local coordinates on C with transition functions all represented as Möbius transformations. Indeed, given two linearly independent solutions χ_i of $(\epsilon_1^2\partial_u^2 + t(u))\chi_i = 0$, $i = 1,2$, one may show that $y(u) = \chi_1/\chi_2$ satisfies $2t(u) = \epsilon_1^2\{y,u\}$. Using y as a new local coordinate, one therefore has $\tilde{t}(y) \equiv 0$. The Möbius transformations are the only allowed transition functions in an atlas formed by a collection of local charts with $\tilde{t}(y) \equiv 0$. Background on projective structures relevant for us is reviewed in [29, 30].

5.2.1. Single-valuedness

In the form (5.40), it becomes much easier to analyse the condition that $\Phi_{\mathbf{E}}(\mathbf{u},\bar{\mathbf{u}})$ should be single-valued. This will be the case iff the function $\phi_{\mathbf{E}}(u,\bar{u})$, which can be decomposed into linearly independent solutions of the differential equations (4.38) as

$$\phi_{\mathbf{E}}(u,\bar{u}) = \sum_{i,j=1}^{2} C_{ij}\chi_i(u)\bar{\chi}_j(\bar{u}), \tag{5.41}$$

has the property of being single-valued. By a change of basis in the space of solutions, one may always bring the matrix C_{ij} in (5.41) into diagonal form. By a rescaling and multiplication of C_{ij} by an overall phase, one may assume that the diagonal matrix elements are contained in $\{-1, 0, 1\}$. The matrix $C_{ij} = \delta_{ij}$ and diagonal matrices C_{ij} having a vanishing diagonal matrix element are invariant under representations $\rho : \pi_1(C) \to \mathrm{SL}(2, \mathbb{C})$, which never occur as holonomies of projective structures [31]. For the discussion of the remaining case $C = \mathrm{diag}(1, -1)$, we may may further transform C_{ij} to the form $C_{ij} = \epsilon_{ij}$, where $\epsilon_{12} = 1$, $\epsilon_{ij} = -\epsilon_{ji}$, having invariance under $\mathrm{SL}(2, \mathbb{R})$. Projective structures having holonomy in $\mathrm{PGL}(2, \mathbb{R})$ are called *real* projective structures. It follows from the observations above that solutions to the single-valuedness condition correspond to real projective structures.

We may thereby conclude that $\Phi_{\mathbf{E}}(\mathbf{u}, \bar{\mathbf{u}})$ can be a single-valued solution to the system of equations (4.38) only if the monodromy of $\epsilon_1^2 \partial_u^2 + t(u)$ is conjugate to a homomorphism of $\pi_1(C)$ into $\mathrm{SL}(2, \mathbb{R})$. The solution $\Phi_{\mathbf{E}}(\mathbf{u}, \bar{\mathbf{u}})$ can then be represented in the factorized form (5.40).

One solution of the conditions above is well known: for each Riemann surface C, there exists a unique metric $d^2s = e^{2\varphi} dy d\bar{y}$ of constant negative curvature. The corresponding projective connection $t(y) = -\frac{1}{4}(\partial_y \varphi)^2 + \frac{1}{2}\partial_y^2 \varphi$ has the Fuchsian group Γ uniformizing $C \simeq \mathbb{H}/\Gamma$ as its holonomy. We may use it to construct a particular solution $\Phi_{\mathbf{E}_0}(u, \bar{u})$ to the quantization conditions via (5.40) and (5.41). The function $\phi_{\mathbf{E}_0}(u, \bar{u})$ appearing in the factorized representation (5.40) for $\Phi_{\mathbf{E}_0}(u, \bar{u})$ is related to the metric of constant negative curvature as $\phi_{\mathbf{E}_0}(u, \bar{u}) = e^{-\varphi(u, \bar{u})}$.

There exists a construction called grafting, allowing us to construct from a given projective structure with Fuchsian holonomy infinitely many other projective structures with the same holonomy [13]. It was furthermore shown in [13] that *all* projective structures with holonomy being a fixed Fuchsian group are obtained by grafting the projective structure furnished by the uniformization theorem, leading to a classification of the projective structures with Fuchsian holonomy.

There exist real projective structures not having Fuchsian holonomy. Such projective structures can be obtained by a small generalisation of the grafting construction and have also been classified in [13]. A subclass of the real projective structures has holonomy in $\mathrm{PSL}(2, \mathbb{R})$. In work in progress [32] we will describe in more detail which non-Fuchsian real projective structures correspond to single-valued eigenfunctions of the quantised Hitchin Hamiltonians.

> *The quantum SOV method establishes a one-to-one correspondence between single-valued eigenfunctions of the quantized Hamiltonians of the* SL(2) *Hitchin system and projective structures with holonomy in* $\mathrm{PSL}(2, \mathbb{R})$ *on C.*

Projective structures with Fuchsian holonomy not coming from the uniformization of C are called exotic. The classification of real projective structures from [13] may be used to classify the solutions of the quantization conditions of the Hitchin system. The functions $\phi_{\mathbf{E}}(u, \bar{u})$ corresponding to exotic projective structures with Fuchsian holonomy define metrics of constant negative curvature via $e^{2\varphi(u, \bar{u})} = (\phi_{\mathbf{E}_0}(u, \bar{u}))^{-2}$ only away from certain singular loci [33].

5.3. Reformulation in Terms of Complex Fenchel–Nielsen Coordinates

The Riemann–Hilbert correspondence between flat connections $\partial_y + A(y)$ and representations $\rho : \pi_1(C) \to G$ relates the moduli space $\mathcal{M}_{\rm flat}(C)$ of flat connections on C to the so-called character variety $\mathcal{M}_{\rm char}(C) = \mathrm{Hom}(\pi_1(C), \mathrm{SL}(2,\mathbb{C}))/\mathrm{SL}(2,\mathbb{C})$. Useful sets of coordinates for $\mathcal{M}_{\rm flat}(C)$ are given by the trace functions $L_\gamma := \mathrm{tr}\,\rho(\gamma)$ associated to simple closed curves γ on C.

Minimal sets of trace functions that can be used to parametrize $\mathcal{M}_{\rm flat}(C)$ can be identified using pants decompositions. Cutting a surface C of genus g with n punctures along a maximal set $\{\gamma_1,\ldots,\gamma_{3g-3+n}\}$ on non-intersecting simple closed curves produces a surface having connected components of type $C_{0,3}$ only. Cutting C along all but one of the curves in $\{\gamma_1,\ldots,\gamma_{3g-3+n}\}$ produces a surface containing a single connected component of type $C_{0,4}$ or $C_{1,1}$. This component will be denoted as $C^{\dagger_r}$ if γ_r is the curve which was not cut. In order to get a coordinate system for $\mathcal{M}_{\rm char}(C)$, one needs two independent coordinates for each $C^{\dagger_r}$, $r = 1,\ldots,3g-3+n$. This is what we will define next.

5.3.1. Complex Fenchel–Nielsen coordinates

Conjugacy classes of irreducible representations of $\pi_1(C_{0,4})$ are uniquely specified by seven invariants,

$$L_k = \mathrm{Tr}\,M_k = 2\cos 2\pi m_k, \qquad k = 1,\ldots,4, \tag{5.42a}$$

$$L_s = \mathrm{Tr}\,M_1M_2, \qquad L_t = \mathrm{Tr}\,M_1M_3, \qquad L_u = \mathrm{Tr}\,M_2M_3, \tag{5.42b}$$

generating the algebra of invariant polynomial functions on $\mathcal{M}_{\rm char}(C_{0,n})$. The monodromies M_r are associated to the curves χ_r depicted in Figure 1. These trace functions satisfy the quartic equation

$$\begin{aligned} L_1L_2L_3L_4 + L_sL_tL_u + L_s^2 + L_t^2 + L_u^2 + L_1^2 + L_2^2 + L_3^2 + L_4^2 \\ = (L_1L_2 + L_3L_4)\,L_s + (L_1L_3 + L_2L_4)\,L_t + (L_2L_3 + L_1L_4)\,L_u + 4. \end{aligned} \tag{5.43}$$

The affine algebraic variety defined by (5.43) is a concrete representation for the character variety of $C_{0,4}$. For fixed choices of $m_1,\ldots,m_4$ in (5.42a), one may use equation (5.43), to describe the character variety as a cubic surface in $\mathbb{C}^3$. This surface admits a parametrization in terms of coordinates (λ,κ) of the form

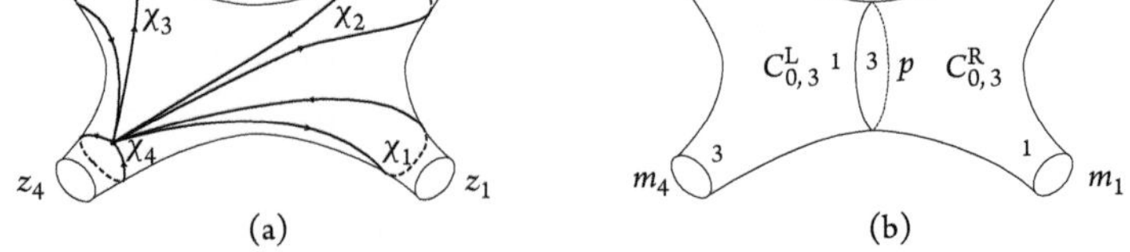

Figure 1 *Basis of loops of $\pi_1(C_{0,4})$ and the decomposition $C_{0,4} = C^L_{0,3} \cup C^R_{0,3}$.*

$$L_s = 2\cos(\lambda/2), \tag{5.44a}$$

$$L_t\big((L_s)^2-4\big) = 2(L_2L_3+L_1L_4)+L_s(L_1L_3+L_2L_4) + 2\cos(\kappa)\sqrt{c_{12}(L_s)c_{34}(L_s)}, \tag{5.44b}$$

$$L_u\big((L_s)^2-4\big) = L_s(L_2L_3+L_1L_4)+2(L_1L_3+L_2L_4) + 2\cos((2\kappa-\lambda)/2)\sqrt{c_{12}(L_s)c_{34}(L_s)}, \tag{5.44c}$$

where $L_i = 2\cos\frac{\lambda_i}{2}$, and $c_{ij}(L_s)$ is defined as $c_{ij}(L_s) = L_s^2+L_i^2+L_j^2+L_sL_iL_j-4$.

For $C \simeq C_{1,1}$ one may similarly parametrize the trace functions along the usual a- and b-cycles on the torus as

$$L_a = 2\cos(\lambda/2), \tag{5.45}$$

$$L_b\big((L_a)^2-4\big)^{\frac{1}{2}} = 2\cos(\kappa/2)\sqrt{(L_a)^2+L_0-2}, \tag{5.46}$$

with L_0 being the trace function associated to the boundary of $C_{1,1}$.

Using pants decompositions as described above, one may define a pair of coordinates (κ_r,λ_r) associated to each cutting curve γ_r, $r=1,\ldots,3g+n-3$. Taken together, the tuples $\mathbf{k} = (\kappa_1,\ldots,\kappa_{3g+n-3})$ and $\mathbf{l} = (\lambda_1,\ldots,\lambda_{3g+n-3})$ form a system of coordinates for $\mathcal{M}_{\text{flat}}(C)$. The coordinates defined above are Darboux coordinates,

$$\Omega = \frac{1}{4\pi}\sum_{r=1}^{3g-3+n} d\kappa_r \wedge d\lambda_r, \tag{5.47}$$

where Ω is the symplectic form on the moduli spaces of flat connections introduced by Goldman and Atiyah–Bott (see [6, 29] and references therein).

5.3.2. Quantization conditions in terms of complex Fenchel–Nielsen coordinates

We had previously reformulated the quantization conditions as the condition that the monodromy of the differential operator $\epsilon_1^2\partial_u^2 + t(u)$ defines a representation of $\pi_1(C)$ in $\mathrm{PSL}(2,\mathbb{R})$. We are now going to reformulate this condition in terms of the coordinates $(\mathbf{l},\mathbf{k})$.

To this end, let us recall some basic facts reviewed in [30]. Fixing an oper represented by a projective connection t_* one may represent generic opers via $t = t_* + \vartheta$, where ϑ is a quadratic differential. Noting that $d := \dim H^0(C,K^2) = 3g-3+n$, we see that the opers represent a half-dimensional subspace $\mathrm{Op}_{\mathfrak{sl}_2}(C)$ of the moduli space of flat $\mathrm{SL}(2,\mathbb{C})$-connections. The monodromy map from $\mathrm{Op}_{\mathfrak{sl}_2}(C)$ to the character variety is locally biholomorphic. Fixing coordinates $\mathbf{E} = (E_1,\ldots,E_d)$ for the vector space $H^0(C,K^2)$, one may therefore use the monodromy map to define $\mathbf{l} = \mathbf{l}(\mathbf{E})$ and $\mathbf{k} = \mathbf{k}(\mathbf{E})$ by analytic continuation in $\mathbf{E}$.

Let us next observe that the coordinates $(\mathbf{l},\mathbf{k})$ as introduced above are closely related to the classical Fenchel–Nielsen coordinates for the Teichmüller component in the character variety $\mathcal{M}^{\mathbb{R}}_{\text{char}}(C)$ parametrizing the Fuchsian groups appearing in the uniformization of

Riemann surfaces. The Fenchel–Nielsen coordinates (l_r, k_r) are given in terms of the coordinates (λ_r, κ_r) simply as $l_r = \mathrm{i}\lambda_r$, and $k_r = \mathrm{i}\kappa_r$. The coordinates (l_r, k_r) are real on Fuchsian groups. Note that the coordinates (λ_r, κ_r) and $(\lambda_r + 4\pi\, n_r, \kappa_r + 2\pi\, \nu_r\, m_r)$ give the same values of the trace functions iff $n_r, m_r \in \mathbb{Z}$, and $\nu_r = 1$ if $C^{\dagger_r}$ is of type $C_{0,4}$, while $\nu_r = 2$ if $C^{\dagger_r}$ is of type $C_{1,1}$.

These observations lead us to reformulate the quantization conditions for the Hitchin system as follows:

To each eigenfunction of the quantized Hamiltonians of the SL(2) *Hitchin system on* C *which is single-valued there exists a projective structure on* C *having holonomy with complex Fenchel–Nielsen coordinates* $(\mathbf{k}, \mathbf{l}) = (\mathbf{l}(\mathbf{E}), \mathbf{k}(\mathbf{E}))$ *satisfying*

$$\mathrm{Re}(\lambda_r) = 2\pi\, n_r, \quad \mathrm{Re}(\kappa_r) = \nu_r\, \pi\, m_r, \quad n_r, m_r \in \mathbb{Z},\ r = 1, \ldots, d, \tag{5.48}$$

where ν_r *was defined above, and* $d := 3g - 3 + n$.

We observe an interesting point: the definition of the coordinates $(\mathbf{l}, \mathbf{k})$ depends on the choice of a pants decomposition. Changing the pants decomposition must relate the integer points defined in (5.48) to each other.

It should be interesting to relate the conditions in (5.48) to Goldman's classification of projective structures with real holonomy [13]. The data classifying such projective structures are collections $\mu(\mathbf{M})$ of disjoint simple closed curves μ_i on C with positive integer weights M_i attached to them. We may represent these projective structures in terms of the sum of a reference oper with a quadratic differential $\vartheta = \vartheta(\mathbf{E}_{\mu(\mathbf{M})})$, defining a discrete set of points in $\mathcal{B}_{\epsilon_1} \simeq \mathrm{Op}_{\mathfrak{sl}_2}(C)$. In this regard, it is suggestive to note that the grafting operation relating different projective structures with Fuchsian holonomy can be represented in terms of the complex Hamiltonian twist flows generated by the complex length functions [29]. What is not clear to us at the moment is how exactly this operation gets represented in terms of the coordinates $\mathbf{l}(\mathbf{E})$.

6. Generating Functions of Varieties of Opers

It follows from the observations above that the quantization conditions are naturally described using the space $\mathcal{P}(S)$ of projective structure on a surface S. A projective structure defines in particular a complex structure. The space of projective structures $\mathcal{P}(S)$ on a surface S is therefore fibred over the moduli space $\mathcal{M}(S)$ of complex structures on S.

The definition of the generating function $\mathcal{W}$ given below will use the results of [34–36] comparing two natural holomorphic symplectic structures on $\mathcal{P}(S)$. The first is defined using the symplectic structure (5.47) on the character variety via the holonomy map. The second comes from the non-canonical isomorphisms of $\mathcal{P}(S)$ to the cotangent bundle $T^*\mathcal{M}(S)$ briefly reviewed below, referring to [34–36] for a more detailed review of the relevant background and further references.

6.1. Generating Function $\mathcal{W}$

Let us first recall that the space of quadratic differentials on a Riemann surface C is canonically isomorphic to the cotangent fibre $T^*\mathcal{M}(S)$ at the point in the moduli space $\mathcal{M}(S)$ represented by the Riemann surface C. This means that $\mathcal{P}(S)$ is non-canonically isomorphic to $T^*\mathcal{M}(S)$. A set of local coordinates $\mathbf{q} = (q_1,\ldots,q_d)$ for $\mathcal{M}(S)$ canonically defines a set $(\mathbf{q},\mathbf{E})$ of coordinates for $T^*\mathcal{M}(S)$, with $\mathbf{E} = (E_1,\ldots,E_d)$ being the coefficients in the expansion of $\vartheta \in H^0(C,K^2)$ with respect to the dual of the basis for the tangent space $T\mathcal{M}(S)$ generated by the vector fields ∂_{q_r}, $r = 1,\ldots,d$. The isomorphisms $\mathcal{P}(S) \simeq T^*\mathcal{M}(S)$ defined by the choice of a reference projective connection allow us to use $(\mathbf{q},\mathbf{E})$ as coordinates for $\mathcal{P}(S)$.

Associating to the projective connection $t(y)$ the holonomy of the connection $(\partial_y + (\begin{smallmatrix} 0 & t \\ 1 & 0 \end{smallmatrix}))dy$ defines a map from $\mathcal{P}(S)$ to the character variety. It follows from the theorems proven in [34–36] that this map relates the natural symplectic structures,

$$\Omega = \frac{1}{\mathrm{i}}\sum_{r=1}^{d} dE_r \wedge dq_r = \frac{1}{4\pi}\sum_{r=1}^{d} d\kappa_r \wedge d\lambda_r. \tag{6.49}$$

The change of Darboux coordinates from $(\mathbf{q},\mathbf{E})$ to $(\mathbf{l},\mathbf{k})$ can be described by a generating function $\mathcal{W}(\mathbf{l},\mathbf{q})$ satisfying

$$\kappa_r(\mathbf{l},\mathbf{q}) = -4\pi\mathrm{i}\frac{\partial}{\partial\lambda_r}\mathcal{W}(\mathbf{l},\mathbf{q}), \qquad \frac{\partial}{\partial q_r}\mathcal{W}(\mathbf{l},\mathbf{q}) = E_r. \tag{6.50}$$

It follows in particular that the subspaces of opers in the moduli spaces of flat connections are Lagrangian, and that the functions $\mathcal{W}$ are the generating functions of this Lagrangian subspace. Note that the functions $\mathcal{W}(\mathbf{l},\mathbf{q})$ satisfying (6.50) depend on the choices of coordinates $(\mathbf{q},\mathbf{E})$ and $(\mathbf{l},\mathbf{k})$. Equation (6.50) furthermore determines $\mathcal{W}(\mathbf{l},\mathbf{q})$ only up to a constant. We will discuss the resulting issues in Section 7 below.

6.2. Quantization Conditions in Terms of the Generating Function $\mathcal{W}$

It remains to observe that the quantization conditions can be reformulated in terms of the function $\mathcal{W}(\mathbf{l},\mathbf{q})$ in a way which resembles the use of the function introduced by Yang and Yang [7] for the description of the quantization conditions in quantum integrable models soluble by the Bethe ansatz method. We immediately find from (5.48) and (6.50) that

for each single-valued eigenfunction of the quantized Hamiltonians of the SL(2) *Hitchin system on* C, *there exist tuples* $\mathbf{n} = (n_1,\ldots,n_d)$, $\mathbf{m} = (m_1,\ldots,m_d)$ *of integers and a solution* $\mathbf{l} = \mathbf{l}(\mathbf{n},\mathbf{m})$ *to the equations*

$$\mathrm{Re}(\lambda_r(\mathbf{n},\mathbf{m})) = 2\pi\, n_r, \quad \mathrm{Re}\left(\frac{\partial}{\partial\lambda_r}\mathcal{Y}(\mathbf{l},\mathbf{q})\right)\Big|_{\mathbf{l}=\mathbf{l}(\mathbf{n},\mathbf{m})} = \nu_r\,\pi\, m_r, \tag{6.51}$$

where $r = 1,\ldots,d$, *and the function* $\mathcal{Y}(\mathbf{l},\mathbf{q})$ *is given as* $\mathcal{Y}(\mathbf{l},\mathbf{q}) = 4\pi\mathrm{i}\,\mathcal{W}(\mathbf{l},\mathbf{q})$.

For a given tuple $(\mathbf{n},\mathbf{m})$ *of integers characterizing a single-valued eigenfunction, one gets the corresponding eigenvalues* $\mathbf{E}=(E_1,\ldots,E_d)$ *as*

$$E_r(\mathbf{n},\mathbf{m})=\frac{\partial}{\partial q_r}\mathcal{W}(\mathbf{l}(\mathbf{n},\mathbf{m}),\mathbf{q}). \tag{6.52}$$

This had previously been observed in [9] for the SL(2,$\mathbb{C}$)-Gaudin model. In a parallel development, it has been proposed in [6] that quantization conditions for the Hitchin system are naturally formulated in terms of the function $\mathcal{W}$. The proposal above completes the proposal from [6] by establishing the precise relation between a particular quantization condition for the quantized Hitchin system to a specific condition formulated in terms of the function $\mathcal{W}$.

7. Global Definition of the Yang's Function

The goal of this section is to clarify which global geometric object is locally represented by the function $\mathcal{W}$. One may note that the local definition for the functions $\mathcal{W}$ given above is sufficient for the goal to formulate the quantization conditions for the Hitchin integrable system. Readers only interested in this aspect can safely skip this section. However, from a mathematical point of view, it seems desirable to clarify if there is a globally defined geometric object on $\mathcal{P}(S)$ locally represented by the functions $\mathcal{W}(\mathbf{l},\mathbf{q})$. We are now going to propose that one can define line bundles on $\mathcal{P}(S)$ having $\mathcal{W}(\mathbf{l},\mathbf{q})$ as their local sections. The proposal can be motivated using the observation that $\mathcal{W}(\mathbf{l},\mathbf{q})$ represents the leading asymptotics of the Virasoro conformal blocks for large central charge. It then follows from known facts on the conformal blocks [37].

7.1. *Global Issues in the Definition of the Functions* $\mathcal{W}$

In order to define the coordinates $(\mathbf{q},\mathbf{E})$, one needs coordinates $\mathbf{q}_\imath$ defined in open sets $\mathcal{U}_\imath\subset\mathcal{M}(S)$ such that $\{\mathcal{U}_\imath;\imath\in\mathcal{J}\}$ forms a cover of $\mathcal{M}(S)$, and a family of reference opers $t_\imath=t_\imath(\mathbf{q}_\imath)$ holomorphic on $\mathcal{U}_\imath$. Coordinates $(\mathbf{q}_\imath,\mathbf{E}_\imath)$ and $(\mathbf{q}_\jmath,\mathbf{E}_\jmath)$ defined on sets $\mathcal{U}_\imath$ and $\mathcal{U}_\jmath$ with non-trivial intersection $\mathcal{U}_{\imath\jmath}=\mathcal{U}_\imath\cap\mathcal{U}_\jmath$ will transform as

$$\sum_{r=1}^{d}E_r^\imath\,dq_r^\imath=\sum_{r=1}^{d}E_r^\jmath\,dq_r^\jmath+df_{\imath\jmath}(\mathbf{q}_\imath), \tag{7.53}$$

with $f_{\imath\jmath}$ being locally defined functions on $\mathcal{U}_{\imath\jmath}$. The functions $f_{\imath\jmath}$, being defined in this way only up to a constant, must satisfy the condition

$$f_{\imath_1\imath_2}+f_{\imath_2\imath_3}=f_{\imath_1\imath_3}+\phi_{\imath_1\imath_2\imath_3}, \tag{7.54}$$

with $\phi_{\imath_1\imath_2\imath_3}$ constant on triple overlaps $\mathcal{U}_{\imath_1}\cap\mathcal{U}_{\imath_2}\cap\mathcal{U}_{\imath_3}$. A collection of functions $f_{\imath\jmath}$ defined on the overlaps of a cover defines what was called a *projective* line bundle in [38].

The dependence of the coordinates $(\mathbf{l},\mathbf{k})$ on the choice of a pants decomposition σ will be made explicit by using the notation $(\mathbf{l}_\sigma,\mathbf{k}_\sigma)$. The generating function for the change of coordinates from $(\mathbf{l}_\sigma,\mathbf{k}_\sigma)$ to $(\mathbf{q}_\imath,\mathbf{E}_\imath)$ will be denoted as $\mathcal{W}_{\sigma\imath}(\mathbf{l}_\sigma,\mathbf{q}_\imath)$.

Changes of the defining coordinate systems are described as follows. A change from coordinates $(\mathbf{l}_{\sigma_1},\mathbf{k}_{\sigma_1})$ to $(\mathbf{l}_{\sigma_2},\mathbf{k}_{\sigma_2})$ is described by a generating function $F_{\sigma_1\sigma_2}(\mathbf{l}_{\sigma_1},\mathbf{l}_{\sigma_2})$ such that

$$\sum_{r=1}^{d}(\kappa_r^{\sigma_1}d\lambda_r^{\sigma_1}-\kappa_r^{\sigma_2}d\lambda_r^{\sigma_2})=dF_{\sigma_1\sigma_2}(\mathbf{l}_{\sigma_1},\mathbf{l}_{\sigma_2}). \tag{7.55}$$

The generating functions must satisfy a condition of the form

$$F_{\sigma_1\sigma_2}(\mathbf{l}_{\sigma_1},\mathbf{l}_{\sigma_2}(\mathbf{l}_{\sigma_1},\mathbf{l}_{\sigma_3}))+F_{\sigma_2\sigma_3}(\mathbf{l}_{\sigma_2}(\mathbf{l}_{\sigma_1},\mathbf{l}_{\sigma_3}),\mathbf{l}_{\sigma_3})=F_{\sigma_1\sigma_3}(\mathbf{l}_{\sigma_1},\mathbf{l}_{\sigma_3})+\Phi_{\sigma_1\sigma_2\sigma_3}, \tag{7.56}$$

with $\Phi_{\sigma_1\sigma_2\sigma_3}$ being constant.

The generating functions transform under changes of coordinates for $\mathcal{P}(S)$ as

$$\mathcal{W}_{\sigma\imath_1}(\mathbf{l}_\sigma,\mathbf{q}_{\imath_1})=\mathcal{W}_{\sigma\imath_2}(\mathbf{l}_\sigma,\mathbf{q}_{\imath_2}(\mathbf{q}_{\imath_1}))+f_{\imath_1\imath_2}(\mathbf{q}_{\imath_1}), \tag{7.57a}$$

$$\mathcal{W}_{\sigma_1\imath}(\mathbf{l}_{\sigma_1},\mathbf{q}_\imath)=\mathcal{W}_{\sigma_2\imath}(\mathbf{l}_{\sigma_2}(\mathbf{l}_{\sigma_1},\mathbf{q}_\imath),\mathbf{q}_\imath)+F_{\sigma_1\sigma_2}(\mathbf{l}_{\sigma_1},\mathbf{l}_{\sigma_2}(\mathbf{l}_{\sigma_1},\mathbf{q}_\imath)). \tag{7.57b}$$

It is known that the $\phi_{\imath_1\imath_2\imath_3}$ define a non-trivial cohomology class [38]. It is therefore not yet clear if there can be any globally defined object locally represented by the functions $\mathcal{W}$ on $\mathcal{P}(S)$.

7.2. *Use of the Glueing Construction*

However, there is a way out. There is a well-known construction defining Riemann surfaces of arbitrary topology by glueing three-punctured spheres. The glueing construction identifies parametrized annular neighbourhoods of two punctures on a possibly disconnected Riemann surface $\check{C}$ to produce a new surface C'. For each cutting curve γ_r, $r=1,\ldots,$ $3g-3+n$, defining a pants decomposition, one may introduce a parameter q_r specifying the identification in the glueing construction in such a way that $q_r\to 0$ corresponds to the nodal degeneration where the length of γ_r vanishes. In this way, one gets families of coordinates $\mathbf{q}_\sigma$ for a neighbourhood $\mathcal{U}_\sigma$ of the boundary component in the moduli space $\overline{\mathcal{M}}(S)$ associated to a pants decomposition σ. By varying the pants decompositions, σ one gets a cover of $\overline{\mathcal{M}}(S)$.

It is possible to choose the identification maps in such a way that all transition functions in the Riemann surface produced by the glueing construction are Möbius transformations. This means that the Riemann surfaces defined in this way come equipped with a natural projective structure. We may use this projective structure to define the coordinates E_r. In this way, we get a family of coordinate systems $(\mathbf{q}_\sigma,\mathbf{E}_\sigma)$ covering $\mathcal{P}(S)$.

We may then consider the functions $\mathcal{W}_\sigma(\mathbf{l}_\sigma,\mathbf{q}_\sigma)$ defined as the generating function for the change of coordinates from $(\mathbf{l}_\sigma,\mathbf{k}_\sigma)$ to $(\mathbf{q}_\sigma,\mathbf{E}_\sigma)$. The functions $\mathcal{W}_\sigma(\mathbf{l}_\sigma,\mathbf{q}_\sigma)$ transform under the changes of the coordinates $(\mathbf{l}_\sigma,\mathbf{k}_\sigma)$ and $(\mathbf{q}_\sigma,\mathbf{E}_\sigma)$ as

$$\mathcal{W}_{\sigma_1}(\mathbf{l}_{\sigma_1},\mathbf{q}_{\sigma_1})=f_{\sigma_1\sigma_2}(\mathbf{q}_{\sigma_1})+F_{\sigma_1\sigma_2}(\mathbf{l}_{\sigma_1},\mathbf{l}_{\sigma_2}(\mathbf{l}_{\sigma_1},\mathbf{q}_{\sigma_1}))+\mathcal{W}_{\sigma_2}(\mathbf{l}_{\sigma_2}(\mathbf{l}_{\sigma_1},\mathbf{q}_{\sigma_1}),\mathbf{q}_{\sigma_2}(\mathbf{q}_{\sigma_1})). \tag{7.58}$$

We are now going to argue that there exists a line bundle over $\mathcal{P}(S)$ having the functions $\mathcal{W}_\sigma(\mathbf{l}_\sigma,\mathbf{q}_\sigma)$ as its local sections. Indeed, our claim must hold if we are able to give an unambiguous definition of the function $\mathcal{W}_\sigma(\mathbf{l}_\sigma,\mathbf{q}_\sigma)$ satisfying (6.50) for the coordinates $(\mathbf{q}_\sigma,\mathbf{E}_\sigma)$ and $(\mathbf{l}_\sigma,\mathbf{k}_\sigma)$ associated to any pants decomposition σ. In overlaps of the respective domains of definition, there exist relations between $(\mathbf{q}_{\sigma_1},\mathbf{E}_{\sigma_1})$ and $(\mathbf{q}_{\sigma_2},\mathbf{E}_{\sigma_2})$, as well as $(\mathbf{l}_{\sigma_1},\mathbf{k}_{\sigma_1})$ and $(\mathbf{l}_{\sigma_2},\mathbf{k}_{\sigma_2})$. It follows that there must exist relations of the form (7.58). The consistency of these relations on triple overlaps then implies a cancellation between the constants $\phi_{\iota_1\iota_2\iota_3}$ and $\Phi_{\iota_1\iota_2\iota_3}$ appearing in the consistency conditions (7.54) and (7.56) for the generating functions $f_{\sigma_1\sigma_2}$ and $F_{\sigma_1\sigma_2}$, respectively.

One way to define the functions $\mathcal{W}_\sigma(\mathbf{l}_\sigma,\mathbf{q}_\sigma)$ unambiguously is to specify the asymptotic behaviour they have at the maximal nodal degeneration of C in the Deligne–Mumford compactification of $\mathcal{M}(S)$. We claim that the following choice does the job:

$$\begin{aligned}\mathcal{W}_\sigma(\mathbf{l}_\sigma,\mathbf{q}_\sigma) \sim & \sum_{r\in I_{0,4}}(\delta(l_r)-\delta(l_{r,1})-\delta(l_{r,2}))\log q_r+\sum_{r\in I_{1,1}}\delta(l_r)\log q_r \\ & +\sum_{v\in\mathbb{P}_\sigma}N(l_{v,1},l_{v,2},l_{v,3})+\mathcal{O}(q_r),\end{aligned} \tag{7.59}$$

where $I_{0,4}$ and $I_{1,1}$ are the subsets of $\{1,\ldots,3g-3+n\}$ for which $C^{\dagger r}\simeq C_{0,4}$ and $C^{\dagger r}\simeq C_{1,1}$, respectively, $\mathbb{P}_\sigma$ is the set of pairs of pants appearing in the pants decomposition σ of C, $l_{v,i}$, $i=1,2,3$, are the complex length coordinates associated to the boundary curves of the pair of pants labelled by $v\in\mathbb{P}_\sigma$, and $N(l_3,l_2,l_1)$ is defined as

$$N(l_3,l_2,l_1)=\frac{1}{2}\sum_{s_1,s_2=\pm}\Upsilon_{\rm cl}\big(\tfrac{1}{2}+\tfrac{i}{4\pi}(s_1l_1+s_2l_2+l_3)\big)-\frac{1}{2}\sum_{i=1}^{3}\mathrm{Re}\big(\Upsilon_{\rm cl}\big(1+\tfrac{i}{2\pi}l_i\big)\big), \tag{7.60}$$

assuming $l_i\in\mathbb{R}$ to simplify the expression and using the notation $\Upsilon_{\rm cl}(x)=\int_{1/2}^{x}du\,\log\frac{\Gamma(u)}{\Gamma(1-u)}$. The proof of this claim is outlined in Section 10.3 and Appendix E of [37]. It is interesting to note that the function $N(l_3,l_2,l_1)$ appears in the semi-classical limit of the Liouville three-point functions [39].

8. Concluding Remarks

8.1. *Semi-classical Limit*

The semi-classical limit of the quantization conditions formulated above is closely related to the geometric picture presented in Section 2.

In order to see this, we need to observe that the leading behaviour of the solutions to the differential equations $(\epsilon_1^2\partial_u^2+t(u))\chi(u)=0$ for $\epsilon_1\to 0$ may be represented as

$$\chi(u) \sim e^{\frac{\mathrm{i}}{\epsilon_1}\int^u du'\, v(u')}, \tag{8.61}$$

where $v(u)$ satisfies $(v(u))^2 = t(u)$. The asymptotic behaviour of the monodromies of the differential equations $(\epsilon_1^2\partial_u^2 + t(u))\chi(u) = 0$ can be expressed in terms of the periods of the canonical 1-form $\lambda = vdu$ on the double cover $\Sigma = \{(u,v); v^2 = t(u)\} \subset T^*C$,

$$L_{r,s} \sim \exp\big(\tfrac{\mathrm{i}}{\epsilon_1}a^r\big), \qquad L_{r,u} \sim \exp\big(\tfrac{\mathrm{i}}{\epsilon_1}a_r^{\mathrm{D}}\big), \tag{8.62}$$

where a^r and a_r^{D} are the periods defined in (2.8) by integrating along a suitable canonical basis for the first odd homology of Σ. It follows immediately from (6.50) that

$$\mathcal{W}\big(\tfrac{1}{\epsilon_1}\mathbf{a},\mathbf{q}\big) \sim \frac{1}{\epsilon_1}\mathcal{F}(\mathbf{a},\mathbf{q}) + \text{regular}, \tag{8.63}$$

where $\mathcal{F}(\mathbf{a},\mathbf{q})$ is the potential appearing in the discussion of the algebraic integrability of the Hitchin system in Section 2.1. The coordinates $(\mathbf{a},\mathbf{a}^{\mathrm{D}})$ form yet another set of Darboux coordinates for $T^*\mathcal{M}(S)$, called homological coordinates in [36], and $\mathcal{F}(\mathbf{a},\mathbf{q})$ is the generating function for the change of coordinates from $(\mathbf{a},\mathbf{a}^{\mathrm{D}})$ to $(\mathbf{q},\mathbf{E})$.

The quantization conditions (6.51) therefore have the following leading asymptotics:

$$\mathrm{Re}(a^r) = \epsilon_1\pi\, n_r, \qquad \mathrm{Re}(a_r^{\mathrm{D}}) = \epsilon_1\pi\, m_r. \tag{8.64}$$

These are the natural Bohr–Sommerfeld quantization conditions for the real action-angle variables introduced in Section 2.1, indicating that the quantization conditions studied in this paper are indeed very natural.

8.2. *Real versus Complex Integrable Systems*

Somewhat different types of conditions expressed in terms of generalized Yang's functions have previously been found in other cases admitting such a formulation [1]. Rather than the pair of real equations (6.51) it was shown in [1] that the quantization conditions for the Toda chain and for the elliptic Calogero–Moser models can be represented as a single complex equation of the form

$$\partial_{a^r}\mathcal{Y}(\mathbf{a},\mathbf{q}) = 2\pi n_r, \qquad r = 1,\ldots,d. \tag{8.65}$$

It seems that the quantization conditions of the type (8.65) are natural in algebraically integrable systems which are complexifications of *real* integrable systems like the Toda chain. In this paper, we have been considering integrable systems which are genuinely complex. The two types of quantization conditions, (8.65) and (6.51), are naturally associated to these two cases, respectively, as is also supported by the semi-classical considerations in Section 8.1 above. Comparing the results of a semi-classical analysis of the quantization conditions for XXX-type spin chains with SL(2, $\mathbb{C}$)-symmetry carried out in [40] with (8.64) indicates that quantization conditions of such spin chains are of the same type as found for the Hitchin system in this paper.

8.3. Relation to Conformal Field Theory

WZW-type conformal field theories can be defined mathematically using the representation theory of the affine Lie algebra $\hat{\mathfrak{g}}_k$ at level k extending a semi-simple finite-dimensional Lie algebra $\mathfrak{g}$. The conformal blocks of WZW-type conformal field theories are defined as elements f in the dual of the vacuum representation V_0 of $\hat{\mathfrak{g}}_k$ invariant under the natural action of the Lie algebra of meromorphic functions allowed to have poles only at a single point $P \in C$. The defining invariance condition may be twisted by families of holomorphic G-bundles $\mathcal{E}_{\mathbf{x}}$, introducing a dependence on a collection of parameters $\mathbf{x} = (x_1, \ldots, x_d)$ representing coordinates on Bun_G.

The conformal blocks can be characterized in terms of the solutions $\mathcal{Z}(\mathbf{x}, \mathbf{q})$ to the KZB equations, taking the form

$$(k + h^\vee)\frac{\partial}{\partial q_r}\mathcal{Z}(\mathbf{x}, \mathbf{q}) = \mathsf{H}_r\,\mathcal{Z}(\mathbf{x}, \mathbf{q}), \tag{8.66}$$

where $\mathbf{q}$ are complex coordinates for the Teichmüller space $\mathcal{T}(S)$, H_r are the quantized Hitchin Hamiltonians, and $h^\vee$ is the dual Coxeter number of $\mathfrak{g}$. In the critical level limit $\epsilon_2 \to 0, \epsilon_2 := -(k + h^\vee)\epsilon_1$, one may solve (8.66) with the ansatz [41] (see [42] for related results)

$$\mathcal{Z}(\mathbf{x}, \mathbf{q}) \sim e^{-\frac{\epsilon_1}{\epsilon_2}\mathcal{S}(\mathbf{q};\epsilon_1)}\Psi(\mathbf{x}, \mathbf{q}; \epsilon_1)(1 + \mathcal{O}(q)), \tag{8.67}$$

where $\mathcal{S}(\mathbf{q}; \epsilon_1)$ and $\Psi(\mathbf{x}, \mathbf{q}; \epsilon_1)$ satisfy

$$\mathsf{H}_r\Psi(\mathbf{x}, \mathbf{q}; \epsilon_1) = E_r\Psi(\mathbf{x}, \mathbf{q}; \epsilon_1), \qquad \frac{\partial}{\partial q_r}\mathcal{S}(\mathbf{q}; \epsilon_1) = E_r. \tag{8.68}$$

This result can be made more precise by using the glueing construction to construct bases of conformal blocks associated to pants decompositions. In the case $\mathfrak{g} = \mathfrak{sl}_2$, one gets solutions $\mathcal{Z}(\mathbf{l}; \mathbf{x}, \mathbf{q})$ to the KZB-equations depending on additional parameters $\mathbf{l} = (\lambda_1, \ldots, \lambda_d)$ parametrizing the intermediate representations[2] used in the glueing construction, one complex number λ_r for each cutting curve γ_r. The analysis above can then be refined by using the Verlinde loop operators in a similar way as was done in [26, 43], giving

$$\kappa_r(\mathbf{l}, \mathbf{q}) = -4\pi \mathrm{i}\frac{\partial}{\partial \lambda_r}\mathcal{S}(\mathbf{l}, \mathbf{q}). \tag{8.69}$$

It follows that the function $\mathcal{S}(\mathbf{l}, \mathbf{q})$ representing the leading term in the asymptotics (8.67) coincides with the function $\mathcal{W}(\mathbf{l}, \mathbf{q})$ studied in this paper.

8.4. Real Geometric Langlands

The results of this note can be reinterpreted as a variant of the geometric Langlands programme. One aspect of the ordinary Langlands programme is the classification of

[2] The parameters λ_r parametrize the weights j_r of the intermediate representations as $j_r = -\frac{1}{2} + \mathrm{i}\lambda_r$.

the (cuspidal) spectrum of the Laplacian on certain locally symmetric spaces. From the point of view of integrable systems, one may view the Laplacian as the 'local' observable one is interested in. Possible degeneracies can be resolved by using additional 'non-local' observables called Hecke operators.

From this point of view, one may interpret the geometric Langlands programme as a conjectural answer to a 'pre-spectral' problem. It describes the natural geometric home for the eigenvalues of the Hitchin Hamiltonians—the variety of opers within the moduli space of local systems.

The natural next step is to define natural quantization conditions defining what might be called cuspidal eigenfunctions of the Hitchin Hamiltonians. In this paper, we propose such a quantization condition. It selects a discrete subset within the variety of opers—a particular subset of the intersections between the variety of opers and the real slice. We propose to view correspondences between real opers and single-valued eigenfunctions of Hitchin Hamiltonians as natural variants of the geometric Langlands correspondence:

$$\boxed{\text{Real } {}^L\mathfrak{g}\text{-opers}} \quad \longleftrightarrow \quad \boxed{\begin{array}{l}\text{Single-valued eigenfunctions}\\ \text{of Hitchin Hamiltonians}\end{array}} \tag{8.70}$$

As opposed to the versions of the geometric Langlands correspondence intensively studied in the literature, this version is not of an algebro-geometric nature; rather, it is based on the relation between the two natural algebraic structures on the moduli of flat connections furnished by the non-algebraic Riemann–Hilbert correspondence. However, the version of the geometric Langlands correspondence proposed above has the virtue of being somewhat closer analogous to the original Langlands programme, in the sense that the single-valued eigenfunctions of the Hitchin-Hamiltonians can be viewed as analogues of the automorphic forms.

Acknowledgements

The author would like to thank Leon Takhtajan and, especially, E. Frenkel for very useful comments on a previous version of the draft.

This work was supported by the Deutsche Forschungsgemeinschaft (DFG) through the collaborative Research Centre SFB 676 'Particles, Strings and the Early Universe', project A10.

REFERENCES

[1] N. A. Nekrasov and S. L. Shatashvili, *Quantization of integrable systems and four dimensional gauge theories*, arXiv:0908.4052 (2009). Published in 16th International Congress on Mathematical Physics, Prague, August 2009, P. Exner, Editor, pp.265–289, World Scientific 2010.

[2] D. Gaiotto, *N=2 dualities*, *JHEP* **08** (2012), no. 8, 034.
[3] D. Gaiotto, A. Neitzke and G. Moore, *Wall-crossing, Hitchin systems, and the WKB approximation*, *Adv. Math.* **234** (2013), 239–403.
[4] N. J. Hitchin, *Stable bundles and integrable systems*, *Duke Math. J.* **54** (1987), no. 1, 91–114.
[5] N. Nekrasov and E. Witten, *The omega deformation, branes, integrability, and Liouville theory*, *JHEP* **1009** (2010), no. 9, 092.
[6] N. Nekrasov, A. Rosly and S. Shatashvili, *Darboux coordinates, Yang–Yang functional, and gauge theory*, *Nucl. Phys. B Proc. Suppl.* **216** (2011), no. 1, 69–93.
[7] C.-N. Yang and C. P. Yang, *Thermodynamics of one-dimensional system of bosons with repulsive delta function interaction*, *J. Math. Phys.* **10** (1969), no. 7, 1115–22.
[8] E. Frenkel, *Lectures on the Langlands program and conformal field theory*, in *Frontiers in number theory, physics, and geometry. II* (P. Cartier, ed.), pp. 387–533. Springer, Berlin, 2007.
[9] J. Teschner, *Quantization of the Hitchin moduli spaces, Liouville theory, and the geometric Langlands correspondence I*, *Adv. Theor. Math. Phys.* **15** (2011), no. 2, 471–564.
[10] A. Balasubramanian and J. Teschner, *Supersymmetric field theories and geometric Langlands: The other side of the coin*, arXiv:1702.0649 (2017). Published in *Proc. Symp. Pure Math.* **98** (2018) 79–106.
[11] E. Frenkel, *Affine algebras, Langlands duality and Bethe ansatz*, in *Proceedings of the International Congress of Mathematical Physics, Paris, 1994* (D. Iagolnitzer, ed.), pp. 606–42. International Press, Cambridge, MA, 1995.
[12] E. Sklyanin, *Separation of variables in the Gaudin model*, *J. Soviet Math.* **47** (1989), no. 2, 2473–88.
[13] W. M. Goldman, *Projective structures with Fuchsian holonomy*, *J. Differential Geom.* **25** (1987), no. 3, 297–326.
[14] D. S. Freed, *Special Kahler manifolds*, *Commun. Math. Phys.* **203** (1999), no. 1, 31–52.
[15] E. Witten, *More on gauge theory and geometric Langlands*, arXiv:1506.0429 (2015). Published in *Advances in Mathematics* **327** (2018), 624–707.
[16] J. C. Hurtubise, *Integrable systems and algebraic surfaces*, *Duke Math. J.* **83** (1996), no. 1, 19–50.
[17] A. Gorsky, N. Nekrasov and V. Rubtsov, *Hilbert schemes, separated variables, and D-branes*, *Comm. Math. Phys.* **222** (2001), no. 2, 299–318.
[18] I. Krichever, *Vector bundles and Lax equations on algebraic curves*, *Commun. Math. Phys.* **229** (2002), no. 2, 229–69.
[19] A. Beilinson and V. Drinfeld, *Quantization of hitchin's integrable system and hecke eigensheaves*, available at http://math.uchicago.edu/~mitya/langlands/hitchin/BD-hitchin.pdf (accessed 15 May 2018) (1991).
[20] B. Feigin and E. Frenkel, *Affine Kac–Moody algebras at the critical level and Gelfand–Dikii algebras*, *Int. J. Mod. Phys.* **7** (1992), Suppl. 1A, 197–215.
[21] B. Feigin, E. Frenkel and N. Reshetikhin, *Gaudin model, Bethe ansatz and correlation functions at the critical level*, *Commun. Math. Phys.* **166** (1994), no. 1, 27–62.
[22] B. Enriquez, B. Feigin and V. Rubtsov, *Separation of variables for Gaudin–Calogero systems*, *Compos. Math.* **110** (1998), no. 1, 1–16.
[23] G. Felder and A. Schorr, *Separation of variables for quantum integrable systems on elliptic curves*, *J. Phys. A* **32** (1999), no. 46, 8001–22.
[24] Y. Hikida and V. Schomerus, *H+(3) WZNW model from Liouville field theory*, *JHEP* **0710** (2007), no. 10, 064.
[25] V. Drinfeld, *Two-dimensional l-adic representations of the fundamental group of curve and automorphic forms on GL(2)*, *Amer. J. Math.* **105** (1985), no. 1, 85–114.
[26] E. Frenkel, S. Gukov and J. Teschner, *Surface operators and separation of variables*, arXiv:1506.0750 (2015). Published in *JHEP* **1601** (2016) 179.

[27] S. Ribault and J. Teschner, *H+(3)-WZNW correlators from Liouville theory, JHEP* **2005** (2005), no 6, 014.

[28] E. Frenkel, *Gauge Theory and Langlands Duality,* talk given at MSRI, Berkeley, in September 2014, available at https://youtu.be/564NQQrFUA0?t=1h8m45s (accessed 4 May 2018).

[29] W. M. Goldman, *The complex-symplectic geometry of* SL(2, $\mathbb{C}$)*-characters over surfaces,* in *Algebraic groups and arithmetic* (S. G. Dani and G. Prasad, eds), pp. 375–407. Tata Inst. Fund. Res., Mumbai, 2004.

[30] D. Dumas, *Complex projective structures,* in *Handbook of Teichmüller Theory, Volume II.* (A. Papadopoulos, ed.), pp. 455–508. European Mathematical Society, Zürich, 2009.

[31] D. Gallo, M. Kapovich and A. Marden, *The monodromy groups of schwarzian equations on closed riemann surfaces, Ann. Math.* **151** (2000), no. 2, 625–704.

[32] T. Figiel and J. Teschner, *in preparation.*

[33] L. A. Takhtajan, *Real projective connections, V. I. Smirnov's approach, and black hole type solutions of the Liouville equation, Theoret. Math. Phys.* **181** (2014), no. 1, 1307–16.

[34] S. Kawai, *The symplectic nature of the space of projective connections on Riemann surfaces, Math. Ann.* **305** (1996), no. 1, 161–82.

[35] B. Loustau, *The complex symplectic geometry of the deformation space of complex projective structures, Geom. Topol.* **19** (2015), no. 3, 1737–75.

[36] M. Bertola, D. Korotkin and C. Norton, *Symplectic geometry of the moduli space of projective structures in homological coordinates, Invent. Math.* **210** (2017), no. 3, 759–814.

[37] J. Teschner and G. S. Vartanov, *Supersymmetric gauge theories, quantization of* $\mathcal{M}_{\text{flat}}$*, and conformal field theory, Adv. Theor. Math. Phys.* **19** (2015), no. 1, 1–135.

[38] D. Friedan and S. H. Shenker, *The analytic geometry of two-dimensional conformal field theory, Nucl. Phys. B* **281** (1987), no. 3–4, 509–45.

[39] A. B. Zamolodchikov and A. B. Zamolodchikov, *Conformal bootstrap in Liouville field theory, Nucl. Phys. B* **477** (1996), no. 2, 577–605.

[40] S. E. Derkachov, G. P. Korchemsky and A. N. Manashov, *Non-compact Heisenberg spin magnets from high-energy QCD. 3. Quasiclassical approach, Nucl. Phys. B* **661** (2003), no. 3, 533–576.

[41] N. Reshetikhin and A. Varchenko, *Quasiclassical asymptotics of solutions to the KZ equations,* in *Geometry, topology, & physics for Raoul Bott,* Conf. Proc. Lecture Notes Geom. Topology, IV (R. Bott and S.-T. Yau, eds), pp. 293–322. Int. Press, Cambridge, MA, 1995.

[42] D. Ben-Zvi and E. Frenkel, *Geometric realization of the Segal–Sugawara construction,* in U. Tillmann, ed., *Topology, geometry and quantum field theory, Proceedings of the 2002 Oxford Symposium in the honour of the 60th birthday of Graeme Segal, London Math. Soc. Lecture Note Ser.,* Vol. 308, pp. 46–97. Cambridge Univ. Press, Cambridge, 2004.

[43] J. Teschner, *Classical conformal blocks and isomonodromic deformations,* arXiv:1707.0796 (2017).